Games, Gambling, and Probability

Textbooks in Mathematics

Series editors:
Al Boggess, Kenneth H. Rosen

Linear Algebra
What You Need to Know
Hugo J. Woerdeman

Introduction to Real Analysis, Third Edition
Manfred Stoll

Discovering Dynamical Systems Through Experiment and Inquiry
Thomas LoFaro, Jeff Ford

Functional Linear Algebra
Hannah Robbins

Introduction to Financial Mathematics
With Computer Applications
Donald R. Chambers, Qin Lu

Linear Algebra
An Inquiry-based Approach
Jeff Suzuki

Mathematical Modeling in the Age of the Pandemic
William P. Fox

Games, Gambling, and Probability
An Introduction to Mathematics
David G. Taylor

Financial Mathematics
A Comprehensive Treatment in Discrete Time
Giuseppe Campolieti, Roman N. Makarov

Linear Algebra and Its Applications with R
Ruriko Yoshida

Maple™ Projects of Differential Equations
Robert P. Gilbert, George C. Hsiao, Robert J. Ronkese

Practical Linear Algebra
A Geometry Toolbox, Fourth Edition
Gerald Farin, Dianne Hansford

An Introduction to Analysis, Third Edition
James R. Kirkwood

Student Solutions Manual for Gallian's Contemporary Abstract Algebra, Tenth Edition
Joseph A. Gallian

Elementary Number Theory
Gove Effinger, Gary L. Mullen

https://www.routledge.com/Textbooks-in-Mathematics/book-series/CANDHTEXBOOMTH

Games, Gambling, and Probability
An Introduction to Mathematics,
Second Edition

David G. Taylor
Roanoke College, Salem, VA
USA

CRC Press
Taylor & Francis Group
Boca Raton London New York

CRC Press is an imprint of the
Taylor & Francis Group, an **informa** business

A CHAPMAN & HALL BOOK

Second edition published 2021
by CRC Press
6000 Broken Sound Parkway NW, Suite 300, Boca Raton, FL 33487-2742

and by CRC Press
2 Park Square, Milton Park, Abingdon, Oxon, OX14 4RN

© 2021 Taylor & Francis Group, LLCFirst edition published by CRC Press 2015

CRC Press is an imprint of Taylor & Francis Group, LLC

Library of Congress Cataloging-in-Publication Data

Names: Taylor, David George, 1979- author.
Title: Games, gambling, and probability : an introduction to mathematics /
David G. Taylor, Roanoke College, Salem, VA.
Other titles: Mathematics of games
Description: Second edition. | Boca Raton : Chapman & Hall/CRC Press, 2021.
| Series: Textbooks in mathematics | Revised edition of: The mathematics
of games. 2015. | Includes bibliographical references and index.
Identifiers: LCCN 2021000947 (print) | LCCN 2021000948 (ebook) | ISBN
9780367820435 (hardback) | ISBN 9781003011583 (ebook)
Subjects: LCSH: Probabilities. | Game theory. | Statistical decision.
Classification: LCC QA273 .T388 2021 (print) | LCC QA273 (ebook) | DDC
519.3--dc23
LC record available at https://lccn.loc.gov/2021000947
LC ebook record available at https://lccn.loc.gov/2021000948

ISBN: 978-0-367-82043-5 (hbk)
ISBN: 978-1-032-01812-6 (pbk)
ISBN: 978-1-003-01158-3 (ebk)

Typeset in CMR10
by KnowledgeWorks Global Ltd.

*To my friends, colleagues, professors,
students, and family, especially my
parents, Tom and Del Taylor
and, though she cannot read,
my loving dog Lilly.*

Contents

List of Figures

List of Tables

Preface

To everyone, you may wonder exactly why I might write a textbook that deals primarily with games of chance, or games involving chance. There are many mathematical topics (and skills) that many might argue are more important than probability. To combat this, I remind everyone that many experiments have shown that the *human brain generally has very serious problems dealing with probability and chance*, and a greater understanding of probability can help develop the intuition necessary to approach risk with the ability to make more informed (and better) decisions [26]. Indeed, since many of the early games discussed are casino games, the study of those games, along with an understanding of the material in Chapter 10, should remind you that gambling is a *bad idea*; you should think of placing bets in a casino as paying for entertainment. Winning can, obviously, be a fun reward, but should not ever be *expected*.

I think that writing a book is a daunting task, and I have always looked in awe at those who are able to do it quite easily. What I hope to do here is to bring my personality, teaching experience, and love of mathematics together into a textbook that aims to show you the beauty and joy of the subject, combined with illustrating its importance. As you will notice throughout the text, before any mathematical topic is presented, it will usually be prefaced with a question.

About This Book

The goal for this textbook is to complement the inquiry-based learning movement. In my mind, concepts and ideas will stick with the reader more when they are motivated in an interesting way. Here, we use questions about various games (not just casino games) to motivate the mathematics, and I would say that the writing emphasizes a "just-in-time" mathematics approach. Topics are presented mathematically as questions about the games themselves are posed.

I really hope you enjoy the book now that it's in its second edition. It should read in a friendly way, and you could very well envision yourself sitting in a classroom listening to me speak as you read. The examples are explained in detail, and there are ample exercises at the end of each chapter for you to practice your newfound knowledge and skills. Interesting topics (to me, at least) that seem tangential within a chapter are located in the appendices and in Chapter 11 so if you are curious about them (and I highly encourage you to be so), check them out! While the book is written primarily to be used as a textbook, it also reads well for anyone who wishes to pick it up (and I suppose by now you are already at that point) and enjoy some mathematics and games.

To the instructor, please take time to develop the material from the first four chapters. Within those, you will find the standard material from an introductory probability course, albeit presented in a much different way and order. The chapters afterward include some discussion of different games, different "ideas" that relate to the law of large numbers, and some more mathematical topics not typically seen in such a book. Feel free, of course, to add your own selection of games as you deem appropriate – the use of games is meant to make the book (and course) feel like fun, and by no means is my list exhaustive! The list is rather my own selection of games used in my own classes, with several others added to show a larger picture.

To the student, please enjoy yourself reading this book, working the exercises, and experiencing the class! I've had a lot of fun teaching courses on which this book is based, and I enjoyed writing the text for the first edition along with adding content and reorganizing material for the second edition. The language used, examples presented, and virtually everything else inside was developed and written with you in mind! I highly encourage you after reading this book or completing a class, to explore mathematics further. One natural choice afterward, if you haven't studied it already, is exploring the world of statistics. That world, while much different from the "absolutes" usually provided by probability theory, is very closely related. Indeed, some people think of statistics as kind of an "applied probability" even though statistics is full of rich theory itself.

Changes for the Second Edition

The first edition of this book was aimed at a special May Term course that I had been teaching at Roanoke College; the goal was to have a book that was friendly to students for whom mathematics was not necessarily their strong point but also was rich enough to introduce a lot of standard probability content. The course itself consists of playing many of the games appearing in this book in order to develop one's intuition by seeing probability in action. Since the first edition was released, I've developed a second course that uses this book in a slightly more traditional course; the new course does not focus on playing games, but rather is a general education course that develops quantitative reasoning while teaching the idea that a knowledge of probability can help anyone make better decisions in life. Teaching this second course led to most of the reorganization of this second edition book along with the ideas for new topics.

Here is a brief description of the changes made for this edition compared to the first.

- The chapter on Blackjack, which was Chapter 4 in the first edition, is now Chapter 8. The game is still fascinating to me and is still one of my favorites, but the mathematics and probability concepts introduced in that chapter did not flow well with the new course that I was teaching, nor did it flow well with a list of "essential" topics for a course on probability.

- In the first edition, Chapter 3 contained a very brief introduction to game theory as a result of a discussion about bluffing in poker; this brief introduction has been expanded into a full chapter, Chapter 5, the contains much of what a student would learn from a first course on game theory from a mathematics and probability point of view.

- Discussing sports provides an opportunity to both introduce some new mathematics as well as use topics from earlier in the book; the new Chapter 7 contains a plethora of different topics in "sports mathematics" that fans of any types of sports should enjoy.

- The first edition had a very long Chapter 5 due to a heavy discussion of the probabilities used at Zombie Dice; in order to make the "more dice" chapter more manageable, Zombie Dice now appears in Section 9.9 and the rest of the chapter appears as Chapter 4 due to reorganization.

- In addition to all of the above additions, new games were added, including Arkham Horror in Section 4.2, Uno in Section 9.3, and Scrabble in Section 9.6.

- In the first edition, there were "mini-excursions" at the end of each of the main chapters; these sections contained interesting things in mathematics and probability chosen by me that were only tangentially related to the chapters in which they appeared; in practice, they got in the way of the main text and were adding to some confusion on where to find things in the book, so they all have been moved to a new Chapter 11.

- Finally, numerous typos were corrected, language changed in many places to make things more clear, and even more exercises were added to the end of each of the first ten chapters.

Thanks

A huge special thanks goes out to my friend and colleague Robert Allen who worked with me to create many of the illustrations that appear throughout this text. His typesetting knowledge and ability to create graphs and figures that exist in my head and barely on paper are phenomenal!

Thanks are also due to my friends and colleagues Adam Childers, Christopher Davis, Roland Minton, Cole Owens, Matthew Pons, Karin Saoub, and Mackenzie Sullivan, who supported me throughout the process and provided valuable feedback. Thanks are especially due to Adam and Christopher who listened to me as I developed initial plans for the book and talked through some computations with them. I would have a lot less hair from pulling it out if not for the time they spent listening! Editing a book that I wrote has been a hard task, and that has been made easier by Adam, Roland, Hannah Robbins, Jeff Spielman, and Claire Staniunas who caught various typos and suggested improvements for awkward phrasing. Roland especially had the unique ability to take what was in my head and say it in the way that I wanted it said; he also provided a lot of valuable feedback and guidance in the development of Chapter 7.

My colleagues Adam and Hannah at Roanoke College used a draft version of this book in their courses many years ago so that I could put the finishing touches on it; many thanks to them, as well as to the students in their classes, for providing their thoughts and ideas as the book was being finalized. In particular, thanks to student Richard Gill, who caught typos and suggested improvements. The second edition is mainly a result of the feedback I received from my own students using the book in my classes along with my own desire to make improvements to the flow and content.

I owe a debt of gratitude to the administration at Roanoke College with extreme thanks to Richard Smith, Dean of the College, who allowed me to delay starting as an associate dean by a semester so that I could keep my planned sabbatical in order to finish this text.

Finally, I would very much like to thank my editor Bob Ross and the entire team at CRC Press, a division of the Taylor & Francis Group, for guiding me through the process and believing in a second edition.

Dr. David G. Taylor
MCSP Department
Roanoke College
Salem, Virginia 24153
taylor@roanoke.edu

1

Mathematics and Probability

1.1 Introduction

If you have access to some dice, go ahead and grab six standard 6-sided dice and roll them. Without even seeing the exact dice that you rolled, I'm pretty sure that you rolled at least two of the same number (for the approximately 1.5% of you that rolled exactly one of each number, I apologize for being incorrect). How did I know that? The behavior of (fair) 6-sided dice is pretty easy to model – if you were talking to a friend, you would probably say that each of the six numbers is equally likely to be rolled. After that, it only takes a modest understanding of the rules of probability to determine the "expected" probability that, when rolling six dice, you will get exactly one of each number. As a side note, the chance of getting exactly one of each number is just slightly more common than the probability of being born a twin (in North America at least) and slightly less common than drawing the A♠ from a thoroughly shuffled deck of cards with jokers included!

If you roll those six dice again, chances are you *still* will not have gotten exactly one of each number. The probability that you do is still the same, but for the 0.024% of you that rolled exactly one of each number on *both* rolls, you might be skeptical. And with good reason! Understanding how the mathematics of probability applies to chance events and actually observing what happens in practice can sometimes be at odds with one another. If you were going merely on "observed" probability, you might think that rolling one of each number is a common event. By the way, you are twice as likely to die by slipping and falling in the shower or bathtub compared to rolling exactly one of each number on both of the rolls!

The problem lies in the fact that repeating the same *experiment* a small number of times will probably not reflect what will happen long-term. If you have time, you might try rolling those six dice ten more times, or maybe a hundred more times. If you're really struggling to find something to do, a thousand or even ten thousand repetitions might convince you that my estimates above are correct (and you may even get numbers close to mine), but you are not likely at all to discover those percentages by only observing what happens!

This chapter is about introducing some of the language of probability, some of the mathematical principles and calculations that can be used right from the start, and some more thoughts about why being able to compute probabilities mathematically doesn't *replace* what you see, but rather compliments it quite well. As a small disclaimer, while the rest of this book focuses on what I consider "fun" games, many of which are probably familiar to you, this chapter strips out the idea of "game" and "rules" in favor of using just dice, coins, and candy to motivate our study. As will be usual throughout this book, though, the mathematics and examples will generally appear *after* the question we are trying to answer. The questions we ask about the various games and situations will drive us to develop the mathematics needed to answer those very questions!

1.2   About Mathematics

Most of this book is about a *subfield* of mathematics known as probability, but before we get into the details, we should spend a few moments talking about mathematics in general. Many of you have had many different "math" classes at various stages of being in school. I use quotes there because the mathematics covered in elementary school through high school is very specific and focused, at least initially, on calculations. Students first learn about numbers and counting, followed by addition, subtraction, multiplication, and division; these topics fall into a very small part of mathematics called arithmetic and are really not topics that mathematicians would think of when discussion the subject at all. Indeed, it is not until students start experiencing algebra, where early equations such as $2 + x = 5$ and $\sqrt{x - 4} \geq 9$ pique our interests. These first equations give us a glimpse into what mathematics *really* is: the science of patterns.

Before we continue, let's take time for a brief side note. It's oftentimes when learning about fractions, or when getting to a first algebra course in middle school or high school, that some students start thinking that they're "not good at math" or that they "don't enjoy math." This feeling can continue throughout someone's schooling and into a majority (or all) of their adult life. I firmly believe that this does not need to be the case, and should not be the case. Some say that there are math people and "not math" people, a notion that I think should be eliminated as much as we possibly can. Sure, mathematics may come more naturally to some people and others may need to put in a bit more time, but this is exactly the same for anything else in life. Consider basketball as an example. Anyone can learn the game and to shoot hoops somewhat well by just watching and playing the game; by putting in more practice, anyone can shoot hoops *better*, and with hours of practice, it's possible for virtually everyone to shoot pretty well and likely play on a high school team (yes, there are other skills in the game other than shooting hoops, but any of the skills can also be learned and practiced). With enough time and effort, anyone can be *good* at basketball; sure, not everyone has the raw talent, time, or desire to become the next NBA star, but the same is true of mathematics. With time and effort, everyone can be good at mathematics, and not everyone may have that time, desire, or even the equivalent of raw talent, to become a professional mathematician. I firmly believe that the world doesn't need to have people that say they're "not good at math" or that they "don't enjoy math."

There are many ways that mathematics is used to explore patterns, and while this book aims to explore only patterns explained by probability, it's worth a moment mentioning some other patterns that are often explained using mathematics. First, fractals are present all around us and are generated using an iterative process; a quick internet search for "Koch snowflake" will show you that the iterative process used starts with an image or idea, and replicates that same image or idea within the existing image. In Figure 1.1, you can see a computer-generated example of how branches and leaves grow on a fern; each "leaf" is really a copy of the original, larger, leaf. Fractals appear often in nature. River deltas, growth spirals on many succulents, and the

FIGURE 1.1: Fractal Example

buds from Romanesco broccoli provide rich examples. The Fibonacci sequence, and the numbers contained in that sequence, also occur in nature quite frequently. Originally used to show the number of total rabbit pairs in a population after a certain number of generations and created by defining the first two terms as $F_1 = 1$ and $F_2 = 1$, along with the relationship that any successive term is obtained by adding together the previous two terms so that $F_n = F_{n-1} + F_{n-2}$ for $n \geq 3$, we obtain the sequence $1, 1, 2, 3, 5, 8, 13, 21, 34, \ldots$. This works in the idealized rabbit pair example where newborn rabbit pairs are placed in a field and each breeding pair mates at the age of one month and at the end of the second month they produce another pair of rabbits. After one month, the original pair mates, but there is still only one pair; at the end of the second month, they produce a new pair, so there are now two pairs in the field. Only the original pair will produce offspring at the end of the third month as the newest pair will wait another month, and so on and so forth. Branching in trees, leaf arrangements of stems, the flowering of artichokes, the spiraling in seashells, and the pattern shown on pine cones are all examples of where Fibonacci numbers show up in nature.

I encourage you to take some time exploring these other famous patterns in mathematics and direct you to [6] to learn more about fractals and to [36] to see more about the Fibonacci sequence and numbers in nature. This book, however, takes an in-depth look at how patterns occur in processes of random chance, and how studying those processes can help us learn more about how chance works, how to make predictions based on those patterns, and how to use the knowledge and field of probability to make better decisions. While author's prerogative creeps in from time to time to explore cool and interesting topics that are only tangentially related to our core focus, probability remains the main cornerstone of this book and, with that, our journey begins.

1.3   Probability

Before we take a look at how dice function, let's step back to something that is even simpler to study: coins. For our purposes, when we talk about coins, we will assume that the coin is fair with a side denoted as "heads" and a side denoted as "tails" (faithful watchers of many sporting events such as the Super Bowl will undoubtedly understand the use of quotations marks there – it is not always easy to decide what image on a coin represents which side). Wait a minute, though! What do we mean by a *fair* coin? Though I think we all have an intuitive sense of what fair means, it's worth taking a moment to highlight what *I* mean by fair.

> **Definition 1.1** A coin, die, or object of chance is said to be *fair* if each of its potential outcomes is in fact equally likely; that is, there is no bias in the object that would cause one of the outcomes to be more favored.

Unless otherwise specified in this text, you can safely assume that any process we talk about, whether it's flipping a coin, rolling dice, or spinning a roulette wheel, is fair. In the real world, however, cheats and con artists will attempt to use unfair devices to take your money or, more generally, take advantage of you in other ways. From a mathematical point of view, unfair coins or dice make for an interesting study in and among themselves; we will occasionally see these unfair games or situations pop up in either the text or exercises, but they will be clearly indicated. To wet your whistle, so to speak, it's actually true that you can use an unbalanced coin, no matter how unbalanced it is, to fairly decide outcomes by changing a coin flip to a pair of coin flips, unless, of course, you're using a two-headed (or two-tailed) coin. You will see how in Exercise 1.13.

To begin our study, let's ask a basic question and then develop the language and mathematics necessary to answer the question. First, it's worth a mention that this particular question will not be answered right now, as we will want to spend some time becoming acclimated to the process involved, but, throughout the book, we will always want to keep the overall question or questions in mind as we develop the words and mathematics needed.

Question

Given four coins, what is the chance that I can flip them all and get exactly three heads?

Language

Officially speaking, performing an action that results in an outcome by chance is called an *experiment*, and the set of possible raw outcomes from that experiment is called the *sample space* for the experiment. A subset of the sample space is usually referred to as a particular *event*, and it is the "event" for which we are interested in the probability, or chance, that it will happen. While the formal language is nice, when we are interested in asking questions and finding the answers later in this book, we won't have the need to be terribly explicit about many of these words; for now, it's worth looking at some basic examples to see how these words relate to situations that we probably already know.

In particular, I should mention that we will frequently use the words "condition" or "situation" instead of "event" and "experiment" throughout this book to keep our discussion more colloquial and less technical, but when the technical language is needed, we will use it.

> **Example 1.1** For the *experiment* of flipping one coin, the *sample space* is the set {heads, tails}. A particular event, or subset of the sample space, could be the set {heads}, which we would describe in English as "flipping a head."

What happens if we add more coins? It should make sense that the set of possible outcomes will grow as we add more coins.

> **Example 1.2** For the *experiment* of flipping two coins, the *sample space* is the set {(heads, heads), (heads, tails), (tails, heads), (tails, tails)}. Note that completely spelling out "heads" and "tails" will become too lengthy and prohibitive soon, so another way to express the sample space would be using the letters H and T. That is, we could also write the space as {HH, HT, TH, TT}, and a possible event, in English, could be "flipping at least one tail" for which the subset of the sample space would be {HT, TH, TT}.

There is a *very* important note to make here. You might be thinking that the sample space elements HT and TH describe the same thing, but they do *not*! Imagine if one of your coins were a quarter and the other were a penny. In that case, flipping heads on the quarter and tails on the penny is a very different *physical* outcome than flipping heads on the *penny* and tails on the *quarter*! Yes, {HT, TH} is an event for this experiment described in English as "flipping a head and a tail," but this event can be physically obtained in these very two distinct ways. Even if the coins were identical (perhaps two pennies freshly minted), you could easily flip one coin with your left hand and the other coin with your right hand; each coin is a distinct entity.

Probability

If you were asked by a friend to determine the probability that heads would be the result after flipping a coin, you would probably answer 1/2 or 50% very quickly. There are two possible raw outcomes when flipping a coin, and the event "obtaining a head" happens to be only one of those raw outcomes. That is, of the sample space $\{H, T\}$ consisting of two elements, *one* element, H, matches the description of the desired event. When we talk about the probability of an event occurring, then, we need only take the number of ways the event could happen (within the entire sample space) and divide by the total number of possible raw outcomes (size of that entire sample space).[1] For instance, for the experiment of flipping a coin and desiring a head, the probability of this event is 1 divided by 2, since there is one way of getting a head, and two possible raw outcomes as we saw. This should hopefully agree with your intuition that flipping a coin and getting a head indeed has probability $\frac{1}{2} = 0.50 = 50\%$. If that seemed easy, then you are ahead of the game already, but be sure to pay attention to the upcoming examples too!

Note here that I've expressed our answer in three different, but equivalent and equal, ways. It is quite common to see probabilities given in any of these forms. Since there are never more ways for an event to happen than there are possible raw outcomes, the answer to *any* probability question will be a fraction (or decimal value) greater than or equal to zero and less than or equal to one, or a percentage greater than or equal to 0% and less than or equal to 100%! Let's put these important notes in a box.

---

[1]Note that this definition does lead to some interesting problems and results if the size of the sample space is infinite; see Appendix A to see some of these interesting examples, or consult a higher-level probability textbook and look for countably infinite or continuous distributions.

Definition 1.2 The *probability* of an event happening is the number of elements in the event's set (remember, this is a subset of the sample space) divided by the number of elements in the entire sample space. If P is the probability of an event happening, then $0 \leq P \leq 1$ or, equivalently, $0\% \leq P \leq 100\%$.

At this point, we have enough information to be able to answer our question. Given four coins and the fact that there are two possibilities for outcomes on each coin, there will be 16 different physical outcomes; that is, there will be 16 elements in our sample space, listed in Table 1.1. Take a moment to verify that these are the *only* 16 outcomes possible.

TABLE 1.1: Outcomes of Flipping 4 Coins

HHHH	HHHT	HHTH	HTHH
THHH	HHTT	HTHT	THHT
HTTH	THTH	TTHH	HTTT
THTT	TTHT	TTTH	TTTT

Example 1.3 Since there are only 4 outcomes with exactly 3 heads, and there are 16 total outcomes, the probability of flipping 4 coins and getting exactly 3 heads is $4/16 = 0.25 = 25\%$.

The same method can definitely be used to answer similar questions for any number of dice, but we will consider a quicker method (involving what is known as a binomial distribution) in Chapter 4. After all, determining the probability of obtaining exactly four heads when flipping ten coins might take awhile to compute, given that there are $2^{10} = 1,024$ possible physical outcomes to flipping those coins, of which 210 of them involve exactly four heads! At the very least, though, knowing this information would lead us to determine that the probability of getting four heads with ten flips would be $210/1024 \approx 0.2051 = 20.51\%$. That probability is possibly a lot larger than you may have thought; at least it was to me the first time I computed it!

TABLE 1.2: Outcomes for Rolling Two Dice

	⚀	⚁	⚂	⚃	⚄	⚅
⚀	⚀⚀	⚀⚁	⚀⚂	⚀⚃	⚀⚄	⚀⚅
⚁	⚁⚀	⚁⚁	⚁⚂	⚁⚃	⚁⚄	⚁⚅
⚂	⚂⚀	⚂⚁	⚂⚂	⚂⚃	⚂⚄	⚂⚅
⚃	⚃⚀	⚃⚁	⚃⚂	⚃⚃	⚃⚄	⚃⚅
⚄	⚄⚀	⚄⚁	⚄⚂	⚄⚃	⚄⚄	⚄⚅
⚅	⚅⚀	⚅⚁	⚅⚂	⚅⚃	⚅⚄	⚅⚅

One of the games we shall study in Chapter 2 is the game of craps. While the game flow and rules will be presented there rather than here, you should, at the very least, know that it involves rolling a pair of standard six-sided dice. Since this probably sounds a bit more exciting than flipping coins (it does to me, at least), let us turn our attention to some probabilities involving dice. Our main question for now will be the following:

Question

What is the probability of rolling a sum of seven on a pair of six-sided dice?

By now, the method for answering this question should be pretty clear. With a list of all physical outcomes available for rolling a pair of dice, we could simply look at all of the possibilities, determine which outcomes lead to a sum of seven, and determine the probability from that.

Take a look at Table 1.2. Here, you can see the 36 different physical outcomes possible (again, to be clear, even if you are rolling two white dice, each die is independent of each other – in the figure, we use white *and* black to represent the two different dice).

Example 1.4 Looking at Figure 1.2, we can see that there are 6 outcomes that result in the sum of the dice equaling 7, namely the outcomes that are shown starting in the lower left (with a white six and black one) proceeding up the diagonal to the right ending at the white one and black six. Since there are 36 total outcomes, and only 6 result in a sum of 7, the probability of rolling a 7 on a single toss of a pair of 6-sided dice is $6/36 = 1/6 \approx 16.67\%$.

Adding Probabilities

Question

What is the probability of rolling two 6-sided dice and obtaining a sum of 7 or 11?

One of the hardest things for a beginning probability student to grasp is determining when we *add* probabilities and when we *multiply* probabilities. For this part of the chapter, we'll be focusing on situations where we add two probabilities together.

The short version of how to remember when to *add* two probabilities together is to remember that addition is used to find the probability that "situation A" *or* "situation B" happens; that is, addition is associated with the word "or." In other words, given two different events, if you're asked to find the chance that either happens, you'll be adding together two numbers (and, for reasons we will discuss, possibility subtracting a number as well). Perhaps this concept is best illustrated through an example or two; for these examples, rather than using a standard 6-sided die, let's use a 10-sided die (pictured in Figure 1.2) instead; faithful players

FIGURE 1.2: 10-Sided Die

of tabletop games such as Dungeons & Dragons should appreciate using what some would call an "irregular" die!

Note that the "0" side on the die can be treated as either 0 or 10. Just to be consistent with the numbering of a 6-sided die, we will treat the "0" side as 10, and consider a 10-sided die (commonly referred to as a "d10") as having the integers 1 through 10, inclusive.

Example 1.5 If we ask for the probability of rolling a number strictly less than 4 or strictly greater than 8 on a 10-sided die, we could of course use the methods we have previously discussed. The outcome set $\{1, 2, 3, 9, 10\}$ describes all of the physical possibilities matching our conditions, so the probability should be $5/10 = 1/2 = 50\%$.

Instead, let's look at each of the conditions separately. The probability of obtaining a number strictly less than 4 is $3/10$ since there are three physical outcomes (1, 2, or 3) that will make us happy, and 10 possible total outcomes. Similarly, the probability of obtaining a number strictly greater than 8 is $2/10$ since the physical outcomes 9 or 10 are the only outcomes that work. Therefore, the probability of getting a result less than 4 or greater than 8 is $3/10 + 2/10 = 5/10 = 50\%$.

That didn't seem so bad, did it? At the moment, it's worth pointing out that the two conditions we had (being less than 4 and being greater than 8) were *disjoint*; that is, there is no outcome possible that fits both conditions at the same time. The result of any roll could be less than 4, just as the result of any roll could be greater than 8, but a result cannot simultaneously be less than 4 *and* greater than 8! Combining conditions where there is indeed overlap is when subtraction is needed, as we see in the next example.

Example 1.6 Now let's determine the probability of rolling a 10-sided die and getting a result that is either a multiple of 3 or (strictly) greater than 5. The multiples of 3 that are possible come from the set $\{3, 6, 9\}$ and the outcomes greater than 5 include $\{6, 7, 8, 9, 10\}$. Combining these together, the set of possible outcomes that satisfy at least one of our conditions is $\{3, 6, 7, 8, 9, 10\}$ (note that we do not list 6 and 9 more than once; a set is simply a list of what's included – once 6 is already a member of the set satisfying our happiness, we don't need another copy of it). For reference, you may have heard of this process of combining two sets together like this as taking the set *union*; indeed you can think of "or" in probability as taking a union. Putting all of this together, there are 6 outcomes we like, and 10 possible outcomes, so the probability we care about is $6/10 = 3/5 = 60\%$.

What happens when we try to use addition here? Well, the "multiple of 3" condition happens with chance $3/10$ and the "greater than 5" condition happens with chance $5/10$, so when we add these together we get $3/10 + 5/10 = 8/10 = 80\%$. What happened?

Our new answer is exactly $2/10$ larger than it should be! Non-coincidentally, $2/10$ is also the probability of obtaining either a 6 or a 9 when we roll a 10-sided die; the moral of the story here, and reproduced below, is that when adding probabilities together for *non-disjoint* events, we must subtract the probability of the overlap.

When there is overlap between events, say A and B, as in Example 1.6, subtracting the overlap is essential if you want to use addition to compute the probability of A or B happening. In Figure 1.3 you can see a visual representation of this phenomenon; if we simply add the probabilities of A and B (the light gray and lighter gray regions, respectively), we are including the dark gray region (the overlap) twice! Subtracting out the probability of the overlap gives us the correct answer (think of this as cutting away the overlap from B to form a moon-like object so that the two events A and moon-B fit together in a nice way). This is formalized below in what we call the additive rule; note that "A and B" will refer to the overlap region or probability.

Theorem 1.3 (The Additive Rule) If A and B are two events (or conditions), the probability that A or B happens, denoted $P(A \text{ or } B)$, is given by

$$P(A \text{ or } B) = P(A) + P(B) - P(A \text{ and } B)$$

where $P(A)$ is the probability of event A, $P(B)$ is the probability of event B, and $P(A \text{ and } B)$ is the probability that both events occur simultaneously (the overlap between A and B).

With this rule, we are now able to use the additive rule in its full form when needed. Note that, for now, many of the "A or B" probabilities seem easier to compute by determining all of the outcomes that satisfy either condition (as we did in the first half of the last two examples), but as the questions become more complicated, it may very well be easier to compute each of $P(A)$, $P(B)$ and $P(A \text{ and } B)$!

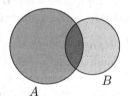

FIGURE 1.3: Additive Rule

It is now time to answer our question about obtaining a sum of 7 or 11 on a roll of two 6-sided dice.

Example 1.7 We have already determined that the probability of getting a sum of 7 on two dice is $1/6$. To obtain a sum of 11, we can see from Figure 1.2 that there are only two ways of doing so; you need to get a 5 on one die, and a 6 on the other die, where the first die could either be the white or black die. It would then seem that the probability of getting an 11 is $2/36 = 1/18$, so to answer our question, the probability of getting a sum of 7 or 11 is $1/6 + 1/18 = 4/18 = 2/9 \approx 22.22\%$ since there is no overlap (you cannot get a sum of 7 and also get a sum of 11 on a single roll of the dice).

Multiplying Probabilities

Question

Suppose you flip a coin and roll a 6-sided die; what is the probability that you get a head on the coin and a 5 or a 6 on the die?

While I doubt that you will encounter too many situations in life where you're asked to flip a coin and roll a die and look for specific outcomes, this question at least will let us get at the heart of when multiplying probabilities is a good idea. More "real world" examples of when we multiply probabilities will follow.

By now, I hope that one of the first things you might consider in your quest to answer this question is the idea of enumerating all of the physical outcomes and then trying to match up which of those fits the conditions that we care about. In this case, the coin flip is completely *independent* of the die roll, so, since we know the coin flip will result in either heads or tails, and the 6-sided die will result in a number between one and six (inclusive), we can list the 12 possible outcomes (twelve since the coin flip has two outcomes, the die roll has six outcomes, and $2 \cdot 6 = 12$).

Looking at Table 1.3 we can see the different results. Note that the outcome, H5, for example, is shorthand for obtaining a head on the coin flip and a five on the die roll. At this point, you can look at the possible outcomes and pick out those that match "head on the coin and a 5 or 6 on the die" and, since only H5 and H6 fit that condition, the probability is $2/12 = 1/6 \approx 16.67\%$.

TABLE 1.3: Flip and Roll Outcomes

H1	H2	H3
H4	H5	H6
T1	T2	T3
T4	T5	T6

While we will consider a fuller definition of what it means for two conditions to be *independent*, for now the easiest way to think about conditions being *independent* is that the probability of one condition happening does not depend at all on the other condition happening (or not). Here it is fairly clear that the result of the die roll doesn't depend at all on the coin flip (and vice versa), so perhaps there is a way to combine the probabilities of the two conditions together.

Theorem 1.4 (Independent Probabilities) For two conditions A and B, if A and B are independent, then the probability that both A and B happen at the same time, denoted $P(A \text{ and } B)$ is

$$P(A \text{ and } B) = P(A) \cdot P(B).$$

How does this work in practice? Let's consider an alternative solution to our question about obtaining a head on a coin flip and a 5 or a 6 on a die roll.

Example 1.8 Obtaining a head on a coin flip happens with probability 1/2, and obtaining a 5 or a 6 on a die roll happens with probability $2/6 = 1/3$. Since the two conditions are independent, the probability of obtaining a head and also a 5 or 6 on the die roll is $(1/2) \cdot (1/3) = 1/6 \approx 16.67\%$, the same as before.

We can also take a look at another example involving independent conditions here.

Example 1.9 Suppose we flip a coin three times in a row. What is the probability that we get a head on the first flip, a head on the second flip, and a tails on the last flip (equivalent to the sequence HHT for flipping a coin three times)? The probability of getting a head on the first flip is 1/2, the probability of getting a head on the second flip (completely independent of the first flip) is 1/2, and the probability of getting a tails on the last flip is 1/2 (again, completely independent of either the first or second flip). Combining these together we see

$$\text{Probability of HHT} = \underbrace{\left(\frac{1}{2}\right)}_{\text{first head}} \cdot \underbrace{\left(\frac{1}{2}\right)}_{\text{second head}} \cdot \underbrace{\left(\frac{1}{2}\right)}_{\text{last tail}} = \frac{1}{8} = 12.5\%.$$

Now we can take time to use the ideas of this chapter to answer a question that players of Yahtzee[2] would like to know (or at least keep in their minds) as they play the game. It should be noted before, though, that *non*-independent probabilities will be examined in Chapter 8.

Question

What is the probability that you obtain a 6 on a roll of a standard 6-sided die if you are allowed to reroll the die once?

While Yahtzee will be discussed in detail in Chapter 4, the question comes from the scenario in which you roll your dice (in Yahtzee you are working with five 6-sided dice), get four sixes on your first roll, and want to try for a Yahtzee (also known as five-of-a-kind). In the game, after the initial roll, you are given the chance to reroll any or all of your dice twice. Here we're asking about our probability of getting that last 6 to complete our Yahtzee.

Example 1.10 There are two ways for us to satisfactorily obtain the much needed 6. The first is simply to roll a 6 on our next roll of that die. This

---

[2]Yahtzee is a registered trademark of the Milton Bradley Company/Hasbro Incorporated.

happens with probability 1/6. The second is for us to need to reroll the die because we did not get a 6, and then roll a 6 on the reroll. This last condition involves two independent events; the first is not rolling a 6 on our next roll, and the second is rolling a 6 on the reroll! This later situation is covered by our multiplication rule for independent events so that probability is

$$\underbrace{\left(\frac{5}{6}\right)}_{\text{roll non-6}} \cdot \underbrace{\left(\frac{1}{6}\right)}_{\text{6 on reroll}} = \frac{5}{36}.$$

Combining these together using the additive rule (since "rolling a 6 on our next roll" and "rolling a non-6 on our next roll and a 6 on the reroll" are disjoint conditions), we obtain our final answer as

$$\underbrace{\left(\frac{1}{6}\right)}_{\text{roll 6}} + \underbrace{\left(\frac{5}{6}\right)}_{\text{roll non-6}} \cdot \underbrace{\left(\frac{1}{6}\right)}_{\text{6 on reroll}} = \frac{11}{36} \approx 30.56\%.$$

Shortcuts

There are many, many questions that can be answered by using the additive rule and independence, along with looking at the raw sample spaces. There are, though, some shortcuts that should be mentioned early so that you can take full advantage of them as your study of probability proceeds.

Question

What is the probability of flipping 20 coins and getting at least 1 head?

Before we answer this question specifically, let's answer a slightly simpler question; look back at Table 1.1 as you think about the probability that we flip 4 coins and try to get at least 1 head. From the 16 outcomes there, 15 of them have at least one head, so the answer is pretty easily determined as $15/16 = 93.75\%$. However, considering the *complement* (or opposite) condition is even simpler! The conditions "get 0 heads" and "get at least 1 head" here are not only disjoint, but exactly one of these conditions must happen!

Theorem 1.5 If A is any condition, and Z is the complement or opposite of A, then it must be true that

$$P(A) = 1 - P(Z).$$

Given the above discussion, this result is easy to understand. Since conditions A and Z given are disjoint, we can use the additive rule and know that there will be absolutely no overlap, so $P(A$ or $Z)$ is given by $P(A) + P(Z)$, but since we know that exactly one of A or Z *must* happen, $P(A$ or $Z) = 1$ (or 100%) How is this useful? By looking at our 4 coin example, note that the opposite of getting at least 1 head is getting no heads, and for 4 coins, this means the very specific physical outcome TTTT. Since the probability of that one outcome is $1/16$, the probability of the opposite of that (getting at least 1 head) is $1 - 1/16 = 15/16$ as we saw before. While this didn't save much time now, let's answer our question.

Example 1.11 There are 1,048,576 possible physical outcomes when flipping 20 coins (since each coin has two possible outcomes, there are 2 outcomes for 1 coin, 4 outcomes for 2 coins, 8 outcomes for 3 coins, continuing until we reach 2^{20} outcomes for 20 coins). The complement of getting at least 1 head is getting 0 heads; that is, getting the one specific outcome of all tails: TTTTTTTTTTTTTTTTTTTT. Since each coin is independent of the others, we can multiply probabilities together with

$$\text{Probability of all tails} = \underbrace{\left(\frac{1}{2}\right)}_{\text{first coin}} \cdot \underbrace{\left(\frac{1}{2}\right)}_{\text{second coin}} \cdots \underbrace{\left(\frac{1}{2}\right)}_{\text{nineteenth coin}} \cdot \underbrace{\left(\frac{1}{2}\right)}_{\text{twentieth coin}}$$

which can be combined together to equal $\left(\frac{1}{2}\right)^{20} = 1/1,048,576$. The answer to the question about getting at least 1 head on 20 coins is then

$$1 - \frac{1}{1,048,576} = \frac{1,048,575}{1,048,576} \approx 99.999905\%.$$

At this point, it would be easy to compute the probability of *not* getting a 6 in our Yahtzee situation previously! On one hand, we could compute it by noticing that to not get a 6 while allowing one reroll, we would have to roll twice and not get a 6 either time, and since the rolls are independent, we would get the probability of not getting the 6 as $(5/6) \cdot (5/6) = 25/36$. On the other hand, using our shortcut, since we already *know* the probability of getting the 6 allowing one reroll is $11/36$, the probability of not getting the 6 is $1 - 11/36 = 25/36$, the same thing!

By now, you should see that there are often multiple ways to compute a specific probability, so you shouldn't be surprised throughout this text if you work an example or exercise problem differently than your classmates (or even me). You will, though, assuming things are done correctly, always get the same answer.

The discussion and examples so far have implied the following easy way to find the total number of outcomes when multiple physical items are used in an experiment. In particular, recall from Table 1.3 that flipping a coin (with

2 outcomes) and rolling a die (with 6 outcomes) at the same time resulting in a sample space of $2 \cdot 6 = 12$ outcomes. This generalizes nicely.

Theorem 1.6 If an experiment or process is created by combining n smaller processes, where the first process has m_1 possible outcomes, and second process has m_2 possible outcomes, and so on, the total number of outcomes for the experiment as a whole is

$$\text{Total Number of Outcomes} = m_1 \cdot m_2 \cdot m_3 \cdots m_n.$$

For example, if we were to roll an eight-sided die and a twenty-sided die, and then flip 2 coins, we would have

$$\text{Total Outcomes} = \underbrace{8}_{\text{die 1}} \cdot \underbrace{20}_{\text{die 2}} \cdot \underbrace{2}_{\text{coin 1}} \cdot \underbrace{2}_{\text{coin 2}} = 640$$

different physical outcomes. Even when the individual pieces have (relatively) few possible outcomes, Theorem 1.6 can give many, many outcomes for the full process!

The Monty Hall Problem

Imagine that you're on a television game show and, as the big winner of the day, get to proceed to the bonus round at the end of the day. The host shows you three large doors on stage; behind two of the doors is a "prize" of little value, such as a live goat, and behind one door is a prize of great value, perhaps a new car. You're asked to select one of those doors; at this point, given no information about where any of the prizes might be, we can agree that the probability of you haphazardly selecting the correct door

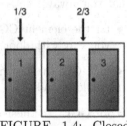

FIGURE 1.4: Closed Doors

(assuming you would like a new car as opposed to a slightly-less-than-new goat) is 1/3 since there are three doors and only one contains a car. However, once you have picked a door, the host opens one of the *other* two doors, revealing a goat, and then immediately asks you if you would prefer to keep your original selection or if you would like to switch your choice to the other unopened door.

Question

Is it more advantageous to keep your original door, switch to the other unopened door, or does it not make a difference?

Take a moment to think about your answer! This particular problem is known as the "Monty Hall Problem," named for the host of the game show *Let's Make a Deal* that was popular throughout the 1960s, 1970s, and 1980s (a more recent version, starring Wayne Brady as the host, debuted in 2009). The bonus round on the show, known as the "Big Deal" or "Super Deal" never aired exactly as the situation above was presented; this particular situation *was* asked of Marilyn vos Savant as part of the "Ask Marilyn" column in *Parade Magazine* in 1990 [45]. What made this problem so famous was that, after Marilyn's (correct) answer was printed, many people, including mathematicians, wrote in to the magazine claiming that her answer was wrong! If your own answer was that it does not matter one bit whether you switch or not, you're among the many people that have been stymied by this brainteaser; the correct answer is that you should most definitely *switch* as doing so doubles your probability of driving off in a new car. Let's see why.

To be fair, there are two small but very reasonable assumptions that need to be made in terms of Monty Hall's behavior; the first is that Monty will always open one of the two doors not selected by you originally. The second is that Monty knows the location of the car and will always open a door to reveal a goat and *not* the car. The elements of our sample space here could be written as ordered triples of the letters G (for goat) and C (for car) such that the triple contains exactly one C and two Gs. That is, the possible (random) configurations for which doors contain which prizes are {CGG, GCG, GGC}, each equally likely. Suppose that your initial choice is door number one and that you always choose to *stay* with that door; then we see that

- for the element CGG, the host will open either door two or door three, you stay with door one, and you win a car,

- for GCG, the host must open door three, and staying with door one makes you lose, and

- for GGC, the host opens door two, for which staying with door one also makes you lose.

Choosing to stick with your original door keeps your chance of winning the same as the initial 1/3; you can verify that if you choose either door two or door three originally, the same argument above applies.

When you choose to *switch*, however, you're (smartly) taking advantage of new information presented about the pair of doors you did *not* choose. Again, suppose you always pick door one and will choose to switch; we see that

- for CGG, the host will open either door two or door three, and switching off of door one will cause you to lose,

- for GCG, the host opens door three, so switching (to door two) will make you win, and

- for GGC, the host opens door two, so switching again makes you win.

Choosing to *switch* from your original door increases your chance of winning the car to 2/3 which can be counterintuitive.

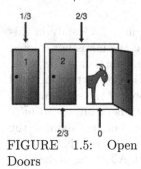

FIGURE 1.5: Open Doors

In other words, looking at Figure 1.4, when you make the first choice of which door you want, you're actually choosing which pair of doors you're interested in learning more about! You *know*, given the assumptions of this problem, that one of those two doors will be opened, so would you rather choose to *stay*, winning with probability 1/3, or *switch*, winning with probability 2/3 by essentially getting two doors for the "price" of one? Indeed, this latter option is the best; at the start, there's a 2/3 probability that the collection of two unselected doors contains the car somewhere, and as we see in Figure 1.5, once Monty Hall opens one of those two doors, the probability the remaining unselected door contains the car must remain as 2/3 (as indicated by the bottom set of numbers in the figure).

There's a distinct possibility right now that you're still not convinced that switching is the correct answer. As one attempt to guide you to be more convinced, suppose instead of three doors, there are *one hundred doors* and, after you select one door at the start, the host of the game show opens 98 doors to reveal 98 cute goats; your initial choice was correct with probability 1 out of 100. The probability that the *other* 99 doors collectively contains a door with a new car behind it is, of course, 99 out of 100 by using Theorem 1.5. After the host opens 98 of those doors, the probability that the 99 doors you did not initial select is *still* 99 out of 100, so you should most definitely switch! If you are still not convinced of the solution to the original problem, find some friends and actually play this game over and over again; you should see that the probability of winning the car by switching is really close to 2/3 in your simulations.

There are many other ways to think about and explain the probabilities involved in this problem; one involves conditional probability and I encourage you to think about the Monty Hall Problem again after reading Chapter 8. For more, consult the plethora of resources available on the internet (including a plethora of simulations and online "games" that show empirically that 2/3 is precisely correct) along with the book "The Monty Hall Problem" by Jason Rosenhouse, found as [41] in the Bibliography. While the situation, as presented and as asked of Marilyn vos Savant in the first place, did not happen on Let's Make a Deal, Monty Hall *did* write back to a mathematician who posed the question earlier, remarking that "once the first [door] is seen to be empty, the contestant cannot" change their door and, "incidentally, after one [door] is seen to be empty, [their] chances are no longer 50/50 but remain what they were in the first place, one out of three."

1.4   Candy (Yum)!

I have to make a slight confession here. You will not find any candy in this
book. The pages are not made out of candy paper, and the ink is not colored
sugar, so you can stop trying to think about eating any portion of the book.
You will also not find a lot of mathematics performed in this section, nor will
we be using the discussion here in the future. The purpose of the section, then,
is to really sell the point that the world of probability isn't always so nice.
Fair dice are nice, as Las Vegas would certainly agree, and perfect shuffles (the
kind that give you a truly randomized deck of cards, not the kind that you can
use to control the layout of a deck) would be wonderful. All of the probability
calculations we do are based on the idea that we know exactly the behavior of
certain processes. For us, that means a coin flip results in heads exactly 50%
of the time and tails exactly 50% of the time. It means that rolling a 6-sided
die results in a 4 exactly 1/6 of the time. Unfortunately, this is not always the
case (and, possibly, never the case).

FIGURE 1.6: M&Ms

My guess is that almost everyone reading this
book has had M&Ms[3] at some point in his or her
life (for those who haven't, Figure 1.6 shows the
classic candy). Who can resist the chocolate candies, covered with a colorful shell, that melt in
your mouth but not in your hand? I also guess
that most everyone has a favorite color among
them all! The question we may ask, though, is
what is the probability that a randomly selected
M&M will be blue (my favorite color)?

There are essentially two ways to answer this question. The first, of course,
would be to ask Mars, Incorporated, the manufacturer of the candy delights!
According to their website, back in 2007, each package of standard milk chocolate M&Ms should have 24% of its candies coated in a lovely blue color.[4] However, the current version of their website does not list the color distributions.

Has the percentage of each color changed since 2007? How would we know?
Is there a secret vault somewhere with the magic numbers for the distribution
of colors? The answers that I have for you for these three questions are

1. possibly,

2. look below, and

3. I have no idea.

The second way to possibly answer the question is to go out to your local
grocery store, purchase a bag of M&Ms, open it up, and count the quantity of

---

[3]M&M is a registered trademark of Mars Incorporated.

[4]For the curious, the same site in 2007 listed the distribution as 24% blue, 14% brown,
16% green, 20% orange, 13% red, and 14% yellow. These numbers are slightly rounded, as
you might imagine, since they add up to more than 100%.

each color that shows up. If the distributions are correct from 2007, then a bag with 200 M&Ms should have about 48 blues (yes, blue is really my favorite color). Maybe you were only able to find a "Fun Size" bag with 20 M&Ms; you should have about 4.8 blue M&Ms (or close to 5, since 0.8 M&Ms are typically not present). A slightly larger bag at the checkout line might have 50 M&Ms and you could work out the expected number of blue candies (12). Hopefully, common sense should tell you that the larger the bag, the closer your results should mirror "reality."

At this point, it sounds like we've moved from probability to statistics! If that's what you were thinking, you're exactly right! The fields are intimately related, and the process of trying to determine what distribution a particular "experiment" (such as color distributions for candies) has can be done with statistical techniques. In statistics, a (usually small) percentage of a population (called a sample) is used to make generalizations about the population itself. Here we are trying to use packages of M&Ms to determine what percentage the factory used to package all of the M&Ms. Larger sample sizes cut down on the amount of error (both from the process of sampling itself and also the use of the sample to make generalizations), so using a large "family size" bag of M&Ms is better than a "Fun Size" bag. Even better would be grabbing large bags of M&Ms from stores in Paris, Tokyo, Dubai, San Diego, and Dauphin, Pennsylvania. The best option would be finding thousands of bags containing M&Ms, randomly selecting a certain number of them, and using those to determine the possible distribution of colors.

If the candy example doesn't convince you that statistics can be important, imagine yourself trying to determine what percentage of voters in your state approve of a certain law. Sure, you could ask everyone in the state directly and determine the exact percentage, but you would be surprised at how much money it would cost and how much time it would take to find the answer to that question (versus choosing a sample of, say 1,000 voters, and asking them).

Most of this book is dedicated to what I like to call "expected probability," the calculable kind for which we can answer many important questions. It's the probability that casino owners rely on to keep the house afloat. However, I firmly believe that "observed probability" can be as important, and in some cases, is the only type of probability available. As you experience the world, try to match up what you see with what you know. Getting a feel for how, say, games of chance work by observing the games and thinking about them can give you a sense of when situations are more favorable for you. Combining what you see with the "expected probability" we will be working with throughout the book will give you an edge in your decision making processes.

Oh, as an instructor of statistics as well, I would encourage you to take a course in statistics or pick up a book on statistics after you finish with this book. Everyone can benefit from an understanding of statistics!

1.5   Exercises

1.1 Consider the "experiment" of flipping three (fair) coins. Determine the set of possible physical outcomes and then find the probability of flipping:
(a) exactly zero heads.   (b) exactly one head.   (c) exactly two heads.
(d) exactly three heads.

1.2 Consider rolling an 8-sided die once.
(a) What are the possible outcomes?
(b) What is the probability of rolling a 3?
(c) What is the probability of rolling a multiple of 3?
(d) What is the probability of rolling an even number or a number greater than 5?
(e) If you just rolled a 6, what is the probability of rolling a 6 on the next roll?

1.3 How many different outcomes are possible if 2 ten-sided dice are rolled? How many outcomes are possible if, after rolling two dice, four coins are flipped?

1.4 Some of the origins of modern gambling can be linked to the writings and studies of Antoine Gombaud, the "Chevalier de Méré" who lived in the 1600s. He was very successful with a game involving four standard six-sided dice where "winning" was having at least one of those dice showing a 6 after the roll. Determine the probability of seeing at least one 6 on rolling four dice; using Theorem 1.5 may be helpful.

1.5 Repeat Exercise 1.4 for three dice and for five dice instead of the four dice used by the Chevalier de Méré; are you surprised at how much the probability changes?

1.6 A player is said to "crap out" on the initial roll for a round of craps if a total of 2, 3, or 12 is rolled on a pair of six-sided dice; determine the probability that a player "craps out."

1.7 For this exercise, suppose that two standard six-sided dice are used, but *one* of those dice has its 6 replaced by an additional 4.
(a) Create a table similar to Table 1.2 for this experiment.
(b) Determine the probability of rolling a sum of 10.
(c) Determine the probability of rolling a 7.
(d) Determine the probability of not obtaining a 4 on either die.
(e) If these dice were used in craps as in Exercise 1.6, what is the probability that a player "craps out"?

1.8 Suppose that you are using two special six-sided dice instead. On one of the dice, the sides are 1, 3, 4, 5, 6, and 8. The other die has 1, 2, 2, 3, 3, and 4 as the sides. Repeat Exercise 1.7 for this pair of special dice and comment on the differences in your answers.

1.9 Instead of rolling just two six-sided dice, imagine rolling *three* six-sided dice.
(a) What is the size of the resulting sample space?
(b) Determine the number of ways and the probability of obtaining a total sum of 3 or 4.
(c) Determine the number of ways and the probability of obtaining a total sum of 5.

(d) Using your work so far, determine the probability of a sum of 6 or more.

1.10 A standard American roulette wheel has 38 spaces consisting of the numbers 1 through 36, a zero, and a double-zero. Assume that all spaces are equally likely to receive the ball when spun.

(a) What is the probability that the number 23 is the result?

(b) What is the probability of an odd number as the result?

(c) Determine the probability that a number from the "first dozen" gets the ball, where the "first dozen" refers to a bet on all of the numbers 1 through 12.

(d) Determine the probability of an odd number or a "first dozen" number winning.

1.11 Most introduction to probability books love discussing blue and red marbles; to celebrate this, we will examine some situations about marbles here. Suppose an urn has 5 blue marbles and 3 red marbles.

(a) Find the probability that a randomly selected marble drawn from the urn is blue.

(b) Assuming that marbles are replaced after being drawn, find the probability that two red marbles are drawn one after another.

(c) Again, assuming that marbles are replaced after being drawn, find the probability that the first two draws consist of one red and one blue marble.

(d) Now, assuming that marbles are *not* replaced after being drawn, find the probability of drawing two blue marbles in a row.

1.12 More generally, suppose that an urn has n blue marbles and m red marbles.

(a) Find an expression in terms of n and m that gives the probability of drawing a blue marble from the urn and find a similar expression for the probability of drawing a red marble from the urn.

(b) Using your expression for the probability of drawing a blue marble, discuss what happens to this probability as m (the number of red marbles) gets larger and larger. Then discuss what happens to the probability as n (the number of blue marbles) gets larger and larger.

(c) Find an expression, in terms of n and m, that gives the probability of drawing two blue marbles in a row, assuming that the marbles are not replaced after being drawn.

(d) Repeat part (c) assuming that the marbles *are* replaced after being drawn.

1.13 Suppose you have an unbalanced coin that favors heads so that heads appears with probability 0.7 and tails appears with probability 0.3 when you flip the coin.

(a) Determine the probability of flipping the coin, obtaining a head, and then flipping the coin again, and obtaining a tail.

(b) Now determine the probability of flipping the coin, obtaining a tail, and then flipping the coin again with a result of a head.

(c) If your friend asks you to flip a coin to determine who pays for lunch, and you instead offer to flip the coin twice and he chooses either "heads then tails" or "tails then heads," is this fair? Note that we would ignore any result of two consecutive flips that come up the same.

(d) Based on your results, could you use this method to fairly decide any situation involving "a coin flip," even with an unbalanced coin?

2

Roulette and Craps: Expected Value

2.1 Roulette

The game of roulette has been around for over 200 years; several games throughout Europe in the 17th and 18th centuries contributed to what has been standardized in the last century. Historians regularly credit mathematician (and inventor, writer, philosopher, and physicist) Blaise Pascal's experiments with perpetual motion as the impetus behind the wheel itself, and the

FIGURE 2.1: Roulette

game that most resembles current roulette started to become famous in Paris during the late 1700s. Indeed, the name itself comes from the French word for "little wheel." A modern roulette wheel appears in Figure 2.1.

The game is rather easy to play, and the mathematics behind the bets will give us a glimpse of exactly how a casino is able to stay in business. If you're lingering around the floor of a casino in Las Vegas, Atlantic City, or anywhere else, it will be very hard to not notice the roulette table! As the ball rolls around the outside of the wheel, the anticipation builds at the table, and as the ball begins to settle into the middle, bouncing around several times and hitting a few spokes, thousands of dollars can be won or lost. Some grumbles and perhaps a lot of excitement can be heard as the ball finally settles into a slot as the croupier yells "thirty-three black!"

We will be working with what we will call American roulette; the exercises at the end of the chapter will explore variations of roulette (in particular European roulette, which is traditionally slightly different than American roulette, as you will see in the exercises). As you can see in Figure 2.2, there are the numbers 1 through 36, separated into half red and half black numbers, and two additional (green) spaces, one called 0 and the other 00 (double zero). Note for the purposes of the "even" bet we'll discuss below, neither 0 nor 00 are considered even during a game of roulette (despite the fact that 0 is an even number to mathematicians and pretty much everyone in the world *but* roulette players). As another historical note, the original game in Europe did include both the 0 and 00, with the 0 space colored red and the 00 space colored black; however, both spaces were considered the "bank's numbers" and did not pay out for the player, even when betting on red or black. The spaces

23

were colored green very shortly after to avoid confusion (the current European version, where 00 is not included, was introduced at the time to compete with casinos that offered roulette with both 0 and 00; the reason that this strategy worked for casinos will become apparent later).

Playing roulette is fairly simple, and the payouts are fairly easy to understand as well, so this early casino game will be a great introduction to some of the mathematical tools we will end up using for most of our game analysis. Roulette is played in rounds, and a round starts by the croupier indicating that it is time to place bets. Shortly after, he or she will give the wheel a spin, start the ball rolling in the opposite direction, and very soon call out "no more bets," at which time the fate of your bets is about to be determined. After the result has been determined, the croupier will take away the losing bets, and pay out on the winning bets. Then a new round will start.

Odds

One example bet that we can use to explain payout odds and typical casino behavior is a bet on the single number 23 (or any other number). What will become especially important in our study of casino games (and the single source of the "house advantage") will be the difference between true mathematical odds and casino payout odds for the various bets.

> **Definition 2.1** The *true mathematical odds* or *true odds* of an event or wager is a pair of numbers, separated by a colon, as in $X : Y$, where the first number X represents the number of ways to not win the wager and the second number Y represents the number of ways to win the wager.

It is easy to think of true odds as (# of ways to lose):(# of ways to win), which will be correct as long as there are no opportunities for the wager to be a push (tie).

> **Example 2.1** For a bet of 23 at roulette, since there are 38 spaces on the wheel and only 1 way to win, there are 37 ways for us *not* to win; the true mathematical odds for a single number bet at American roulette is 37:1 (in Definition 2.1, X is 37 and Y is 1).

Spoken out loud, you would read "37:1" as "thirty-seven to one." One important note here; even though the probability of getting a 23 on one spin of the wheel is indeed $\frac{1}{38}$ (one physical outcome gives success, out of a possible 38 outcomes), the true mathematical odds are 37:1 since, while the ideas are very, very closely related, the meaning is slightly different. To convert true mathematical odds of 3:2 to a probability of winning, note that since there are 3 ways of losing and 2 ways of winning, there are 5 possible outcomes. Then the probability of winning is 2/5 and *not* 2/3.

Theorem 2.2 An event with true mathematical odds $X : Y$ can be converted into a probability of winning as

$$\frac{Y}{X+Y}$$

and an event with probability A/B can be converted into true mathematical odds of $(B - A) : A$.

It's also worth mentioning here that the true mathematical odds of say 4:2 and 2:1 are identical for all purposes that we will use them for (use Theorem 2.2 to convert each of those odds to probabilities and you will get the same value). At this point, you might wonder where the house gets its advantage. If you wager $1.00 on 23 at roulette, and it actually does come up, how much money will you now have? The following definition will help.

Definition 2.3 The *casino payout* or *casino or house odds* for a wager or event, expressed also in the form $X : Y$, means that the casino will pay you winnings of $$X$ for each $$Y$ you wager.

A quick study of roulette payouts will tell you that the casino payout is listed at 35:1, or 35 to 1. This means, using Definition 2.3, that on your $1.00 wager, the casino will give you an additional $35.00 for winning; you also get your original $1.00 back, giving you a total of $36.00 returned on that $1.00 bet. There are two important things to mention here; *first, unless otherwise specified, it is assumed that your original wager will be returned to you in addition to your winnings.* It should also be common sense, for example, that if instead you wager $2.00 on number 23 and win, that the 35:1 house odds means that you will receive $70.00 in winnings, since, as per Definition 2.3, you receive winnings of $35.00 for each $1.00 that you bet.

Let's now take a look at all of the types of bets and payouts for American roulette as depicted in Figure 2.2. A bet on a single number, as we discussed above, is pictured as item ① and has a 35:1 payout. Item ② is a pair of adjacent numbers, such as betting on 19 and 20, or 7 and 10, or, as pictured, 2 and 3. The payout for a pair of numbers is 17:1 as is often called a "split" bet. A wager of three numbers in a row, such as ③ in the figure, has a payout of 11:1 and is called taking the "street." Choosing a bet on four numbers of a square, such as 17, 18, 20, and 21 pictured as bet ④ is called a "corner" bet and has a 8:1 payout from the casino. Item number ⑤ is the "first five" bet, with house odds of 6:1, and, as we will see, is the single worst bet that can be made at the game of roulette. Bet ⑥ is an example of taking a pair of adjacent streets, often called a "six line" bet that pays 5:1. Two types of bets, ⑦ and ⑧, have a 2:1 payout. These wagers consist of first dozen, second dozen, or third dozen (bets identified by ⑦), and each of the three columns of numbers depicted as ⑧ (one example is the collection of numbers 1, 4, 7, 10, 13, 16, 19, 22, 25, 28, 31, and 34). Finally, there are the three different types of even money bets (defined as having a payout of 1:1): first 18 or last 18 (⑨ and ⑩), red or black (⑪), and even or odd (⑫). A summary of this information is provided in Table 2.1.

What questions can we ask about roulette? Determining the probabilities of any of these bets should not be difficult at this point. With only 38 spaces on the American roulette wheel, all of these can be computed directly without the need of the additive rule or other "heavy" machinery from Chapter 1. However, as you hopefully have experienced in previous mathematics classes, practice is the key to true understanding. It is definitely worth the time to make sure that computing the probability of any of these bets is straightforward and simple. We will do one together, and the rest are left for you as an exercise at the end of the chapter.

FIGURE 2.2: Roulette Bets

Example 2.2 To determine the probability of the "dozens" bet for American roulette, we recall that there are 12 possibilities for our bet to be successful (for example, for the "first dozens" bet, any of the numbers from 1 to 12 will make us happy) out of a possible 38 slots on the wheel. This gives us a $\frac{12}{38} \approx 31.58\%$ chance of success.

At this point, we are ready to ask the next question of interest to us in our study of roulette, and a question that will be of interest to us for any casino game we may study.

Question

What is the house advantage for a bet on a single number in roulette?

Expected Winnings (Expected Value)

Before we answer our specific question, imagine yourself in a scenario where you are about to flip a coin, and before you do so, a friend offers you a "friendly" bet. If the coin comes up heads, he will pay you $1.00, and if the coin comes up tails, you have to pay him $1.00. Is this a fair bet? Most people would look at this and quickly agree that it is indeed a fair bet, but the better question is: why? When I ask my students, I usually get a response such as "since half of the time you win a dollar and half of the time you lose a dollar, in the long run no one makes (or loses) money." If you had a similar thought, you're well ahead of the game! The idea here is that we could easily compute the *average* winnings, which tells us a lot about this bet.

$$\text{Average Winnings: } \underbrace{\left(\frac{1}{2}\right) \cdot \$1.00}_{\text{heads}} + \underbrace{\left(\frac{1}{2}\right) \cdot (-\$1.00)}_{\text{tails}} = \$0.00$$

TABLE 2.1: Roulette Bets

Bet	Payout
Single Number	35:1
Adjacent Numbers	17:1
Three Numbers	11:1
Four Numbers	8:1
Five Numbers	6:1
Six Numbers	5:1
Dozens	2:1
Columns	2:1
First/Last 18	1:1
Red/Black	1:1
Even/Odd	1:1

This tells us that the *expected winnings* (or *expected value*) for this bet is $0.00, meaning no advantage for you or for your friend. How do things change when the situation isn't evenly balanced (as with a coin)? Instead of using a simple average, we can use a *weighted average* to determine the expected winnings. My guess is that you are already familiar with this concept but perhaps the name doesn't ring a bell; in the classroom, when computing grades, oftentimes "exams" will count more than "participation," so your final grade average in the course is computed by applying a higher weight for exams than participation. We do the same here to answer our question.

Example 2.3 To determine the house advantage on a $1.00 single number bet, say 34, we need to know the probability of winning and losing. Since only the number 34 will lead us to a win, the probability of hitting that single number (along with any other single number) is $\frac{1}{38}$. Any of the other 37 results on the wheel will cause us to lose, with probability $\frac{37}{38}$ (or, as a faithful reader from Chapter 1 will remember, this is also $1 - \frac{1}{38}$ since in this scenario, it is only possible to win or lose; there are no ties). If we win, the payout is 35:1, so we will win $35.00. If we lose, we simply lose our $1.00 wager.

$$\text{Expected Winnings: } \underbrace{\left(\frac{1}{38}\right) \cdot \$35.00}_{\text{win}} + \underbrace{\left(\frac{37}{38}\right) \cdot (-\$1.00)}_{\text{lose}} \approx -\$0.0526$$

From our calculation, it appears that, in the long run, we lose about 5.26

cents per single number wager if we bet $1.00. Looking at the expected winnings calculation, you should notice that if we wagered $10.00 instead of $1.00, each number in the formula would be multiplied by 10, including the answer. We can then safely say that the house advantage on a single number bet in roulette is $0.0526 per dollar, or 5.26% if you prefer.

Note that, even though our original $1.00 wager is returned to us, it doesn't show up in the expecting *winnings* calculation except as a possible loss if we lose. We do not consider the money that was actually wagered to be part of the winnings, only as that possible loss.

While most of the other house advantage values will be left to you to finish as an exercise, I would like to do two more. First, going from one extreme to the other, let's compute the house advantage (expecting winnings) for a bet on red (since the bet amount scales the expected winnings, it's enough for us to compute expecting winnings for a $1.00 wager and express our answer as a "per dollar" result or as a percentage).

Example 2.4 From looking at the roulette wheel or betting area, we see that 18 out of the 38 spaces are red. Our chance of winning is $\frac{18}{38}$ and our chance of losing is the opposite $\frac{20}{38}$. With 1:1 casino odds, we will only win $1.00 if our bet is successful, so the expected value computation yields the below.

$$\text{Expected Winnings: } \underbrace{\left(\frac{18}{38}\right) \cdot \$1.00}_{\text{win}} + \underbrace{\left(\frac{20}{38}\right) \cdot (-\$1.00)}_{\text{lose}} \approx -\$0.0562$$

So far you may be wondering if *all* bets in roulette have the same house advantage! It certainly looks that way! However, if you were reading carefully when the bet types were discussed, you'll remember that the "first five" bet, depicted as ⑤ in Figure 2.2, was described as being the worst bet in roulette. Let's see why.

Example 2.5 For the "first five" wager, we need either the 0, 00, 1, 2, or 3 to come up, for a total of 5 possibilities out of the 38 spaces. We can see that the probability of winning is $\frac{5}{38}$ and the probability of losing is $\frac{33}{38}$. With a payout of 6:1, we can now proceed.

$$\text{Expected Winnings: } \underbrace{\left(\frac{5}{38}\right) \cdot \$6.00}_{\text{win}} + \underbrace{\left(\frac{33}{38}\right) \cdot (-\$1.00)}_{\text{lose}} \approx -\$0.0789$$

Since you will see that all of the other wagers give a 5.62% house advantage, the 7.89% house advantage for the "first five" bet makes it significantly better for the casino, and thus significantly worse for you, the player.

Hopefully you can see the importance of the expected value calculation, so named because it gives us an idea of what to "expect" in the long run. However, it's worth noting here that oftentimes the numerical value computed for an expected value is not *obtainable* in that, for example, you will never actually lose $0.0562 on a single bet at roulette; people often call expected value the (weighted) average to stress this phenomenon.

A very small trip into the world of game design now seems appropriate.

Question

How does a casino maintain an edge on new games or wagers?

As a small example, we will consider the idea of adding a new bet to roulette that wins whenever a prime number comes up on the wheel. You may remember that a prime number is a positive integer greater than one whose only divisors are itself and one. From the collection of numbers 1 through 36, the set of prime numbers consists of 2, 3, 5, 7, 11, 13, 17, 19, 23, and 29, for a total of 10 numbers. How much should the casino pay out for this wager?

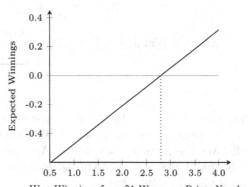

W = Winnings for a $1 Wager on Prime Number

FIGURE 2.3: Expected Winnings on Prime Number

Example 2.6 Given that there are 10 ways to win, out of the 38 possible outcomes for American roulette, there must be 28 ways to lose. This gives us true mathematical odds of 28:10, or, by dividing both numbers by 10, equivalent true mathematical odds of about 2.8:1. To determine the expected winnings, consider a payout of W on a $1.00 wager for this bet. We then get

$$\text{Expected Winnings} = \underbrace{\left(\frac{10}{38}\right) \cdot \$W}_{\text{win}} + \underbrace{\left(\frac{28}{38}\right) \cdot (-\$1.00)}_{\text{lose}} = \frac{10W - 28}{38}$$

where we drop the dollar sign on the right to keep the formula from looking too ugly. Pictured in Figure 2.3 is this quantity, plotted for values of W ranging from $0.50 to $4.00. Do you find it rather curious that the expected winnings are $0.00 when $W = 2.8$? I would imagine by now that you don't find that surprising! If the casino wants to guarantee a house advantage for a game, any payout that is below the true mathematical odds will do.

To see how the casino can really affect their bottom line, we try two values of W that seem plausible. Note again that for true mathematical odds of X:1, as long as the casino payout is set at Y:1 where $Y < X$, the casino will make money. For a payout of 5:2 (which could also be written as 2.5:1), we get

$$\text{Expected Winnings } (W = 2.5) = \underbrace{\left(\frac{10}{38}\right) \cdot \$2.50}_{\text{win}} + \underbrace{\left(\frac{28}{38}\right) \cdot (-\$1.00)}_{\text{lose}} = -\$0.079$$

and for a payout of 9:4 (the same as 2.25:1), we get

$$\text{Expected Winnings } (W = 2.25) = \underbrace{\left(\frac{10}{38}\right) \cdot \$2.25}_{\text{win}} + \underbrace{\left(\frac{28}{38}\right) \cdot (-\$1.00)}_{\text{lose}}$$
$$= -\$0.145.$$

Players, of course, would prefer the first choice, and the casino might actually use the payout of 5:2 because it is simultaneously better for the player but still gives the house an advantage. A payout of 11:4 or 12:5 would be even better for the player while still keeping the house happy, but those payouts look a bit silly, don't they?

Before we move on to craps, it's worth formalizing this notion of expected value so that it can be used more in the future, especially when there are options other than "winning" and "losing."[1]

> **Definition 2.4** If values $W_1, W_2, \ldots, W_n$ can be obtained from an experiment with respective probabilities $p_1, p_2, \ldots, p_n$, then the expected value of the experiment is given by
>
> $$\text{Expected Value } = p_1 \cdot W_1 + p_2 \cdot W_2 + \cdots + p_n \cdot W_n.$$

To illustrate Definition 2.4 where more than two pieces are added, consider a slight modification to the "red" bet a roulette where a player wins when a red number comes up, loses when a black number or 0 comes up, and *pushes* when double zero (00) comes up; that is, a 00 result on the wheel is a tie for the player on this bet and the player's wager is simply returned.

---

[1]The definition, when using an infinite number of possible values, again can get interesting; see Appendix B for an interesting byproduct of infinity and expected value.

Example 2.7 For the modified red bet, note that there are still 18 results out of the 38 that result in a win, but there are only 19 results that count as a loss (any black number along with 0). With probability 1/38 the wager, $1.00 for our example, is returned to the player for a net profit (or loss) of $0.00. The expected value (expected winnings) for this bet then are

$$\text{EV} = \underbrace{\left(\frac{18}{38}\right) \cdot \$1.00}_{\text{win}} + \underbrace{\left(\frac{1}{38}\right) \cdot \$0.00}_{\text{push}} + \underbrace{\left(\frac{19}{38}\right) \cdot (-\$1.00)}_{\text{lose}} \approx -\$0.0263,$$

slightly better than the usual house advantage, but a house advantage nonetheless!

2.2 Summations

We've avoided using summation notation for formulas and computations so far, but discussing expected value gives us a nice opportunity to introduce the notation which will appear later in the book. If you consider the following *long* list of things to be added together, you might wish that there were a simpler way to express the same thing. Adding up the first 10 squares, one would obtain

$$1^2 + 2^2 + 3^2 + 4^2 + 5^2 + 6^2 + 7^2 + 8^2 + 9^2 + 10^2 = 385,$$

but if you wanted to add together the first 100 squares, writing each term out would be a waste of space, not to mention also very tedious. We could use shorthand such as

$$1^2 + 2^2 + 3^2 + \cdots + 99^2 + 100^2 = 338,350$$

where the use of $\cdots$ is fairly clear, but when the expressions become more and more complicated, even better shorthand is needed.

Definition 2.5 (Summation Notation, Version 1) The repeated addition of predictable terms can be represented in *summation notation*, where the shorthand on the left side of

$$\sum_{i=a}^{n} f(i) = f(a) + f(a+1) + f(a+2) + \cdots + f(n-1) + f(n)$$

represents the addition on the right, and is verbalized as "the sum from *i* equals *a* to *i* equals *n* of $f(i)$" where $f(i)$ represents some function in terms of *i* (called the *indexing variable* for the summation).

The benefits of Definition 2.5 should be immediately clear. Adding together the first 100 squares is

$$\sum_{i=1}^{100} i^2 = 1^2 + 2^2 + 3^2 + \cdots + 99^2 + 100^2$$

where the left-hand side is concise and clear. We can also list other small examples such as

$$\sum_{i=5}^{10} (2i+1) = (2 \cdot 5 + 1) + (2 \cdot 6 + 1) + (2 \cdot 7 + 1) + (2 \cdot 8 + 1)$$

$$+ (2 \cdot 9 + 1) + (2 \cdot 10 + 1) = 96$$

and

$$\sum_{s=-2}^{3} s \cdot (s-1) = -2 \cdot (-2-1) + -1 \cdot (-1-1) + 0 \cdot (0-1) + 1 \cdot (1-1)$$

$$+ 2 \cdot (2-1) + 3 \cdot (3-1) = 16$$

where the second example here illustrates that we need not always use the variable *i*; the meaning should still be clear from how Definition 2.5 is posed.

What's even nicer about summation notation is that we can take things one step further. Suppose that *S* is a *set* of ordered pairs (or triples, and so on) given in the form (x, y). For example, let

$$S = \{(0,3), (1,2), (2,1), (3,0)\}$$

so that we can now illustrate summations performed on a set, as we have in

$$\sum_{(x,y)\in S} x^2 + y = (0^2 + 3) + (1^2 + 2) + (2^2 + 1) + (3^2 + 0) = 20,$$

noting that the fancy symbol $\in$ means "in the set." The summation process here takes *every* element of the set *S*, computes $x^2 + y$, and then adds the resulting numbers. Same idea, but a bit more general! We make that formal with the following.

Definition 2.6 (Summation Notation, Version 2) The *summation notation* over a set S is written as

$$\sum_{s \in S} f(s)$$

where each element of the set S is inserted into the function to obtain values for $f(s)$; these values are all added together to obtain the final result.

Now that we have this notation available to us, take a quick look back at Definition 2.4. The definition for expected value can be expressed is summation notation as

$$p_1 W_1 + p_2 W_2 + \cdots + p_n W_n = \sum_{i=1}^{n} p_i \cdot W_i$$

and, using the general form, we can find the expected sum when rolling two dice.

Example 2.8 To determine the average value for the experiment of rolling two dice and taking the sum of the faces, let the set S consist of all possible pairs obtained when rolling two dice. That is,

$$S = \{(1,1), (1,2), \ldots, (1,6), (2,1), \ldots, (2,6), \cdots, (6,1), (6,2), \ldots, (6,6)\}$$

so that we can express the expected value as

$$\text{Expected Sum Two Dice} = \sum_{s \in S} \underbrace{\left(\frac{1}{36}\right)}_{\text{probability}} \cdot \underbrace{(x+y)}_{\text{sum}} = 7$$

where for each ordered pair $s = (x, y)$ we take $x + y$ as the sum. Note that there are 36 terms in this calculation, and the explicit calculation is omitted.

To end our coverage of summation notation, for now, let's take a look at a silly example for roulette.

Example 2.9 Suppose the a bettor wants to wager \$1.00 on each of the "regular" numbers for a round of roulette, with "regular" meaning that no wager will be placed on the zero or double zero spaces. What does this do to the expected winnings calculation that we've done before? Using summation notation, we get

$$\text{Expected Winnings} = \sum_{i=1}^{36} \underbrace{\left(\left(\frac{1}{38}\right) \cdot \$35 + \left(\frac{37}{38}\right) \cdot (-\$1)\right)}_{\text{winnings for bet on number } i} \approx 0\$1.89$$

which seems quite large. However, remember that we need to spend a total of $36.00 in order to make all of the wagers, and as a fraction of the amount wagered, the house advantage is just $1.89/36 = 0.0526$ or about 5.26%, the same as it was for any one of these 36 wagers!

2.3   Craps

Like roulette, a craps table in a casino can be the source of a ton of energy! The suspense as the shooter prepares to throw the dice, trying to roll a 7 or 11, or perhaps make their point, leads up to the cheers as the table wins, or groans as the shooter fails and the dice pass to the next shooter. While the rules of craps can seem daunting

at first, and the game definitely has a culture behind it (you will hear terms such as "yo-leven," "seven out," and "easy eight"), the excitement, possibility of winning (or, of course, losing) a lot of money, and ambience provide for a fun experience. Who knew that two dice could provide that much entertainment?

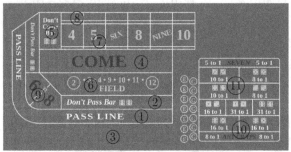

FIGURE 2.4: Craps Bets

While craps does have a good amount of "single roll" bets, the key to understanding how craps works is knowing that essentially it is played in rounds, with many of the bets on the table staying put until the round comes to an end. Many of the various bets are pictured in Figure 2.4, but for now we will focus on two of the basic bets, the "pass line" and "don't pass line" labeled by ① and ② respectively.

At the start of a round, the on/off button at the table (seen in Figure 2.5) will be placed in the "off" position, signaling that the table is ready for a *come out roll*. On this initial roll, the shooter's goal is to roll the dice and obtain a total of 7 or 11, for a pass line win at a 1:1 payout. A total of 2, 3, or 12 for the come out roll results

FIGURE 2.5: On/Off Button

in a loss of the pass line bet, and those totals are commonly referred to as "craps" during this roll. On any other value (4, 5, 6, 8, 9, or 10), the *point* is established, the on/off button is turned to the "on" side and placed on the point total rolled, and the game proceeds into the second phase of the round.

For the second phase, the shooter's goal is to roll the point before they rolls a 7. Rolling the point results in a win for the pass line, and rolling a 7 results in a loss. If any other total is rolled during this phase, as far as the pass and don't pass lines are concerned, the bets stay on the table and continue. Only after either a 7 or the point is rolled does the on/off button revert to the "off" side and bets are resolved for those two types of bets. At this time, a new round is about to begin, and possibly a new shooter is determined (a shooter traditionally retains the dice as long as he or she continues to roll 7s or 11s, or makes the point; failing to make the point by rolling a 7 in the second phase is "sevening out" and will move the dice to the next shooter).

Question

What is the house's advantage on the pass line bet at craps?

By now, hopefully the first tool that comes to mind to answer this question is the expected value method we employed for roulette! Of course, not all expected value computations will be as short as those we saw earlier, and our craps calculations will require some work. The key here is that we need to somehow determine the probability that a pass line bet will win. Recall that Figure 1.2 lists all 36 possibilities for rolling two dice, so we can consult that to help us out.

TABLE 2.2: Pass Line/Don't Pass Line

Initial Roll	Pass Line Win	Don't Pass Line Win
Natural (7 or 11)	8/36	–
Craps (2, 3, or 12)	–	4/36
4	1/36	1/18
5	2/45	1/15
6	25/396	5/66
8	25/396	5/66
9	2/45	1/15
10	1/36	1/18
Total	244/495	251/495

There are many ways that a pass line bet will win. Perhaps the easiest is looking at the come out roll, and determining when that first roll will end the round by the shooter rolling a total of 7 or 11. Using Figure 1.2 as a guide, we note that there are 6 ways in which to roll a total of 7, and 2 ways in which to roll a total of 11, so there are 8 ways (out of a possible 36) for the pass line to win on the first roll. This happens with probability 8/36.

However, if a point is established, things change. Suppose 4 is established as the point (note that this happens on the come out roll with probability 3/36 since there are 3 ways to roll a total of 4). During the second phase of the round, only a roll of 4 or a roll of 7 are important; *all other rolls are irrelevant to the outcome of the pass line bet.* Since we don't care about any other result than those, let's take a look at the number of ways to roll just a 4 or a 7; there are 3 ways to roll a four, and 6 ways to roll a seven, so we only have a total of 9 outcomes that matter, and 3 of those result in a win. Thus *once a point of four is established,* there is a probability of 3/9 that the pass line will win (3 outcomes help us out of the 9 outcomes we care about). We then see that the probability of establishing the point as 4 on the come out roll, and then winning by rolling a 4 during phase two, is

$$\underbrace{\left(\frac{3}{36}\right)}_{\substack{\text{establish 4} \\ \text{as the point}}} \cdot \underbrace{\left(\frac{3}{9}\right)}_{\substack{\text{roll four} \\ \text{before seven}}} = \frac{1}{36} \approx 0.0278$$

since the come out roll is completely independent of the phase two rolls; we can multiply the two probabilities as we have done.

While Table 2.2 includes the probabilities for each of the pass line point possibilities (along with the don't pass probabilities, which we will discuss soon), let us take a look at one more of these possibilities (you will be asked later to verify some of the other numbers in the table). Namely, suppose that the come out roll results in a point of 8 being established. This happens with probability 5/36 since there are 5 ways to roll a total of 8 with two dice. Afterwards, we only care about rolling an 8 or rolling a 7; since there are 5 ways to roll an eight, and 6 ways to roll a seven, we have 11 outcomes that we care about, so the probability that, once a point of eight is established, the pass line wins is 5/11. Combining this together we have

$$\underbrace{\left(\frac{5}{36}\right)}_{\substack{\text{establish 8} \\ \text{as the point}}} \cdot \underbrace{\left(\frac{5}{11}\right)}_{\substack{\text{roll eight} \\ \text{before seven}}} = \frac{25}{396} \approx 0.0631$$

which is what appears in Table 2.2.

We can now compute the expected value of the pass line bet.

Example 2.10 From Table 2.2, we see that the probability of winning a pass line bet of, say, $1.00, is 244/495. Since, in this case, the only other option is losing, we lose with probability $1 - 244/495 = 251/495$. Then we have

$$\text{Expected Winnings} = \underbrace{\left(\frac{244}{495}\right) \cdot \$1.00}_{\text{win}} + \underbrace{\left(\frac{251}{495}\right) \cdot (-\$1.00)}_{\text{lose}} = -\$0.01414$$

for a house advantage of 1.414%.

At this point, I'm sure that you're wondering about the "don't pass line" bet at craps; Table 2.2 seems to imply that it's the exact opposite of the pass line wager, and for *the most part* it is. Let's imagine that the rules for winning the don't pass line wager were exactly the same as the *losing* conditions for the pass line (that is, the don't pass line wager will win on a come out roll of craps (2, 3, or 12), or, if a point is established, will win if a 7 is rolled before the shooter makes his or her point). The probability of this is 251/495, so let's compute the expected value,

$$\text{Expected Winnings} = \underbrace{\left(\frac{251}{495}\right) \cdot \$1.00}_{\text{win}} + \underbrace{\left(\frac{244}{495}\right) \cdot (-\$1.00)}_{\text{lose}} = \$0.01414,$$

which implies that our expected winnings are *positive*, indicating a house *disadvantage*! Surely this is not the case!

Example 2.11 To fix this, the casino changes the rules ever-so-slightly; instead of a come out roll of 12 being a win for the don't pass line bet, it is a *push*. That is, there is no win or loss there; your bet is simply returned to you (you win, or lose, $0.00). The probability of a don't pass line win is therefore reduced to $251/495 - 1/36 = 949/1980$ since the chance of getting a 12 on the come out roll is 1/36. Then we have our expected winnings as

$$\underbrace{\left(\frac{949}{1980}\right) \cdot \$1.00}_{\text{win}} + \underbrace{\left(\frac{1}{36}\right) \cdot \$0.00}_{\text{push}} + \underbrace{\left(\frac{244}{495}\right) \cdot (-\$1.00)}_{\text{lose}} = -\$0.01364$$

for a house advantage of 1.364%.

Note that since pushes result in a win (or loss) of $0.00, we really need not include them in our expected winnings calculations. We will tend to not list these explicitly in the future unless it would be instructive to see them. Also note that this very subtle change to the game creates an advantage for the house, albeit smaller than their advantage for the pass line bet. The don't pass line bet is slightly more favorable to the player than the pass line bet, so it is mathematically the better bet, but traditionally this is considered "betting against the shooter" which, at a community table game, could make others see you in a slightly negative light!

Free Odds

There's one more small "twist" to discuss before we touch base on some of the other wagers in craps. If the come out roll results in
a point being established, both the pass line
bettor and the don't pass line bettor are allowed to *increase* their wagers by placing a

TABLE 2.3: Odds Payouts

Point	Pass	Don't Pass
4 or 10	2:1	1:2
5 or 9	3:2	2:3
6 or 8	6:5	5:6

separate bet (pictured as ③ in Figure 2.4 for the pass line version; various
casinos will place the don't pass line bet differently) called "taking the free
odds" or simply "taking odds."

The payouts for free odds are listed in Table 2.3. You'll notice that the
payout depends upon not just whether your bet is placed on the pass line
or don't pass line, but also on the actual point that is determined from the
come out roll. Before we proceed, I should mention that the *original* wager
is still rewarded at a 1:1 payout; these new payouts are only from the extra
wager placed as free odds. In addition, it is common for casinos to allow you
to place more on the free odds wager than just your initial bet! While "five
times" odds, or 5x odds are common, you will see casinos occasionally offering
10x, 50x, or even 100x odds! Odds of up to 100x means that even if you place
only $1.00 on the original bet, you could put up to $100.00 on free odds. Note
that a payout of 6:5, for example, is equivalent to a payout of 1.2:1 (the latter
could be more useful when determining exact payouts).

Question

What makes free odds so interesting mathematically, and are they a
good idea?

The answer to this question lies in the subtle naming differences between
true mathematical odds (Definition 2.1) and casino odds (Definition 2.3). If
you take a look back at Example 2.3 regarding a $1.00 wager on the number
34 at roulette, you might ask the question about what would happen if the
casino payout were 37:1 (the true mathematical odds for a single number bet,
since there is one way to win and thirty-seven ways to lose)? In that case, we
would get

$$\text{Expected Winnings: } \underbrace{\left(\frac{1}{38}\right) \cdot \$37.00}_{\text{win}} + \underbrace{\left(\frac{37}{38}\right) \cdot (-\$1.00)}_{\text{lose}} = \$0.00,$$

a rather uninteresting result if you are the casino and trying to have an advantage.

Here, though, let's take a look at what happens if you wager $1.00 on the
pass line, the point is established as 10, and you decide to place an additional
$1.00 as your free odds bet. Note that this example will only examine what
happens with the free odds bet itself.

Example 2.12 Since the payout for the free odds bet on the pass line with a point of 10 is 2:1, we can use the ordinary expected value calculation to determine the house advantage. Note that there are 3 ways to roll a ten, and 6 ways to roll a seven, so, with 9 total outcomes we care about, the probability of winning is 3/9 and that of losing is 6/9. We get

$$\text{Expected Winnings: } \underbrace{\left(\frac{3}{9}\right) \cdot \$2.00}_{\text{win}} + \underbrace{\left(\frac{6}{9}\right) \cdot (-\$1.00)}_{\text{lose}} = \$0.00,$$

a house advantage of exactly 0%!

That was definitely interesting! Let's do one more, just for practice. Suppose that we bet $1.00 on the don't pass line, the point is established as 8, and we decide to take advantage of higher odds available at our table to place an additional free odds wager of $6.00.

Example 2.13 The payout for the free odds bet for the don't pass line and a point of 8 is 5:6, our bet of $6.00 will net us winnings of $5.00 should we win. There are 6 ways to roll a seven to for us to win, and 5 ways to roll an eight for us to lose, so we have

$$\text{Expected Winnings: } \underbrace{\left(\frac{6}{11}\right) \cdot \$5.00}_{\text{win}} + \underbrace{\left(\frac{5}{11}\right) \cdot (-\$6.00)}_{\text{lose}} = \$0.00,$$

another 0% advantage for the house!

You will have the chance to compute the house advantage for the other possibilities involving free odds in the exercises, but the short version of the story here is that they will all result in a house advantage of 0%! They are called "free" odds because the casino, in the long run, makes no money on these bets. Another way to think about that is that these bets are completely fair; the only advantage the casino has is on the initial pass line (or don't pass line) bet. Mathematically, it is always a good idea to take full advantage of this by wagering the most that you possibly can on free odds, though you must always remember that your money is completely at risk, and you should never gamble expecting to make money; wager only what you can reasonably afford for entertainment purposes!

Question

How do the free odds affect the house advantage for the pass line and don't pass line bets overall?

At this point, using the probabilities from Table 2.2 and odds from Table 2.3 will help us answer the question. For simplicity, we will assume that a $10.00 wager is placed on the pass line, mainly to make our numbers look better, but also to be more realistic. Casinos will rarely pay out winnings less than $1.00, and those winnings under $1.00 for us would be lost forever, resulting in more of an advantage for the house. We will also be cautious, and merely offer up an additional $10.00 for free odds when available.

Example 2.14 We only have possible wins (and losses) of $10.00 on the come out roll; assuming a point is established and we move to phase 2 of a craps round, then our additional $10.00 wager for free odds gives us potential losses of $20.00. Wins at this point will vary according to Table 2.3, but in each case we win $10.00 on the initial wager, and odds payouts on the extra $10.00 wager, so we will end up with, using probabilities from Table 2.2,

$$\text{Expected Value} = \underbrace{\left(\frac{8}{36}\right) \cdot \$10}_{\substack{\text{win come} \\ \text{out roll}}} + \underbrace{\left(\frac{4}{36}\right) \cdot -\$10}_{\substack{\text{lose come} \\ \text{out roll}}} + \underbrace{\left(\frac{1}{36}\right) \cdot \$30}_{\substack{\text{win point} \\ \text{of four}}} + \underbrace{\left(\frac{1}{18}\right) \cdot -\$20}_{\substack{\text{lose point} \\ \text{of four}}}$$

$$+ \underbrace{\left(\frac{2}{45}\right) \cdot \$25}_{\substack{\text{win point} \\ \text{of five}}} + \underbrace{\left(\frac{1}{15}\right) \cdot -\$20}_{\substack{\text{lose point} \\ \text{of five}}} + \underbrace{\left(\frac{25}{396}\right) \cdot \$22}_{\substack{\text{win point} \\ \text{of six}}} + \underbrace{\left(\frac{5}{66}\right) \cdot -\$20}_{\substack{\text{lose point} \\ \text{of six}}}$$

$$+ \underbrace{\left(\frac{25}{396}\right) \cdot \$22}_{\substack{\text{win point} \\ \text{of eight}}} + \underbrace{\left(\frac{5}{66}\right) \cdot -\$20}_{\substack{\text{lose point} \\ \text{of eight}}} + \underbrace{\left(\frac{2}{45}\right) \cdot \$25}_{\substack{\text{win point} \\ \text{of nine}}} + \underbrace{\left(\frac{1}{15}\right) \cdot -\$20}_{\substack{\text{lose point} \\ \text{of nine}}}$$

$$+ \underbrace{\left(\frac{1}{36}\right) \cdot \$30}_{\substack{\text{win point} \\ \text{of ten}}} + \underbrace{\left(\frac{1}{18}\right) \cdot -\$20}_{\substack{\text{lose point} \\ \text{of ten}}} = -\$0.1414$$

which seems the same as before, considering that on a $10.00 wager, this results in a house advantage of 1.414% as in Example 2.10! However, since we wager a total of $10.00 when the come out roll is 2, 3, 7, 11, or 12 (which happens with probability 12/36) and a total of $20.00 when a point is established (with probability 24/36), our *expected wager* on average is

$$\text{Expected Wager} = \left(\frac{12}{36}\right) \cdot \$10 + \left(\frac{24}{36}\right) \cdot \$20 \approx \$16.67,$$

indicating that the house advantage in this situation is *really*

$$\frac{-\$0.1414}{16.67} \approx -\$0.00848,$$

a house advantage of 0.848%, notably *less than* 1%!

It's amazing how much these free odds can affect the overall house advantage at craps; this is the primary reason that many regard the game of craps, using the pass line or don't pass line bet and taking free odds, as having the lowest house advantage in the casino! While card counting at Blackjack can give the player an advantage, it can be risky, hard to perform, and take awhile to have a large effect. Indeed, with 5x odds available, and taking maximum odds, the house advantage becomes 0.326%, and with 100x odds, that advantage shrinks to 0.0209%, essentially 0%!

Note that we haven't discussed the don't pass line wager with free odds much, as you will do that yourself soon.

Other Wagers

It is now time to take a look at some of the other wagers in craps. This will not be an exhaustive list, and, if you do feel like playing craps at some point, and want to explore other wagers than the pass line and don't pass line, consult other resources or your friendly craps dealer.

The *come* bet and *don't come* bet, pictured in Figure 2.4 as ④ and ⑤, are virtually identical to the pass line and don't pass line, except you can place these wagers at any time (and I tend to think of this bet as starting your own little round of craps, where the next roll will be your own come out roll, and if a point is established, your bet will be placed on the corresponding number to keep track of it). Mathematically, there are no interesting questions here since the house advantage mirrors that of the pass and don't pass line wagers.

The *field* bet, identified in the layout as ⑥, is very commonly played. It is a single-roll bet (that is, it can be placed before any roll, and will be paid out after that roll), and wins on a 2, 3, 4, 9, 10, 11, or 12. The casino payout is 2:1 if the roll is a total of 2 or 12, and otherwise the payout is 1:1.

Bets ⑦ and ⑧ represent buy, place, and lay bets. These are multi-roll bets similar to the pass and don't pass bets, but let the bettor choose the specific number that they are buying or placing (trying to win by having that number rolled before a seven), or laying to have a seven rolled before that specific number. Payouts are similar to the pass line and don't pass line, but are somewhat lower than the true mathematical odds. Some bets (namely for buying as opposed to placing) have a commission charged on wins, and most of these bets are always active (continue to stay on the board and paid out until they lose) except on come out rolls. We will not discuss these at all.

Big 6 and Big 8 are pictured as ⑨ and, again, are multi-roll bets that the 6 (or 8 for Big 8) will be rolled before a 7 is rolled. Payout is at 1:1, and since a bettor could do a buy or place for the same number, and these payouts are closer to the true mathematical odds, experienced craps players will never wager on Big 6 or Big 8 (and, in some places such as Atlantic City, it is illegal for casinos to offer this bet or even have it on their craps tables).

The craps bet ⑩ and its related bet in red, the seven bet, are single roll bets that a 2, 3 or 12 (or for the seven bet, a seven) will be rolled on the next roll of the dice. The payout for the craps bet is 7:1, and the payout for the seven bet is 4:1.

The bets near ⑪ are varied in nature. The bets are pictured on the table, and the 2, 3, 11, and 12 bets are single roll bets that the specific number will come up on the next roll. Both the 2 and the 12 pay 30:1, and the 3 and eleven pay 15:1. The other bets nearby are the "hard" bets. Hard 4 and Hard 10 both pay 7:1 and win if a total of 4 (or 10) is rolled "the hard way," meaning as a pair of twos (or pair of fives), before a seven or the corresponding total is rolled any other way. Hard 6 and Hard 8 are similar, with a payout of 9:1 (a higher payout because there are more ways to get 6 or 8 the easy way).

Before you're on your own, let's take a look at two of these by computing the house advantage.

Example 2.15 For the single roll bet of seven, note that of the 36 different outcomes for rolling two dice, only 6 of those result in a win. With a payout of 4:1, we have

$$\text{Expecting Winnings} = \underbrace{\left(\frac{6}{36}\right) \cdot \$4.00}_{\text{win}} + \underbrace{\left(\frac{30}{36}\right) \cdot -\$1.00}_{\text{lose}} \approx -\$0.16667$$

for a house advantage of about 16.67%!

Example 2.16 For the multi-roll bet of Hard 4, there is only 1 outcome that will help us win (double twos); for losses, we will only lose if a seven is rolled (with 6 possible ways) or if four is rolled the easy way (with a one and a three on the dice, with 2 possible ways). This makes a total of 9 outcomes we care about, and since only one of those ways is a win, the probability for Hard 4 winning is 1/9. A payout of 7:1 lets us proceed, obtaining

$$\text{Expecting Winnings} = \underbrace{\left(\frac{1}{9}\right) \cdot \$7.00}_{\text{win}} + \underbrace{\left(\frac{8}{9}\right) \cdot -\$1.00}_{\text{lose}} \approx -\$0.11111$$

for a house advantage of just over 11.11%.

2.4   Exercises

2.1 Determine the probability of each bet for American roulette listed in Table 2.1.

2.2 Determine the true mathematical odds of each bet for American roulette listed in Table 2.1.

2.3 Show that the house advantage for any American roulette bet (except the "first five" bet) is always 5.26%, by computing the expected winnings for the wagers not computed in Examples 2.3 through 2.5.

2.4 Imagine a new bet for American roulette called the Fibonacci bet. This is a bet that one of the numbers 1, 2, 3, 5, 8, 13, 21, or 34 will hit.

(a) Determine the true mathematical odds of the Fibonacci bet.

(b) If you were the casino, what house odds would you set for winning this bet, assuming you want the house to have some advantage?

(c) Compute the expected winnings of the Fibonacci bet using your house odds.

2.5 Consider American roulette with the 0 and 00 spaces removed from the roulette wheel.

(a) What do you think the expected winnings for betting on the number 31 would be? Why?

(b) Calculate the expected winnings for a bet on the number 31.

(c) Calculate the expected winnings for a bet on red.

2.6 For this exercise, refer to the calculations of Example 2.6 and the discussion that follows.

(a) What payout for the "prime number" bet should the casino offer so that the house advantage for the bet is about 5.26% (the same as for most other bets)?

(b) What payout should the casino offer for the "first five" bet so that the house advantage is about 5.26%? Why do you think the casino offers a payout of 6:1 instead?

2.7 Suppose a "greedy" casino decides to add a triple zero (000) space to their roulette wheel. Consider the following questions, assuming that the casino payouts stay the same.

(a) Determine the expected winnings for a bet on the number 17.

(b) Determine the expected winnings for a bet on odd.

(c) Comment on how much the house advantage increases by adding the 000 space.

2.8 European Roulette does not have the double zero (00) space that American Roulette has, but the casino payouts remain the same. For each of the bets in Table 2.1, determine the true mathematical odds, along with the expected value, for that bet.

2.9 For the "prime number" bet introduced in Example 2.6, determine the payout needed, assuming a European Roulette table, for the expected value to equal about 2.7%.

2.10 European casinos often feature a "half-back" rule on all even-money wagers (such as even or odd) that gives the player *half* of his or her money back if 0 is the outcome of a spin; compute the expected value on even-money bets for European Roulette with this change.

2.11 Other European casinos let even money wagers "ride" for one more spin of the wheel should 0 be the outcome; the wager is returned to the player if the bet wins the second time. That is, if a wager is placed on red, and 0 comes up, the bet remains on red for the next spin where, on this second spin, an outcome of red returns the wager to the player while anything other than red loses. Determine the expected value for even-money bets with this rule and comment on this rule's effect compared to standard rules.

2.12 For each point total of 5, 6, 9, and 10, determine the probability calculations for establishing that point total and then winning on the pass line to obtain the values in Table 2.2.

2.13 For each point total of 4, 5, 6, 8, 9, and 10, determine the probability calculations for establishing that point and then winning on the don't pass line to obtain the values in Table 2.2.

2.14 Some casinos modify craps so that the don't pass line bet is a push on a come out roll of 3 *instead* of a come out roll of 12. Determine the expected value (house advantage) of the don't pass line wager under this modified rule, and comment on your answer versus the usual advantage of 1.364%. Ignore free odds.

2.15 Repeat Example 2.14 for the don't pass line wager; you might use a base bet of $6.00 (and an additional $6.00 for free odds when allowed) to help with calculations.

2.16 Verify that the house advantage for craps, using the pass line wager and free odds, is 0.326% using 5x maximum odds and 0.0209% using 100x maximum odds.

2.17 Free odds at many casinos recently have been labeled as "3-4-5x" odds, meaning that the bettor can wager, as free odds, up to 3 times their initial bet on a point total of 4 or 10, up to 4 times their initial bet on a point of 5 or 9, and up to 5 times their initial bet on a point of 6 or 8; compute the house advantage for the pass line bet using these odds, assuming you bet the maximum allowed. Why do you think these casinos use this structure for free odds?

2.18 Determine the house advantage for the following bets at craps (note that the payouts are given in Section 2.3).

  (a) the field bet   **(b)** craps   **(c)** Hard 8   **(d)** double six (12)

2.19 Many casinos offer a game called the "Wheel of Fortune" (sometimes known as Big 6 or the Money Wheel) that features a large wheel with various money symbols on it. A fairly standard version has 54 spaces, with 24 spaces labeled $1, 15 spaces labeled $2, 7 spaces labeled $5, 4 spaces labeled $10, 2 spaces labeled $20, 1 space with a Joker, and 1 space with another special symbol. Players can wager that any of these space types will come up on the next spin, and the $5 space, for example, pays 5:1 odds. Both special spaces pay 40:1, but the joker space *only* wins if the joker is spun, and the same is true for the special symbol.

  (a) Determine the true mathematical odds for each of the bet types.

  (b) Determine the expected value for each of the bet types; which wager offers the lowest house advantage?

  (c) Some casinos change the payout for both the joker space and special symbol space to 45:1 instead of 40:1; how does this change the expected value for these wagers?

  (d) Other casinos replace one of the $1 spaces with another $5 space; recompute the expected values for $1 and $5 and comment on how this small change affects the advice you would give to bettors that choose to play the game.

3

Counting: Poker Hands

3.1 Cards and Counting

First, I should make sure that the title of this section is not confused with
card counting which is a technique used
when playing Blackjack in a casino in
order to obtain an edge over the house
advantage; this topic will be discussed
in Chapter 8. With that out of the way,
it's now time to take a look at some
questions regarding the game of poker.

FIGURE 3.1: Royal Straight Flush

At first, we'll concern ourselves with versions that give the player an initial
hand of five cards, and study the distribution and probabilities of those five
cards. Then we'll spend some time on some variants.

For those who aren't familiar with poker, some background is necessary.
A standard deck of playing cards consists of 52 cards, arranged into 13 cards
of each suit (there are four suits, ♣ (clubs), ♦ (diamonds), ♥ (hearts), and
♠ (spades)). Those 13 cards are the ace, the two, the three, up to the ten,
followed by the *face cards* Jack, Queen, and King. The *rank* of a numbered
card is its value, and those cards are ordered in the usual way (for example, a
three has a lower rank than a six). The Jack is ranked above a ten, with the
Queen ranked above a Jack, and the King is above the Queen. For virtually
all versions of poker, the ace can be treated as low (essentially behaving like
a one) or high (ranked immediately above a King) in order to make the best
hand of cards that you can.

Poker Hands

Possibly the first thing that a poker player learns, after knowing the cards and
suits, are the different "named hands" that are possible with five cards. Before
we count the ways of obtaining the particular hands and then determine the
probabilities, let's review these hands and make sure that we're all on the
same page. The information is summarized in Table 3.1 for easy reference.

A *straight flush* (see Figure 3.1) is a hand consisting of five cards in a sequence, all of the same suit. The name comes from the fact that this hand meets the criteria for two of the named hands below (namely the straight and the flush), and examples of a straight flush include the hands 8♥ 9♥ 10♥ J♥ Q♥ and A♠ 2♠ 3♠ 4♠ 5♠. Some people, and especially video poker machines, distinguish between an ordinary straight flush and a *royal straight flush*, or simply *royal flush*, when the hand involves the 10, Jack, Queen, King, and Ace, as in 10♦ J♦ Q♦ K♦ A♦. Note that, as with the straight below, two hands with a straight flush are ranked according to their high card; higher is better. Two straight flushes with the same high card are equal in value.

A *four of a kind* is fairly self-explanatory; it consists of the four cards of the same rank, and a fifth card of different rank. The rank of the four matching cards is used to break ties (again, higher is always better), and in games with multiple decks, wildcards, or community cards, ties are broken by using the unmatched card. An example of a four of a kind is 4♣ 4♦ 4♥ 4♠ 7♦.

The *full house* is a hand that consists of three matching cards of one rank, and two matching cards of a different rank (in other words, using the language below, a hand that has both a three of a kind and a pair). For example, you might have the hand 5♦ 5♥ 5♠ J♣ J♠, which would commonly be called out as "fives over Jacks" or "fives full of Jacks." Ties are broken by the rank of the three matching cards (and in games where that could be tied, the rank of the pair would be used).

The next hand we've already partially discussed; the regular *flush* consists of five cards that have the same suit, not in sequence; for example, 4♣ 8♣ 10♣ Q♣ A♣. To break ties, only the highest card is considered, unless those too are tied, in which case you look at the second highest card, and, if those are the same, the third highest, and so on and so forth.

A *straight* consists of five cards in a sequence, with at least one card of a different suit than the others. As with the straight flush, ties are broken by comparing the rank of the high card (suit does not matter). An example of a straight that is not a straight flush is 5♥ 6♦ 7♣ 8♦ 9♠.

Three cards of one rank, and two different unmatched cards, forms a *three of a kind*; for example, both 9♦ 9♥ 9♠ 2♣ A♠ and Q♣ Q♥ Q♠ 2♦ 3♣ are hands with three of a kind. The second is ranked higher, because the rank of the three matched cards is checked first (even though the first hand has an Ace). In games with wildcards or other added rules, the rank of the highest unmatched card can be used to break ties if two players have the same matched three cards, and if that unmatched card is the same, consider the lower of the two unmatched cards.

A hand such as 8♦ 8♠ 2♥ 2♠ K♣ or 7♦ 7♥ 6♣ 6♠ A♠ would be called a *two pair* since there are two cards of the same rank, two other cards of matching rank, and a third card that matches neither of the paired ranks. The first is better than the second example because the rank of the highest pair is considered first (even though the sixes and the ace in the second example are better than the twos and king of the first); if those are tied, the rank

TABLE 3.1: Poker Hands

Name	Description	Example
Royal Flush	cards in sequence, suits match, ace high	10♦ J♦ Q♦ K♦ A♦
Straight Flush	cards in sequence, suits match	8♥ 9♥ 10♥ J♥ Q♥
Four of a Kind	four cards, same rank	4♣ 4♦ 4♥ 4♠ 7♦
Full House	three of a kind and a pair	5♦ 5♥ 5♠ J♣ J♠
Flush	non-sequenced, suits match	4♣ 8♣ 10♣ Q♣ A♣
Straight	sequenced, suits non-matching	5♥ 6♦ 7♣ 8♦ 9♠
Three of a Kind	three cards, same rank	Q♣ Q♥ Q♠ 2♦ 3♣
Two Pair	two cards, same rank, twice	8♦ 8♠ 2♥ 2♠ K♣
Pair	two cards, same rank	7♦ 7♠ K♠ 5♥ 2♥
High Card	none of the above	K♠ Q♠ 10♦ 5♣ 2♥

of the other pair is considered, and the unmatched card is used as the last tiebreaker.

The "last" named hand we'll discuss is *one pair*, a hand that has two cards of the same rank, and three cards of different rank that are not the same rank as each other (or the pair itself). Examples would be 7♦ 7♠ K♠ 5♥ 2♥ and 7♣ 7♥ K♦ 6♥ 2♠, where the second example is better than the first (because the rank of the matched cards, checked first, is the same, as is the highest unmatched card; the six in the second example beats the five in the first).

Finally, a hand such as K♠ Q♠ 10♦ 5♣ 2♥ that doesn't match any of the above is a hand that many would call *high card*. For hands in this category, to break ties, consider the rank of the highest card, then second highest, and so on.

Now that we have that out of the way, we can get to some of the mathematics.

Question

How many different five card hands are there in poker? How many different combinations of each named hand are possible?

Combinations (Binomial Coefficients)

The *question* itself may need a little explanation. We don't mean *types* of hands, we mean different physical five card hands. Counting these may seem easy at first, but none of the tricks of Chapter 1 will help. Each card is not independent of the others. You might imagine the process of dealing a hand of cards as receiving your first card, your second card, your third card, and so on, to form your hand of five cards, but for poker and classification purposes, receiving your cards in the order 5♦ Q♠ 3♥ J♥ A♣ or in the order 3♥ J♥ Q♠ A♣ 5♦ results in the same hand, even though the order in which they were received is different. This is because, for purposes of five-card poker, *order doesn't matter*. We'll need a new way of counting, and from this we'll be able to start counting more difficult scenarios.

Suppose that instead of five cards, we're only dealing with *one card* poker hands (a very boring game though). How many different one card hands are there? If you answered 52, you got it right; since there are 52 distinct cards in the deck, there are exactly 52 possibilities for your one card. That wasn't terribly interesting.

How do things change if we instead ask how many *two card* poker hands exist? Now things get a bit more interesting. Since there are 52 choices for your first card, and 51 choices for your second card, it would seem that $52 \cdot 51 = 2,652$ would be the answer, but that is only a good start. Both 4♠ J♦ and J♦ 4♠ would appear on your list of 2,652 hands, but both of these possibilities represent the *same* hand (remember, for us here, *order doesn't matter*)! In fact, for any possible two-card hand, say 2♣ 3♠, you will have counted it *twice* in your list of 2,652 hands, once as 2♣ 3♠ and once as 3♠ 2♣, so it should make sense that the true quantity of distinct two-card poker hands is $2652/2 = 1,326$ since we counted *ordered* hands and then removed half (the duplicates).

This can, of course, continue! For *three* card poker hands, there are $52 \cdot 51 \cdot 50 = 132,600$ possible *ordered* hands, and now we have to count the number of ways we can order three cards such as 5♦ 7♥ J♠. In the

TABLE 3.2: The 6 Orderings

5♦ 7♥ J♠	5♦ J♠ 7♥
7♥ 5♦ J♠	7♥ J♠ 5♦
J♠ 5♦ 7♥	J♠ 7♥ 5♦

two card case, we divided by 2 because there were 2 ways for us to order two pre-chosen objects (two choices for the first object, and then only one choice to choose the second object since there is only one left). Given our three cards, there are three choices to select the first card, then only two choices to select the second card, and one choice for the last card once the first two are selected. This gives us a total of $3 \cdot 2 \cdot 1 = 6$. Since the numbers are still relatively small, we can formally list the orderings of 5♦ 7♥ J♠, given in Table 3.2. Let's formalize this in an example.

Example 3.1 Since there are $52 \cdot 51 \cdot 50 = 132,600$ possible ordered three card poker hands, and $3 \cdot 2 \cdot 1 = 6$ possible orderings for any given subset of three cards, there are $132600/6 = 22,100$ possible distinct three card poker hands.

It should be starting to become clear how we will generalize. First, many of you will probably recognize $3 \cdot 2 \cdot 1 = 3!$ as the *factorial* function. For those that don't, we'll introduce that here.

Definition 3.1 The *factorial* of a nonnegative integer n, written as $n!$, is defined as the product of all distinct integers between and including n and 1. That is,

$$n! = n \cdot (n-1) \cdot (n-2) \cdots 3 \cdot 2 \cdot 1.$$

When n is 0, we define $0! = 1$.

We will look at several examples, but it is worth mentioning again that $0!$ is defined to be equal to 1.[1] Also note that most (scientific or graphing) calculators have this function built-in; consult your manual or look through your calculator's "Catalog" of functions to find it.

Example 3.2 Here we show and compute $5!$ and $9!$.

$$5! = 5 \cdot 4 \cdot 3 \cdot 2 \cdot 1 = 120$$
$$9! = 9 \cdot 8 \cdot 7 \cdot 6 \cdot 5 \cdot 4 \cdot 3 \cdot 2 \cdot 1 = 362,880$$

Note that the factorial function grows quite fast! If you and eight friends are deciding how to order yourselves (perhaps choosing who gets which of nine seats on the front row at a concert), there are 362,880 different ways for you to sit! At this point, we can now answer our main question about five card poker hands.

Example 3.3 There are $52 \cdot 51 \cdot 50 \cdot 49 \cdot 48 = 311,875,200$ ordered five card poker hands, and $5! = 120$ different orders, so there are $311875200/120 = 2,598,960$ distinct possible five card poker hands.

This idea of counting possibilities when we have 52 objects (say cards) and want to choose some, say five, of them is very useful. In fact, we have a special notation and special tool available to us for just this case. The notations

$$\binom{52}{5} = {}_{52}C_5 = 2,598,960$$

are used to represent this, and both notations are read as "52 choose 5." While both notations are common, we will use the first notation almost exclusively. Since, for example,

$$8 \cdot 7 \cdot 6 = \frac{8 \cdot 7 \cdot 6 \cdot 5 \cdot 4 \cdot 3 \cdot 2 \cdot 1}{5 \cdot 4 \cdot 3 \cdot 2 \cdot 1} = \frac{8!}{5!}$$

---

[1]There are many good reasons as to why $0! = 1$ that are outside the scope of this book; for now, perhaps the best two are that "it works" and, if you think about the number of ways to order 0 objects, there is exactly 1 way to order nothing.

a very compact way to express "8 choose 3" would be as

$$\binom{8}{3} = \frac{8 \cdot 7 \cdot 6}{3!} = \frac{\frac{8!}{5!}}{3!} = \frac{8!}{3!5!},$$

leading to the next definition.

Definition 3.2 The *combination* or *binomial coefficient* for "n choose r" represents the number of ways of selecting r distinct objects from a collection of n objects, where order doesn't matter. It is given by the formula

$$\binom{n}{r} = \frac{n!}{r!(n-r)!}.$$

Example 3.4 The number of five card poker hands, since we are choosing 5 cards from a collection of 52 cards, is

$$\binom{52}{5} = \frac{52!}{5!(52-5)!} = \frac{52!}{5!47!} = 2,598,960.$$

Your calculator should be able to do these "choose" calculations easily, assuming you're working with a scientific or graphing calculator. Look in your "catalog" or manual for "nCr" and give that a try, either by entering "52 nCr 5" or "nCr(52,5)" depending on the syntax of the calculator.

Now we can start answering questions about the different named hands. First, let's take a look at the *straight flush* and count those. The best advice I can give for working with named hands and determining exactly how to count them is to determine what choices need to be made in completely describing a particular hand. For the straight flush, there are two bits of information that completely determine the five cards. First, there's the suit, and second, there's the high card of the straight.

Example 3.5 To count the number of straight flushes, first note that of the four suits, one of which must be chosen for our hand, there are also ten possible high cards for a straight (starting with the five and higher ranks, any card can be the high card for a straight; only two, three, and four are ineligible), of which one must be chosen. At this point, the specific straight flush has been chosen. Since choosing which suit and which high card we will use is independent, we can multiply; there are

$$\underbrace{\binom{4}{1}}_{\text{suit}} \cdot \underbrace{\binom{10}{1}}_{\text{high card}} = 4 \cdot 10 = 40$$

different straight flushes (of which only 4 are royal flushes). This puts the probability of obtaining a straight flush from a shuffled deck upon drawing five cards at

$$\frac{40}{2,598,960} \approx 0.00001539 = 0.001539\%$$

which is a 1 in 64,974 chance.

Before we compute some more, recall that a probability of 1 in 64,974 gives true mathematical odds (recall Definition 2.1) of 64,973:1; a casino could reasonably pay out $60,000.00 on a $1.00 wager at a game where the goal was to select five random cards from a deck and receive a straight flush. The payout sounds very high, the minimum wager could be set very low, and the expected winnings of

$$\left(\frac{1}{64974}\right) \cdot \$60,000 + \left(\frac{64973}{64974}\right) \cdot (-\$1) = -\$0.0765$$

show that the game is still very favorable to the house.

Now, let's take a look at some other named hands. Specifically, we will calculate the number of *flushes* that we could possibly get. Note that the number of possible hands for each named type is listed in Table 3.3. Besides the flush, three of a kind, and two pair to follow, you will be asked to determine how to obtain these numbers as an exercise.

TABLE 3.3: Poker Hand Counts

Name	Number	Probability
Straight Flush	40	0.00001539
Four of a Kind	624	0.0002401
Full House	3,744	0.001441
Flush	5,108	0.001965
Straight	10,200	0.003940
Three of a Kind	54,912	0.02113
Two Pair	123,552	0.0475
Pair	1,098,240	0.4226
High Card	1,302,540	0.5012

Example 3.6 Much as with the last example, there are four different suits, one of which must be chosen. Afterwards, the exact make-up of the cards is (mostly) irrelevant. Of the 13 possible cards of the selected suit, we need to choose 5. Again, we will multiply, since once the suit is selected, the 5 cards are selected independently. We will get

$$\underbrace{\binom{4}{1}}_{\text{suit}} \cdot \underbrace{\binom{13}{5}}_{\text{five cards}} = 4 \cdot 1,287 = 5,148$$

different hands of entirely one suit. However, we have also included the straight flushes in this count, so if we subtract those 40, we will give 5,108 different flushes (since flushes require that the five cards be non-sequenced). The probability of a flush is then

$$\frac{5108}{2,598,960} \approx 0.001965 = 0.1965\%.$$

Now, for something a little different, let's determine the number of hands that qualify as a *three of a kind*. Recall for this hand we need three matching cards by rank, and then two other cards that are unmatched and not the same as the three matching cards. We do not need to worry about suit here (though games where you have more than five cards and more than one deck, where you select the best five card hand, do cause some difficulties).

Example 3.7 First, we need to select the rank for the three matching cards. With 13 ranks available, we choose 1 of those. Once the rank is selected, though, we need to choose 3 of the possible 4 cards (since we're looking for a three of a kind). What about the other two cards? They need to not match the already selected rank, so there are now only 12 ranks available, and since we need the two cards to also not form a pair (we are not counting full houses here), we need to choose 2 of those ranks. Finally, for each of the 2 selected ranks, we will need to choose 1 of the 4 cards available of those ranks. We'll get

$$\underbrace{\binom{13}{1}}_{\substack{\text{triplet} \\ \text{rank}}} \cdot \underbrace{\binom{4}{3}}_{\substack{\text{cards for} \\ \text{triplet}}} \cdot \underbrace{\binom{12}{2}}_{\substack{\text{unmatched} \\ \text{cards' rank}}} \cdot \underbrace{\binom{4}{1} \cdot \binom{4}{1}}_{\substack{\text{unmatched} \\ \text{cards}}} = 13 \cdot 4 \cdot 66 \cdot 4 \cdot 4 = 54,912$$

for a probability of 0.0211 or 2.11%.

Now let's take a look at hands classified as *two pair* and count these.

Example 3.8 To count the number of possible two pair hands, we again look at what the essence of the hand actually is. We need two different pairs of cards, and a third that matches none of those two pairs. With 13 ranks, we choose 2 to be the ranks for the pairs (note that, when dealing with the full house on your own, you will need to select a rank for the three of a kind and a rank for the two of a kind separately). For each of those, we need to choose 2 of the possible 4 cards of that rank; afterwards, of the 11 remaining ranks, choose 1 for the unmatched fifth card, and from that rank, choose 1 of the possible 4 cards, obtaining

$$\underbrace{\binom{13}{2}}_{\substack{\text{ranks for} \\ \text{two pair}}} \cdot \underbrace{\binom{4}{2} \cdot \binom{4}{2}}_{\substack{\text{cards for} \\ \text{both pair}}} \cdot \underbrace{\binom{11}{1}}_{\substack{\text{unmatched} \\ \text{card's rank}}} \cdot \underbrace{\binom{4}{1}}_{\substack{\text{unmatched} \\ \text{card}}} = 78 \cdot 6 \cdot 6 \cdot 11 \cdot 4 = 123,552$$

or a probability of 0.0475 = 4.75%.

You'll be asked to verify the named hand counts of Table 3.3 not covered in Examples 3.5, 3.6, 3.7, and 3.8 as an exercise later. Note that once all of the "named hands" are counted, determining the count for *high card* is simply a matter of subtracting the number of "named hands" from 2,598,960, the total number of possible five card poker hands.

3.2 Seven Card Pokers

There are a variety of different poker games available that use seven cards instead of just five. These include, for example, seven card stud, Omaha Poker, and, perhaps the most famous version of poker available today, Texas Hold'Em. Rules for these various games are readily available online, and, except for Texas Hold'Em, which we'll discuss in the next section, the plan now is not to get into the specific rules of each of these games, but rather to use our newfound counting skills to work with the idea of using seven cards to make the best five card poker hand (something common, in some fashion or another, to all of these).

The questions we'll address are similar to before; if we're about to be given a set of seven cards from which to make the best five card poker hand, what are the probabilities that we'll get, say, a full house? I should note that in some of these poker games, all seven cards belong to you, while for some others, you have a specific number of cards in your hand, and the others are "community cards" available to be used by all.

Example 3.9 What is the probability that we'll be able to construct a royal (straight) flush if seven cards are randomly available to us? Much as before, we'll have to choose 1 of the available 4 suits, say diamonds, and this already fixes five of the cards we need (the A♦ K♦ Q♦ J♦ 10♦). The other two cards in our seven-card hand can be any of the other 47 cards, since at this point no card will make our hand any better. We'll arrive at

$$\text{Probability of Royal Flush} = \frac{\overbrace{\binom{4}{1}}^{\text{suit}} \cdot \overbrace{\binom{47}{2}}^{\text{other cards}}}{\underbrace{\binom{52}{7}}_{\text{seven card hands}}} = \frac{4 \cdot 1,081}{133,784,560} \approx 0.000032$$

which is a good bit better than the five-card version (given in Table 3.3, dividing all straight flushes by 10 gives us the probability of a royal flush there as 0.000001359).

We do need to be very careful with our logic, though. If we want to consider *any* straight flush, the presence of extra cards to choose from throws a few monkey wrenches into the mix! Just as in Example 3.5 earlier, we still need to choose 1 of the 4 suits, and we need to choose 1 of the 10 possible "high cards" for the straight. This, however, is where things get interesting. Suppose our selection is clubs for the suit, and the nine for "high card." This gives us the five cards **9♣ 8♣ 7♣ 6♣ 5♣**. Are we allowed to choose from any of the remaining 47 cards for the other two? Not quite; if we've selected **9♣** as our high card, then we are not allowed to have **10♣** among our collection of cards (since now the 10 would be the high card for the straight). However, if the ace is selected as the high card, then, as we saw in Example 3.9, it doesn't matter. We have two cases; if we select any of the *nine* cards five through king as the high card, we only have 46 cards to choose from for the extra 2 cards. If the ace is chosen as the high card, we have the full set of 47 from which to select. This gives us

$$
\text{Probability of Straight Flush} = \frac{\overbrace{\binom{4}{1}}^{\text{suit}} \cdot \overbrace{\binom{9}{1}}^{\text{high card}} \cdot \overbrace{\binom{46}{2}}^{\text{other cards}} + \overbrace{\binom{4}{1}}^{\text{suit}} \cdot \overbrace{\binom{47}{2}}^{\text{other cards}}}{\underbrace{\binom{52}{7}}_{\text{seven card hands}}}
$$

$$
= \frac{4 \cdot 9 \cdot 1{,}035 + 4 \cdot 1{,}081}{133{,}784{,}560} \approx 0.000311.
$$

As this shows, when a pool of seven cards is involved, determining the number of (or probability of) most named hands will require similar caution. We consider two more here, and some will be left for you to do in the exercises. Let's take a look at the full house.

Example 3.10 Unlike the five-card version where a full house is simply a three of a kind combined with a pair, there are now three distinct ways to form a five-card full house from a group of seven cards. First, it's possible that the seven cards give us a three of a kind, a pair, and two other unmatched cards (computed as ① below). Next, we could very well have a three of a kind and *two* pair, where the higher of the pair will be used to form our full house (computed as ②). Finally, we could have two three of a kinds, and an unmatched card, where two of the cards from the lower ranked three of a kind will be used for the "pair" part of the full house (shown as ③). Note that a hand that contains a three of a kind and a *four of a kind* is not included here, as that hand will be counted in the better four of a kind category. We count each of these possibilities separately, then combine.

$$① = \binom{13}{1} \cdot \binom{4}{3} \cdot \binom{12}{1} \cdot \binom{4}{2} \cdot \binom{11}{2} \cdot \binom{4}{1} \cdot \binom{4}{1} = 3,294,720$$

rank for   cards for   rank for   cards for   ranks for   unmatched
triple   triple   pair   pair   unmatched   cards

$$② = \binom{13}{1} \cdot \binom{4}{3} \cdot \binom{12}{2} \cdot \binom{4}{2} \cdot \binom{4}{2} = 123,552$$

rank for   cards for   ranks for   cards for
triple   triple   pairs   pairs

$$③ = \binom{13}{2} \cdot \binom{4}{3} \binom{4}{3} \cdot \binom{11}{1} \cdot \binom{4}{1} = 54,912$$

ranks for   cards for   rank for   unmatched
triples   triples   unmatched   card

Adding all of these together gives us 3,473,184 ways to get a full house, and out of the total of 133,784,560 possible seven card collections, a probability of $3,473,184/133,784,560 \approx 0.025961$, a good bit higher than the five-card only version.

Our last example here to demonstrate the importance of careful thinking will be the hand known as two pair. Pay special attention while working through this and notice how careful we must be with the suits when seven cards are involved; we must be careful, when counting two pairs, to make sure that no flushes are created.

Example 3.11 There are two general ways to form a two pair hand given a collection of seven cards from which to work. The first, which ends up being the easiest, is to have *three pairs* among those seven, along with one other unmatched card, giving a pattern of AABBCCD, where each distinct letter represents a card value. A quick thought exercise will convince you that in this scenario, there will be at most four cards that can share the same suit (each suit could appear at most once in each of the pairs, so spaces, for example, could assigned to only one card for AA, BB, and so on), avoiding a flush. We count, as usual, finding

$$\binom{13}{3} \cdot \binom{4}{2} \cdot \binom{4}{2} \cdot \binom{4}{2} \cdot \binom{10}{1} \cdot \binom{4}{1} = 2,471,040.$$

ranks for   cards for   rank for   unmatched
pairs   pairs   unmatched   card

Now things get a little more difficult because, if our collection consists of two pairs and three unmatched cards (with the pattern AABBCDE), flushes are possible. The two pairs can consist of either 2, 3, or 4 different suits (for examples of these situations, consider 5♥ 5♣ 7♥ 7♣ for 2 suits, 6♦ 6♠ J♠ J♣ for 3 suits, and 3♥ 3♠ 9♣ 9♦ for 4 suits). For each two-card pair, there are six ways (four choose two) to select suits, so for *both* pairs, there are $6 \cdot 6 = 36$ different ways to select suits. For the three *unmatched* cards, there are $4 \cdot 4 \cdot 4 = 64$ different ways to select suits, so now we just need to be careful combining these.

Of the 36 ways to select suits for the pairs, exactly 6 of them consist of 4 different suits; here, for the AABB portion of the hand, we choose 2 of the possible 4 suits for the pair AA (there are 6 combinations), and the other two suits become the suits for pair BB as well. In this case, the unmatched cards *cannot* help form a flush, since at most we will have 4 cards of one suit. Here, with these 6 possibilities, any of the 64 suit combinations for the unmatched cards are possible. Now, again of those 36 ways to select suits for the pairs, exactly 6 of them consist of 2 different suits, such as 5♥ 5♣ 7♥ 7♣, since for the pattern AABB, we choose 2 of the possible 4 suits for pair AA, and pair BB then contains the *same* two suits. However, in this scenario, having the three unmatched cards be *all* hearts or *all* clubs would be an issue; the CDE portion of the hand cannot consist of three cards that all share the same suit as either of the suits of the pairs. Of the 64 suit combinations for the pattern CDE, the 2 with each unmatched card having the same suit as either of the pairs' suits is therefore not allowed. That is, for example, with the pairs as 5♥ 5♣ 7♥ 7♣, we cannot have all three unmatched cards be hearts, nor could we have all three unmatched cards be clubs. Finally, of the 36 ways to select suits for the pairs, there are only 24 of them remaining, and these must have 3 different suits, as in 6♦ 6♠ J♠ J♣. Since the pairs (the AABB portion of the hand) consist of four cards and three suits, one suit must be repeated. Of the 64 ways to select suits for the unmatched cards, then, we need to disallow the possibility that all three (the CDE portion of the pattern) match that shared suit of the pairs (spades for example). This leaves

$$\underbrace{6 \cdot (64 - 2)}_{\substack{\text{two suits} \\ \text{for pairs}}} + \underbrace{24 \cdot (64 - 1)}_{\substack{\text{three suits} \\ \text{for pairs}}} + \underbrace{6 \cdot 64}_{\substack{\text{four suits} \\ \text{for pairs}}} = 2,268$$

possible suit combinations that do *not* yield flushes when two pairs and three unmatched cards are involved.

Now we can get around to counting the two pair hands with three un-matched cards; of the 13 ranks available, we need to choose 5 of them, and then remove the 10 possibilities of getting a straight. Of those 5 ranks, we can choose 2 to represent the ranks for the pairs (the remaining 3 ranks will serve as the unmatched card ranks). Finally, we can choose 1 of the 2,268 ways to select the suits for the cards, yielding

$$\underbrace{\left(\binom{13}{5} - 10 \right)}_{\substack{\text{ranks for} \\ \text{all cards}}} \cdot \underbrace{\binom{5}{2}}_{\substack{\text{pick ranks} \\ \text{for pairs}}} \cdot \underbrace{\binom{2,268}{1}}_{\substack{\text{pick suit} \\ \text{combination}}} = 28,962,360.$$

If we add this to the number of two pair hands coming from three pairs and an unmatched card, we get 31,433,400 possible two pair hands, for a probability of about 0.23496.

By now you can see that working with counting possibilities when we are trying to form the best five-card hand from a larger collection can be difficult but not impossible. If we wanted to count just the raw hands that *could* be classified as two pair without regard to higher-ranked hands, the process would become much simpler. However, if you *did* end up with a flush that happened to also contain two pairs, I would hope that you would play the hand as a flush and *not* as the lower ranked hand!

3.3 Texas Hold'Em

The form of poker that you are most likely familiar with, based on its television coverage the last few years, is Texas Hold'Em. It can be quite exciting to play when one player pushes all of his or her chips into the pot, yelling "all in," and letting their very existence in the tournament ride on whether his or her opponents call the bet *and* have a better hand. Matches determined on the last revealed card can cause the audience to go wild, and at least one player may go wild as well!

The premise is fairly simple. Each player receives two cards to call their very own (no one other than that player sees these cards, except perhaps the television audience), and over the course of the round, five more cards are dealt face up to the table as community cards (in order, a collection of three cards called *the flop*, then one card called *the turn* or *fourth street*, and a final card called *the river* or *fifth street*). If at least two players remain at the end, the cards of each player are revealed and the higher ranked poker hand takes the pot.

Assuming that Adam, Betty, Chris, Dawn, and Ethan are sitting around a table in clockwise order, play begins with the dealer, say Adam, shuffling the deck of cards and both Betty and Chris placing the initial bets; in order to guarantee at least *some* action on the hand, Betty is required to place a bet called the *small blind* and Chris is required to place a bet called the *big blind*. Typically, the big blind amount is the current minimum bet at the table, and the small blind is half of that. Once the cards are dealt, the first round of betting proceeds with Dawn either folding or calling the big bet amount (or raising, but we won't get into the specifics of betting rules here). Around the table, each player makes the same decision until all players have either folded or put the same amount into the pot. One card is placed face down in a discard pile (called burning a card) and the top three cards are revealed face up as *the flop*, and then another round of betting starts with Betty (or the first player to the left of the dealer who is still active in the hand). Another card is burned and one more card, *the turn*, is revealed face up, with another round of betting, and, after another burn card, *the river* card being revealed face up, the last round of betting occurs before hands are revealed. For more about the intricacies of the game, feel free to consult any of the numerous online resources available on the rules of the game.

Question

How can we apply probability to help us play better at Texas Hold'Em?

One note about the mathematics of the "burn" card is that, assuming a completely randomized deck, it is technically unnecessary and can be considered to be part of the tradition behind the game. In practice, however, as much as one tries to shuffle a deck of cards, invariably clumps of cards are likely to stay together while shuffling. Using burn cards is one way to cut down on the effect of these clumps. At this point, we can now take a look at a few scenarios to answer our question.

> **Example 3.12** Suppose that your initial two cards are 5♦ 5♠, and the flop is revealed as Q♠ 6♥ 2♣. What is the probability that you'll make three of a kind off of the turn? There are 47 unknown cards left for us, since we must ignore the burn cards along with any cards in the hands of our opponents (we have no idea what these may be). Of those 47, there are two fives that will help us get three of a kind. Therefore, the probability we make three of a kind is $2/47 \approx 0.04255$ or 4.255%.

It's worth noting at this point that if you were instead watching this match on television, and a player had the two fives and the flop was as in Example 3.12, chances are the probability displayed for making his hand would be much different. Probabilities displayed on televised matches *do* take into account the cards of other players and which hands are likely to win against the others. The calculations done here are the basis for these probabilities.

You are very unlikely to see Texas Hold'Em players using calculators at the poker table! If we don't need the exact probability and, instead, an estimate will do, there is a nice way to do this in your hand. At any point in the game after the flop, there will either be 47 or 46 unknown cards to you when you are asked to place a wager (before the turn or the river, respectively). If, as in Example 3.12, you need one of two possible cards to complete the hand you are looking for, the probability of making the hand on the next card at those times is roughly $2/47 \approx 0.04255$ and $2/46 \approx 0.04348$. The values are *very close* to the same value if we assume 50 cards in the deck, at $2/50 = 0.04$. Indeed, the common rule of thought during Texas Hold'Em is given by Theorem 3.3, where the addition of one to the percent is done to offset the error we introduced by overestimating the cards left in the deck.

> **Theorem 3.3 (Simple Texas Hold'Em Probability)** To determine an approximate probability that the next card you see at Texas Hold'Em will complete a particular named poker hand, count the number of unseen cards that will make the hand, multiply by two, and add one; this number is very close to the percent chance that you will make your hand.

For Example 3.12, the simple probability rule given in Theorem 3.3 gives 5%, since there are two cards that can help us. Let's look at one more example, using this result.

Example 3.13 Suppose you hold two spades, say **7♠ 3♠**, and the flop along with the turn yield a total of two more spades, as in A♥ A♠ 6♦ 2♠. What is the approximate probability of making a flush on the river? Since 4 of the possible 13 spades have been seen, there are 9 unknown to us, and therefore using Theorem 3.3 gives

$$9 \cdot 2 + 1 = 19,$$

giving us an approximate 19% chance of making our flush (versus 19.57%, which is the more correct probability obtained as 9/46).

For a quick and fast way to approximate probabilities during actual game play, Theorem 3.3 works pretty well!

At this point, you may be wondering about the strength of your initial two-card hand and whether or not a specific starting hand is worth playing; this is quite hard to show mathematically without resorting to computer simulations. I should add that I am not against using them at all, and will frequently do so in my work. Unfortunately, it does mean that working out examples showing *why* a particular hand is ranked very high versus very low would be beyond this book. Several online sources have these results; as you might guess, a pair of aces, then kings, then queens form the top three hands (with probabilities of about 31%, 26%, and 23%, respectively, of winning with those cards). An ace and a king of the same suit beats out a pair of jacks for the next spot, and for the rest, you can search online.

3.4 Exercises

3.1 Your initial cards for a hand of poker are 3♦, 4♥, 5♠, 6♠, and 9♣. Determine the probability that, after discarding the **9♣**, you complete your straight upon drawing one additional card.

3.2 Now, suppose that your initial cards for a hand of poker are 3♠, 7♠, 8♥, 10♠, and J♥. What is the probability that you discard the 3♠ and complete your straight with an additional card? Why does this value differ from the answer in Exercise 3.1?

3.3 For each of the following, repeat Exercise 3.1 but assume that, through some casual peeking, you know the person sitting to your left at the table has the given cards.

(a) 3♣ 4♣ 9♥ J♦ A♣
(b) 2♥ 3♠ J♦ J♥ Q♦
(c) 2♠ 2♥ 5♥ 5♣ 7♦

3.4 Verify the counts given in Table 3.3 listed below (these were not computed in this chapter).

(a) four of a kind   (b) full house   (c) straight   (d) pair

3.5 Imagine that a new named poker hand is created called the *semi-flush* that consists of five cards of the same color. For example, 4♦ 6♦ 9♥ J♥ Q♦ would qualify as a semi-flush since all of the cards are red.

(a) Determine the number of five-card hands that qualify as a semi-flush but *not* as a regular flush (don't worry about whether a hand could count as a straight or not).

(b) How does the "rarity" of a semi-flush compare to the other named hands of Table 3.3? Do you think it would be worth having this hand?

3.6 Determine the number of distinct *four card* poker hands.

3.7 For *four card* poker, determine the number of and probability of the following poker hands.

(a) four of a kind   (b) straight flush   (c) three of a kind   (d) flush
(e) straight   (f) two pair   (g) one pair   (h) high card

3.8 For *three card* poker, recall that the total number of possible hands is 22,100, computed in Example 3.1. Determine the number of and probability of the following poker hands.

(a) straight flush   (b) three of a kind   (c) straight   (d) flush
(e) one pair   (f) high card

3.9 Imagine *six-card* poker where all six cards matter. For each of the following types of possible poker hands, determine the number of and probability; as a hint, with the exception of "high card," the hands are ordered by increasing counts (and, therefore, probability).

(a) straight flush   (b) four of a kind and a pair   (c) two three of a kinds   (d) flush   (e) four of a kind an no pair
(f) straight   (g) three pair   (h) three of a kind and a pair
(i) three of a kind and no pair   (j) two pair   (k) one pair
(l) high card

3.10 For seven-card poker where the best five-card hand matters, determine the number of and probability of the following poker hands.

(a) four of a kind   (b) flush   (c) three of a kind   (d) one pair

3.11 For poker as in Exercise 3.10, determining the number of straights can be difficult. Do so, using the following questions as a guide.

(a) For 7 cards of distinct rank, determine the number of suit combinations that do *not* result in a flush.

(b) Use this information to determine the number of straights formed when 7 distinct ranks are involved; be careful, as we were with the straight flush, with what happens when an ace is and is not the high card for the straight.

(c) Being careful like above, determine the number of straights formed when 6 distinct ranks are involved (this means the seven cards contain a pair).

(d) Continuing, determine the number of straights formed when 5 distinct ranks are involved in two cases. One case is that the seven cards contain a three of a kind involved, and the other case is that it has two pairs.

(e) Combine the above to determine the total number of and probability of a straight.

4

More Dice: Counting and Combinations, and Statistics

If you have walked down the board game aisle at your local department store, super store, or toy store, it doesn't take long to realize that there are a boatload of different dice games available. Since dice are very portable, easy to use and understand, and can have custom dice faces to represent many different things, it's quite understandable why. In this chapter, we take a look at

FIGURE 4.1: Five Dice

a few of my favorite games that have some mathematical questions (and answers), and we also use this opportunity to reinforce many of the ideas from the first few chapters.

4.1 Liar's Dice

Liar's dice is a game that has been around since at least the 1500s; commercial versions have become widely available since the 1970s, and the game made an appearance in popular culture with the movie *Pirates of the Caribbean: Dead Man's Chest* along with the video game *Red Dead Redemption*. While there are many versions of the game, with different rules for bidding and dice representation and presentation, the version described below may be the most common (and definitely the version of the game I still frequently play with my colleagues).

Each player uses a cup and five normal 6-sided dice in order to play (see Figure 4.1). Because of this, the number of players is theoretically unlimited, subject only to the number of dice (and I suppose cups) present. Games with too few players (two or three) as well as games with too many players (more than, say eight), do not really give the "true feel" for the game as it starts. However, you are certainly free to try the game with any number of players to get a feel for it. In our example below, we assume that there are 7 players.

Once one player is chosen to go first, each player rolls their dice and keeps their dice concealed from the other players using the cup (perhaps "shaking the cup" is more descriptively correct than "rolling" the dice in the cup). At this point, the starting player makes a bid reflecting the quantity of a certain die face appearing among all players' dice. A common rule that is added to the game to make it slightly more exciting is that ones are *wild* (that is, they match whatever die face is the current bid), and ones may not be bid. For instance, the first player, in a game with seven players, may see their dice faces as ⚀ ⚀ ⚂ ⚂ ⚄ and decide to bid that there are at least 13 fours at the table. The next player (to their left) may choose to

- *challenge* that bid if they think that there are fewer than 13 fours at the table,

- *raise* that bid with a higher bid of their own (bidding rules are discussed below), or

- yell *spot on* if they believe that the player's bid is *exactly* correct.

When challenged, if the bid is indeed correct (of, say, at least 13 fours (and ones) at the table), the challenger loses one of their dice; if the bid is wrong, the bidder loses the die instead. If "spot on" is called, the player calling it loses a die if there are not exactly the bid number of the named dice at the table; otherwise, the bidder loses a die, and the caller receives one die back (up to a maximum of five dice). For raising, the new bid must increase either the *quantity* of dice involved in the bid, or the numeric *value* of the die face involved, or both. For the initial bid of 13 fours, for example, 13 fives or 13 sixes would be an appropriate raise, as would 14 twos. However, 13 twos would *not* be a valid bid.

The opportunity to raise, challenge, or call "spot on" goes around the table clockwise until someone decides to challenge or call "spot on." At that point, after the outcome is decided, the player to the left of the initial bidder starts the next round, and play continues until all players but one are eliminated (have no dice remaining). While it is definitely possible to play it safe early in the game with six opponents, strategy and bluffing play key roles in the end game. Early in the game, knowledge of the expected number of dice plays an important part of the game, and inspires our next question.

Question

What are the probabilities for the quantity of named faces at Liar's Dice, and what is the expected quantity?

How can we determine this? For our group of seven players, there are a total of 35 dice at the table. Of course, we already *know* what our 5 dice are, so there are 30 unknown dice at the table. Let's suppose that the opening bid is as indicated; that is, the bidder says there are 13 fours at the table. Suppose also that we have ⚀ ⚀ ⚂ ⚃ ⚄, and we want to determine the probability of different amounts of fours or ones at the table. Since there are a total of 30 unknown dice, if we want to compute the probability of having 13 fours or ones, we need to choose 10 of those 30 dice to be fours or ones (since we already have one four and two ones).

Example 4.1 Since each of the dice is independent of the rest, the probability of any *specific set* of 10 dice being fours or ones can be determined by using multiplication, combined with the probability of 1/3 for a die being a four or a one (let's call that a *success*), and a probability of 2/3 for a die being something other than a four or a one (let's call that a *failure*). We will need 10 dice with the 1/3 probability of success, and the remaining 20 dice with the 2/3 probability of failure. From Chapter 3 we recall that the binomial coefficient 30 choose 10 will give us the *number* of ways to select which of the 30 dice result in success. Combining this all together we'll get

$$\text{Probability of 10 4s or 1s} = \underbrace{\binom{30}{10}}_{\substack{\text{ways to choose} \\ \text{success dice}}} \cdot \underbrace{\left(\frac{1}{3}\right)^{10}}_{\substack{\text{ten fours} \\ \text{or ones}}} \cdot \underbrace{\left(\frac{2}{3}\right)^{20}}_{\substack{\text{twenty non-} \\ \text{fours or ones}}} \approx 0.15301$$

as the probability of there being exactly 10 fours or ones among the 30 unknown dice.

Binomial Distribution

Of course, Example 4.1 is only an interesting example if calling "spot on" was of interest to you; we need more information if we're to determine how good (or how bad) the bid of 13 fours could be for us! There was nothing special about the number 10 in the last example, and it leads us to the following definition.

Definition 4.1 (Binomial Distribution) Suppose we have n trials (for example, dice) of an experiment, where the probability of success on any single trial is p (for example, 1/3 for the above) and the probability of failure on any single trial is $q = 1 - p$ (for example, $1 - 1/3 = 2/3$ for the above). Assuming the trials are independent, the probability of exactly x successes in those n trials is given by the *binomial distribution* as

$$\text{Probability of } x \text{ Successes} = \binom{n}{x} \cdot p^x \cdot q^{n-x}.$$

Determining the probability of x dice showing fours or ones out of the 30 total is now pretty easy; we can plug in different values of x and see what happens, as I have done for Figure 4.2. Note that in that figure, x is indeed along the traditional x-axis, and since the probability of getting 20 or more fours or ones is essentially zero, I have only plotted this graph for x values ranging from zero to twenty. You might also guess that we did not end up looking at $x = 10$ in Example 4.1 by happenstance; it appears from the distribution here that 10 gives the highest probability, and you might call it the "average" value for this setup. If you also find this shape familiar and looking like a normal distribution or bell curve, that's not far off either.[1] Let's formalize this.

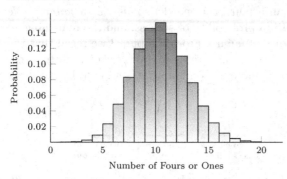

FIGURE 4.2: Binomial Distribution

The concept of expected value has been used a lot so far in this book, and we will continue to use it even more! Throughout Chapter 2 we primarily used it to compute expected winnings in order to determine a house advantage and we will do that again in Chapter 8. Here, however, we'll use it in a broader sense. Let's reduce the number of dice to make our calculations more explicit and instructive; consider rolling 6 standard dice, and we want to simply count the number of threes rolled. Using Definition 4.1, you should recognize this setup with $n = 6$ and $p = 1/6$ (and thus $q = 5/6$).

What is the expected number of threes when rolling 6 dice? Certainly the possibilities are that you can get anywhere from zero to six threes, so we'll need the probability of getting each of these using the binomial distribution. For practice, let's do 0, 1, and 2 threes together (the rest you should practice yourself and make sure you get the same approximate numbers as I do). Using the formula, with $n = 6$, $p = 1/6$, and $q = 5/6$, we get

$$\text{Probability of 0 Threes} = \binom{6}{0}\left(\frac{1}{6}\right)^0\left(\frac{5}{6}\right)^6 \approx 0.334898$$

$$\text{Probability of 1 Three} = \binom{6}{1}\left(\frac{1}{6}\right)^1\left(\frac{5}{6}\right)^5 \approx 0.401878$$

$$\text{Probability of 2 Threes} = \binom{6}{2}\left(\frac{1}{6}\right)^2\left(\frac{5}{6}\right)^4 \approx 0.200939.$$

The other probabilities, for 3, 4, 5, and 6 threes, respectively, are 0.053584, 0.008038, 0.000634, and 0.000021.

---

[1]For more about the connection between the binomial distribution and the normal distribution, check out Appendix C; while they are not the same, normal distributions can be used to approximate binomial distributions quite well.

Example 4.2 The expected number of threes when rolling 6 dice can be obtained from the expected value formula, which, by now, we have used quite a lot. If you like, you can imagine being paid $1.00 for each three that you get, in which case the computation is exactly the same as an expected winnings formula, except we can't possible lose! The expected number (average) number of threes is then

$$\text{Average} = \underbrace{0.334989 \cdot 0}_{0 \text{ threes}} + \underbrace{0.401878 \cdot 1}_{1 \text{ three}} + \underbrace{0.200939 \cdot 2}_{2 \text{ threes}} + \underbrace{0.053584 \cdot 3}_{3 \text{ threes}}$$
$$+ \underbrace{0.008038 \cdot 4}_{4 \text{ threes}} + \underbrace{0.000643 \cdot 5}_{5 \text{ threes}} + \underbrace{0.000021 \cdot 6}_{6 \text{ threes}} \approx 1.00,$$

which actually makes a lot of sense! If there's a 1/6 chance of getting a three, and you roll 6 dice, on average doesn't it make sense that you would get one three?

In fact, that phenomenon is not unique to this particular choice of numbers! The binomial distribution, which has a lot of applications beyond just rolling dice, has this easy formula for the expected number of successes.[2]

Theorem 4.2 For a binomial distribution with n trials and probability of success p, the expected number of successes is

$$\text{Expected Value} = n \cdot p.$$

Using this, our situation with 30 unknown dice and wanting to know the expected number of fours or ones becomes easy; since there is a 1/3 chance of getting a four or a one, we can use this result with $n = 30$ and $p = 1/3$ to get an expected quantity of fours and ones as 10; exactly what Figure 4.2 seems to indicate.

Of course, this result can be used for any experiment that fits the framework of the binomial distribution! This doesn't generally work well with playing cards (unless cards are drawn and then immediately returned to the deck before the next draw) since drawing a J♦ as your first card definitely affects the probability of getting a Jack or a diamond on the next card (along with affecting the probability of everything else as well). However, it does work well for coins; if you flip 10 coins, the expected number of heads is five (using $n = 10$ and $p = 1/2$). Using the binomial distribution itself, we can also compute the probability of flipping 10 coins and getting exactly 4 heads (as

---

[2]You should feel free to consult an advanced probability or statistics text on why this is always true, but hopefully your intuition and our example will convince you.

alluded to in Chapter 1).

$$\text{Probability of 4 Heads} = \underbrace{\binom{10}{4}}_{\substack{\text{which coins} \\ \text{are heads}}} \cdot \underbrace{\left(\frac{1}{2}\right)^4}_{\substack{\text{probability} \\ \text{four heads}}} \cdot \underbrace{\left(\frac{1}{2}\right)^6}_{\substack{\text{probability} \\ \text{six tails}}} \approx 0.2051$$

Question

How can we evaluate bids at Liar's Dice to successfully challenge a bid?

First, I should mention that this is a *very* tough question to completely answer. In our example where we have ⚀ ⚀ ⚂ ⚄ ⚅ and the opening bid (made by someone else) is 13 fours (or ones), we have to remember that the person making the bid *has information we do not have*, namely their own dice! In fact, one could make the argument that the person placing the bid has looked at their own dice, noticed that they also had 3 fours or ones, computed the expected value of the 30 dice that they don't know, and bid the expected total of fours or ones as 13. On one hand, that would be a very smart (perhaps bold) opening bid; on the other hand, they could be *bluffing*, which, as we will see further in Chapter 5, adds a dimension to the game that is very hard to analyze mathematically.

Of course, just because 13 seems like a safe number, we need to remember that a bid of 13 fours or ones means that the bidder is confident (hopefully) that there are *at least* 13 fours or ones; how likely is it that there are *more* than 13 of those dice faces? It seems that not only do we need to consider the expected value in order to play and bid wisely, but we also need to consider the likely *spread* of the quantity of dice faces.

Spread

You will find a fuller definition and study of the idea of spread and *standard deviation* in a full statistics or advanced probability book; here, we need only some of the basic ideas in order to look at the range of likely values of the number of fours or ones at the Liar's Dice table. Yes, to figure out the probability of having at least 13 fours or ones present you could add together the probabilities that there are exactly 13, exactly 14, exactly 15, exactly 16, and so on, the whole way up to exactly 30 (though a lot of these later probabilities will essentially be zero). This could take time, be cumbersome, and almost require that you have a calculator or computer in front of you playing Liar's Dice, which just might take away slightly from the fun!

Instead, consider the following result which approximates the spread of a binomial distribution quite well, due to the fact that binomial distributions can be very well approximated by normal curves.

Theorem 4.3 For a binomial distribution with n trials, probability of success p, and probability of failure $q = 1 - p$, the *standard deviation* for the number of successes is

$$\text{Standard Deviation} = \sqrt{n \cdot p \cdot q}.$$

If μ is the expected value and σ is the standard deviation for a binomial distribution, then approximately 68% of the time, the number of successes will be between $\mu - \sigma$ and $\mu + \sigma$. In addition, the number of successes will be between $\mu - 2 \cdot \sigma$ and $\mu + 2 \cdot \sigma$ about 95% of the time and between $\mu - 3 \cdot \sigma$ and $\mu + 3 \cdot \sigma$ about 99.7% of the time.

The value σ^2, the square of the standard deviation, is referred to as the *variance* of the distribution.

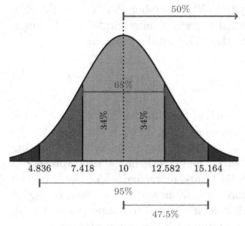

FIGURE 4.3: Spread

How can this be useful? Let's take a look. Supposing that we have 30 unknown dice, what are the likely ranges for the quantity of fours or ones? Well, we have seen earlier (and we can recompute) that the expected number of fours or ones is

$$\mu = n \cdot p = 30 \cdot \left(\frac{1}{3}\right) = 10,$$

and, using our new results for spread, we can compute the standard deviation as

$$\text{Standard Deviation} = \sigma = \sqrt{n \cdot p \cdot q} = \sqrt{30 \cdot \left(\frac{1}{3}\right) \cdot \left(\frac{2}{3}\right)} \approx 2.5820.$$

Using Theorem 4.3, we can now predict that about 68% of the time, the quantity of fours or ones will be in the range of 10 ± 2.5820, or 7.418 to 12.582. Even better, about 95% of the time, the quantity will be in the range of $10 \pm 2 \cdot 2.5820$, resulting in the range 4.836 to 15.164. Since the distribution for the number of fours and ones is *symmetric*, as seen for example in Figure 4.3, about 50% of the time the quantity will be 10 or greater, but we also know that 34% of the time, the quantity will be between 10 and 12.582, and 47.5% of the time the quantity will be between 10 and 15.164, giving us useful information about how likely certain situations can be when large numbers of dice are involved.

You might be wondering if this information is good in practice. When I play Liar's Dice, the expected value is always something that enters my mind (and, when playing with students, it becomes common practice for someone to shout out the new expected value once someone loses a die) when making or evaluating bids. I'm not sure that my students do this, but I definitely take the standard deviation and spread into account when playing, especially when large numbers of dice are involved.[3]

As we quickly saw, the binomial distribution can be used with coins as well (while another use could be randomly guessing at multiple-choice quizzes and predicting your score, I chose to stick with coins). You might flip a coin 100 times and get 45 heads and 55 tails; is this unusual? Would you think the coin is unfair? What if you get 40 heads and 60 tails? Let's practice once more with the idea of spread.

> **Example 4.3** With 100 coins, and considering a "head" as a success, we're working with a binomial distribution (flips are independent and identical) with $n = 100$, $p = 1/2$, and $q = 1/2$. The expected number of successes, of course, is $\mu = n \cdot p = 100 \cdot (1/2) = 50$, a number that you could have predicted without the fancy binomial expected value formula of Theorem 4.2! However, we can also compute the standard deviation as
>
> $$\text{Standard Deviation} = \sigma = \sqrt{n \cdot p \cdot q} = \sqrt{100 \cdot \frac{1}{2} \cdot \frac{1}{2}} = 5$$
>
> so with Theorem 4.3 in mind, about 68% of the time you should see $50 - 5 = 45$ to $50 + 5 = 55$ heads. Another way of looking at this is the other 32% of the time, you should get less than 45 heads or more than 55 heads! Getting 45 isn't a rare event at all! Even better, about 95% of the time you should see $50 - 2 \cdot 5 = 40$ to $50 + 2 \cdot 5 = 60$ heads, or rather 5% of the time you will see fewer than 40 heads or more than 60 heads! On its own, seeing 40 heads doesn't prove the coin is unfair, as it's bound to happen from time to time; however, if it starts happening often with the same coin, the fairness of the coin could be in question.

Using Tables

For Liar's Dice, since the probability of getting a one or any other single value on a die is fixed at $p = 1/3$, we can create tables of values that can be useful while playing the game. These tables are given at the end of the book in the Tables section as Tables T.1 and T.2. The question here is how one can use these tables to make better decisions while playing.

The first thing that you will have to remember to use the tables is that the value of n used is *not* the current number of dice at the table! Since you already know exactly what *your* dice are, the correct row to use in either table

---

[3]For the curious, the (normal) approximation we are using for the intervals of Theorem 4.3 are valid as long as you are using at least 18 dice; in general the approximation works well as long as the interval $(\mu - 3\sigma, \mu + 3\sigma)$ is entirely contained in the interval $(0, n)$.

is the row corresponding to the number of dice, n, that your opponents have. Since both tables include values of n from 1 to 30, the tables accommodate games that start with seven players.

The first table, Table T.1, gives the probability that *exactly* X dice match any specific value (when ones are also included). For instance, if there are $n = 20$ dice other than yours at the table, the probability that among those 20 dice there are 7 twos and ones is about 0.18. This also means that the probability that, for those 20 dice, there are 7 fives and ones is about 0.18. Make sure that you can find this number in Table T.1. While this table is not all that useful, it does help when thinking about "spot on" calls and should be used primarily for that purpose. Note that the numbers in this table could be computed by hand, since the value in row n and column X is simply

$$\text{Row } n, \text{ Column } X \text{ Value} = \binom{n}{X} \cdot \left(\frac{1}{3}\right)^X \cdot \left(\frac{2}{3}\right)^{n-X}$$

via the binomial distribution of Definition 4.1.

Dashes are included in Tables T.1 and T.2 for entries that don't make sense; for instance, you would never ask for the probability of having 14 sixes and ones if there were only 9 dice being rolled. For reference, the expected value μ and standard deviation σ are given for each n possibility in Table T.1.

The second table, Table T.2, can be much more useful during game play. This table gives the probability that *at least* X dice match a specific value (when ones are included) for the desired number n of other dice present at the table. For the $n = 20$ dice that your opponents hold, the table tells us that the probability of at least 7 twos and ones among their dice is about 0.52. This table is very useful for making the initial bid; once others have made some sort of bid, you'll have to decide if they've been bluffing or telling the truth and use your instinct to modify how to use the table. Once someone else bids some fours, for example, assuming they're not bluffing, you can be reasonably sure that the number of fours (and ones) at the table is a little higher than you may have thought.

Values in both tables have been truncated to two decimal places in order to keep probabilities that are *almost* a sure thing from looking like a sure thing. For instance, when the others have 30 dice, the probability that there's at least 1 three or one among those dice is given in Table T.2 as 0.99 even though that value is much closer to 1.00 (it's roughly 0.9999948). It's not a guarantee that there's at least 1 three or one for those 30 dice, but it's *very* unlikely.

If we use the shorthand notation $P(X = k)$ to represent the probability of having exactly k dice that match either a one or the named value, then what Table T.2 gives on row 13 and column 5, for example, is

$$\text{Row 13, Column 5} = P(X = 5) + P(X = 6) + \cdots + P(X = 12) + P(X = 13)$$

which, since the probability of having between 0 and 13 "successes" here must be 100%, is the same as

Row 13, Column $5 = 1 - (P(X = 0) + P(X = 1) + \cdots + P(X = 4))$

where the term inside the parentheses is what mathematicians would call the *cumulative probability distribution*.

We could go into more detail here, using *summation notation* to express this nicely, but since we will not use the cumulative probability distribution further, discussion of the shorthand notation for sums will happen later in this chapter when absolutely needed.

How exactly can these tables be useful? Here's an example of how I would use (and have used) them.

Example 4.4 When playing the game, I like to put the person after me in an awkward position of having to raise my bid or challenge my call; making a call that is relatively safe while still attractive for the next opponent to challenge means riding that fine line on Table T.2. Suppose against 4 other players, it's my turn to make the first bid. In this example, say they collectively have 17 dice, and I still have all five. Whatever value I choose to call, the chance that the other players have at least 6 of them (combined with ones) is 0.52, and the chance that they have at least 7 (combined with ones) is only 0.32. This is what I mean by riding the fine line; if I have ⊡ ⊡ ⠿ ⊠ ⊞, I might open with a bid of 9 sixes (or ones). With the chance of my opponents having at least 6 of them greater than 1/2, combined with my own 3, I feel safer than not that the table has at least 9.

If you do end up playing a lot of Liar's Dice, a neat variant you can play is having the person who makes the first bid *also* pick the value for the wildcard dice (rather than having it be one all of the time). With a roll such as the ⊡ ⊡ ⠿ ⊠ ⊞ situation of Example 4.4, making twos wild and calling 8 threes (or ones) would be an exciting bluff since among your opponent's dice, there's only a 0.17 chance of 8 threes or ones, but your opponents also would assume that by choosing two as the wild and bidding threes that *you* actually have some.

4.2   Arkham Horror

Set in the world of H.P. Lovecraft's fictional universe, containing powerful elder gods such as Cthulhu and Nyarlathotep, the board game Arkham Horror[4] is a cooperative game where investigators in the (fictional) city of Arkham, Massachusetts work to find clues, defeat monsters, and enter the unknown

---

[4]Arkham Horror is a registered trademark of Fantasy Flight Games.

"other worlds" to stop evil from prevailing. During game play, the investigators need to perform skill tests of varying types, and these skill tests are all carried out using regular six-sided dice. For example, if a character has a Lore value of 3, when that character is asked to do a Lore Test, they would roll three dice, and the test passes if at least one success (a 5 or 6) is rolled. Many tests have a modifier to them to make passing the test more or less difficult. Let's look at an example.

Example 4.5 The investigator Dexter Drake is asked to make a Lore -2 test during a game of Arkham Horror; from his investigator sheet, his current Lore value is four, so, with the skill test modifier of -2, Dexter rolls only two dice.

We could use the binomial distribution and add the probability of getting one success (a 5 or a 6) to the probability of getting two successes (two 5s or 6s) which gives us the total probability of passing the Lore test, but a better way to compute the probability of passing the test is to use Theorem 1.5; that is, let's subtract the probability of getting *zero* successes from 1 to get

$$\text{Probability Pass Lore Test} = 1 - \underbrace{\left(\frac{2}{3}\right)^2}_{\substack{\text{two non-}\\\text{successes}}} = \frac{5}{9} \approx 55.6\%$$

as the probability of passing the test.

Sometimes, a skill test is made even harder by requiring more than one success in order to pass. For example, a player might be asked to do a Strength $+1$ [3] test in order to move a heavy bookcase out of the way in search of a clue.

Example 4.6 Investigator Daniela Reyes, with a base Strength of 3, is being asked to do a Strength $+1$ [3] test while exploring the Black Cave in an attempt to move a large rock. The modifier of $+1$ means that she gets to roll four dice, but the [3] requirement means that she needs to get 3 successes definitely will make the probability of passing much lower. We can do this one of two ways. First, we can simply add together the probability of getting 3 successes and of getting 4 successes, obtaining

$$\text{Prob. Pass Strength Test} = \underbrace{\binom{4}{3}\left(\frac{1}{3}\right)^3\left(\frac{2}{3}\right)^1}_{\text{3 successes, 1 failure}} + \underbrace{\binom{4}{4}\left(\frac{1}{3}\right)^4\left(\frac{2}{3}\right)^0}_{\text{4 successes, 0 failures}} = \frac{1}{9},$$

much lower than it would be if we only needed one success! The other way to compute this would be to subtract the probability of getting 0, 1, or 2 successes from 1, getting

$$\text{Prob. Pass Strength Test} = 1 - \left(\underbrace{\left(\frac{2}{3}\right)^4}_{0 \text{ successes}} + \underbrace{\binom{4}{1}\left(\frac{2}{3}\right)^3\left(\frac{1}{3}\right)^1}_{1 \text{ success}} \right.$$

$$\left. + \underbrace{\binom{4}{2}\left(\frac{2}{3}\right)^2\left(\frac{1}{3}\right)^2}_{2 \text{ successes}} \right) = \frac{1}{9}$$

which is refreshing to see since both answers *should* be the same!

As long as we keep one method of computing these, we can generalize; using the "subtracting from one" method because of its simplicity when fewer successes are needed, we find that the probability of passing a skill test, when x successes are needed and n dice are rolled (where n is the final number after a skill test modified is taken into account), is given by

$$\text{Pass Skill Test Probability} = 1 - \sum_{j=0}^{x-1} \binom{n}{j}\left(\frac{2}{3}\right)^{n-j}\left(\frac{1}{3}\right)^{j}$$

and you can match this up with the computations done in Examples 4.5 and 4.6.

TABLE 4.1: Skill Test Pass Probabilities

Successes Needed	Dice Rolled 1	2	3	4
1	0.333333	0.555556	0.703704	0.802469
2		0.111111	0.259259	0.407407
3			0.037037	0.111111
4				0.012346

Successes Needed	Dice Rolled 5	6	7	8
1	0.868313	0.912209	0.941472	0.960982
2	0.539095	0.648834	0.736626	0.804908
3	0.209877	0.319616	0.429355	0.531779
4	0.045268	0.100137	0.173297	0.258650

With a bit of programming, we can automate this computation easily to give you Table 4.1 which shows the probability of passing a skill test, with the number of successes needed given as the rows and the number of dice you roll, up to eight, given as the columns. As you would hopefully guess and as the table easily illustrates, rolling more dice certainly does increase the probability of passing a given skill test!

4.3   Yahtzee

The dice game of Yahtzee[5] is probably one of my all-time favorites. It is not uncommon for some of my colleagues to stop by my office during lunch for a round or two of the game.

On the surface, the game appears to be entirely luck-based. Rolling dice is something that cannot really be controlled. While history has certainly seen its fair share of cheaters at craps where some players are able to roll the dice with more accuracy regarding the results than anything close to random, putting dice in a cup and rolling from the cup provides a way to counteract the ability to cheat in Yahtzee. However, once the scoring is explained and explored, you may realize that the game has a rich strategy for choosing in which category a particular round of rolling should be scored. Optimal play, if it can be discovered, will of course turn the game into one of luck and following a strategy, but opponents who are not aware of such a strategy will be at a disadvantage.

Yahtzee uses five dice, often placed in a cup so that the cup can be shaken (to randomize the dice) and then poured out to result in the first roll of the dice. Based on what you see and what scoring categories may already be used, you may choose to keep some or none of the dice and reroll the others. After this second roll, you again have the option to keep some or none of the dice and reroll the others (including any dice you kept after the first roll that you now, for whatever reason, wish to roll again). After this third and last roll, you must choose one of the 13 categories in which to score that roll; for your second round, there will only be 12 categories to select, then 11 for the third round of the game, and so on. The game ends after each player has played 13 rounds, and the winner is the player with the highest total score. The various categories are divided into the "upper section" and "lower section" categories. The six upper section categories are

- *aces* or ones, where the score entered is the sum of the dice showing a one; that is, for this category, the same as the *number* of dice showing a one,

- *twos* where the score entered is the sum of the dice showing a two; this is equivalent here to *twice* the number of dice showing a two,

---

[5]Yahtzee is a registered trademark of the Milton Bradley Company/Hasbro Incorporated.

- *threes* where the score entered is the sum of the dice showing a three, or *three* times the number of dice showing a three,

- *fours* where, as you can imagine, the score entered is four times the number of fours, or the sum of the dice showing a four,

- *fives* where the sum of dice showing a five, or five times the quantity of dice showing a five is scored, and

- *sixes* where dice showing a six are scored as the sum of those die's faces, or six times the quantity of those dice.

An "upper bonus" is awarded to the player if the total score for the upper section is 63 or greater. For the lower section, "poker-like" categories are featured, with those seven categories being

- *three of a kind* where you need at least three of the same die face present to score; scoring is taking the sum of *all* five dice,

- *four of a kind* where you need at least four of the same die face; score by taking the sum, again, of *all* five dice,

- *full house* where you need two dice of the same face and the other three dice of a different face, scored at a flat 25 points,

- *small straight* where you need at least four of the dice appearing in sequential order, scored at 30 points,

- *large straight* where you need all five dice appearing in sequential order, scored at 40 points,

- *Yahtzee* (or five of a kind) where you need all five dice showing the same die face, scored at 50 points, and

- *chance* where you simply score the total sum of all five dice.

Note that these categories are summarized, with example rolls, in Table 4.2. A special thing happens if you roll a Yahtzee after you have already scored 50 points in the Yahtzee category; you automatically receive a 100-point bonus (indicated by a check mark on the score sheet, or a bonus chip in older versions of the game) and then must still score your roll in one of the categories. If the appropriate box in the upper section is available, it must be scored there. Otherwise, you may score it in any of the bottom categories (including full house, small straight, or large straight, at full points as a "wildcard") if possible. If the only available category happens to be a non-matching upper section category, unfortunately, a 0 must be scored for one of those. Hopefully the 100 points is enough to quickly erase any pain that taking a 0 gives you.

There are many questions that we can ask mathematically about Yahtzee; the last that we'll study, of course, will concern best game play. For now, though, we can use some of the mathematics we have already explored to answer some questions, and you'll be responsible for some others in the exercises. For us, a list of distinct Yahtzee rolls would mean a list of all possible results after the first roll (or second or third rolls, but including the dice that you chose not to reroll). Here, order does not matter one bit, so we would consider the rolls of ⚀ ⚀ ⚀ ⚃ ⚄ and ⚀ ⚃ ⚄ ⚀ ⚀ the same.

TABLE 4.2: Yahtzee Categories

Category	Example	Score
Aces	⚀ ⚀ ⚁ ⚃ ⚄	2
Twos	⚁ ⚁ ⚁ ⚁ ⚃	6
Threes	⚂ ⚂ ⚂ ⚂ ⚂	12
Fours	⚀ ⚂ ⚂ ⚃ ⚄	8
Fives	⚃ ⚃ ⚃ ⚃ ⚄	20
Sixes	⚀ ⚂ ⚅ ⚅ ⚅	18
Three of a Kind	⚁ ⚂ ⚃ ⚃ ⚃	22
Four of a Kind	⚀ ⚃ ⚃ ⚃ ⚃	18
Full House	⚀ ⚀ ⚀ ⚃ ⚂	25
Small Straight	⚀ ⚂ ⚃ ⚃ ⚄	30
Large Straight	⚀ ⚀ ⚂ ⚃ ⚄	40
Yahtzee	⚂ ⚂ ⚂ ⚂ ⚂	50
Chance	⚀ ⚃ ⚃ ⚅ ⚅	24

Question

How many distinct Yahtzee "rolls" are possible?

To count the number of distinct rolls, we can employ the methods of poker hand counting that we discussed in Chapter 3. To start, note that there are very few cases for us to consider, since for now we're only interested in the total number of possible rolls. Either our roll has a five of a kind, four of a kind, three of a kind, a pair, or no matching dice. We'll consider each of these separately.

One of the easiest to consider is the five of a kind; here, in order to completely describe the roll, we need only determine the rank each of the five dice will have. As with poker, we'll use "rank" to mean the die face involved. Since there are six possibilities, choose one of them to describe the exact five of a kind to obtain

$$\binom{6}{1} = 6$$

possible five of a kinds. Continuing, for the possible four of a kinds, note that we still need to choose one of the possible six ranks for the 4 matching dice, but then we also need to choose a different rank for the fifth die. We can do this in one of five possible ways, resulting in

$$\binom{6}{1} \cdot \binom{5}{1} = 6 \cdot 5 = 30$$

different ways. Counting the possible three of a kinds can be done in several ways, and I present the two-handed approach here to illustrate one of the ways of thinking about this category. First, we could have a three of a kind and then a pair; to count this, we can choose two of the possible six ranks to select the ranks involved, and then choose *which* of our two ranks becomes the triple (done in one of two possible ways). This gives us

$$\binom{6}{2} \cdot \binom{2}{1} = 15 \cdot 2 = 30$$

so-called full houses, and then we also need to count the situations in which we have a three of a kind and two unmatched dice. For this, we select three of the possible six ranks, and then choose one of those three to represent the actual triple,[6] in

$$\binom{6}{3} \cdot \binom{3}{1} = 20 \cdot 3 = 60$$

ways, for a total of 90 possible three of a kinds.

As an exercise, you will show that there are 120 possible rolls with at most only pairs involved (this includes two pair and one pair rolls), so to finish our count, for no matching dice, note that from the six ranks, we need to choose five of them, for

$$\binom{6}{5} = 6$$

TABLE 4.3: Roll Type Counts

Type	Number
Five of a Kind	6
Four of a Kind	30
Three of a Kind	90
Pairs	120
No Matching	6
Total	252

possible rolls with no match.

This information is summarized in Table 4.3, and you can see there the final tally for distinct Yahtzee rolls, the sum of all of these cases, gives us a grand total of 252. Note that this is *far* fewer than the number of *ordered* Yahtzee rolls, as that total would be $6^5 = 7,776$ since each die results in one of six possible faces.

You might find it interesting that there are the same number of possible Yahtzee (five of a kind) configurations as there are no matching dice configurations! In fact, since the only possible large straights are ⚀ ⚁ ⚂ ⚃ ⚄ and ⚁ ⚂ ⚃ ⚄ ⚅, there are far fewer, by a factor of three, combinations that yield a large straight than a Yahtzee! However, rolling a Yahtzee with one roll of the dice is far less common than rolling a large straight, so let's compute the probabilities and see why.

---

[6]As an illustration that there are different ways to think about this, note that we could also say that we'll choose one of the possible six ranks to represent the triple, and then from the five remaining ranks, choose two to pick the unmatched dice. This gives us $\binom{6}{1} \cdot \binom{5}{2} = 6 \cdot 10 = 60$ as well.

Example 4.7 Figuring out the probability of getting a Yahtzee with one roll is actually fairly easy. We already know that there are six possible rolls that result in the five of a kind, so once the particular roll (that is, rank) is determined, *each* die rolled needs to result in that rank. For each of the dice involved, there's a 1/6 chance of getting the right value, so the probability of a Yahtzee is then

$$\text{Probability of Yahtzee} = \underbrace{\binom{6}{1}}_{\text{rank}} \cdot \underbrace{\left(\frac{1}{6}\right) \cdot \left(\frac{1}{6}\right) \cdot \left(\frac{1}{6}\right) \cdot \left(\frac{1}{6}\right) \cdot \left(\frac{1}{6}\right)}_{\text{roll the rank on each die}} = 6 \cdot \left(\frac{1}{6}\right)^5$$

$$= \frac{1}{1,296} \approx 0.000772.$$

What changes when we want to talk about the large straight? Well, note that for a Yahtzee of fours the only possible roll that we would be happy with is ⚃ ⚃ ⚃ ⚃ ⚃, but for a large straight, starting at 1 and ending at 5, we would be equally happy with ⚀ ⚁ ⚂ ⚃ ⚄ or ⚄ ⚃ ⚂ ⚁ ⚀. The lesson here is that while there was only one ordering for the Yahtzee, there are many orderings for a large straight. Once we've determined whether the large straight begins with either a 1 or a 2, there are five choices for the first die, four choices for the second die, three choices for the third, two for the fourth, and one for the last, for a total of

$$5 \cdot 4 \cdot 3 \cdot 2 \cdot 1 = 5! = 120$$

different orderings. Once an ordering is selected, though, there's a 1/6 chance that a particular die will result in the face we need. This gives us

$$\text{Prob. of Lg. Str.} = \underbrace{\binom{2}{1}}_{\substack{\text{start} \\ \text{rank}}} \cdot \underbrace{5 \cdot 4 \cdot 3 \cdot 2 \cdot 1}_{\text{orderings}} \cdot \underbrace{\left(\frac{1}{6}\right) \cdot \left(\frac{1}{6}\right) \cdot \left(\frac{1}{6}\right) \cdot \left(\frac{1}{6}\right) \cdot \left(\frac{1}{6}\right)}_{\text{roll needed rank on each die}}$$

$$= \binom{2}{1} \cdot 5! \cdot \left(\frac{1}{6}\right)^5 = 2 \cdot 120 \cdot \left(\frac{1}{6}\right)^5 = \frac{5}{162} \approx 0.030864$$

for the chance of rolling a large straight with a single roll. This is exactly 40 times more likely than rolling a Yahtzee on one roll!

Note here the presence of the $(1/6)^5$ term; in this case, and our cases to follow, once we have selected a specific outcome for the dice (by choosing the *type* of roll as well as the *ordering* of the dice), the probability of that specific dice roll is always $(1/6)^5$. Each die face is precisely controlled by our counting techniques, making the probability part itself much simpler. Example 4.7 also illustrates a new concept for us different from the notion of combinations given by Definition 3.2. The number of possible orderings, or *permutations*, of a set can be determined using factorials. Let's see a fuller definition that applies in more circumstances.

Permutations

Definition 4.4 The number of *permutations* of r objects from a collection of n objects is given by the formula

$$_nP_r = n \cdot (n-1) \cdots (n-r+2) \cdot (n-r+1) = \frac{n!}{(n-r)!}$$

and, when $n = r$, this turns into the nice formula

$$_nP_n = n \cdot (n-1) \cdots 2 \cdot 1 = n!.$$

Most of the time our use of permutations will be clear enough to avoid using the notation. In particular, when we are completely ordering a collection of n objects, we will write $n!$ instead of $_nP_n$ as this happens often enough to avoid confusion. The need to order only a subcollection for us will be minimal, but for an example, consider the following.

Example 4.8 Oftentimes disagreement can reign when I have friends over and we're about to choose what board game we want to play. While we do end up playing several and enjoying our time together, on at least two occasions I've asked everyone to write down their top three game choices, thereby giving me information on what everyone may want to play. Assuming that I have 10 games (in reality, I have over 70 choices, but let's keep the numbers down for this example), how many different top three orderings are possible? Using Definition 4.4 provides the quick and easy answer, using $n = 10$ since there are 10 game choices and $r = 3$ since I've asked for an ordered top three. We get

$$\text{Number of Top Three Lists} = \frac{10!}{(10-3)!} = \frac{10!}{7!} = 10 \cdot 9 \cdot 8 = 720$$

different top three lists. For the curious, there are 328,440 top three lists if 70 games are present!

By now, you might be wondering what changes need to be made to determine other probabilities at Yahtzee when the option to reroll dice up to two times takes effect. Before we get into best strategy at the game, which will take awhile on its own, it's worth taking the time to answer a question that many may ask you when they learn that you are studying the mathematics of games.

Question

What is the probability of getting a Yahtzee?

There are essentially three ways to get a Yahtzee; the first, and arguable the "luckiest" is simply rolling a five of a kind on the first roll. While this doesn't happen often, Example 4.7 from earlier gives us this probability as $1/1,296$ or approximately 0.08%. Alternatively, we could use *two* rolls to get a Yahtzee. For example, on our first roll, we might get four of a kind, and then be lucky in completing the Yahtzee on the next roll (the first of two possible rerolls). There are other ways to roll a five of a kind in two rolls, of course, so let us analyze this case before we move on to the case where all three rolls are needed to get the Yahtzee.

Example 4.9 Here we wish to determine the probability of rolling a Yahtzee in two rolls after possibly keeping some dice after the first roll. Suppose that the first roll results in four of a kind; common sense should tell you that it makes sense to keep the four of a kind, and reroll hoping to match the rank of that quadruple. To determine the probability, recall from Table 4.3 that there are 30 possible four of a kinds. Once we have chosen one of these, we need only choose which of the five dice yields the unmatched die. Then, recalling that the probability of any specific ordering of five dice is $(1/6)^5$, multiplying by a final $1/6$ to successfully match the rank for the four of a kind on the reroll is all that's necessary. This gives us a probability[7] of

$$4 \text{ Then } 1 = \underbrace{\binom{30}{1} \cdot \binom{5}{1}}_{\substack{\text{dice ranks} \\ \text{and ordering}}} \cdot \underbrace{\left(\frac{1}{6}\right)^5}_{\text{probability}} \cdot \underbrace{\left(\frac{1}{6}\right)}_{\text{reroll}} \approx 0.003215.$$

For a concrete example of this situation, suppose the first roll were ⚁ ⚁ ⚁ ⚁ ⚄, where you would choose to reroll the ⚄, and hope to obtain a ⚁ to complete the Yahtzee.

The case where you obtain a three of a kind (with two other dice different from the three of a kind's rank, possibly a pair, possibly not) and then successfully reroll the two dice and obtain a Yahtzee will be left for you as an exercise to show that the probability is approximately 0.005358. Instead,

---

[7]Note that we could also compute this probability as $\binom{6}{1}\binom{5}{1}\left(\frac{1}{6}\right)^4\left(\frac{5}{6}\right) \cdot \left(\frac{1}{6}\right)$ where we first choose the rank for the four of a kind, then choose which position the unmatched die occurs, and finally give the probability of getting 4 "successes" and 1 "failure" on the first roll, followed by a "success" on the reroll. I chose *not* to use this method because we already have computed in Table 4.3 the counts for the various four of a kinds, and our method in Example 4.9 yields a simpler-looking probability model.

let us consider for this example what happens if our first roll results in a pair and we reroll the remaining three dice (not all matching, as this case would have been considered a three of a kind) hoping for a Yahtzee. Of the 120 dice choices for a pair from Table 4.3, exactly 60 of them come from having a pair and three other completely different dice and 60 come from having two pair and an unmatched die. For the case of having two pair, we choose 1 of the possible 60 die selections, then for the specific ordering of dice, we choose 1 of the 5 positions to represent the unmatched die, and then choose 2 of the remaining 4 positions for the location of the first pair (we could also then choose 2 of the of remaining 2 positions for the last pair, but there is only one way to do this anyway). Combining this together, we'll get

$$\text{2 Pair, Then 3} = \underbrace{\binom{60}{1} \cdot \binom{5}{1} \cdot \binom{4}{2}}_{\substack{\text{dice ranks} \\ \text{and ordering}}} \cdot \underbrace{\left(\frac{1}{6}\right)^5}_{\text{probability}} \cdot \underbrace{\left(\frac{1}{6}\right)^3}_{\text{reroll}} \approx 0.001072$$

where the $(1/6)^3$ term comes from needing to get all three dice on the reroll to match the desired rank for the Yahtzee. For the 60 cases with one pair and all other dice unmatched, even among themselves, we select one of those sixty cases, then choose 2 of the 5 positions for the pair, and for the remaining unmatched dice, there are 3! orderings (we are permuting those three dice, as per Definition 4.4), yielding

$$\text{1 Pair, Then 3} = \underbrace{\binom{60}{1} \cdot \binom{5}{2} \cdot 3!}_{\substack{\text{dice ranks} \\ \text{and ordering}}} \cdot \underbrace{\left(\frac{1}{6}\right)^5}_{\text{probability}} \cdot \underbrace{\left(\frac{1}{6}\right)^3}_{\text{reroll}} \approx 0.002143$$

as the probability.

Finally, recall that from Table 4.3 there are 6 possibilities for getting our five dice to have *no* matching faces; in this case, we select one of these six cases, and among the five unmatched dice, there are 5! orderings. Note here that we will choose to reroll *all* of the dice,[8] obtaining a probability of

$$\text{All Different, Then 5} = \underbrace{\binom{6}{1} \cdot 5!}_{\substack{\text{dice ranks} \\ \text{and ordering}}} \cdot \underbrace{\left(\frac{1}{6}\right)^5}_{\text{probability}} \cdot \underbrace{\binom{6}{1} \cdot \left(\frac{1}{6}\right)^5}_{\text{reroll}} \approx 0.000071.$$

By putting everything from Example 4.9 together, we now know that the probability we will obtain a Yahtzee in exactly two rolls (this means not getting a Yahtzee on the first roll, and, whatever we choose to do optimally from the first roll, obtaining it on the second) is about 0.011859. Compared to Example 4.7, where the probability of Yahtzee on the first roll is 0.000772, we see that we're over 15 times more likely to get a Yahtzee with just one reroll than we are on a single roll.

Multinomial Distribution

Before we continue, we should take a quick pause to see if the computations of Example 4.9 can be "cleaned up" a little; that is, perhaps there's a tool that can be used to help out a little bit with the orders. Remember that the binomial distribution of Definition 4.1 provides a way to talk about "success" and "failure" in matching particular faces to specific dice (among other things of course), so perhaps some generalization of that can give us a way to determine the number of orderings for each scenario discussed.

There were 60 rolls of the 252 total that resulting in something we would call two pair, described in Example 4.9 as the "2 Pair, Then 3" scenario. Sure, there are 60 ways to choose 1 of the possible two pair situations, but once one of these (for example, having 2s and 4s for the two pairs, and a 5 as the unmatched die) is selected, how many orderings are possible? In the example, we chose one of the five dice positions to receive the unmatched die, then chose two of the remaining four to receive the first pair, completely describing all possibilities. Notice, though, that

$$\binom{5}{1} \cdot \binom{4}{2} = 30 = \frac{5!}{2! \cdot 2! \cdot 1!} = \binom{5}{2, 2, 1}$$

where the notation on the right is new and will be explained soon. How can the appearance of each factorial term be explained? Using the permutations of Definition 4.4, we know there are 5! ways to arrange five dice in general. We must be careful, however, when counting ways to arrange the five dice and have a pair, for while, using a black die for emphasis, the rolls ⚁ ⚂ ⚃ ⚄ ⚄ and ⚄ ⚂ ⚃ ⚃ ⚄ are *physically* different, they represent the *exact same* ordering for game purposes. Dividing by 2! removes this double counting since there are 2! ways to order the 2s, and then dividing by a second 2! removes the double counting of the 4s.

The notation on the right side of the equation above comes from the following definition.

---

[8]From a mathematical point of view, the case where you choose to keep *one* of the dice from the first roll and try to get a four of a kind on the second roll results in exactly the same probability since when keeping no dice, you can roll any of the 6 possible Yahtzee rolls, and when keeping one die, you need to roll the specific Yahtzee that matches. You'll be asked to show that keeping one die results in the same probability as an exercise.

Definition 4.5 (Multinomial Coefficient) The *multinomial coefficient*, given by the formula

$$\binom{n}{x_1, x_2, \ldots, x_k} = \frac{n!}{x_1! \cdot x_2! \cdots x_k!}$$

counts the number of ways choosing from n objects a collection of k types of objects, where there are x_1 of one type, x_2 of another type, continuing up to x_k of the last type. Note that we require $x_1 + x_2 + \cdots + x_k = n$ so that each of the n objects appears as one of the types.

Using Definition 4.5 is fairly simple once you have had a chance to practice. Appealing again to the "2 Pair, Then 3" situation, once we know that we want a pair of 2s, a pair of 4s, and an unmatched 5, the number of orderings is given by the multinomial coefficient

$$\binom{5}{2, 2, 1} = 30$$

since we have five slots available, and we need two 2s, two 4s, and one 5. This inspires the following generalization of the binomial distribution.

Definition 4.6 (Multinomial Distribution) The *multinomial distribution* describes an experiment with n trials where each trial results in any number, say k, of different results; the trials are identical in that the probability of a specific outcome (or result), say R_i, remains the same for each repetition. If the probability of outcome R_i is given as p_i, then the probability that the experiment yields an x_1 number of outcome R_1, an x_2 number of outcome R_2, and so on, is given by

$$\text{Probability } (R_1 = x_1, \cdots, R_k = x_k) = \binom{n}{x_1, x_2, \ldots, x_k} \cdot p_1^{x_1} \cdot p_2^{x_2} \cdots p_k^{x_k}$$

where we require that $x_1 + x_2 + \cdots + x_k = n$ (each of the n trials gives one of the R_is as its outcome) and $p_1 + p_2 + \cdots + p_k = 1$ (the list of k possible results for a trial is exhaustive). Note that when we write the shorthand notation $R_i = x_i$, we mean that x_i of the trials ended up being result R_i.

As you're reading through Definition 4.6, the symbols may get in the way of a good working knowledge of how it can be useful. For our regular six-sided die, for example, we would use $k = 6$ since there are six different results, and R_1 might represent the outcome of rolling a one, R_2 could represent the outcome of rolling a two, and so on. However, if we do want a pair of 2s, a pair of 4s, and the unmatched 5 as before, it may be easier to ignore anything but 2s, 4s, and 5s, and instead use Definition 4.6 with $k = 3$ and the three outcomes R_1, R_2, and R_3 representing 2s, 4s, and 5s respectively. Luckily for us, until we get to the study of Zombie Dice in Chapter 9, we won't need to be this explicit. The multinomial coefficient of Definition 4.5 will suffice.

You should, however, see some similarities between Definition 4.6 and Definition 4.1. In particular, when there are only two possible outcomes for a trial (that is, $k = 2$ and only R_1 and R_2 exist), they agree! The requirement that $x_1 + x_2 = n$ means that, if R_1 is a "success," then R_2 might as well be called a failure. Since here $p_1 + p_2 = 1$ implies that, by writing $p_1 = p$ and $p_2 = q$, the only thing in Definition 4.6 that appears much different is the multinomial coefficient. In this two-outcome situation, we can see that, by a little bit of playing around,

$$\binom{n}{x_1, x_2} = \frac{n!}{x_1! x_2!} = \frac{n!}{x_1!(n - x_1)!} = \binom{n}{x_1}$$

from before (since, again, $x_1 + x_2 = n$). Neat! Of course, you need not remember *why* this gives you the same result, but I think it's always nice to know that the generalization gives you the exact computations as before!

Again, our use here of Definition 4.6 will be minimal and we will rely on past experience and the multinomial coefficients themselves (Definition 4.5) to guide our way. However, you should be thinking that each of our rolls, once ranks are selected for the type of roll we are investigating, is truly a multinomial distribution where we have the presence of 5 trials (at the start, then possibly fewer on rerolls) and probability $p_i = 1/6$ for each of the die faces. In fact, we could call this situation luxurious here because the term $p_1^{x_1} p_2^{x_2} \cdot p_k^{x_k}$ in Definition 4.6 is reduced to $\left(\frac{1}{6}\right)^n$ for us, where n is the number of dice being rolled. This "luxurious" situation ends, as I have already warned, when you encounter Zombie Dice later in this book.

Example 4.10 To recompute the probability of getting two pair on your initial roll and then completing your Yahtzee on the second roll is given by

$$2 \text{ Pair, Then } 3 = \underbrace{\binom{60}{1}}_{\text{ranks}} \cdot \underbrace{\binom{5}{2, 2, 1}}_{\text{ordering}} \cdot \underbrace{\left(\frac{1}{6}\right)^5}_{\text{probability}} \cdot \underbrace{\binom{3}{3}}_{\text{ordering}} \cdot \underbrace{\left(\frac{1}{6}\right)^3}_{\text{probability}} \approx 0.001072,$$
$$\underbrace{\phantom{\binom{3}{3} \cdot \left(\frac{1}{6}\right)^3}}_{\text{reroll}}$$

the same value as before. Again, note that once the ranks are selected and an ordering given, the probability of getting that ordering is given by $\left(\frac{1}{6}\right)^5$ at the start. On the reroll, there are only three dice to roll, so we choose 3 of the 3 slots available to be of one specific type (in this case, the binomial and multinomial coefficients are the same when only one type is needed).

Returning to our overall questions about the probability of a Yahtzee, what happens when we allow the second possible reroll? As you might imagine, there are many cases to consider when asking about the probability of getting a Yahtzee with all three rolls. In order to be as clear as possible, we will take care of these possibilities in separate examples rather than one example spanning many pages. The first possibility is that we roll a four of a kind on the first roll, do not complete our Yahtzee on the second roll, and then successfully match that last die on the third roll. Remember, again, at this point we only care about the situations where it does indeed take all three rolls to get a Yahtzee; the one roll and two roll versions were covered in Examples 4.7 and 4.9.

Example 4.11 To get a Yahtzee in three rolls, with the first roll resulting in a four of a kind, the second roll *not* matching, and the third roll matching (with probability $1/6$), the probability can be calculated as

$$4, 0, \text{Then } 1 = \underbrace{\binom{30}{1}}_{\text{ranks}} \cdot \underbrace{\binom{5}{4,1}}_{\text{ordering}} \cdot \underbrace{\left(\frac{1}{6}\right)^5}_{\text{probability}} \cdot \underbrace{\binom{5}{1}}_{\text{rank}} \cdot \underbrace{\binom{1}{1}}_{\text{ordering}} \cdot \underbrace{\left(\frac{1}{6}\right)}_{\text{probability}} \cdot \underbrace{\left(\frac{1}{6}\right)}_{\text{third roll}}$$

$$\underbrace{}_{\text{first roll}} \quad \underbrace{}_{\text{second roll}}$$

$$\approx 0.002679$$

where the term for the first roll appears slightly different from Example 4.9. The multinomial coefficient is used to count the orderings of four dice of one rank and one die of a different rank. Note that for the second roll, we choose one of the five unmatching ranks to be our result, and then, on rolling 1 die, there is only 1 ordering possible as the "1 choose 1" computes.

As you might guess, the next scenario to consider is what happens if the first roll of the dice results in a three of a kind. Note again that, like Example 4.9, we will need to consider the effect of rolling a three of a kind in a "full house" formation or with two other unmatched dice. While Example 4.11 is very explicit with its terms, what appears in the examples below is somewhat less explicit in my use of braces.

Example 4.12 There are two ways in which we can get three of a kind and then complete our Yahtzee in two more rolls; if we initially roll, say, ⚁ ⚁ ⚁ ⚂ ⚄, we would definitely choose to reroll the ⚂ and ⚄ if we're going for a Yahtzee. Our next roll, if we're completing the Yahtzee while needing both rerolls, could give us either 1 or 0 more twos. The case where we start with three of a kind, obtain one more on the first reroll, and finish the Yahtzee on roll three, we'll consider here; the case where the first reroll results in zero matching dice is left for you later as an exercise.

Regardless of whether the first roll gives you a full house or not, assuming that we do have a three of a kind, we will choose to reroll the other two dice, hoping for a Yahtzee in two rolls. Assuming we get one additional matching die on the first reroll (giving us a four of a kind leading into the final roll), we note that the two rerolled dice can be ordered in 2 ways, using the multinomial coefficient $\binom{2}{1,1}$. For the final roll we need to complete the Yahtzee by matching the desired rank with probability 1/6. Combined with the first roll probabilities for a full house versus a three of a kind with two unmatched dice, we'll get

$$3 \text{ (FH)}, 1; 1 = \underbrace{\binom{30}{1} \cdot \binom{5}{3,2} \cdot \left(\frac{1}{6}\right)^5}_{\text{first roll}} \cdot \underbrace{\binom{5}{1} \cdot \binom{2}{1,1} \cdot \left(\frac{1}{6}\right)^2}_{\text{second roll}} \cdot \underbrace{\left(\frac{1}{6}\right)}_{\text{third roll}}$$

$$\approx 0.001786$$

and

$$3 \text{ (non-FH)}, 1; 1 = \underbrace{\binom{60}{1} \cdot \binom{5}{3,1,1} \cdot \left(\frac{1}{6}\right)^5}_{\text{first roll}} \cdot \underbrace{\binom{5}{1} \cdot \binom{2}{1,1} \cdot \left(\frac{1}{6}\right)^2}_{\text{second roll}} \cdot \underbrace{\left(\frac{1}{6}\right)}_{\text{third roll}}$$

$$\approx 0.007144$$

where I should mention that the probability for the full house (or three of a kind and two unmatched) on the first roll given here is part of Exercise 4.11.

With two of the cases done, it would seem like our journey through computing the probability of a Yahtzee in exactly three rolls is almost complete. There are only two other cases to consider; the first happens when the initial roll results in a most a pair (or two), and the other happens when the first roll has no matching dice at all. While the latter is fairly easy to work with, the case where we have at most two matching dice is the most involved.

Note that I continue to use the multinomial coefficients of Definition 4.5 here for most of these computations; you are, of course, encouraged to think about other ways to count orderings and compute probabilities that result in the same result. There is at least one other way to do so that you've already encountered.

Example 4.13 One possibility for getting a Yahtzee when the first roll has only a pair or two is rerolling three dice, obtaining 2 matching dice that match the desired rank, and then getting the fifth die to match on the second reroll. For instance, if the first roll were ⊡ ⊡ ⊡ ⊠ ⊞, you would reroll the ⊡, ⊠, and ⊞, and maybe obtain ⊡ ⊡ ⊞ with those three dice. At this point, rerolling the ⊞ and getting ⊡ would be great! Dividing the probability calculation up into two pieces (for two pair versus only a single pair) as in Example 4.9, for the two pair version we get

$$2 \ (2P), \ 2, \ \text{Then} \ 1 = \underbrace{\binom{60}{1} \cdot \binom{5}{2,2,1} \cdot \left(\frac{1}{6}\right)^5}_{\text{first roll}} \cdot \underbrace{\binom{5}{1} \cdot \binom{3}{2,1} \left(\frac{1}{6}\right)^3}_{\text{second roll}} \cdot \underbrace{\left(\frac{1}{6}\right)}_{\text{third roll}}$$

$$\approx 0.002679$$

where, in the second roll, we have a term that says if we need to get two matching dice and one unmatched die, we need to choose one of the three dice (or positions) to be the unmatched. Only the "first roll" computation changes when a single pair is rolled for the first roll, obtaining

$$2 \ (1P), \ 2, \ \text{Then} \ 1 = \underbrace{\binom{60}{1} \cdot \binom{5}{2,1,1,1} \cdot \left(\frac{1}{6}\right)^5}_{\text{first roll}} \cdot \underbrace{\binom{5}{1} \cdot \binom{3}{2,1} \left(\frac{1}{6}\right)^3}_{\text{second roll}} \cdot \underbrace{\left(\frac{1}{6}\right)}_{\substack{\text{third} \\ \text{roll}}}$$

$$\approx 0.005358$$

as that probability.

   There are other ways, of course, to get a Yahtzee in your third roll assuming a pair at the start; the case when the second roll results in matching one additional die, and then needing two on the last roll will be left to you (these cases I would call the "2 (2P), 1, Then 2" and "2 (1P), 1, Then 2" situations).

   The perhaps more interesting situation occurs when the first roll results in something like ⊡ ⊡ ⊡ ⊠ 𝟙 and, after rerolling the non-threes, you have ⊠ ⊡ ⊡ ⊠ ⊠. Immediately you should notice that to have the best chance of rolling a Yahtzee on the last roll, you should *keep the three of a kind* and reroll the threes! The probability of getting a three of a kind on three dice is

$$\text{Probability 3 of a Kind with 3 Dice} = \underbrace{\binom{5}{1}}_{\text{rank}} \cdot \underbrace{\left(\frac{1}{6}\right)}_{\text{probability}} = \frac{5}{216}$$

where we have only five ranks to choose for the three of a kind (in our example, if we chose *three* as the rank, we would have gotten a Yahtzee with two rolls, and that case was covered in Example 4.9). Our chance of getting a pair, then a different three of a kind, and finally matching with the two rerolled dice for an initial two pair is

$$2 \ (2P), \ \text{"3,"} \ \text{Then} \ 2 = \underbrace{\binom{60}{1} \cdot \binom{5}{2,2,1} \cdot \left(\frac{1}{6}\right)^5}_{\text{first roll}} \cdot \underbrace{\left(\frac{5}{216}\right)}_{\substack{\text{second roll} \\ \text{three of a kind}}} \cdot \underbrace{\left(\frac{1}{6}\right)^2}_{\text{third roll}}$$

$$\approx 0.000149$$

and for an initial one pair only is

$$2\ (1P),\ \text{``3,''}\ \text{Then}\ 2 = \underbrace{\binom{60}{1} \cdot \binom{5}{2,1,1,1} \cdot \left(\frac{1}{6}\right)^5}_{\substack{\text{first roll}}} \cdot \underbrace{\left(\frac{5}{216}\right)}_{\substack{\text{second roll} \\ \text{three of a kind}}} \cdot \underbrace{\left(\frac{1}{6}\right)^2}_{\substack{\text{third roll}}}$$

$$\approx 0.000298.$$

To complete all possibilities for what happens when the initial roll consists of one or two pairs, we need only determine the probability of having no dice on the first reroll matching our desired rank and also not forming a three of a kind. Since the probability of rolling three dice different from our desired rank is $\left(\frac{5}{6}\right)^3$, we just need to subtract, in the appropriate place, the 5/216 used above as the chance we get the three of a kind. We obtain

$$2\ (2P),\ 0,\ \text{Then}\ 2 = \underbrace{\binom{60}{1} \cdot \binom{5}{2,2,1} \cdot \left(\frac{1}{6}\right)^5}_{\substack{\text{first roll}}} \cdot \underbrace{\left(\left(\frac{5}{6}\right)^3 - \left(\frac{5}{216}\right)\right)}_{\substack{\text{second roll, no} \\ \text{three of a kind}}} \cdot \underbrace{\left(\frac{1}{6}\right)^2}_{\substack{\text{third roll}}}$$

$$\approx 0.000595$$

for two pairs at the start, and for only one pair at the start,

$$2\ (1P),\ 0,\ \text{Then}\ 2 = \underbrace{\binom{60}{1} \cdot \binom{5}{2,1,1,1} \cdot \left(\frac{1}{6}\right)^5}_{\substack{\text{first roll}}} \cdot \underbrace{\left(\left(\frac{5}{6}\right)^3 - \left(\frac{5}{216}\right)\right)}_{\substack{\text{second roll, no} \\ \text{three of a kind}}} \cdot \underbrace{\left(\frac{1}{6}\right)^2}_{\substack{\text{third roll}}}$$

$$\approx 0.001191,$$

completing this example.

As our final case, we could certainly roll all of the dice, get *no* matches, and then need to rely on getting a Yahtzee with only the two remaining rolls. Take a moment to convince yourself that it does not matter whether we keep one die and try to match it, or whether we simply reroll *all* five dice. To ease computation, we will do the latter.

Example 4.14 The probability of rolling all five dice, getting no matches, and then using the remaining two rolls to get Yahtzee is easy because we have already done most of the work! Example 4.9 covers the probability of getting a Yahtzee in exactly two rolls. The probability that we roll all five dice, get no matches, then reroll all of the them, working to get our five of a kind is

$$0,\ \text{Then Yahtzee in Two Rolls} = \underbrace{\binom{6}{1} \cdot 5! \cdot \left(\frac{1}{6}\right)^5}_{\substack{\text{first roll}}} \cdot \underbrace{0.011859}_{\substack{\text{from} \\ \text{Example 4.9}}} \approx 0.001098$$

since we can reuse that result now. Note that we could have instead used a multinomial coefficient here, too, since

$$5! = \binom{5}{1,1,1,1,1}.$$

After a somewhat long journey, we are finally able to put all of these numbers together. From Examples 4.11, 4.12, 4.13, and 4.14, along with Exercises 4.12 and 4.13, the probability that we obtain a Yahtzee using all three rolls is approximately 0.033396, and adding to this the results from Examples 4.7 and 4.9 (the probabilities of getting a Yahtzee in exactly one and two rolls, respectively), the overall probability of getting a Yahtzee is about 0.046027, or about 4.6%.

If you were wondering how the chance for Yahtzee would be affected if *three* rerolls were possible (that is, if you had four possible rolls of the dice), a different method for computing these probabilities will be explored in Chapter 6. While this new method may make Examples 4.7, 4.9, 4.12, 4.13, and 4.14 seem unnecessary, the methods developed and examples of counting and probabilities are invaluable and will definitely help us in the future.

Given our probability calculation, it seems that we can expect a Yahtzee to happen, on average, 1 in 21.73 rounds of the game assuming, of course, that you *only* try for a Yahtzee on every turn! Since there are more categories than simply "Yahtzee" in the game, in practice, where you might be aiming to roll a large straight (or something else), the occurrence of the five of a kind in this dice game is a bit rarer than 1 in 21.73 rounds. Some of you might try for an "All Yahtzee" strategy, trying to obtain the largest possible score at the game. The probability of getting all 13 rounds to result in a Yahtzee can be determined using the binomial distribution of Definition 4.1 with $p = 0.046027$ and $n = 13$, resulting in

$$\text{Probability of 13 Yahtzee Rolls} = \binom{13}{13} \cdot (0.046027)^{13} \cdot (1 - 0.046027)^0$$

$$\approx 0.00000000000000000000416$$

so don't get your hopes up too high! In general, your average score with this strategy will be a good bit less[9] than optimal Yahtzee strategy. This, of course, begs the following question.

Question

How can we make the best decision at each stage in Yahtzee?

---

[9]The average score using the "All Yahtzee" strategy is about 171.52, while that of optimal Yahtzee play is about 254.59.

As you might imagine, with 13 categories to complete and up to three rolls allowed per round, determining the best play for Yahtzee is very non-trivial. Having ⚀ ⚁ ⚂ ⚃ ⚄ as your first roll of the game is very good; getting the 40 points for the large straight that early is wonderful! However, getting that same roll as the first roll of your *last* round is terrible if the only thing left for you to complete is, say, Yahtzee. You may also know that one key to a successful Yahtzee game is making sure to score at least 63 points in the "upper section" so that you get the 35 point bonus. To keep track of how your progress is going toward getting that bonus is fairly simple, since exactly three ones, three twos, three threes, three fours, three fives, and three sixes, scored in the upper section, adds up to exactly 63 points; if you get less than three for any of those categories, you'll need to balance that with getting more than three for at least one category (missing a six, as you might guess, is way worse than missing a one).

The way to determine best play for Yahtzee, in general, is first to see what methods we can develop for a much simpler game.

Mini-Yahtzee

Using the name "Mini-Yahtzee" is probably too generous to describe the game that I'm about to tell you about. While it does involve using dice (well, technically, one die) and categories (only two), there are no rerolls involved. My intent here is to highlight the process involved in looking at the possibly *ends* of a game to decide what decisions should be made to get there.

The game of Mini-Yahtzee involves only two rounds. The current player rolls a single six-sided die, and then must score the result in one of two possible categories: double or square. Scoring in the *double* category involves taking the die face, doubling that value, and recording the result on your score sheet. The *square* category involves taking the die face, squaring the value,

TABLE 4.4: Mini-Yahtzee Card

Category	Score
Double	
Square	
Total	

and recording that. The player then rolls the die and scores the result, as appropriate, in the remaining category. A scorecard can be seen in Table 4.4.

Some strategy decisions are pretty clear; if your first roll is ⚀, you're better off scoring this in the double category, because for any roll except a one, your next roll will be better squaring (at possibilities of 4, 9, 16, 25, or 36) versus squaring the one, scoring that one, and shooting for, at best, a 2, 4, 6, 8, 10, or 12 on the last roll. Similarly, your intuition should tell you that if you roll ⚅ at the start, you should square this number and proudly score a 36 on your scorecard, entirely due to luck.

Figure 4.4 shows part of a *decision tree* for this game. The first level of branching describes what happens when you roll the die for the first time. Since each of the numbers one through six are equally likely, we assume that the unlabeled edges each have the value 1/6 (they have been omitted in order

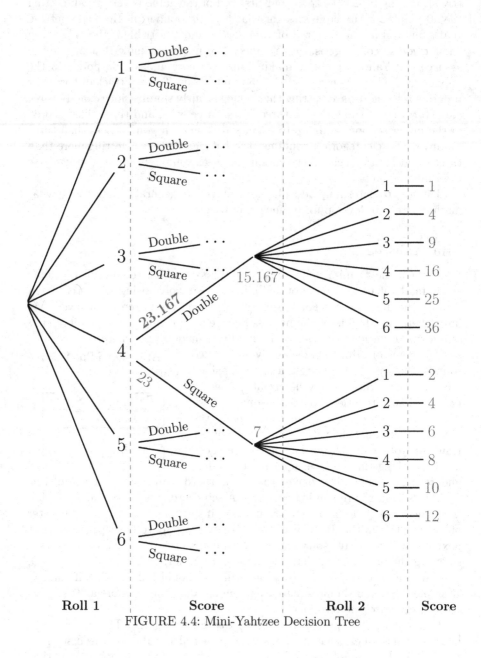

FIGURE 4.4: Mini-Yahtzee Decision Tree

to keep the tree uncluttered). At this point, a decision needs to be made on whether to double the result or instead square it; each of the six die results has two branches coming off of it, one representing the path to follow if doubling the result, and the other to follow for squaring. Then, the next branch, for each decision possibility, splits into six possibilities for the second die roll, followed by a single branch for each result, since the decision for roll two is forced. The decision tree here is very closely related to the *probability tree* of Definition 8.2 that we'll see later; the only difference here is the added "branching" where we actually need to make a decision. Also, note that Figure 4.4 only shows the second roll and subsequent decision branch for a starting roll of ⊡ in order to save space. How can this be helpful? Let's explore the gray numbers of Figure 4.4 through an example.

Example 4.15 What should you do if the first roll of the die is ⊡? The answer lies is considering what may happen on the *second* roll. Suppose that you decide to double the four and score an 8 in the doubling category. You're forced to square the next die roll, and since each die face occurs, equally likely, with probability 1/6, your expected score for "square" on the next roll is

$$\text{Expected Square} = \left(\frac{1}{6}\right) \cdot 1^2 + \left(\frac{1}{6}\right) \cdot 2^2 + \left(\frac{1}{6}\right) \cdot 3^2 + \left(\frac{1}{6}\right) \cdot 4^2$$
$$+ \left(\frac{1}{6}\right) \cdot 5^2 + \left(\frac{1}{6}\right) \cdot 6^2 \approx 15.167,$$

which can also be written using summation notation as

$$\text{Expected Square} = \sum_{n=1}^{6} \left(\frac{1}{6}\right) \cdot n^2 = 15.167$$

so that your expected score, heading down the double branch after rolling ⊡, is $8 + 15.167 = 23.167$. Note that the "double" branch of Figure 4.4 has a gray 15.167 there, indicating the value of that path, and the double path itself is labeled with 23.167, your expected score for this decision.

Similarly, suppose you decide to square the four, scoring 16 in that category. The expected score for "double" on the next roll is

$$\text{Expected Double} = \sum_{n=1}^{6} \left(\frac{1}{6}\right) \cdot (2n) = 7$$

and, adding the 16 from squaring to that, gives an expected score of the square branch for ⊡ as $16 + 7 = 23$.

Choosing to score ⊡ on the first roll in the double category is mathematically the best option.

Of course, there was nothing special about using ⊡ in Example 4.15 except for the fact that the difference between choosing to score it as a double versus scoring it as a square is very close. You can repeat this same process for each of the other possible results from the first roll, and Table 4.5 gives the expected scores (at the end of the game) for each of the possible starting result and each decision. From the table, it appears that you should score only a five or a six as a square from the first roll and otherwise

TABLE 4.5: Mini-Yahtzee Values

Roll	Double	Square
1	17.167	8
2	19.167	11
3	21.167	16
4	23.167	23
5	25.167	32
6	27.167	43

score the result as a double. This strategy may very well be one of the simplest for a game! Now, if only the game were more fun.

The way to find the best strategy for a game like Yahtzee is virtually the same as what we did using Figure 4.4 and Example 4.15 albeit on a much, much larger scale. Even starting with our Mini-Yahtzee game, note that adding just one more category will add both another decision branch and roll branch on the right; in addition, the first decision point would have three choices instead of two, and the second would have two instead of one. The game of Yahtzee involves thirteen categories, and instead of one die with six possibilities, we have five dice with 252 possibilities (remember Table 4.3 has the counts for roll types). For each of the 252 roll types, at the start, there are 13 choices of category, followed by another 252 possible roll results, and then 12 choices for a category, and this process continues. Without any trimming, there will be

$$102,917,054,120,865,385,267,539,244,701,579,843,993,600$$

paths through this tree. That's definitely *way* too many to include in this book. By the way, this also does *not* include rerolls, so one would hope that trimming the tree can reduce this number to a more manageable quantity.

Thankfully, some mathematicians and computer scientists have studied this problem and have somewhat recently developed the full strategy with the use of computers. A short article by Phil Woodward appeared in *Chance* in 2003, and researchers James Glenn and Tom Verhoeff both independently solved the Yahtzee problem shortly before that time. The latter of these two, as of this writing, has a website with a "Yahtzee Proficiency Test" that will let you play the game of Yahtzee and provide you with feedback about where your decisions are less than optimal. A quick web search for the quoted name should guide you there, or, of course, you can check this book's Bibliography!

Included in the Tables section at the end of the book, as Table T.3, is an almost optimal strategy that you can use in Yahtzee play. Each possible result with five dice is listed in the table with a 13-character string associated to it which specifies, in order, which category that result is best scored in. The numerical characters mean to fill in the corresponding box in the upper section of your score sheet; the others are:

- T for three of a kind,
- F for four of a kind,
- H for full house,
- S for small straight,
- L for large straight,
- Y for Yahtzee, and
- C for chance

Even though these rankings are based on computations for just the *first* round of a game, you can still use this table as you play to make good decisions later (if the leftmost character's category is already used on your card, use the next leftmost character and so on). While not perfect, it will work very well for you as long as you remember that once the upper section bonus is obtained, you can be more willing to score a zero up top, and once a Yahtzee is obtained, you can use further Yahtzee results as wildcards as per the rules.

> **Example 4.16** For a quick example that shows Table T.3 in action, suppose that you roll ⚅ ⚅ ⚅ ⚂ ⚀. The corresponding string from the table is "6TC12FHY3L45S" which means that, assuming "sixes" is free as a category, you'll be scoring this result there. Of course, with two rolls left, it makes sense to improve your round, so we would roll the ⚂ and ⚀ again; if after the third roll we are left with ⚅ ⚅ ⚅ ⚅ ⚃, the new string "6FTC12HY43L5S" tells us to prioritize sixes over four of a kind, and four of a kind over three of a kind, and so on; here, if sixes, four of a kind, and three of a kind are already used, we would score 28 in Chance and start the next turn.

We do need to keep in mind, of course, if you're not playing a solitaire game and instead are trying to beat your friends, late in the game you may need to change your own personal strategy. For instance, if Yahtzee and Small Straight are the only categories left, and you need to beat your friend who has 35 more points than you, you need to go for the Yahtzee no matter what, even though the Small Straight is more likely!

4.4 Exercises

4.1 Suppose a single coin is flipped 6 times; use the Binomial Distribution found in Definition 4.1 to find the probability that x heads are flipped for x ranging from 0 to 6.

4.2 Use Definition 4.1 to explain the calculation, found in Example 1.11, of determining the probability that zero heads are flipped on 20 coins.

4.3 Suppose 100 standard 6-sided dice are rolled.

(a) Find the probability that 15 fours are rolled.

(b) Determine the expected number of fours that should be rolled when these 100 dice are used.

(c) What is the standard deviation for the number of fours that are rolled on 100 dice?

(d) Using the notion of spread, determine an interval for the number of fours you would expect to be showing about 68% of the time 100 dice are rolled.

4.4 When a skill test in Arkham Horror is failed, the investigator can spend a clue token that they have earned in order to reroll one of the failed dice used for the test; repeat Example 4.5 if the investigator has a clue token to spend if the initial skill test roll does not pass.

4.5 In Arkham Horror, it is possible for players to gain a condition called "Blessed" which modifies the rules for conducting a skill test; rolls of 4, 5, or 6 are now considered successes instead of just 5s and 6s.

(a) Repeat Examples 4.5 and 4.6 to determine the probability of passing those skill tests when an investigator is Blessed.

(b) Recompute Table 4.1 for an investigator that is currently Blessed.

4.6 Similar to the previous exercise, in Arkham Horror, it is also possible for players to gain a condition called "Cursed" where only rolls of a 6 will count as a success when doing a skill test (instead of both 5s and 6s).

(a) Repeat Examples 4.5 and 4.6 to determine the probability of passing those skill tests when an investigator is Cursed.

(b) Recompute Table 4.1 for an investigator that is currently Cursed.

4.7 Another cooperative board game that my best friend and I often play is Shadows of Brimstone.[10] When on a map tile with no enemies, players can scavenge that tile (look for items); the player rolls three six-sided dice, and for each six rolled, draws a card from the scavenge deck as a reward.

(a) Determine the probability of getting to draw zero cards from the scavenge deck; repeat for one card and for two cards.

(b) What is the expected number of cards you draw from the scavenge deck?

(c) After rolling the dice, the player can spend a "grit" token to reroll any number of dice just rolled; assuming that the player will only reroll non-sixes, repeat parts (a) and (b) if the player uses a "grit" token during their scavenge roll.

4.8 Also in Shadows of Brimstone, the weapon that I use to attack enemies uses eight-sided dice to determine how many hits are made on an attack, as well as the number of critical hits. The weapon hits on a roll of 3+, critically hits on a roll of 6+, and misses on a 1 or a 2. Assuming that I roll 6 dice, determine the expected number of critical hits, regular hits, and misses.

4.9 With good gear in Shadows of Brimstone, even attacks from a large enemy can be mitigated fairly well. If Beli'al attacks me, it rolls six regular 6-sided dice and scores a hit against me for each roll of 5+. Since my defense is listed as 3+, for each successful hit from Beli'al, I get to roll a 6-sided die and on a roll of three or higher, that hit is blocked. Each unblocked hit to my character does six damage to me, but a point of damage can be ignored using my armor of 4+ (for each point of damage, I roll a 6-sided die and that point is ignored on a roll of four or higher). Determine the expected number of damage points I take from an attack round by Beli'al.

---

[10]Shadows of Brimstone is a registered trademark of Flying Frog Productions.

4.10 Determine the number of possible initial rolls in Yahtzee where the roll contains two pairs. Then, do the same for an initial roll that contains one pair and three unmatched dice. This completes the count of 252 distinct Yahtzee rolls given on page 78.

4.11 Complete Example 4.9 by showing that the probability of rolling a three of a kind, rerolling the other two dice, and obtaining a Yahtzee is approximately 0.005358. Note that you should consider the full house and non-full house cases separately.

4.12 Complete Example 4.12 by determining the probability of rolling a three of a kind, rolling no matching dice on the first reroll, and then rolling both matching dice to complete the Yahtzee on the last roll. As in Example 4.12, you will want to consider the full house and non-full house versions separately for both the initial roll *and* the first reroll.

4.13 Complete Example 4.13 by determining the probability of rolling a pair (or two pair), rolling one matching die on the first reroll, and then rolling both matching dice to complete the Yahtzee on the last roll. Remember to treat the two pair and one pair cases separately, and now the reroll could result in a pair or not a pair!

4.14 A footnote to Example 4.9 remarks that, when rolling five unmatched dice on your first roll in Yahtzee, the probability of getting a Yahtzee on your second roll does not depend on whether you choose to keep one die and roll for a four of a kind or you choose not to keep any dice and roll for a five of a kind. Show that this is true by determining the probability of getting a Yahtzee on your second roll assuming that you roll five unmatched dice on the first roll and choose to keep one of the dice, and then comparing this to the probability calculated in Example 4.9.

4.15 Consider a different version of Mini-Yahtzee where instead of "double" and "square" as the scoring options, you have "triple" and "cube" as the options. Create a decision tree similar to Figure 4.4 for one possible first roll and then create a table similar to Table 4.5 so that you can determine optimal strategy on how to score your first roll.

5

Game Theory: Poker Bluffing and Other Games

Imagine playing a round of Texas Hold'Em with a **6♠** and 9♦ in your hand; marching though the flop, turn, river reveals the community cards shown in Figure 5.1. Your hand isn't *actually* worth anything, but you *could* choose to bluff. That is, since your opponents do not know what you have in your hand for sure, you could pretend to have something that you do not. If your 9♦ were actually a 8♦, you'd have a nice straight!

FIGURE 5.1: Poker

At the core, the metagame of bluffing involves you making one of two decisions, either to bluff or not to bluff, and your opponent also making one of two decisions, either to call your bluff or to fold. This type of situation can be modeled by a branch of mathematics called game theory. We consider some of the basics of game theory here, and for a more complete coverage of the topic, you can look at a book completed devoted to the subject, such as [49], [59], or [4]. Before we get to the game theory basics and explore its uses beyond poker, let's stick to discussing poker and develop a bit of the motivation behind game theory.

5.1 Bluffing

As any good poker player will tell you, bluffing is an important aspect of the game. For those that aren't too aware of the practice, *bluffing* is the art of pretending to have a better hand of cards than you actually do. The most common way a bluff is expressed is when a player makes or raises a bet when the player's hand is unlikely to win against the other players' hands (a fact the player is perhaps well aware of). The bluffer is hoping to steal the pot by causing all other players to fold their hands. In some cases, the bluffer is also hoping to trick his or her opponents into contributing a little bit more into the pot before the bluffer really lays it on thick and encourages others to drop out. Bluffs, though, can occur in many other avenues, including bidding during games such as Liar's Dice, which we saw in Chapter 4, and auctions, driving up the price of an object when the bidder has no intent to be the last bidder (this can be seen, and heard, on the popular television show *Storage Wars*).

As important as bluffing is, amateur players often bluff more than is needed. If you watch television coverage of the *World Series of Poker* events, you'll rarely, if at all, see any professional poker player running forward with a hand in Texas Hold'Em of, say, 2♦ 7♠. Mostly because of its inability to form a straight or flush with three additional cards, this hand (and any other unsuited version of it), probabilistically, is the worst possible starting hand for the game. It is not uncommon, though, for players to attempt to run this hand locally at their Friday Night Poker nights. I have been guilty of this in the past, attempting to bluff my way forward to steal the pot. I will not detail my poor results at doing so. It should be noted, though, that this hand *can* be viable if you're the big blind, and at least get to see the flop (the first three community cards revealed); more information is always better in making an informed decision.

Question

How often should we bluff?

This very innocent question is fun to examine because of the interesting connections we can make to game theory. We will see how that works soon, but for now, let's take a look at how bluffing can be mathematically sound. For our study, we will limit ourselves to the game of Texas Hold'Em; see Section 3.3, if you haven't already, to learn about that game. We will also make copious use of the simple probabilities for Texas Hold'Em given in Theorem 3.3, mainly to ease calculations, but also since accurately determining when your opponent is bluffing is not an exact science either.

Consider what happens in the following scenario. The pot currently has $50 in it, and you have a hand that is essentially worthless, perhaps K♣ Q♦ 6♣ 4♠ 2♠. You're fairly confident that your opponent, Bob, has a better hand. Is bluffing profitable? Assuming that you are willing to bet $25 as a bluff, hoping that Bob folds and you steal the $50 currently in the pot, the answer completely depends on the probability that Bob will fold. Assuming that he folds 30% of the time, we can use expected value to determine your expected winnings. Note that any money you already have in the pot is irrelevant; the question is only about whether you should wager an additional $25 to have a chance at getting the $50 available. We get

$$\text{Expected Winnings} = \underbrace{0.30 \cdot (\$50)}_{\text{Bob folds}} + \underbrace{0.70 \cdot (-\$25)}_{\text{Bob calls}} = -\$2.50$$

which is a losing proposition for you; you should not attempt to bluff. However, suppose instead that Bob folds 40% of the time. Then we have

$$\text{Expected Winnings} = \underbrace{0.40 \cdot (\$50)}_{\text{Bob folds}} + \underbrace{0.60 \cdot (-\$25)}_{\text{Bob calls}} = +\$5.00,$$

which means that we should bluff.

You might wonder what the tipping point is; what value for the chance that Bob folds turns a losing proposition for our bluff into a positive one? We can determine this using expected values and a variable for Bob's probability.

Example 5.1 If we let p represent the probability that Bob folds, we can perform the above expected value computation again; the tipping point occurs when the expected winnings are exactly $0.00, since at this probability, it doesn't matter whether we bluff or not. We should bluff for whatever values of p give a positive value. We get

$$\text{Expected Winnings (\$)} = \underbrace{p \cdot (50)}_{\text{Bob folds}} + \underbrace{(1-p) \cdot (-25)}_{\text{Bob calls}} = 75p - 25$$

so we should bluff whenever $75p - 25 > 0$, which happens when $p > 1/3 \approx 0.33$.

Note that, as far as the dollar amounts are involved, only the proportion of your bet to the current pot size matters. You would still find a value of $p = 1/3$ for the tipping point if the pot were, say, $100 and you wanted to risk $50 by bluffing. Take a moment to verify this statement; look at Example 5.1 and repeat it with these new dollar figures and convince yourself that any scaling does not affect the

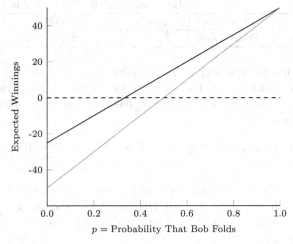

FIGURE 5.2: Expected Value for a $25 Bluff (Black) and $50 Bluff (Gray)

tipping point. Suppose now that instead of wagering $25 to attempt to steal a $50 pot, you're willing to wager $50 (maybe you want to stand up at the table and proclaim "let's double that pot" at this point). In Figure 5.2, you can see the expected value graph for both of these situations; note that Bob needs to fold more than a whopping 50% of the time for a "double the pot" bluff to be profitable.

Optimal Bluffing

So far, we haven't actually addressed our question at all; our analysis was based entirely on your opponent's choice to call your bluff or not. To get a better handle on what you should do, we need to look at situations with a bit more depth.

Suppose in a game of Texas Hold'Em, you have 3♥ 3♠, and the community cards on the table so far are J♦ J♣ 10♠ 9♥. The pot currently stands at $50, and before the river is revealed, you have the opportunity to raise an additional $50 (we are keeping things fairly simple here in terms of options) as a bluff, or move into the next round. You're again fairly sure that Bob has a better hand at the moment (and later, perhaps through television coverage, you find out that he had Q♥ 8♣ for a straight).

You have the option to check or bluff; Bob also has two options, either calling your bet, or folding or checking depending on whether you bluff or not, respectively. We ignore the possibility that Bob can raise here for simplicity. Since there are four cards that will make our hand (a full house, getting any of the remaining threes or Jacks on the river) better than Bob's, we can approximate the probability of that happening as about 9% (taking four, doubling it, and adding one via Theorem 3.3).

TABLE 5.1: Payoff Matrix

	Bob Calls	Bob Checks/Folds
You Bluff	−$36.50	$50.00
You Don't Bluff	$9.00	$4.50

If you decide to bluff, and Bob believes you and folds, you pick up winnings of $50 (the current pot). Otherwise, if you decide to bluff and Bob calls your raise, then there are now $100 of winnings available to you (the original $50 in the pot, plus Bob's calling of your bet). The expected winnings here, based on the probability of 0.09 for making your hand, are

$$\text{Expected Winnings Bluff, Bob Calls} = \underbrace{0.09 \cdot \$100}_{\text{beat Bob}} + \underbrace{0.91 \cdot (-\$50)}_{\text{Bob wins}} = -\$36.50$$

which doesn't look too nice for you. On the other hand, if you decide *not* to bluff and instead check *this* round, you can wager your $50 *after* seeing the river, or simply check. If you make your hand, you bet $50, and Bob calls, you have

$$\text{Expected Winnings Wait, then Bet} = \underbrace{0.09 \cdot \$100}_{\text{make hand}} + \underbrace{0.91 \cdot \$0}_{\text{not made}} = \$9.00$$

and if you decide not to bet, and simply check, with Bob checking, you have

$$\text{Expected Winnings Wait, then Check} = \underbrace{0.09 \cdot \$50}_{\text{make hand}} + \underbrace{0.91 \cdot \$0}_{\text{not made}} = \$4.50.$$

These last two computations imply that if you don't bluff and don't make your hand, you will fold and not risk any additional money. This information is summarized in Table 5.1.

How can we use this information to dictate how we play? Our biggest reward comes from bluffing, but that also comes with the risk of the biggest loss. The expected winnings for not bluffing can never be negative, and that sounds attractive. Assuming that Bob is smart and knows the way that you play, it is in Bob's best interest to choose the option that makes your expected winnings the worst; if we assume that Bob can read you correctly, if you bluff, he'll call, giving us expected winnings of −$36.50. If you don't bluff, he'll check or fold, holding you to expected winnings of only $4.50. Other than choosing dumber people to play against, the only way around this is remembering that we need not choose to always bluff (or not bluff) in this scenario! Given repeated play, where we're likely to encounter this situation again, perhaps the best idea is to haphazardly decide how often we bluff using some probability.

Example 5.2 If *you* decide to bluff about 5% of the time, for a probability of bluffing of 5%, then what happens if Bob sticks to a strategy of *always* calling? Using Table 5.1, we can compute

$$\text{Expected Winnings Bob Call} = \underbrace{0.05 \cdot (-\$36.50)}_{\text{you bluff}} + \underbrace{0.95 \cdot \$9.00}_{\text{no bluff}} \approx \$6.73$$

and if Bob instead sticks to the strategy of always checking or folding (as appropriate), we have

$$\text{Expected Winnings Bob Check/Fold} = \underbrace{0.05 \cdot \$50.00}_{\text{you bluff}} + \underbrace{0.95 \cdot \$4.50}_{\text{no bluff}} \approx \$6.78$$

which, assuming Bob chooses to always call given this scenario (as your opponent, he definitely wants to keep your expected winnings small, since your winnings are his losses), means your outcome is better at $6.73 than the $4.50 given the pure strategy of not bluffing and Bob checking.

Of course, what you should do is determine the probability p that you should bluff that *equalizes* these expected winnings. Then, *no matter what Bob does*, your expected winnings will be the same. Looking at Example 5.2, replacing 0.05 with p and 0.95 with $1 - p$, we can set these two expected winnings equal, to (leaving off the dollar signs) obtain

$$-36.5p + 9(1 - p) = 50p + 4.5(1 - p)$$

for which algebra yields the solution $p \approx 0.04945$, so we should bluff just *a little* less than 5% of the time.

If that's not enough, Bob can do the same thing! He can determine the probability for which he should call by considering what happens if you always bluff versus never bluffing; that is, looking at your expected winnings for each row of Table 5.1.

Example 5.3 Assuming that you always bluff, the expected value for you based on Bob calling with probability p is

$$\text{Expected Winnings (You Bluff)} = \underbrace{p \cdot (-\$36.50)}_{\text{Bob calls}} + \underbrace{(1-p) \cdot \$50.00}_{\text{Bob folds}}$$

$$= \$50.00 - \$86.50p$$

and the expected value for you assuming that you never bluff is

$$\text{Expected Winnings (You Don't Bluff)} = \underbrace{p \cdot (\$9.00)}_{\text{Bob calls}} + \underbrace{(1-p) \cdot \$4.50}_{\text{Bob checks}}$$

$$= \$4.50 + \$4.50p.$$

Again, some algebra yields the solution $p = 0.50$, so Bob should call half of the time.

What does all of this mean? If you bluff just shy of 5% of the time, no matter what Bob does, your expected winnings will be the same. If Bob decides to call 50% of the time, then *no matter what you do,* your expected winnings (which are also Bob's expected losses) will be the same. If *both of you* adopt these strategies, then each of you is playing optimally; this choice of strategies is the *equilibrium* for this particular situation, and is what rational players should do.

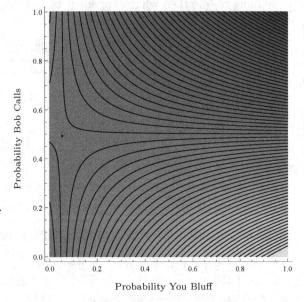

FIGURE 5.3: Effect of Bluffing and Calling Probabilities

Figure 5.3 shows the effect of the probabilities for you bluffing and for Bob calling, where darker shades represent smaller numbers, such as the worst-case for you of $-\$36.50$ at the upper right, and lighter shades represent larger numbers, such as the best-case for you of $\$50.00$ at the lower right. The dot at the point $(0.04945, 0.50)$ is the equilibrium solution that "balances" out those two cases. It is better for you than the "great for Bob, terrible for you" case in the upper-right, and better for Bob than the "terrible for Bob, great for you" of the lower-right!

Note from Figure 5.3 that *any* deviation of Bob from his "call" probability of about 55% gives *you* a higher expected value, assuming you stick with your own "bluff" probability. Surely we can generalize this idea and come up with a solution method that doesn't rely on computing so many things each time that we want to determine optimal bluffing (and calling) strategy. Indeed, this is the early focus in the branch of mathematics known as game theory.

5.2 Game Theory Basics

Game theory came into prominence in the 1950s and 1960s as a method for determining the best decisions for a wide array of applications. Pioneered by John Nash, the main character portrayed in the movie *A Beautiful Mind*, it involves more than just the application to bluffing that we have seen. Economists and military strategists have used game theory to answer difficult questions, where rather than poker players, the players involved in a "game" could be stores trying to price items high or low, or countries deciding to go to war or not go to war.

The model we will see here mirrors our study of bluffing in that we make several assumptions about a game in order to present a model that may be most useful to you. Our games will involve only two players at first, but general game theory models do incorporate methods for games with more than two, in which typically the study of coalitions between players is possible. In addition, for this section, we will assume a *zero sum* game, where one player's winnings are exactly equal to the other player's losses. Game theory does study non-zero sum games, but the solutions can be much messier; we consider those in the next section. As a reminder, if you wish to study game theory in a more in-depth way, please feel free to consult a book dedicated to the subject, such as the book by Peter Straffin found as [49] in the Bibliography.

A game begins as a *payoff matrix* such as the one we saw earlier in Table 5.1. In general, for a two-choice game between two players, we can represent the matrix with variables as in Table 5.2. One player, called the row player, has to make a choice between row A and row B. The

TABLE 5.2: Payoff Matrix

	C	D
A	$(x, -x)$	$(y, -y)$
B	$(z, -z)$	$(w, -w)$

other player, the "opponent," called the column player, chooses between column C and column D. Entries in the matrix are pairs of numbers, where a pair such as (a, b) represents a payoff of a for the row player and a payoff of b for the column player. In Table 5.2, you'll note that the pairs are symmetric in that the payoff for one player is the negative of the other; this is a byproduct of us assuming a zero sum game for this section, since for a zero sum game, each payoff pair should add to zero.

For this section, we'll abbreviate our payoff matrices; since we do have the zero sum model assumption, we can assume that the column player's number will be the negative version of the row player's number and only write the row player's number in the matrix, as seen in Table 5.3. In this model, it's important to remember that the row player likes bigger

TABLE 5.3: Simpler Payoff Matrix

	C	D
A	x	y
B	z	w

numbers, and the column player likes smaller numbers; the column player, in fact, *loves* negative numbers, since they represent a positive amount for the column player. It's also important to note that both players know the values in the payoff matrix, but make their choice independent of each other, and before either player has revealed his or her choice to the other.

Dominance and Saddle Points

TABLE 5.4: Game 1

	C	D
A	−1	2
B	1	3

There are two main strategies to solve a two-choice game. The first, which *must be checked first*, is to see if any *pure strategy* solutions exist; that is, is there a singular choice of A or B for the row player and singular choice of C or D for the column player that results in *equilibrium*, meaning that neither player could take advantage of the other *if* they knew what his or her opponent was going to choose. For instance, in Game 1 pictured in Table 5.4, look what happens if the row player chooses B and the column player chooses C.

Even if the row player *knew* that the column player was going to choose C, they would want to pick B since a payoff of 1 is better than a payoff of −1. Similarly, even if the column player *knew* that the row player was choosing B, they would pick C since a payoff of 1 to the row player is better than a payoff of 3 to the row player. This is what is meant by an equilibrium solution; in this case, both players have a pure strategy in that the row player should always choose B and the column player should always choose C. Such a point, called a *saddle point* for the game, is a solution that is simultaneously the smallest entry in its row (good for the column player) and the largest entry in its column (good for the row player). Game theory dictates that, if present, rational players should always play a saddle point solution. The result, or *value*, of Game 1 is a payoff of 1 to the row player, a loss of 1 for the column player.

> **Definition 5.1** In a zero sum game, a *saddle point* in a game matrix is location where the matrix value is simultaneously the smallest entry in its row and the largest entry in its column.

How can we easily find saddle points? There is a nice process that can be used regardless of matrix size that doesn't involve manually checking each value to see if it satisfies Definition 5.1.

> **Theorem 5.2 (Finding Saddle Points)** To find a saddle point, consider a three-step process.
>
> 1. First, find the smallest value of each row and underline the largest of these; this is called the *maximin*. If more than one value is tied for the largest, underline them all.
>
> 2. Next, find the largest value of each column and underline the smallest of these; this is called the *minimax*. Again, if more than one value is tied for the smallest, underline them all.
>
> 3. Finally, if the *value* of the maximin and minimax are exactly the same, then that value is the saddle point value, and the location of the saddle point in the matrix is at intersection of the row with the maximin and the column with the minimax (if there are multiple intersection points satisfying this, then any of those locations is a saddle point and, as we'll see soon, those values must be the same).

Consider this process applied to the game in Table 5.4. Now, in Table 5.5, we've added the row minimums, column maximums, and followed the process of Theorem 5.2. Row A has row minimum -1 while row B has minimum 1, so we underline the 1 as our maximin; similarly, column C has column maximum 1 and column D has maximum 3, so we underline

TABLE 5.5: Game 1 Marked Up

	C	D	
A	-1	2	-1
B	<u>1</u>	3	<u>1</u>
	<u>1</u>	3	

the 1 as our minimax. Since the value of the maximim and minimax are the same (with value 1), that value becomes our saddle point, and since B and C are the only row and column involved, the row player should pick B and the column player should pick C.

Example 5.4 Now, take a look at the game below which now has three rows and three columns.

	D	E	F	
A	1	3	-2	-2
B	0	-1	4	-1
C	<u>2</u>	4	3	<u>2</u>
	<u>2</u>	4	4	

The process of Theorem 5.2 has been carried out for you already. In this case, note that row C and column A provide the location of the saddle point, which you can also verify satisfies Definition 5.1 as being simultaneously the smallest value in the row and largest value in the column. You should take a minute to verify that the process of Theorem 5.2 matches what you would expect to happen.

Consider what happens in Game 2 which appears in Table 5.6. Take a minute to see what happens when you try to find a saddle point. You should realize that the maximin is 3 and the minimax is 4, so Theorem 5.2 does not yield a saddle point solution. In fact, you can manually verify that no entry in the

TABLE 5.6: Game 2

	C	D
A	−1	5
B	4	3

matrix is simultaneously the smallest entry in its row and largest entry in its column; while we will still be able to find a solution, the equilibrium solution that we find will be a *mixed strategy* solution, which will be covered in the next subsection. First, though, we have a little more discussion of saddle points along with a short discussion about dominance.

There was a brief comment in Theorem 5.2 that if multiple rows or columns were involved in the maximin and minimax calculation, there should be more than one saddle point. Let's take a moment to explore this idea. In the following, suppose that a, b, c, and d are the values and locations of four saddle points for a game, as expressed below.

$$
\begin{array}{ccccc}
\ddots & \vdots & & \vdots & \cdot \\
\cdots & a & \cdots & b & \cdots \\
& \vdots & & \vdots & \\
\cdots & c & \cdots & d & \cdots \\
\cdot & \vdots & & \vdots & \ddots
\end{array}
$$

Since a is supposed to be a saddle point, we know that a is the smallest value in the row and largest in the column, giving us that $c \leq a \leq b$ as a relationship. Similarly, since d is supposed to be a saddle point, it is also the smallest value in its row and the largest in the column, giving us the inequality $b \leq d \leq c$. Combining these inequalities together gives us that $c \leq a \leq b \leq d \leq c$ and since c cannot have two different values, this means that $c = a = b = d$ so that the value of all saddle points must be the same (if there are more than four, you can perform a very similar argument to arrive at the fact that all saddle points must have the same value). In this case, players would be free to play any of their respective rows or columns that correspond to saddle point locations.

A game could also be made simpler by using the concept of *dominance*. Take another look at Game 1 in Table 5.4 and you'll notice that the row player would be better off *always* playing row B since that player prefers that row no matter what the column player decides to do; if the column player prefers C, then the row player prefers the 1 of row B to the −1 of row A and, if the column player prefers D, the row player prefers the 3 of row B to the 2 of row A. Both players are aware of this fact, so it makes sense that both players know that the row player will pick B, and then the column player, preferring 1 to 3, since the column player prefers smaller numbers (remember, the payoffs represent the payoff to the row player only), will select column C.

Definition 5.3 If at least one value of row M in a game matrix is larger than its corresponding value in row N of the same matrix and all other values of row M are no smaller than their corresponding values in row N, we say that row M *dominates* row N or that row N is *dominated by* row M. Similarly, if at least one value of column P in a game matrix is smaller than its corresponding value in column Q of the same matrix and all other values of column P are no larger than their corresponding values in column Q, we say that column P *dominates* column Q or that column Q is *dominated by* column P.

In looking back at Game 1, we would say that row A is dominated by row B (or that row B dominates row A); row A should not be considered, as a rational player would select a row that would give them an outcome that is always at least as good as the dominated row. With the removal of row A, we could then say that D is dominated by column C (or that column C dominates column D) and should also be removed, leaving us with the same solution as before.

You might be curious about the interaction between the concept of dominance and the existence of saddle points. Since a saddle point value must the largest in its column, the only way in which a row containing a potential saddle point could be removed via dominance would be if that same value exists in another row; similarly, since a saddle point value is also the smallest in its row, the only way a column containing a potential saddle point would disappear using dominance would be in that same value also is present in another column. In either case, or both cases, the saddle point would remain.

Mixed Strategies

Remember that we were unable to solve Game 2 in Table 5.6 by finding a saddle point? Since no pure strategy solution existed for that game, we'll need to look for another way to solve this, similar to what we did when looking at bluffing with the game in Table 5.1. The way to proceed uses expected values. Let's look at this from the perspective of the row player first when looking at Game 2. If the row player picks row A with probability p and row B with probability $1 - p$ (since the row player has to pick a row), we can look at the expected values of each of the two column choices. We get, for column C

$$\text{Expected Value (C)} = \underbrace{p \cdot (-1)}_{\text{row A}} + \underbrace{(1 - p) \cdot 4}_{\text{row B}} = 4 - 5p$$

and for column D, we get

$$\text{Expected Value (D)} = \underbrace{p \cdot 5}_{\text{row A}} + \underbrace{(1-p) \cdot 3}_{\text{row B}} = 3 + 2p$$

as our expected values. Being a crafty row player, we decide that our value of p should be chosen so that, *no matter what the column player chooses*, the expected value of the column player should be the same. In other words, if our value of p is such that the column expected values are the same, the column player cannot take advantage of us since their payoff will be the same regardless of their choice. We can set our expected values equal to each other and solve for p, and solving $4 - 5p = 3 + 2p$ gives $p = 1/7$. That means the row player should play row A with probability $1/7$ and row B with probability $6/7$.

Similarly, the column player will also have the same internal discussion about determining their probability of playing C versus playing D. If the probability of playing C is given by p, we get

$$\text{Expected Value (A)} = \underbrace{p \cdot (-1)}_{\text{column C}} + \underbrace{(1-p) \cdot 5}_{\text{column D}} = 5 - 6p$$

for the expected value of the first row and

$$\text{Expected Value (B)} = \underbrace{p \cdot 4}_{\text{column C}} + \underbrace{(1-p) \cdot 3}_{\text{column D}} = 3 + p$$

for the second. Again, the column player will choose p so that the row player's choice of row A versus row B is moot (the expected values of the two rows are the same), so we solve $5 - 6p = 3 + p$ and get a value of $p = 2/7$, so the column player will play C with probability $2/7$ and D with probability $5/7$.

What about the *value* of the game? Since we've equated all of the expected values, to determine the value of the game, we need only plug in the value of p for one of the players into one of the expected value computations they have already done. Using $p = 2/7$ and considering the expected value of row A, we get

$$\text{Value of the Game} = \underbrace{\left(\frac{2}{7}\right) \cdot (-1)}_{\text{column C}} + \underbrace{\left(\frac{5}{7}\right) \cdot 5}_{\text{column D}} = \frac{23}{7} \approx 3.286$$

for a payoff of about 3.286 units to the row player. As a final solution to the game, we say that the row player should play mixed strategy $(\frac{1}{7}A, \frac{6}{7}B)$, the column player should play mixed strategy $(\frac{2}{7}C, \frac{5}{7}D)$, and the game has value $\frac{23}{7}$. Note that each player could "make" their choice by rolling a seven-sided die, or using a random number generator. You should take a moment to verify that the expected values for row B, column C, and column D also give $\frac{23}{7}$ as the value.

As an **important warning**, the method of equating expected values, or the simplified methods below for obtaining mixed strategy solutions, are only valid when no pure strategy solution exists. You should always use dominance to eliminate rows or columns and look for saddle points **before** proceeding to find any mixed strategy solution!

For a two-by-two game, where each player only has two choices, we can simplify this process by using a method called *oddments*. By using variables instead of numbers, we can give the general result, and you are encouraged to look at [49] or [59] for the details of obtaining this result.

Theorem 5.4 (Game Solution for Two-by-Two) To determine a mixed strategy solution for a two-by-two game with no pure strategy solution, consulting a general game matrix given in Table 5.2, the row player should use a mixed strategy choosing between row A and row B with the following probabilities

$$\text{Row A Probability} = \frac{|z - w|}{|x - y| + |z - w|}$$

$$\text{Row B Probability} = \frac{|x - y|}{|x - y| + |z - w|}$$

and the column player should use a mixed strategy of the following probabilities

$$\text{Column C Probability} = \frac{|y - w|}{|x - z| + |y - w|}$$

$$\text{Column D Probability} = \frac{|x - z|}{|x - z| + |y - w|}$$

where $|a|$ represents the absolute value of quantity a.

Using this, we can now verify the mixed strategy solution for Game 2 that we developed using expected values. Using the numbers from Table 5.6, we get, for the row player, the probabilities

$$\text{Row A Probability} = \frac{|4 - 3|}{|-1 - 5| + |4 - 3|} = \frac{1}{7}$$

$$\text{Row B Probability} = \frac{|-1 - 5|}{|-1 - 5| + |4 - 3|} = \frac{6}{7}.$$

As the row player, we should play row A with probability 1/7 of the time, and row B with probability 6/7 to equalize the expected value of the two column-player choices. Note that the two probabilities do, in fact, add up to 100% as we definitely require. It's refreshing to see that Theorem 5.4 does indeed give us the same solution as we saw earlier! Repeating this to determine the column player's probabilities, you should get

$$\text{Column C Probability} = \frac{|5-3|}{|-1-4|+|5-3|} = \frac{2}{7}$$

$$\text{Column D Probability} = \frac{|-1-4|}{|-1-4|+|5-3|} = \frac{5}{7}.$$

An exercise for you later will be to show that the probabilities in Theorem 5.4 arise from the method of equating expected values!

With this, it makes sense to quickly revisit Examples 5.2 and 5.3 by applying our new and quick way of determining probabilities.

Example 5.5 Using the game matrix given in Table 5.1, we may determine the probabilities that we should bluff or not bluff. Using Theorem 5.4, we can proceed, getting

$$\text{Bluff Probability} = \frac{|9.00-4.50|}{|-36.50-50.00|+|9.00-4.50|} = \frac{4.50}{91.00} \approx 0.04945$$

as we did before. The probability that we should not bluff is given by the shortcut $1 - 0.04945 = 0.95055$, or, by

$$\text{Don't Bluff Probability} = \frac{|-36.50-50.00|}{|-36.50-50.00|+|9.00-4.50|} = \frac{86.50}{91.00}$$
$$\approx 0.95055$$

as we expected. I love it when a plan comes together!

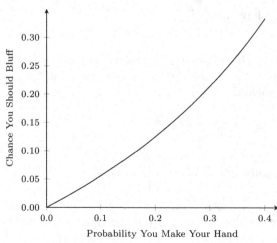

FIGURE 5.4: When to Bluff

I should mention that I do realize it might be quite hard to do all of these calculations live at a poker game when you're playing. The intent isn't for you to use them at the table, but rather both see how mathematics and game theory can help out, and also give you a sense that if you changed, say, the probability that you make your full house to that of trying to get one of the eight cards available to complete an open-ended straight, or changed the amount of your raise when bluffing, you could get a rough sense of how the probability you *should* bluff changes.

In Figure 5.4, I show how the probability you should bluff changes based on the probability you make your hand. As an exercise later you will show that when the probability of making your hand is between about 39.3% and 66.7%, you bluffing and Bob calling becomes a saddle point, and when the probability of making your hand is 66.7% or higher, you bluffing and Bob folding is the saddle point. Why did I include this? I found the question interesting, and you are certainly well-equipped now to figure it for yourself anyway.

Before we move on to larger games, one more example of finding mixed strategies for a 2×2 game is in order, borrowed from [59].

Example 5.6 Imagine two people sitting around a bar at the airport, waiting for their next flight. Rose proposes a game to Colin and says to him "let's point fingers at each other with either one or two finger and, if we match with one finger, you buy me a drink and if we match with two fingers, you buy me two drinks" and, not to let the unmatched possibilities go unnoticed, she also says "if we don't match, I'll let you off with just paying me a dollar." She claims that the game will help pass the time.

Colin replies that it sounds like a very dull game, at least in the early stages, but says "if you'd care to pay me $4.11 before each game as partial compensation for all of those $5.50 drinks I'll have to buy you, then I'd be happy to pass the time with you." Rose counters with a side payment of just $4.10, and Colin finally gives in and plays the game, but comments that "you really should pay $4.11, at least once in every 30 games, but I suppose it won't last that long."

How does this game go? Why did Colin ask for the specific side payment of $4.11 in the first place? Let's model this with a game matrix, using the notation R1 and R2 for Rose putting out one or two fingers, and C1 and C2 for Colin putting out one or two fingers.

	C1	C2
R1	$5.50	$1.00
R2	$1.00	$11.00

Check that you agree with this matrix. Since drinks cost $5.50, the cases when the number of fingers are matched are the total cost for either one drink or two drinks, and the unmatched cases are when Rose just gets $1.00 from Colin. In this case, Rose will be choosing among her two rows, and Colin will be choosing among her two columns; apologies for the bad puns here, courtesy of [49]. While I do enjoy puns and would have used Rose and Colin throughout this chapter, I thought you might tire of them.

At any rate, take a moment to verify that no dominance exists among rows or among columns, and that no saddle points exist. Our solution will involve a mixed strategy, found by equating expected values, which is made easy by Theorem 5.4. We get

$$\text{R1 Probability} = \frac{|1.00 - 11.00|}{|5.50 - 1.00| + |1.00 - 11.00|} = \frac{20}{29} \approx 0.69$$

and

$$\text{R2 Probability} = \frac{|5.50 - 1.00|}{|5.50 - 1.00| + |1.00 - 11.00|} = \frac{9}{29} \approx 0.31$$

for Rose, and for Colin we get

$$\text{C1 Probability} = \frac{|1.00 - 11.00|}{|5.50 - 1.00| + |1.00 - 11.00|} = \frac{20}{29} \approx 0.69$$

and

$$\text{C2 Probability} = \frac{|5.50 - 1.00|}{|5.50 - 1.00| + |1.00 - 11.00|} = \frac{9}{29} \approx 0.31$$

which means that both players should point one finger about 69% of the time and point two fingers about 31% of the time. What about the value of the game? Remember, we need only compute one expected value to get the value of the game to Rose, so let's use R1 to see that

$$\text{Value of the Game} = \underbrace{\left(\frac{20}{29}\right)}_{C1} \cdot \$5.50 + \underbrace{\left(\frac{9}{29}\right)}_{C2} \cdot \$1.00 \approx \$4.1034$$

is the value.

Colin's side payment request becomes apparent here; in the long run, Rose gets a little more than \$4.10 each game, so by asking for \$4.11, Colin is trying to get *a little more* from Rose than the game is worth to her, while Rose countered with something just a little less how much she values the game. These players are sneaky! Note that you could have subtracted the \$4.10 that Colin agreed to from *every* entry of the matrix to solve this game, which would give the same mixed strategy solution, with an adjusted value of just \$0.0034; see Exercise 5.7 to see why this always works.

Larger Games: $2 \times m$ and $n \times 2$

Of course, the examples we have seen so far involve either two-by-two games, where if dominance and saddle points are not present, Theorem 5.4 comes to our rescue, or larger games where the only method we have to solve the game is by using dominance and saddle points. What do we do if there are more options? Let's work through an example, continuing our use of the bluffing example provided by Table 5.1.

Question

How often should we bluff if our opponent could also raise instead of just calling our bet or checking/folding?

TABLE 5.7: Payoff Matrix, Version 2

	Bob Calls	Bob Checks/Folds	Bob Raises
You Bluff	−$36.50	$50.00	−$77.50
You Don't Bluff	$9.00	$4.50	$0.00

So far, we assumed that the only decision Bob had in our poker example was to call our bet (and therefore calling our bluff) or choose to check or fold, depending on what action we took. Bob could also choose to call our bluff in a stronger way but *raising* our bet and putting the decision back on us. Imagine a new column for Bob that addresses what happens when Bob raises our bet by an additional $50. If we are bluffing, we're likely going to roll with that bluff and keep going, so we call Bob's raise and put in our own additional $50; if we end up winning, we've earned $150 (the initial pot of $50, Bob's call of $50 along with his additional $50 raise), giving us

$$\text{Expected Winnings Bluff, Bob Raise} = \underbrace{0.09 \cdot \$150}_{\text{beat Bob}} + \underbrace{0.91 \cdot (-\$100)}_{\text{bob Wins}} = -\$77.50$$

using that same 9% chance of our hand beating Bob's hand as before. On the other hand, if we're playing it cool and not intending to bluff and Bob raises, we might fold, giving us no losses (the money already in the pot is already committed, and, as a reminder, we are only looking for the expected value of bluffing from this point on) for an expected winnings for "wait, then fold when Bob raises" of $0; this is reflected in Table 5.7.

In this case, note that the "Bob Raises" column dominates *both* of his other columns, so Bob will choose to raise, and then "don't bluff" is a clear choice for us, so the outcome would be that we choose not to bluff, Bob chooses to raise, and we fold, resulting in a value of $0 for the game. In fact, if you have been paying close attention and also checked this modified game matrix for saddle points, you would have noticed that $0.00 is actually the saddle point for this game as well! Again, dominance and saddle points do play nice together.

Now, imagine that with Bob choosing to raise, we might actually choose to do a slight bluff at that point; in a way, this is a game within a game, a metagame if you will, and we'll leave out the details here, but imagine mentally flipping a coin to decide to bluff after Bob raises that has the effect of giving

TABLE 5.8: Payoff Matrix, Version 3

	Bob Calls	Bob Checks/Folds	Bob Raises
You Bluff	−$36.50	$50.00	−$77.50
You Don't Bluff	$9.00	$4.50	$13.50

us a 9% chance overall of choosing to bluff after the raise and then winning our hand. Our expected winnings would then be

$$\text{Expected Winnings Wait, Bob Raises} = \underbrace{0.09 \cdot \$150}_{\text{make hand}} + \underbrace{0.91 \cdot \$0}_{\text{not made}} = \$13.50$$

which is reflected in Table 5.8. Now, if you check for dominance, you'll notice that neither row dominates the other, nor does any column dominate any other column. In addition, the maximin is $4.50 and the minimax is $9.00, so there is no saddle point either. What do we do?

If dominance and saddle points are not present, the solution is "easy" since it is known that the solution to any $2 \times m$ game (with two rows and m columns, where $m > 2$) or $n \times 2$ game (with n rows, where $n > 2$ and 2 columns) comes from a 2×2 *subgame*! There are only three possible subgames here (one-by-one, delete a column and solve the resulting game), so it's not too onerous to solve each of them for an equilibrium solution. The key is to check the solution from each 2×2 game in the larger game; if the mixed strategy solution of the 2×2 game holds up against all of the choices in the original game, then that 2×2 solution *is* the overall solution. Here, "holds up" means that the two-strategy player "does at least as well (usually better) against any of the other fellow's strategies as he does against the pair that appear" in the subgame from which we developed our solution [59].

Of course if m or n are fairly large, then trying to solve each subgame and then looking at how the solution performs against all of the rows or columns of the player with many choices could take awhile. A better method is to look at which choices for the player with more than two choices are dominated by some *combination* of other choices; for instance, in Table 5.8, it's actually the case that "Bob Raises" is dominated by a combination of "Bob Calls" and "Bob Checks/Folds" but the mathematical way to determine this is beyond the scope of this book (if you're interested in learning more about how to solve this directly, consult a book on linear algebra, such as [39]). A simpler way is to do this graphically. The player with two choices has their choices put on the two sides of a horizontal axis, and a vertical axis is placed above each choice (with a range going from the smallest entry in the matrix to the largest). Then, each choice for the player with multiple choices, such as Bob in this case, is represented by a line on the graph. For instance, "Bob Calls" becomes a line going from −$36.50 on the left (the "You Bluff" option for us) to $9.00 on the right (the "You Don't Bluff" option for us).

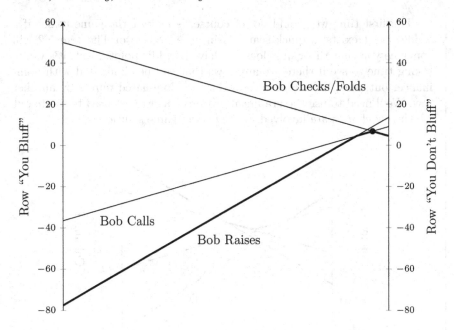

FIGURE 5.5: Payoff Diagram for Bluffing

Because Bob is the focal player here, the player we're trying to narrow down to just two options, and since Bob prefers smaller numbers, we will look only at the lower envelope of the graph (the part that would form a ceiling, if we imagine the area that would be coated by green spray paint if we pointed a can upward from the bottom of the graph and sprayed); this part appears in bold in Figure 5.5. Further, since *we* prefer larger numbers, the resulting solution will come from Bob's choices that intersect at the *highest* part of this lower envelope, which happens at the dot in the figure. Put simpler, to narrow down a column player's choices, on the resulting payoff diagram, find the highest point of the lower envelope (and, as we'll see, to narrow down a row player's choices, we'll pick the lowest point on the upper envelope of the diagram). Here, the intersection only involves "Bob Calls" and "Bob Checks/Folds" so the overall solution will be found by eliminating "Bob Raises," which for us means that the solution found in Example 5.5 still applies.

Example 5.7 The game matrix for this example, taken from [49] is a 4×2 matrix, which will have its solution as a 2×2 subgame.

	E	F
A	−3	5
B	−1	3
C	2	−2
D	3	−6

The first time we should do, of course, is look at the game to see if a saddle point exists; a quick search using the process of Theorem 5.2 will convince you that this game does not have a saddle point. Next, it's worth taking time to see if there are any rows that can be eliminated with dominance, but, alas, there are no rows that are dominated purely by another row. We'll need to use the graphical process that we just used to figure out which row choices are involved in the overall game solution.

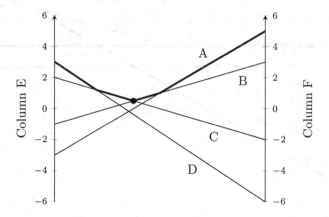

The graph above shows the result of putting the column player's choices, E and F, as the left and right vertical axes and lines graphed for each of the row player's choices. Since the row player needs to eliminate some of their strategies, and the row player enjoys larger numbers, we look at the upper envelope this time (the boundary of where water would collect if it were raining from the top of the graph). Because the column player enjoys smaller numbers, the row choices involved in the final solution will be those formed by the intersection of lines at the lowest point on this upper envelope. In this case, that means only B and C will be involved. We'll therefore solve the subgame below:

	E	F
B	−1	3
C	2	−2

Using Theorem 5.4, we see that the row player should play B with probability $\frac{1}{2}$ and C with probability $\frac{1}{2}$; the column player should play E with probability $\frac{5}{8}$ and F with probability $\frac{3}{8}$. This results in a value of the game, using B, of

$$\text{Value of Game (Row B)} = \underbrace{\left(\frac{5}{8}\right) \cdot (-1)}_{\text{column E}} + \underbrace{\left(\frac{3}{8}\right) \cdot 3}_{\text{column F}} = \frac{1}{2}$$

to the row player.

To close out this part of our discussion, we should verify that claim that the solution found in Example 5.7 does result in a better outcome for the row player than if they attempted to play excluded options A and D. Using the probabilities for E and F established in the example, we would get

$$\text{Expected Value, A} = \underbrace{\left(\frac{5}{8}\right) \cdot (-3)}_{\text{column E}} + \underbrace{\left(\frac{3}{8}\right) \cdot 5}_{\text{column F}} = 0$$

and

$$\text{Expected Value, D} = \underbrace{\left(\frac{5}{8}\right) \cdot 3}_{\text{column E}} + \underbrace{\left(\frac{3}{8}\right) \cdot (-6)}_{\text{column F}} = -\frac{3}{8}$$

which are both worse than the value of the game, $\frac{1}{2}$, found in Example 5.7, which *is* the expected value for both B and C.

Larger Games: 3×3 and Beyond

There are many board games that involve playing cards from your hand and, based on the choice of card, the result of a specific card will create food, gold, or meeples on the board. We discuss the following example in the abstract, which you could adapt to your favorite game. Imagine that the row player can play card A, B, or C, and the column player can

TABLE 5.9: Game 4

	D	E	F
A	-4	1	0
B	2	-6	5
C	1	3	-3

choose between cards D, E, and F. The numbers in Game 5, which appears in Table 5.9, are somewhat arbitrary, but we could say that if the result is AD, the row player gets a food item, valued at 1 "end game" point and the column player gets a meeple, valued at 5 points, for a net total of -4 to the row player towards the final score. The other matrix entries could be obtained by similar reasoning, based on whatever game you might be playing (the exact values here have no specific meaning to any specific game). This particular assignment of numbers and example comes from [23].

Question

How can the methods of game theory determine the best card to play?

The question here is simple and innocent; we also know how to start solving the game of Table 5.9 by using our earlier methods. First, we should check for dominance; a quick check of A versus B, A versus C, and B versus C should convince you that no row dominates any other row, and looking at the columns in a similar way yields that no column dominates any other. To check for saddle points, note that the row minimums are -4, -6, and -3, so the maximin is -3, and the column maximums are 2, 3, and 5, with a minimax of 2. Since the maximin and minimax have different values, there are no saddle points.

The method of equating expected values to determine a mixed strategy will be our only way to proceed. If we assume that the column player plays D, E, and F with probabilities p_1, p_2, and $1 - p_1 - p_2$, respectively, we can begin, noting that since the player must choose *something*, only the probabilities of D and E are variables, as the probability of F is completely determined by those for D and E. We get

$$\text{Row A} = \underbrace{p_1 \cdot (-4)}_{\text{column D}} + \underbrace{p_2 \cdot 1}_{\text{column E}} + \underbrace{(1 - p_1 - p_2) \cdot 0}_{\text{column F}} = -4p_1 + p_2$$

for the first row and

$$\text{Row B} = \underbrace{p_1 \cdot 2}_{\text{column D}} + \underbrace{p_2 \cdot (-6)}_{\text{column E}} + \underbrace{(1 - p_1 - p_2) \cdot 5}_{\text{column F}} = -3p_1 - 11p_2 + 5$$

for the second row. These must be equal, so we can set these equal to each other, to get that

$$-4p_1 + p_2 = -3p_1 + -11p_2 + 5$$

or, by moving the variables to the left side,

$$-p_1 + 12p_2 = 5. \tag{5.1}$$

This is not enough information for us to determine p_1 and p_2, but we also have not yet used anything about row C, which has

$$\text{Row C} = \underbrace{p_1 \cdot 1}_{\text{column D}} + \underbrace{p_2 \cdot 3}_{\text{column E}} + \underbrace{(1 - p_1 - p_2) \cdot (-3)}_{\text{column F}} = 4p_1 + 6p_2 - 3$$

as its expected value. Equating the expected values for rows B and C gives

$$-3p_1 - 11p_2 + 5 = 4p_1 + 6p_2 - 3$$

or, again by moving the variables to the left side,

$$-7p_1 - 17p_2 = -8 \tag{5.2}$$

is the result.

We *could* also get an additional equation by equating the expected values for rows A and C, but that does not give us any additional information, since if A and B have equal expected values, and B and C have equal expected values, we already know that A and C will also have equal expected values. Equations (5.1) and (5.2) form a system of two equations and two unknown variables, and we can solve Equation (5.1) for p_1, plug this into Equation (5.2) to get a value for p_2, and then use that value back in Equation (5.1) to get the value for p_1. You should get $p_1 = 11/101$ and $p_2 = 43/101$, meaning that the probability of playing D is $11/101$, the probability for E is $43/101$, and that for F is $1 - \frac{11}{101} - \frac{43}{101} = 47/101$.

Instead, if you have seen a bit of linear algebra or wish to consult a good linear algebra book such as [39], Equations (5.1) and (5.2) can be solved by techniques in linear algebra by solving the matrix equation:

$$\begin{bmatrix} -1 & 12 \\ -7 & -17 \end{bmatrix} \cdot \begin{bmatrix} p_1 \\ p_2 \end{bmatrix} = \begin{bmatrix} 5 \\ -8 \end{bmatrix}$$

Take a few minutes to repeat this process to determine the probabilities for the row player by using the expected values for columns D, E, and F. You should get $26/101$, $28/101$, and $47/101$ as the probabilities for rows A, B, and C, respectively. Computing the value should get you $-1/101$, a fairly fair game as far as game theory games go!

By noticing that Equation (5.1) could also have been written as

$$-4p_1 + p_2 + 0 \cdot (1 - p_1 - p_2) = 2p_1 - 6p_2 + 5(1 - p_1 - p_2)$$

without expanding the last term of each side and combing like terms, also without expanding the last term of each side, we could have also written

$$(-4 - 2)p_1 + (1 - 6)p_2 + (0 - 5)(1 - p_1 - p_2) = 0$$

to observe that the *differences* in matrix entries is what's important, which is also evident in Theorem 5.4. This leads us to the following *almost* general solution for a 3×3 game.[1]

Theorem 5.5 (Game Solution for Three-by-Three) Suppose that a game matrix with three choices per player is given by

	D	E	F
A	a	b	c
B	d	e	f
C	g	h	i

and that no saddle point exists, and no rows or columns are dominated by any others. Then for the row player, compute the following values

$$j = |(d - e)(h - i) - (e - f)(g - h)|$$
$$k = |(a - b)(h - i) - (b - c)(g - h)|$$
$$l = |(a - b)(e - f) - (b - c)(d - e)|.$$

---

[1] A word of warning is in order here. It is known that the solution to a 3×3 game will either be a solution to some 2×2 subgame or a solution to the original 3×3 game. It is possible, as we saw with $2 \times m$ and $n \times 2$ games that some *pair* of rows (or columns) can dominate the other row (or column) of a 3×3 game; this is hard to determine, and this theorem only applies if no dominance of any sort exists. For more details about how difficult it can be to isolate these cases, see the "More on Dominance" section of the "Miscellany" chapter of [59].

Then, the row player should play their rows with probabilities

$$\text{Row A Probability} = \frac{j}{j+k+l}$$

$$\text{Row B Probability} = \frac{k}{j+k+l}$$

$$\text{Row C Probability} = \frac{l}{j+k+l}.$$

Similarly, for the column player, compute

$$x = |(b-e)(f-i) - (c-f)(e-h)|$$
$$y = |(a-d)(f-i) - (c-f)(d-g)|$$
$$z = |(a-d)(e-h) - (b-e)(d-g)|$$

so that the probability the column player should play their columns is given by

$$\text{Column D Probability} = \frac{x}{x+y+z}$$

$$\text{Column E Probability} = \frac{y}{x+y+z}$$

$$\text{Column F Probability} = \frac{z}{x+y+z}.$$

By my own admission, Theorem 5.5 is unwieldy and not terribly pretty, but it is the generalization of Thereom 5.4 for three-by-three games. It is also a straightforward application of linear algebra; from the original matrix, create a new row by subtracting row B from row A, and call this row AB. Then, subtract row C from row B to get a new row, called BC. We use the matrix with the two rows AB and BC to do our computations. For the value x, corresponding to the probability for column D, in Theorem 5.5, we remove "column D" from this new matrix, and x is the absolute value of the determinant of the resulting 2 matrix. Then, y is the absolute value of determinant of this new matrix with "column E" removed, and z is found similarly. The denominator of the probability is just the sum of these values, also called oddments.

We will not use Theorem 5.5 here, but you are encouraged to use it in the exercises whenever relevant. Note that this process can generalize even further; for a 4×4 game (with no saddle points and no dominance of any type), you form the column probabilities by getting the four oddments similar to what we just did. Form a new matrix with rows called AB, BC, and CD (assuming the row choices are A, B, C, and D of course), and get the oddment for the first column by removing the first column from the new matrix and computing the resulting determinant, in absolute value. If w, x, y, and z are the four column oddments, then the probability of playing the first column is

$\frac{w}{w+x+y+z}$, the second column has probability $\frac{x}{w+x+y+z}$, and so on. The process for a 5×5, or even an $n \times n$ for $n \geq 3$, is very similar.

Of course, you could instead use the expected value process that we did above!

5.3 Non-Zero Sum Games

Thus far, our discussion about matrix games has revolved around zero sum games; recall, a game is a zero sum game if the payoff to one player is the negative of the payoff for the other player. As we have seen, there is a rich toolbox for solving zero sum games, and it is known that every zero sum game for two players must have a solution as either a pure strategy or a mixed strategy. Sure, with larger games, those strategies may be hard to find in terms of the time involved to find them, but it is always possible. We shall soon see that this is *not* the case with non-zero sum games – a game of this type will take more analysis, and there will be a few different ways to develop solutions that are not equivalent with one another, not optimal according to what I consider to be a reasonable set of criteria. It is in our best interest to identify when a game that, on the surface, appears to be non-zero sum is actually equivalent to a zero sum game.

TABLE 5.10: Game 1: Non-Zero Sum

	C	D
A	(20, −3)	(10, 2)
B	(12, 1)	(16, −1)

TABLE 5.11: Game 1: Non-Zero Sum Adjusted

	C	D
A	(3, −3)	(−2, 2)
B	(−1, 1)	(1, −1)

Consider the game represented in Table 5.10. At first glance, it is easy to see that this does not fit our definition of a zero sum game; if the row player picks A and the column player picks C, the row player gets 20 units and the column player gets −3 units, and these are clearly not negation opposites of each other. However, if you transform the row player's numbers (first entry in each payoff ordered pair) using the function $f(x) = \frac{1}{2}(x - 14)$, so that the 20 becomes a 3, the 10 becomes a −2, and so on, you get the modified game presented in Table 5.11 which *is* zero sum! Considering only the row player's numbers, does it seem reasonable that we could solve this new game and obtain a solution for the original name? The subtraction of 14 units could be thought of as the row player having a natural talent or pre-existing funds worth 14 units to them, with those units not available to the column player at all. The halving could also make sense as a difference of scale; perhaps the row player is using a different currency than the column player, and scaling their units down should have no effect on the strategy we recommend, as using U.S. Dollars versus Mexican Pesos or any other currency as the matrix entries should not change the outcome (other than the value of the game). In fact, Exercises 5.7 and 5.8 have you show that adding the same constant value to matrix entries does not change the strategy recommended by our methods, nor does multiplying matrix entries by the same positive constant.

Using the methods of this chapter, you should be able to determine that the row player should play $(\frac{2}{7}A, \frac{5}{7}B)$ and the column player should play $(\frac{3}{7}C, \frac{4}{7}D)$, for a value of $\frac{1}{7}$ to the row player for the game in Table 5.11. Each player should still play those strategies for the original game in Table 5.10, resulting in a loss of $\frac{1}{7}$ to the column player, long term, and a win of $14\frac{2}{7}$ for the row player (using the inverse of the function used to find the zero sum equivalent game). How can you tell if a game that appears to be non-zero sum is equivalent to a zero sum game?

The graph in Figure 5.6 plots each of the payoffs of Table 5.10 as an ordered pair on a standard axis. The first observation that you probably will make is that all four of these points lie on the same line of negative slope. This is, by far, the easiest method to determine whether or not a game with payoff ordered pairs that are not zero sum

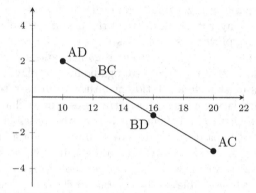

FIGURE 5.6: Payoff Ordered Pairs for Game 1

can be adjusted in order to make the game equivalent to a zero sum game. When this is the case, one of the player's payoffs can be adjusted by a positive linear function to obtain the negative value of the matching payoff for the other player.

Theorem 5.6 (Equivalence of Non-Zero Sum to Zero Sum) If, when plotted as ordered pairs on a standard axis, all payoffs of a game lie on the same line, which also has negative slope, then that game is equivalent to a zero sum game. In that case, use two of the ordered pairs to determine the positive linear function of the form $f(x) = a(x-b)$, where $a > 0$ that will transform one player's payoffs into the negative version of the other player's payoffs. The recommended pure or mixed strategy for the new game will be the recommended strategy for the original game, and the value of the game to the player whose payoffs were adjusted can be determined by taking the value in the new game and adjusting it back using the inverse of $f(x)$.

The requirement that the slope of the line be negative (and the value of a being positive) in Theorem 5.6 ensures that the two players originally are diametrically opposed in their interests (in both the original game and the adjusted game); any game for which the line connecting all payoff ordered pairs does not have negative slope cannot be equivalent to a zero sum game.

Movement Diagrams

TABLE 5.12: The Movie Game

	DMay	DDecember	DMarch
WMay	$(100, 100)$	$(600, 400)$	$(600, 200)$
WDecember	$(400, 600)$	$(0, 0)$	$(400, 200)$
WMarch	$(200, 600)$	$(200, 400)$	$(-100, -100)$

Courtesy of [19], we have the "game" in Table 5.12 that shows two companies, Warner Brothers and Disney, deciding when they should release their new movies, *Wonder Woman 2* and *Black Panther 2*, respectively, into the theaters. The numbers shown represent millions of dollars of profit. For example, if Warner Brothers releases their movie in December (row WDecember) and Disney releases in May (column DMay), then Warner Brothers earns $400 million in profit and Disney earns $600 million in profit. Certainly, as can be seen in the game matrix, both movie companies want to avoid releasing in the same month, as they will experience lower profits (and, if released in a moviegoer light month such as March, could compete with each other so much that both studios realize a loss).

TABLE 5.13: The Movie Game

	DMay		DDecember		DMarch
WMay	$(100, 100)$	$\rightarrow$	$(600, 400)$	$\leftarrow$	$(600, 200)$
	$\downarrow$		$\uparrow$		$\uparrow$
WDecember	$(400, 600)$	$\leftarrow$	$(0, 0)$	$\rightarrow$	$(400, 200)$
	$\uparrow$		$\downarrow$		$\uparrow$
WMarch	$(200, 600)$	$\leftarrow$	$(200, 400)$	$\leftarrow$	$(-100, -100)$

To examine first how the studios might decide to act, we consider a *movement diagram*. Between each row, we insert an arrow pointing to the studio's preferred outcome; since the row player decides which row to choose, the row player will draw an arrow toward the higher number in each column. Similarly, between each column, the column player will draw an arrow pointing to the higher number in each row. See Table 5.13 for how the arrows should point for our example.

Note that the $(400, 600)$ and $(600, 400)$ entries only have arrows pointing *into* them; we'll soon see that these entries are called equilibrium points. If either entry is currently what both players are deciding to do, neither player wants to move off of that entry (at the $(400, 600)$ location, for instance, if Disney is picking May, Warner Brothers would not want to move off of December, and if Warner Brothers is picking December, Disney would not want to move off of May). Let's use some smaller examples to really explore movement diagrams and true non-zero sum games.

Perhaps the most famous non-zero sum two player game in history is the Prisoner's Dilemma, represented in Table 5.14. Originally created by Melvin Dresher and Merrill Flood at the RAND Corporation in 1950, with a story added years later by Albert Tucker at Stanford University, the setup goes something like this.

TABLE 5.14: Prisoner's Dilemma

	C	D
A	$(0, 0)$	$(-2, 1)$
B	$(1, -2)$	$(-1, -1)$

Imagine that two people are arrested for a joint crime and they are each being interrogated in separate rooms. The investigating detectives tell the players that if one of them confesses and the other does not, the confessor gets a reward (payoff of one unit) and the partner gets a heavy sentence (payoff of -2 units). However, if both confess, each will get a light sentence (payoff of -1 unit) and if neither confess, both go free (zero payoff for both). In Table 5.14, A and C represent "don't confess" for the row player and column player, respectively, and B and D likewise represent "confess" for the respective players.

What happens when we analyze this game? If you only look at the row player's numbers, you'll notice that row B (strongly) dominates row A since in the first column, the row player prefers 1 to 0 and in the second column the row player prefers -1 to -2; using zero sum analysis on this, we would recommend row B (confess). The same analysis applies if you look only at the column player's numbers, where column D (strongly) dominates column C, so we would recommend column D (confess), resulting in a payoff ordered pair of $(-1, -1)$ at the solution BD. However, *both* players would prefer the outcome of $(0, 0)$ at AC when neither player confesses! The *movement diagram* for this game shows what happens.

	C	D
A	$(0, 0)$ $\rightarrow$	$(-2, 1)$
	$\downarrow$	$\downarrow$
B	$(1, -2)$ $\rightarrow$	$(-1, -1)$

For the row player, for each column, we again draw arrows between each row pointing to the preferred (larger) outcome for the row player; both arrows point down here since between each row, we see the bottom number in row B is preferred. Similarly, for the column player as we saw before, for each row, we draw arrows between each column pointing to the preferred outcome for the column player; here, both arrows point to the right since column D is preferred. Again, we notice that BD is the "natural" outcome of the game since all flow through the game matrix, so matter the starting point, ends up at BD. Officially, this stable point is called an *equilibrium*, or *Nash equilibrium* in honor of John Nash. We will revisit the Prisoner's Dilemma later in Section 5.3 and, for now, we will explore movement diagrams a bit more.

TABLE 5.15: Movement Diagram Game 2

	C		D
A	$(2,3)$	$\leftarrow$	$(3,2)$
	$\uparrow$		$\uparrow$
B	$(1,0)$	$\rightarrow$	$(0,1)$

The following examples are taken from [49] and both provide examples and practice for movement diagrams while also showing how tricky non-zero sum games can be to find "good" solutions. Looking at Game 2 in Table 5.15, note that the row player prefers row A in all cases, so the arrows both point up, while the column player prefers D when row B is played, resulting in a right arrow, and prefers C when row A is played, leading to the left arrow. Note that all paths in this game matrix lead to AC as the solution, which is the equilibrium point for the game.

In Game 3, given in Table 5.16, note that there is no row or column dominance at all, so we have an arrow of each direction for rows and for columns. Verify that you agree with the direction of the arrows. Note in this game that both BC and AD are equilibria since both locations have only arrows pointing into them. However, you can clearly see that these equilibria are not equivalent or interchangeable; this is another difference between zero sum games and non-zero sum games. Previously, whenever two saddle points existed, they had the same value for both players. Now, it is possible for a game to have two or more non-equivalent and non-interchangeable equilibrium points.

TABLE 5.16: Movement Diagram Game 3

	C		D
A	$(1,1)$	$\rightarrow$	$(2,5)$
	$\downarrow$		$\uparrow$
B	$(5,2)$	$\leftarrow$	$(-1,-1)$

TABLE 5.17: Movement Diagram Game 4

	C		D
A	$(2,4)$	$\leftarrow$	$(1,0)$
	$\downarrow$		$\uparrow$
B	$(3,1)$	$\rightarrow$	$(0,4)$

Finally, consider Game 4 in Table 5.17 and take a moment to check the direction of the arrows. Just like in the Game 3, we do not see any dominance among rows or columns, so we yet again see an arrow of each direction for rows and columns. In this game, however, each location has one arrow pointing to it and one arrow pointing away. There is no pure strategy equilibrium in Game 4, and we'll need to explore other methods than just movement diagrams to come up with a possible solution. As a reminder, though, non-zero sum games are messy, and even the solutions presented for the Prisoner's Dilemma and Game 3 are not ideal by some measures (as we will see, the solution for Game 2 actually satisfies a reasonable definition of optimality).

Nash Equilibrium Explored

Providing a solution for Game 4 does not require much more work than we've done before. In order to find the Nash equilibrium, since a pure strategy solution does not exist, we appeal to our earlier work and a desire to have expected values be the same when a player looks for their individual solution.

Consider the point of view for the row player. If we adapt our prior methods to the row player in order to determine their solution, we wish to play A and B with probabilities chosen so that the column player cannot gain an advantage; we do this by equating expected values for the *column player* using the *column player's numbers only*. Similarly, the column player wants to choose probabilities for C and D to make the *row player's expected values* the same. Let's explore this.

Example 5.8 First, we should split the game so that each player's numbers are in their own matrix.

	C	D
A	2	1
B	3	0

Row Player's Numbers

	C	D
A	4	0
B	1	4

Column Player's Numbers

Looking at the game on the right, which uses only the column player's numbers from Table 5.17, and using Theorem 5.4, you should be able to compute that the row player should play $(\frac{3}{7}A, \frac{4}{7}B)$ which results in a value of $\frac{16}{7}$ for the column player for both C and D. The column player's probabilities using just their numbers is irrelevant, but we do know that the column player can do no better than $\frac{16}{7}$ on average regardless of the strategy they decide to use.

For the column player, we use the left side game which has only the row player's numbers from Table 5.17. This game results in the column player playing $(\frac{1}{2}C, \frac{1}{2}D)$ which results in a value of $\frac{3}{2}$ to the row player. Again, no matter what the row player decides to do (though we already know what they will do based on the previous paragraph), they will do no better than $\frac{1}{2}$.

What actually happens in terms of the value of the game? We need to use the ordered pairs of Table 5.17 along with the combined probabilities given by the mixed strategy solution to figure this out, obtaining

$$\text{Value} = \underbrace{\left(\frac{3}{7}\right)\left(\frac{1}{2}\right) \cdot (2,4)}_{AC} + \underbrace{\left(\frac{3}{7}\right)\left(\frac{1}{2}\right) \cdot (1,0)}_{AD} + \underbrace{\left(\frac{4}{7}\right)\left(\frac{1}{2}\right) \cdot (3,1)}_{BC}$$

$$+ \underbrace{\left(\frac{4}{7}\right)\left(\frac{1}{2}\right) \cdot (0,4)}_{AC} = \left(\frac{3}{2}, \frac{16}{7}\right)$$

where using ordered pairs in the calculation is a nice placeholder for computing the value for the row player (by using just the first number in each ordered pair) and also the column player (by using the second number). The players get their individual values just fine in this case; by having the row player choose a strategy based on equating the column player's expected values, the column player gets that expected payout regardless of what the column player chooses to do (and the same applies if you switch "row" and "column in this sentence").

In a way, this could complete the study of two-person games when cooperation and discussion is not allowed. While we only discussed 2×2 games here for non-zero sum games, since the method for a player to find their solution is to solve the game, with their opponent's numbers, using zero sum strategies, the same methods that we used previously for $2 \times m$, $n \times 2$, 3×3, and larger games, will work; if a pure strategy equilibrium does not exist, then a mixed strategy solution exists in the game or a subgame. John Nash formalized this in 1950.

> **Theorem 5.7 (Nash Equilibrium Theorem)** Every two-person matrix game, where players make simultaneous decisions independent of one another, has at least one Nash equilibrium in either pure strategies or mixed strategies.

We did not discuss how *good* the mixed strategy solutions found in the manner of Example 5.8. This will be formally discussed in Section 5.3, but for now you might be wondering if it might be better for, say, the row player to look at *their own numbers* to find a strategy that they may like better.

Prudential Strategies

A more prudent approach to determine what the row player should do, perhaps, is to consider their own numbers. Looking at the left side of the table in Example 5.8, which you will remember contains only the row player's numbers, we can see that the value of 1 is a saddle point for this game (since the row player is using their own numbers to determine their strategy, we should use dominance and saddle points before looking for a mixed strategy solution). The row player should play A as their *prudential strategy* which gives a value of 1 to them; here, we will also now say that 1 is the row player's *security level*. What does this mean? Since, if the row player always plays A and will be guaranteed either a payoff of 1 or 2 units, the value 1 is their security level since they will get at least 1 unit by using this strategy.

For the column player, look at the right side of the table in Example 5.8.

This game has no dominance or saddle point for the column player.[2] Using Theorem 5.4 gives $(\frac{4}{7}C, \frac{3}{7}D)$ as the recommended strategy for the column player, with security level, or value, to them of $\frac{16}{7}$.

What actually happens if the two players both play their prudential strategies? We can use the same "shorthand" calculation as we did in Example 5.8 to figure out the combined value. Note that since the row player always plays row A, we use a probability of 1 for the first probability whenever row A is involved, and a probability of 0 for the second whenever row B is involved. Doing so, we get

$$\text{Value} = \underbrace{1 \cdot \left(\frac{4}{7}\right) \cdot (2,4)}_{\text{AC}} + \underbrace{1 \cdot \left(\frac{3}{7}\right) \cdot (1,0)}_{\text{AD}} + \underbrace{0 \cdot \left(\frac{3}{7}\right) \cdot (3,1)}_{\text{BC}}$$

$$+ \underbrace{0 \cdot \left(\frac{4}{7}\right) \cdot (0,4)}_{\text{AC}} = \left(\frac{11}{7}, \frac{16}{7}\right)$$

so it seems that while the column player indeed gets their security level, the row player actually does a bit better, getting $\frac{4}{7}$ units more than their security level of 1.

If this hasn't gotten complicated already, with two ways of thinking about how to play a non-zero sum game with two players where a pure strategy does not exist, consider the following. If the column player truly believes that the row player is going to use their prudential strategy of always playing row A, then they really should play column C and *not* their own prudential strategy, as this would result in the payoff at AC of $(2,4)$, which makes the column player much happier (and, it also makes the row player happier). In this case, the column player would be playing their *counter-prudential strategy* – their best response to their opponent's prudential strategy. Similarly, if the row player believes that the column player will indeed play their prudential strategy of $(\frac{4}{7}C, \frac{3}{7}D)$, the row player should look at their own expected values of

$$\text{Expected Value, Row A} = \underbrace{\left(\frac{4}{7}\right) \cdot 2}_{\text{column C}} + \underbrace{\left(\frac{3}{7}\right) \cdot 1}_{\text{column D}} = \frac{11}{7}$$

and

$$\text{Expected Value, Row B} = \underbrace{\left(\frac{4}{7}\right) \cdot 3}_{\text{column C}} + \underbrace{\left(\frac{3}{7}\right) \cdot 0}_{\text{column D}} = \frac{12}{7}$$

---

[2]There's an important note here that needs to be mentioned. Our definition of saddle point was written in the context of a zero sum game, where the column player prefers *smaller* numbers; here, because the column player has their own payoff values in the matrix and now prefers larger numbers, to determine if the column player has a saddle point using their own numbers, switch the roles of minimax and maximin from before, or look for an entry that is simultaneously the largest value in the row and smallest value in the column.

and choose to play row B as their counter-prudential strategy! And, if both players decide to play their counter-prudential strategies at the same time, BC, or $(3,1)$, will be the outcome.

Definition 5.8 (Prudential and Counter-Prudential Strategies) In a non-zero sum game, a player's recommended strategy, in terms of a pure or mixed strategy, in their own game using their own numbers is called their *prudential strategy*. The value to that player, in that game, is their *security level*. A player's *counter-prudential strategy* is their best response to their opponent's prudential strategy.

Example 5.9 Before moving on to talk about the optimality of some of these solutions, let's explore finding prudential strategies in a different game.

	C	D
A	$(1,1)$	$(2,4)$
B	$(5,2)$	$(-1,0)$

Consider the game above. Again, to determine the prudential strategy, security level, and counter-prudential strategy for each player, it makes sense to split this game into the game for the row player and the game for the column player, as we see below:

	C	D
A	1	2
B	5	−1

Row Player's Numbers

	C	D
A	1	4
B	2	0

Column Player's Numbers

For the row player, neither row dominates the other, and no saddle point exists. Using Theorem 5.4 produces the prudential strategy, as a mixed strategy, of $(\frac{6}{7}A, \frac{1}{7}B)$ and a security level of $\frac{11}{7}$. For the column player, again, no column is dominated, and no saddle point (remember the footnote from earlier, it would be the largest value in a row that's also the smallest value in a column) exists. Theorem 5.4 gives $(\frac{4}{5}C, \frac{1}{5}D)$ as the prudential strategy, with security level of $\frac{8}{5}$.

What actually happens if both players play prudentially? By computing the combined expected value using both sets of probabilities, we get

$$\text{Value} = \underbrace{\left(\frac{6}{7}\right)\left(\frac{4}{5}\right) \cdot (1,1)}_{AC} + \underbrace{\left(\frac{6}{7}\right)\left(\frac{1}{5}\right) \cdot (2,4)}_{AD} + \underbrace{\left(\frac{1}{7}\right)\left(\frac{4}{5}\right) \cdot (5,2)}_{BC}$$

$$+ \underbrace{\left(\frac{1}{7}\right)\left(\frac{1}{5}\right) \cdot (-1,0)}_{AC} = \left(\frac{11}{7}, \frac{8}{5}\right)$$

as the combined value, so each player gets their own security level when both players use their prudential strategies.

To figure out the row player's counter-prudential strategy, we will assume that the column player does play $(\frac{4}{5}C, \frac{1}{5}D)$; looking at expected values, we see

$$\text{Expected Value, Row A} = \underbrace{\left(\frac{4}{5}\right) \cdot 1}_{\text{column C}} + \underbrace{\left(\frac{1}{5}\right) \cdot 2}_{\text{column D}} = \frac{6}{5}$$

and

$$\text{Expected Value, Row B} = \underbrace{\left(\frac{4}{5}\right) \cdot 5}_{\text{column C}} + \underbrace{\left(\frac{1}{5}\right) \cdot (-1)}_{\text{column D}} = \frac{19}{5}$$

are the results, so the row player's counter-prudential strategy is to pick row B. Similarly, assuming that the row player uses $(\frac{6}{7}A, \frac{1}{7}B)$ as their strategy, the column player has expected values of

$$\text{Expected Value, Column C} = \underbrace{\left(\frac{6}{7}\right) \cdot 1}_{\text{row A}} + \underbrace{\left(\frac{1}{7}\right) \cdot 2}_{\text{row B}} = \frac{8}{7}$$

and

$$\text{Expected Value, Column D} = \underbrace{\left(\frac{6}{7}\right) \cdot 4}_{\text{row A}} + \underbrace{\left(\frac{1}{7}\right) \cdot 0}_{\text{row B}} = \frac{24}{7}$$

so the column player has column D as their counter-prudential strategy.

The theory for non-zero sum games is simply not well-defined in terms of what will be best. In [49], Straffin writes that "the result of our discussion must be that the solution theory for zero sum games does not carry over to non-zero sum games, and in fact that there is no cogent general solution concept for non-zero sum games" and that one "simply cannot give a general prescription for how to play all such games when communication between players is not allowed." It does make sense, however, to discuss one possible way to find *better* solutions, or at least a way to evaluate the goodness of a solution based on rational criteria.

Pareto Optimality

With the continued caveat that non-zero sum games do not give rise to a single "correct" way to play a game, we can at least give a way to figure out solutions that meet some version of being "better" than others. Consider again the Prisoner's Dilemma game, presented in Table 5.14. The equilibrium point discovered using the movement diagram was the $(-1, -1)$ entry, but *both* players recognize that $(0, 0)$ would be a better outcome.

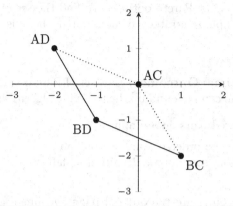

FIGURE 5.7: Pareto Diagram for Table 5.14

Figure 5.7 shows the four payoff ordered pairs on a graph, connected by lines in order to form what we call a convex polygon.[3] The line segments that would be illuminated if a light were placed in the upper-right of the graph are dotted. Why, you might ask? For the point $(0,0)$, moving either directly right or directly up takes you out of the polygon; the same is true of any point along the dashed lines, including $(-2,1)$ and $(1,-2)$. Since the polygon (solid border, dotted border, and the interior) represents all possible pure (the points shown as dots) and mixed (the rest of the border along with the interior) strategies for the players along with their outcomes, the dotted line segments are "better" outcomes that rational players might consider the best options available. Not being able to move to the right and stay within the game means that there are no outcomes better for the row player (since we cannot move right) that keep the column player at the same, or better, payoff. Similarly, not being able to move up from a location on the dotted line segment means there are no outcomes that are better for the column player (since we cannot move up) that keep the row player at their same or better payoff. Players could be happy, or at least happier, at any of these points, since no outcome exists that is better for at least one player and at least the same for the other. These outcomes from the set of *Pareto optimal* outcomes, named for Vilfredo Paredo, renowned Italian economist and engineer.

> **Definition 5.9** An outcome in a matrix game is said to be *Pareto optimal* if there exists no other outcome in the game that would either be better for both players or would be better for one player and exactly the same for the other.

Determining if a specific outcome is Pareto optimal or not is not hard; given an outcome (X, Y), look for any other entries that have a larger value than X for the row player such that the value for the column player is at least Y or an entry that has a larger value than Y for the column player with the row payoff at least X; if none exist, (X, Y) is Pareto optimal, but if such an

---

[3]A convex polygon is formed so that no two sides create a "cave" that would make part of the graph concave; a strict, mathematical, definition, is that a convex polygon is one such that if you were select any pair of points inside of the polygon, the line segment that connects those two points always lies inside of the polygon itself.

outcome does exist, then (X, Y) is not Pareto optimal. To find the set of all Pareto optimal outcomes, the graphing strategy demonstrated above is the easiest way to do so.

Theorem 5.10 (Pareto Optimal Outcomes) The set of Pareto optimal outcomes for a matrix game can be found by the following algorithm:

    1. Plot the payoff ordered pairs on a graph.

    2. Connect as many of the ordered pairs as you can in order to form a convex polygon; some ordered pairs may fall in the interior of the polygon.

The set of Pareto optimal outcomes are the ordered pairs and line segments that face the upper-right (would be illuminated by a light in the upper-right corner of the graph).

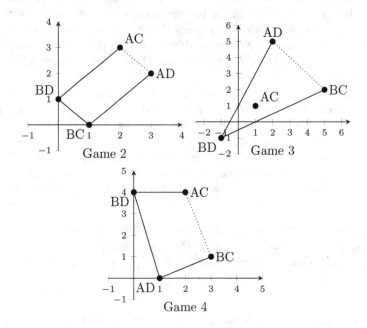

FIGURE 5.8: Pareto Diagram for Tables 5.15, 5.16, and 5.17

We use Theorem 5.10 on the games in Tables 5.15, 5.16, and 5.17 and the results are in Figure 5.8. For Game 2 from Table 5.15, note that the pure strategy equilibrium point, $(2, 3)$, is indeed Pareto optimal (a good thing). The outcome $(3, 2)$ is also Pareto optimal, but not stable as we saw in the movement diagram. Looking at Game 3 which appears in Table 5.16, we see that both stable equilibrium points are Pareto optimal and that one outcome, $(1, 1)$, appears in the interior of the polygon in order to keep it convex. Finally, in Game 4, which comes from Table 5.17, note that the Nash equilibrium we found by equating expected values in Section 5.3 is *not* Pareto optimal.

Again, while the notion of Pareto optimality does not work to provide a single solution to a non-zero sum game, it does give a way to evaluate the "goodness" of potential outcomes; rational players may very well consider that the only acceptable outcomes to a non-zero sum game should be Pareto optimal.

Prisoner's Dilemma Revisited

Our methods for (trying to) solve games have assumed that each game was meant to played exactly once between the two players. For our return to looking at the prisoner's dilemma, we'll first talk about repeated play – that is, the same two individuals playing the same game more than one time.

For non-zero sum games, this is an interesting question, as each time you see what your opponent has selected, you gain a little more information about how they play. In a way, that player is communicating *something* to you about their thought process, and if the game is played several times, your opponent's play on round six, for example, is a direct response to what you played in the first five rounds.

Table 5.18 is a generalized version of the prisoner's dilemma found in [49]. The C and D refer to cooperating (with each other) and defecting; the value of R is the reward for cooperation (for example, set to 0 in our version of the prisoner's dilemma), S represents the "sucker payoff" for when you cooperate but your opponent defects, T is the temptation

TABLE 5.18: Theoretical Prisoner's Dilemma

	C	D
C	(R, R)	(S, T)
D	(T, S)	(U, U)

value that can allure you to defect with the hopes that your opponent cooperates, and U is the payoff for complete noncooperation. In order to guarantee the row and column dominance required by the prisoner's dilemma along with ensuring that DD stays non-Pareto optimal compared to CC, the values must satisfy $T > R > U > S$ and, to keep CD and DC as other Pareto optimal outcomes, R no less than $\frac{S+T}{2}$.

It would be great if both players can, without communicating, decide to cooperate so that the mutually benefiting outcome of (R, R) is realized; perhaps if one player starts cooperating, the other will do so at some point after repeated play. However, imagine playing five rounds of this game; in the fifth game, both players know that strategy D dominates strategy C, so the outcome of the fifth game will be DD. Once that game is decided, the fourth game is effectively the last game, so, for the same reason, DD will be the outcome. Working backwards, it's clear that DD will be the outcome of the first and all successive games.

The interesting scenario happens when the player's do not know how many times they will be playing. Presented in [49], suppose that the players are guaranteed to play at least once; the probability that after one round of the game another will happen is given by p, so the probability of playing the first game is 1, the probability of playing the second game is p, the probability of playing the third game is p^2 (since playing the third game is contingent on playing the second), and so on. If our opponent decides to cooperate until we defect, after which they will also defect, our payoff can be computed as

$$R + pR + p^2 R + \cdots + p^{m-1}R + p^m T + p^{m+1}U + p^{m+2}U + \cdots$$

where m represents the number of the game where we first choose to defect and shoot for that temptation payoff. Using geometric sum formulas (not overly important here, but presented as Theorem 10.1 later in this book) and some algebra, this simplifies to

$$\frac{R - p^m R + (1-p)p^m T + p^{m+1}U}{1-p} \tag{5.3}$$

as the long-term expected value should we defect on game m.

On the other hand, if we always decide to cooperate, the payoff will be

$$R + pR + p^2 R + p^3 R + \cdots$$

which can also be simplified using a geometric series formula (see Theorem 7.2) to

$$\frac{R}{1-p} \tag{5.4}$$

which provides us a way to determine when it makes sense to cooperate.

If expression (5.4) is greater than expression (5.3), then it makes sense that we should always cooperate, which leads to our opponent always cooperating! A bit of algebra shows that this happens when

$$p > \frac{T - R}{T - U}$$

which, in our version of the prisoner's dilemma in Table 5.14, is the case when $p > 1/2$. That is, as long as the probability of playing a successive round of the game is larger than 50%, we should cooperate all of the time.

In practice, however, it's probably unreasonable to assume that the only way to play is to start by cooperating and then only defect after a certain point. Even when the number of rounds is known at the start, players that try the prisoner's dilemma rarely always end up in at outcome DD all of the time. In the early 1980s, Robert Axelrod [4] invited game theorists to submit computer programs to him that would play prisoner's dilemma against each other in a tournament; despite many programs being complex, the winner, among 14 entries, was a program called "Tit for Tat" that would always start by choosing C and for any successive round would simply choose what its opponent did in the previous round. After reporting the results of the tournament along with a detailed analysis of how each algorithm did, Alexrod invited entries for new prisoner's dilemma tournament. This time, over 60 programs were entered, including some that were specifically designed to beat "Tit for Tat" but, surprisingly, "Tit for Tat" won again; the programs designed to beat it did well against it, but did not do well at all against each other. Axelrod's tournament gives some mathematical evidence that cooperating and doing good is better than the alternative.

5.4   Three-Player Game Theory

As we wind down our coverage of game theory, let's take a moment to think about a matrix game involving three players. We still have the row player, who we shall call Rose for this section in order to make things clearer and to again use our silly pun, along with the column player, Colin, but now we have a player who will choose among a third dimension, or layer, and call him Larry. In Table 5.19 appears a constant sum game for three players; payouts are in ordered triples in the order of Rose, Colin, and then Larry. First, note that from Exercise 5.12, constant sum games are equivalent to zero sum games, and this remains true when three players are involved; what follows here is applicable to zero sum games just fine. Also, recall that all players will make their decision independently of each other and simultaneously, so that if Rose decides to go with her A, depending on what the other two players do, the result could be any of $(4, 3, 3)$, $(1, 2, 7)$, $(3, 6, 1)$, or $(2, 5, 3)$.

TABLE 5.19: Three Player Game Matrix

E

	C	D
A	$(4, 3, 3)$	$(1, 2, 7)$
B	$(3, 5, 2)$	$(0, 4, 6)$

F

	C	D
A	$(3, 6, 1)$	$(2, 5, 3)$
B	$(2, 7, 1)$	$(1, 6, 3)$

If you excel at drawing things in three dimensions, you can draw the movement diagram for this game to see if there are any equilibrium points; if not, you can draw the usual movement diagram for each of layer E and layer F, and then draw arrows "out of a layer" or "into a layer" as needed to determine which outcome Larry might prefer. For instance, looking at the AC option in both E, which is $(4, 3, 3)$, and F, which is $(3, 6, 1)$, since Larry would prefer a payout of three units as opposed to one, an arrow would point *to* $(4, 3, 3)$ and an arrow would point *out of* $(3, 6, 1)$. Thankfully, here, the use of dominance helps us out. In all cases, Rose prefers A to B, so she will always pick A; similarly, for Colin, he prefers his C to his D in all cases, so he will always play C. Finally, in all cases the payout for Larry is better in E than it is in F, so the outcome of this game should be ACE, or $(4, 3, 3)$. However, it could be the case, in an arbitrary three player zero sum or constant sum game, that more than one equilibrium point may exist, and they may not be interchangeable.

Furthermore, a new possibility exists that could not happen in a two player game; two of the players could decide to enter into a coalition and jointly decide to play a certain strategy! For example, Colin and Larry could, rather than independently choosing between their own options, try to turn this into essentially a two player game of Rose against the group of Colin and Larry; in this case, Colin and Larry, together, would treat CE, CF, DE, DF as the "group" choices, resulting in a 2×4 two-player constant sum game. Let's see how each of the possible coalition games goes.

Example 5.10 Using Rose as the row player in our adjusted game, using the payouts from Table 5.19 and only listing payouts to the row player in our adjusted game, we get

	CE	CF	DE	DF
A	4	3	1	2
B	3	2	0	1

which we can analyze using methods earlier in this chapter, quite easy to do given that row dominates row B, so Rose will pick A, and then, since we have payouts listed only for the row player, the coalition of Colin and Larry will pick DE, the saddle point for this game, for a final outcome of ADE $= (1, 2, 7)$ so that Rose only gets one unit, Colin gets two units, and Larry gets seven units.

For a coalition of Rose and Larry, where Colin becomes the lone player, we will have rows in our adjusted game correspond to Colin's choices C and D along with payouts for Colin (the middle number in the ordered triples of Table 5.19). We get

	AE	AF	BE	BF
C	3	6	5	7
D	2	5	4	6

as the adjusted game with Rose and Larry playing together, which has a saddle point at ACE $= (4, 3, 3)$, the equilibrium for the game when coalitions are not considered. Remember, be careful when creating these adjusted matrices; the numbers will correspond to the player that is playing by themselves and not as part of the coalition.

Finally, for a coalition of Rose and Colin against Larry, where the adjusted game rows correspond to Larry and his choices and numbers, we get

	AC	AD	BC	BD
E	3	7	2	6
F	1	3	1	3

which has a saddle point at BCE = $(3, 5, 2)$. In this coalition, Rose gets three units, Colin gets five units, and Larry only gets two.

It's worth noting here that Example 5.10 worked out quite nicely since each of the adjusted games had a saddle point; this need not always be the case, and these adjusted games could be solved by dominance and mixed strategies, as needed. Another noteworthy observation is when we consider who each person favors as their "partner" in a coalition. When Rose joins with Larry, she gets four units, and when she works jointly with Colin, she only gets three units; Rose prefers to be in a group with Larry. Looking at Colin, with Larry he gets two units and with Rose he gets five units, so he prefers to work with Rose (and, interestingly enough here is that Colin does *worse* pairing up with Larry than working alone). Finally, Larry gets seven units with Colin and three units with Rose, so he prefers a coalition with Colin.

There's no clear coalition that would form here, since no group of two players prefers each other as a partner, so we might think that the equilibrium point of ACE = $(4, 3, 3)$ would be the outcome. However, if Larry talked to Colin and offered him a side payment of three units (changing the outcome of ADE to $(1, 5, 4)$), Colin might be happy to join with Larry and Rose would only end up with one unit. And, if other possible side payments are considered, we might find ourselves running in circles; for more about coalition theory using characteristic functions, refer to [49]. For now, though, you can appreciate that even zero sum and constant sum games, when a third player is added, can become complicated. The nice, easy (relatively), theory of pure and mixed strategies and true solutions only exists for two player games in a zero or constant sum situation. With that, our coverage of game theory comes to an end, as long as we don't count some examples that appear in Chapter 7.

5.5 Exercises

5.1 Use Theorem 5.4 to solve the following two-by-two matrix games; that is, find either pure or mixed strategies for each player that equates expected values.

(a)
	C	D
A	−1	4
B	0	2

(b)
	C	D
A	7	4
B	2	5

(c)
	C	D
A	−2	3
B	4	−1

5.2 Repeat Example 5.5 to show that the probability that Bob should call is about 0.55488.

5.3 For a two-by-two zero-sum matrix game, is it possible that one player has a pure strategy solution while the other player is forced to use a mixed strategy solution? Why or why not?

5.4 Solve each of the following zero sum games; that is, find the pure or mixed strategy to be played by each player along with the value of the game.

(a)

	C	D
A	-1	1
B	1	-4

(b)

	D	E	F
A	-4	4	-3
B	-1	-1	0
C	2	-1	2

(c)

	D	E	F
A	3	2	4
B	-1	1	5
C	4	0	-1

(d)

	C	D	E	F
A	3	2	4	-1
B	-1	1	-4	4

5.5 Consider the following zero sum game that has one of the entries replaced by a variable t.

	C	D
A	0	4
B	t	3

(a) Determine the range of values for t for which row A will dominate row B.

(b) Determine the range of values for t that will make the value t (outcome BC) a saddle point for the game.

(c) Determine the range of values for which no dominance or saddle points will exist in the game.

5.6 Rose currently holds a black 2 and a red 9 in her hand, and Colin holds a red 3 and a black 8 in his hand; the two players are about to play a game whereby each player simultaneously chooses a card from their hand and places it face down on the table. At that time, the cards are revealed. If the played cards have the same color, Rose wins, but if the played cards have different colors, Colin wins. The amount won is equal to the number displayed on the card played by the winning player. Model this game as a two-by-two zero sum game. Then, solve the game and give the value.

5.7 For an arbitrary game matrix of size $m \times n$, show that adding the same constant value, d, to every matrix entry does not change the solution that the row player and column player should follow, but does change the value by adding d to the value of the original game.

5.8 For an arbitrary game matrix of size $m \times n$, show that multiplying every matrix entry by the same positive constant value, $d > 0$, does not change the solution that the row player and column player should follow, but does change the value by multiplying the original game value by d.

5.9 If the value d in Exercise 5.8 where instead a negative value (so that $d < 0$), what do you think would happen to the recommended pure or mixed strategy solution by our game theory methods? Form a guess, test your hypothesis with a specific example, and show that your hypothesis is correct.

5.10 Show that the probabilities given in Theorem 5.4 are correct and come from applying the method of equating expected values (as in Example 5.3) to the game matrix shown in Table 5.2.

5.11 Consider the following *constant sum* game where each payoff ordered pair adds to 10. Show that this game is equivalent to a zero sum game, and solve it.

	C	D
A	$(13, -3)$	$(4, 6)$
B	$(8, 2)$	$(11, -1)$

5.12 Show that any *constant sum* game, where each payoff pair of the form (x, y) satisfies $x + y = d$ for the same constant d, such as one that appears in Exercise 5.11, is equivalent to a zero sum game.

5.13 For each of the following games, determine if the game is equivalent to a zero sum game using Theorem 5.6.

(a)

	C	D
A	$(-1, 2)$	$(2, -2)$
B	$(5, -3)$	$(1, 1)$

(b)

	C	D
A	$(8, 4)$	$(12, 2)$
B	$(10, 3)$	$(2, 7)$

5.14 For each of the following non-zero sum games, draw the movement diagram, determine all pure strategy equilibria, and determine whether or not each outcome is Pareto optimal.

(a)

	C	D
A	$(0, 0)$	$(1, 2)$
B	$(2, 1)$	$(-1, -1)$

(b)

	C	D
A	$(1, 1)$	$(0, 5)$
B	$(5, 0)$	$(4, 4)$

5.15 Determine the mixed strategy Nash equilibrium solution for the following game.

	C	D
A	$(0, 3)$	$(3, 0)$
B	$(1, 1)$	$(2, -3)$

5.16 For the following game, for each player, find their prudential strategy, security level, and counter-prudential strategy.

	C	D
A	$(1, 1)$	$(2, 4)$
B	$(5, 2)$	$(-1, -1)$

5.17 Consider the following constant sum game between Rose, Colin, and Larry.

E

	C	D
A	$(4, 3, 3)$	$(1, 2, 7)$
B	$(3, 5, 2)$	$(0, 4, 6)$

F

	C	D
A	$(3, 6, 1)$	$(2, 5, 3)$
B	$(2, 7, 1)$	$(1, 6, 3)$

(a) Determine the pure strategy equilibrium point in this game by drawing a movement diagram.

(b) If Colin and Larry form a coalition against Rose, what will the solution to the game be?

(c) Similarly, if Rose and Larry form a coalition against Colin, what happens?

(d) How about if Rose and Colin form a coalition against Larry?

(e) Determine each player's preferred coalition partner.

6

Probability/Stochastic Matrices: Board Game Movement

I'm a huge fan of board games; my closet at the moment has at least thirty true board games among a collection of over seventy games. I would wager that most of you are familiar with some of the very classic board games (see Figure 6.1), such as Monopoly,[1] and have spent many nights playing them with friends and family members. Since a good num-

FIGURE 6.1: Board Games

ber of these games involve rolling some dice for movement, you should immediately recognize that probability can play an important role in studying how these games actually work.

6.1 Board Game Movement

The main focus of this chapter will be answering some questions about the real estate trading game from Parker Brothers called Monopoly. There are many different ways to phrase the main question of interest because, in a way, all other questions that I find interesting are based off of a single thought process.

Question

Is there a way to predict the long term importance of particular spaces on board games?

I'll give a short answer here; yes there is! Our journey to explore why and how, and how this information can be useful, though, will take a careful study of the mathematics involved along the way. Instead of Monopoly, let's consider

---

[1] Monopoly is a registered trademark of Parker Brothers/Hasbro Incorporated.

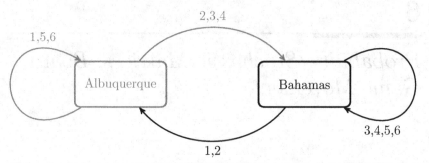

FIGURE 6.2: Simple Game

a much smaller game. In fact, let's make it a very small game, perhaps with what you might consider to be silly rules. Figure 6.2 represents a board game where there are only two spaces. I've called these Albuquerque and Bahamas, partially after two places that I happened to have visited during the year 2013, but also to avoid being too boring and calling them A and B. The diagram indicates what happens when you roll a six-sided die. If you're currently in the Bahamas, for example, and you roll a 1 or a 2, you'll end up in Albuquerque, but otherwise, on a 3, 4, 5, or 6, you get to stay in the Bahamas. This gives you *every* possibility of what happens by starting in the Bahamas; that is, if you are currently there, there's a 2/6 chance you will be sent to Albuquerque and a 4/6 chance you stay put, and these two together describe 100% of the options.

TABLE 6.1: Simple Game Table Form

Start \ End	Albuquerque	Bahamas
Albuquerque	3/6	3/6
Bahamas	2/6	4/6

Of course, we can put this information into tabular form and more or less dispense with the actual "game board" for further study. Table 6.1 represents the same information as Figure 6.2, but, at least to me as a mathematician, is a bit more pleasing. It also looks a lot like the mathematical object called a matrix that some of you may be familiar with. Indeed, let me strip away the labeling from Table 6.1 and give this matrix a name.

$$P_{6.1} = \begin{bmatrix} 3/6 & 3/6 \\ 2/6 & 4/6 \end{bmatrix}$$

This matrix is an example of a type of matrix in probability called a *probability matrix*, or, as some call it, a *Stochastic* matrix or *transition* matrix.

Probability Matrices

Definition 6.1 A square matrix with n rows and n columns of the form

$$P = \begin{bmatrix} p_{1,1} & p_{1,2} & \cdots & p_{1,n} \\ p_{2,1} & p_{2,2} & \cdots & p_{2,n} \\ \vdots & \vdots & \ddots & \vdots \\ p_{n,1} & p_{n,2} & \cdots & p_{n,n} \end{bmatrix}$$

is called a *probability matrix* (or *Stochastic* or *transition* matrix), with the number $p_{i,j}$ representing the probability of starting at space i and ending at space j after one iteration of the game process, and where

    1. For each row, the sum of the entries adds up to 1 (the row completely lists all possibilities of what happens if you start on the space represented by that row); in symbols, for each i, $p_{i,1} + p_{i,2} + \cdots + p_{i,n} = 1$, and

    2. Each entry is a number between 0 and 1, inclusive (the entries are indeed probabilities); in symbols, for each i and each j, $0 \le p_{i,j} \le 1$.

We can verify that our matrix $P_{6.1}$ above, based on the data in Table 6.1, fits Definition 6.1. If you look at each row individually, the entries do indeed add up to 1, and each entry of matrix $P_{6.1}$ is a number between 0 and 1.

What happens if we multiply matrix $P_{6.1}$ by itself? Let's take a look, recalling from our mathematical past how to multiply matrices (no need to fret here; while a minimal demonstration is included below, check out Appendix D for more details on matrix multiplication).

$$P_{6.1}^2 = P_{6.1} \cdot P_{6.1} = \begin{bmatrix} 3/6 & 3/6 \\ 2/6 & 4/6 \end{bmatrix} \cdot \begin{bmatrix} 3/6 & 3/6 \\ 2/6 & 4/6 \end{bmatrix} = \begin{bmatrix} 15/36 & 21/36 \\ 14/36 & 22/36 \end{bmatrix}$$

In this multiplication, we take the gray row of the matrix on the left and multiply by the gray column on the right to get the gray entry, 14/36, of the result on the right, as in

$$\frac{2}{6} \cdot \frac{3}{6} + \frac{4}{6} \cdot \frac{2}{6} = \frac{14}{36},$$

while the other entries follow in a similar manner; again, see the appendix for more of a refresher on matrix multiplication. For now, we want to talk about what this multiplication *does* and *means*.

Let's repeat that calculation again, but I will remove the colors and, instead, offer a reminder of what each number means in terms of the number itself. This will shed some light on exactly what 14/36 represents.

$$\underbrace{\left(\frac{2}{6}\right)}_{\substack{\text{start Bahamas}\\ \text{end Albuquerque}}} \cdot \underbrace{\left(\frac{3}{6}\right)}_{\substack{\text{start Albuquerque}\\ \text{end Albuquerque}}} + \underbrace{\left(\frac{4}{6}\right)}_{\substack{\text{start Bahamas}\\ \text{end Bahamas}}} \cdot \underbrace{\left(\frac{2}{6}\right)}_{\substack{\text{start Bahamas}\\ \text{end Albuquerque}}} = \frac{14}{36}$$

$$\underbrace{\text{Bahamas} \rightarrow \text{Albuquerque} \rightarrow \text{Albuquerque}}_{} \qquad \underbrace{\text{Bahamas} \rightarrow \text{Bahamas} \rightarrow \text{Albuquerque}}_{}$$

It should now make sense, then, if I told you that the 14/36 actually represents the probability of starting in the Bahamas and ending up in Albuquerque after *two* rolls of the die! The other three numbers in the resulting matrix $P_{6.1}^2$ represent similar quantities. For example, the 21/36 in the upper-right is the probability of starting in Albuquerque and ending up in the Bahamas after two moves. This is the major reason why some mathematicians call a matrix that fits Definition 6.1 a *transition* matrix.

It's worth noting here that if I asked about the probability of starting in the Bahamas and ending up in Albuquerque after two moves, this is the same calculation that you would want to do. From the Bahamas, either I can stay there for one turn and then move to Albuquerque, or I can move to Albuquerque first and then stay there for the second roll. These are represented in the matrix multiplication obtaining 14/36 as we saw, but the matrix multiplication method is nice because it does the *other three* possibilities at the same time. For a "board game" with only two squares, that savings aren't that great, but for a game with, say, only 10 spaces, there would be 100 probability calculations to do; the matrix method there would definitely be a huge time-saver.

At this point, you might be thinking that $P_{6.1}^3$ and $P_{6.1}^4$ would represent the probability of starting on a particular space and then ending up on other spaces after 3 and 4 rolls of the dice, respectively. You would be exactly right! Here are the original matrix $P_{6.1}$ and the matrices for what happens after 2, 3, and 4 rolls again, with entries converted to decimals.

$$P_{6.1} = \begin{bmatrix} 0.500000 & 0.500000 \\ 0.333333 & 0.666667 \end{bmatrix} \qquad P_{6.1}^2 = \begin{bmatrix} 0.416667 & 0.583333 \\ 0.388889 & 0.611111 \end{bmatrix}$$

$$P_{6.1}^3 = \begin{bmatrix} 0.402778 & 0.597222 \\ 0.398148 & 0.601852 \end{bmatrix} \qquad P_{6.1}^4 = \begin{bmatrix} 0.400463 & 0.599537 \\ 0.399691 & 0.600309 \end{bmatrix}$$

If we continue, the multiplications become very interesting, as you will find that

$$P_{6.1}^8 = \begin{bmatrix} 0.4 & 0.6 \\ 0.4 & 0.6 \end{bmatrix} = P_{6.1}^9 = P_{6.1}^{10} = P_{6.1}^{11} = \cdots = P_{6.1}^k$$

for all values of $k \geq 8$. What does this mean? If you are thinking about the *long-term behavior* of this game, after 8 or so rolls of the dice, no matter what space you originally started on, you have a 40% chance of being in Albuquerque and a 60% chance of being in the Bahamas! That is, the *steady state* of this

FIGURE 6.3: Another Game

system is that players spend 40% of the time in Albuquerque and 60% of the time in the Bahamas. If this were a real estate game, I would definitely prefer to own the Bahamas assuming, at least, the rent obtained from other players is high enough. More on that later.

Steady-States

You might wonder if this phenomenon will always happen. If you have taken or will take a course in linear algebra, the answer, you may or will know, is a resounding yes! The Perron-Frobenius theorem, not repeated in this book because we do not need its full power, states that any probability matrix (satisfying Definition 6.1) has one as a left eigenvector and a unique solution to the matrix equation $x \cdot P = x$ where x is the steady-state vector representing long-term behavior of the system, typically represented as

$$\begin{bmatrix} p_1 & p_2 & \cdots & p_n \end{bmatrix}$$

where each p_i is the proportion of time spent on the space represented by i in long-term play of the game. In this case,

$$x = \begin{bmatrix} 0.4 & 0.6 \end{bmatrix}$$

and, if you perform the matrix multiplication $x \cdot P_{6.1}$, you will see that x does not change.

For the linear algebra enthusiast, solving the matrix equation $x \cdot A = x$ above is equivalent to solving the matrix equation $x \cdot (A - I) = 0$ (you may be used to the vector x on the right; we have put it on the left here so that vectors can be represented on the page more pleasingly as horizontal vectors, but the methods you may have seen previously still apply). Since small examples can be examined by hand with matrix multiplication and observing what happens after 2, 3, 4, or more rolls, and large examples will need computational power, we do not dwell here on techniques of solving these systems by hand. You are encouraged to consult your calculator's instruction manual on how to input and work with matrices, or use computer software or computing websites to assist you.

Now let's take a look at the game board presented in Figure 6.3. Players start on the space represented by S; rolling a single 6-sided dice, players move a number of spaces represented by the roll moving to the right, and once the player reached the End space, the game is over for them. In this version, if a player rolls a 4 and is currently on space D, they can proceed to the End space. Sounds like fun, right? Before we continue, you should take a moment to think about what the long-term steady-state for this game would be; can you logically determine, later in the game, on what spaces (or space) you spend the largest amount of time?

The probabilities of moving from one space to another are shown in Figure 6.4; note that this probability *diagram* is very similar to a probability tree (see Definition 8.2 appearing later in this book). The main difference is that a "tree" in mathematics can only have one edge coming *into* any specific node; the diagram presented here can be considered a *probability graph* since, in mathematics, trees are part of a general class of objects called graphs.

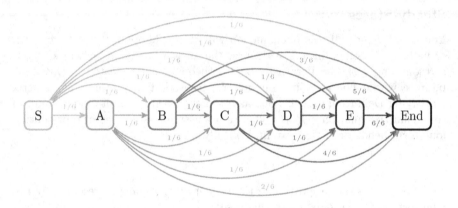

FIGURE 6.4: Another Game's Probabilities

Example 6.1 To create the probability matrix for this example, consider the probabilities labeled in Figure 6.4. From the S space, you have a 1/6 chance of reaching any of the spaces, A, B, C, D, E, or End; you cannot remain on space S, which we will represent in our matrix as the first row and first column (the other rows and columns follow in game order, so the second row represents starting at A, the third row represents starting at B, and so on), so the first row of the matrix would be

$$\begin{bmatrix} 0 & 1/6 & 1/6 & 1/6 & 1/6 & 1/6 & 1/6 \end{bmatrix},$$

and, you can probably determine that from the space labeled A, you can reach B, C, D, E, or End, but the probabilities are not all the same. Either a 5 or 6 will let you move from A to End, so this row of the matrix becomes

$$\begin{bmatrix} 0 & 0 & 1/6 & 1/6 & 1/6 & 1/6 & 2/6 \end{bmatrix}.$$

If you're currently on space E, *any* roll will take you to the End space, and if you are currently on the End space, if you wanted to roll, you would be forced to stay at End, so the last two rows of our probability matrix will be

$$\begin{bmatrix} 0 & 0 & 0 & 0 & 0 & 0 & 1 \end{bmatrix},$$

resulting in the following overall probability (or transition) matrix.

$$P_{6.4} = \begin{bmatrix} 0 & 1/6 & 1/6 & 1/6 & 1/6 & 1/6 & 1/6 \\ 0 & 0 & 1/6 & 1/6 & 1/6 & 1/6 & 2/6 \\ 0 & 0 & 0 & 1/6 & 1/6 & 1/6 & 3/6 \\ 0 & 0 & 0 & 0 & 1/6 & 1/6 & 4/6 \\ 0 & 0 & 0 & 0 & 0 & 1/6 & 5/6 \\ 0 & 0 & 0 & 0 & 0 & 0 & 1 \\ 0 & 0 & 0 & 0 & 0 & 0 & 1 \end{bmatrix}$$

If your guess earlier was that you would end up at the End spot permanently, resulting in a steady-state vector

$$\begin{bmatrix} 0 & 0 & 0 & 0 & 0 & 0 & 1 \end{bmatrix}$$

of all zeroes except a one in the last spot, you would be right! And, since the worst-case scenario playing this game, perhaps besides actually having to play the game, is rolling a 1 each time, it would take at most 6 moves to advance from S to End. You can compute powers of $P_{6.4}$ yourself (with a lot of zeroes involved it really isn't bad), and you can confirm that

$$P_{6.4}^6 = \begin{bmatrix} 0 & 0 & 0 & 0 & 0 & 0 & 1 \\ 0 & 0 & 0 & 0 & 0 & 0 & 1 \\ 0 & 0 & 0 & 0 & 0 & 0 & 1 \\ 0 & 0 & 0 & 0 & 0 & 0 & 1 \\ 0 & 0 & 0 & 0 & 0 & 0 & 1 \\ 0 & 0 & 0 & 0 & 0 & 0 & 1 \\ 0 & 0 & 0 & 0 & 0 & 0 & 1 \end{bmatrix}$$

resulting in our "logical" conclusion as well. It's good to know that our mathematical methods work and agree with what we know must be true!

Some of you may be thinking that the game in Figure 6.3 would be ever-so-slightly more exciting if players were required to reach the End space by an exact roll. If, for example, you were at space D and rolled a 3, you would have to stay at D and try to reach End on your next turn. Of course, if you were at D and rolled a 1, you could happily proceed to space E and try to roll a 1 on the next turn. How does this change things? Will it affect the final steady-state vector? The probability diagram is in Figure 6.5; take a moment to think about your answer before we move on to this next example.

Example 6.2 How does the probability matrix change for the situation presented in Figure 6.5? Since there's no chance of a "wasted" roll from the start space S, the first row remains the same. However, if you start your turn on space A and roll a 6, you will remain exactly where you are on space A. This gives us the second row, corresponding to starting at space

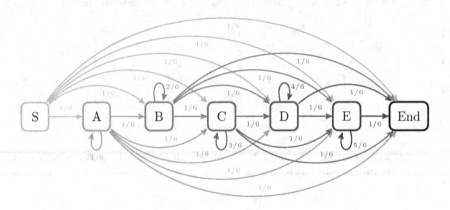

FIGURE 6.5: Another Game's Probabilities with Rule Variation

A, in our probability matrix as

$$\begin{bmatrix} 0 & 1/6 & 1/6 & 1/6 & 1/6 & 1/6 & 1/6 \end{bmatrix}.$$

Continuing on, the third row would end up being

$$\begin{bmatrix} 0 & 0 & 2/6 & 1/6 & 1/6 & 1/6 & 1/6 \end{bmatrix}$$

since rolling a 5 or a 6 at this point leaves you on space B. Continuing in this fashion, you can reason that the matrix we'll get is the following.

$$P_{6.5} = \begin{bmatrix} 0 & 1/6 & 1/6 & 1/6 & 1/6 & 1/6 & 1/6 \\ 0 & 1/6 & 1/6 & 1/6 & 1/6 & 1/6 & 1/6 \\ 0 & 0 & 2/6 & 1/6 & 1/6 & 1/6 & 1/6 \\ 0 & 0 & 0 & 3/6 & 1/6 & 1/6 & 1/6 \\ 0 & 0 & 0 & 0 & 4/6 & 1/6 & 1/6 \\ 0 & 0 & 0 & 0 & 0 & 5/6 & 1/6 \\ 0 & 0 & 0 & 0 & 0 & 0 & 1 \end{bmatrix}$$

To compare this to what happened in Example 6.1, let's see what the matrix looks like for starting in particular spaces and rolling 6 times (recall in Example 6.1 this resulted in us always being at the End space). There are less zeroes in the matrix, so by-hand computations aren't nearly as pleasant as before. Using my computer, I get

$$P_{6.5}^6 = \begin{bmatrix} 0.00000 & 0.00002 & 0.00135 & 0.01425 & 0.07217 & 0.24711 & 0.66510 \\ 0.00000 & 0.00002 & 0.00135 & 0.01425 & 0.07217 & 0.24711 & 0.66510 \\ 0.00000 & 0.00000 & 0.00137 & 0.01425 & 0.07217 & 0.24711 & 0.66510 \\ 0.00000 & 0.00000 & 0.00000 & 0.01563 & 0.07217 & 0.24711 & 0.66510 \\ 0.00000 & 0.00000 & 0.00000 & 0.00000 & 0.08779 & 0.24711 & 0.66510 \\ 0.00000 & 0.00000 & 0.00000 & 0.00000 & 0.00000 & 0.33499 & 0.66510 \\ 0.00000 & 0.00000 & 0.00000 & 0.00000 & 0.00000 & 0.00000 & 1.00000 \end{bmatrix}.$$

You can see that the probability of being on spaces A and B after 6 rolls is pretty low, but since we need to land on End by exact count, we haven't quite reached a steady-state yet. In fact, it takes 49 rolls to have a 99.999% chance of being on the End space; that is, with 5 digits after the decimal point of accuracy,

$$P_{6.5}^{49} \approx \begin{bmatrix} 0 & 0 & 0 & 0 & 0 & 0 & 1 \\ 0 & 0 & 0 & 0 & 0 & 0 & 1 \\ 0 & 0 & 0 & 0 & 0 & 0 & 1 \\ 0 & 0 & 0 & 0 & 0 & 0 & 1 \\ 0 & 0 & 0 & 0 & 0 & 0 & 1 \\ 0 & 0 & 0 & 0 & 0 & 0 & 1 \\ 0 & 0 & 0 & 0 & 0 & 0 & 1 \end{bmatrix},$$

and with linear algebra computations, the steady-state vector is

$$\begin{bmatrix} 0 & 0 & 0 & 0 & 0 & 0 & 1 \end{bmatrix}.$$

Yahtzee, Revisited

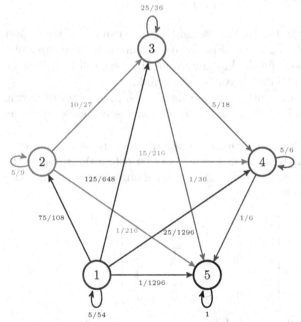

FIGURE 6.6: Board Game Yahtzee

You might remember that when we looked at Yahtzee[2] in Chapter 4, I made a comment about a second method to compute the probability of getting a Yahtzee. At this point, that method has already been introduced to you. We can treat the game of Yahtzee as a "board game" by considering the spaces on the board as locations representing whether you currently have five, four, three, two, or even one "of a kind" and the opportunity to roll or reroll moves you from space to space. This sort of thinking is pictured in Figure 6.6 where the

---

[2]Yahtzee is a registered trademark of the Milton Bradley Company/Hasbro Incorporated.

spaces indicate a roll's largest amount of matching dice. If you're currently at a particular space, the probabilities indicated are those of getting to the next space (including possibly the same space where you are currently) on the very next roll.

The probabilities given in Figure 6.6 are all found throughout the Yahtzee section of Chapter 4 and, more specifically, in Examples 4.7, 4.9, 4.12, 4.13, and 4.14. You are encouraged to verify that all of these probabilities are indeed correct; for now, we will only check three of them.

First, note that if you currently have "one of a kind" (that is, nothing good as far as trying for a Yahtzee goes), the probability that you will end up with four of a kind after rolling all five dice again is given by

$$\text{Four of a Kind} = \underbrace{\binom{30}{1}}_{\text{rank}} \cdot \underbrace{\binom{5}{1}}_{\text{ordering}} \cdot \underbrace{\left(\frac{1}{6}\right)^5}_{\text{probability}} = \frac{25}{1,296}$$

which appears as the non-reroll computation in Example 4.9. If you currently have a pair, the probability of ending up after rolling the other three dice and staying at a pair is

$$\text{Pair, then Pair} = \left(\frac{5}{6}\right)^3 - \frac{5}{216} = \frac{5}{9}$$

which appears in Example 4.13 as the "second roll, no three of a kind" part of the scenario "2 (2P), 0, Then 22." Finally, if you currently have three of a kind, the chance that you'll finish the Yahtzee on the next roll is $1/36$ since both remaining dice need to match your desired rank.

Note here that if you currently have a Yahtzee, the probability of having Yahtzee after one more roll is 100%, since you should wisely choose to roll no more dice!

The information in Figure 6.6 can be summarized nicely as the following probability matrix, noting that the first row (and first column) represents the "one of a kind" state, the second row (and second column) represents the "two of a kind" state, and so on:

$$P_{6.6} = \begin{bmatrix} \frac{5}{54} & \frac{75}{108} & \frac{125}{648} & \frac{25}{1296} & \frac{1}{1296} \\ 0 & \frac{5}{9} & \frac{10}{27} & \frac{15}{216} & \frac{1}{216} \\ 0 & 0 & \frac{25}{36} & \frac{5}{18} & \frac{1}{36} \\ 0 & 0 & 0 & \frac{5}{6} & \frac{1}{6} \\ 0 & 0 & 0 & 0 & 1 \end{bmatrix}$$

By now, having seen how probability matrices work and knowing that Yahtzee cuts you off at a maximum of three rolls, the direction should be clear. The third power of matrix $P_{6.6}$ will tell us the probability, in at most three rolls, of ending up at a particular space given a certain starting space.

My matrix-multiplying device gives

$$P_{6.6}^3 = \begin{bmatrix} 0.000794 & 0.256011 & 0.452402 & 0.244765 & 0.046029 \\ 0 & 0.171468 & 0.435814 & 0.316144 & 0.076575 \\ 0 & 0 & 0.334898 & 0.487611 & 0.177491 \\ 0 & 0 & 0 & 0.578704 & 0.421296 \\ 0 & 0 & 0 & 0 & 1 \end{bmatrix}$$

telling us that the probability of getting a Yahtzee in three rolls (first row, last column) is about 0.046029, the same value as we found in Chapter 4 (the subtle change in the sixth digit after the decimal is due to rounding results *before* the addition that we did back then).

How would allowing another reroll affect the probability of getting a Yahtzee, as is allowed once per game (or on all rounds by purchase) on the popular *Dice with Buddies* social networking game? We need only look at

$$P_{6.6}^4 = \begin{bmatrix} 0.000074 & 0.142780 & 0.409140 & 0.347432 & 0.100575 \\ 0 & 0.095260 & 0.366155 & 0.396420 & 0.142165 \\ 0 & 0 & 0.232568 & 0.499370 & 0.268062 \\ 0 & 0 & 0 & 0.482253 & 0.517747 \\ 0 & 0 & 0 & 0 & 1 \end{bmatrix}$$

which shows us that the probability of ending with a Yahtzee with three rerolls, for a total of four rolls altogether, is slightly more than 10%. That extra reroll more than doubles our chance of getting the elusive five of a kind. For the curious, it takes a round of at least *ten* rolls to make the probability of getting a Yahtzee more than 50%.

Now that $P_{6.6}$ is available to us, the question that might be on your mind is about the steady-state vector or equilibrium. While *no* finite amount of rolls will guarantee a Yahtzee (if you've already made a billion rolls, and have come close to getting a Yahtzee and currently have a four of a kind, the probability of matching that last die is still 1/6, so it's never guaranteed), it should not be too surprising that

$$v = \begin{bmatrix} 0 & 0 & 0 & 0 & 1 \end{bmatrix}$$

is the steady-state vector. If you want to either have fun or practice with matrix multiplication, verify that $v \cdot P_{6.6} = v$ to see that I'm not making this up; just like the other "games" we've discussed which have a final "resting" space.

Cyclic Boards

Examples 6.1 and 6.2, along with our journey through Yahtzee again, all illustrate a concept that should appeal to your sense of logic. Games that have a clearly defined End point will have a long-term steady-state vector that puts the End spot as having a 100% chance of reaching it. Even games for which a small cycle exists (for example, if you were forced to move from

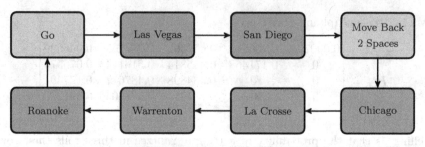

FIGURE 6.7: More Fun Game

Start \ End	Go	LV	SD	Move	C	LC	W	R
Go	0	2/6	1/6	0	1/6	1/6	1/6	0
Las Vegas	0	1/6	1/6	0	1/6	1/6	1/6	1/6
San Diego	1/6	1/6	0	0	1/6	1/6	1/6	1/6
Move Back 2 Spaces	0	0	0	0	0	0	0	0
Chicago	1/6	1/6	1/6	0	0	1/6	1/6	1/6
La Crosse	1/6	2/6	1/6	0	0	0	1/6	1/6
Warrenton	1/6	2/6	1/6	0	1/6	0	0	1/6
Roanoke	1/6	2/6	1/6	0	1/6	1/6	0	0

TABLE 6.2: Probability Table for Game of Figure 6.7

space D to space B in Figure 6.3), if an End exists, with enough turns, the probability of being permanently stuck in the cycle is so low that it becomes negligible. Instead, what we need to consider is a game, such as Monopoly, where the goal of the game isn't necessarily to reach the end, but rather has a board that involves moving around and around until a goal is reached. We'll consider one more "fake" board game before we study two actual board games.

In the game represented by Figure 6.7, eight spaces are on the game board. Six of the spaces represent properties available for purchase by players, the Go space represents the starting spot and a space that, when you pass over the space on future rolls, rewards you greatly, and the "Move Back 2 Spaces" space does exactly what you think it would do. We'll use a standard 6-sided die for this game.

Rolling a 2, 4, 5, or 6 while starting at Go is somewhat uninteresting. These rolls place you in San Diego, Chicago, La Crosse, or Warrenton, respectively. Note that rolling either a 1 or a 3, however, sends you to Las Vegas, since rolling a 3 will move you back two spaces to Las Vegas as though you had a rolled a 1 in the first place. At this point, when game boards are larger and larger and computer calculations are necessary, the goal for us is to make sure we construct the probabilities correctly and accurately represent the game in a probability matrix. Take a few minutes to convince yourself that the probabilities in Table 6.2 are correct.

Since it is not possible to end your turn on the "Move Back 2 Spaces" square, it is also not possible to *start* on that square. If we want to make a probability matrix out of this, we do need to be careful. Definition 6.1 requires that the entries of *every* row add up to 1, so we can do away with the need to include the "Move Back 2 Spaces" idea in our official matrix, given here as

$$P_{6.7} = \begin{bmatrix} 0 & 2/6 & 1/6 & 1/6 & 1/6 & 1/6 & 0 \\ 0 & 1/6 & 1/6 & 1/6 & 1/6 & 1/6 & 1/6 \\ 1/6 & 1/6 & 0 & 1/6 & 1/6 & 1/6 & 1/6 \\ 1/6 & 1/6 & 1/6 & 0 & 1/6 & 1/6 & 1/6 \\ 1/6 & 2/6 & 1/6 & 0 & 0 & 1/6 & 1/6 \\ 1/6 & 2/6 & 1/6 & 1/6 & 0 & 0 & 1/6 \\ 1/6 & 2/6 & 1/6 & 1/6 & 1/6 & 0 & 0 \end{bmatrix}.$$

At this point you are more than welcome to compute powers of the matrix $P_{6.7}$ by hand to try and find the steady-state vector. This won't be too bad, since you will have a good idea of what happens rather quickly, and it takes only computing $P_{6.7}^{11}$ to see the full picture. I obtain

$$[0.107507 \quad 0.247450 \quad 0.142857 \quad 0.124993 \quad 0.125051 \quad 0.124643 \quad 0.127499]$$

as the steady-state, indicating that players spend about 24% of the time in Las Vegas. As you may guess, this increased proportion for Las Vegas compared to the other spaces is entirely due to the effect of "Move Back 2 Spaces," and this effect also increases your time in San Diego as well. Relatively speaking, Chicago, La Crosse, Warrenton, and Roanoke have about the same proportion, and Go has the least (due to the inability to reach Go on one roll of the die if you're starting on the most popular space of Las Vegas).

6.2 Pay Day (The Board Game)

Pay Day[3] is perhaps one of my favorite board games from when I was a child; though there really isn't too much strategy involved, since your fate in the game is mostly decided by the roll of the die. There are, of course, some decisions to be made along the way, and your decisions can definitely be positively influenced by the healthy knowledge of probability and analysis you are learning from this book.

The game itself was first published in 1975 and was the board shown in Figure 6.8. Since then, the game has been released again several times, with an updated version in 1994 and, more recently, the all-new "Big Pay Day." The game that we will discuss and analyze, partly for nostalgic reasons but also for simplicity reasons,[4] will be the original 1975 edition. Since our primary

---

[3]Pay Day is a registered trademark of Parker Brothers/Hasbro Incorporated.

[4]The 1994 version contains a square on the board where players decide whether to move to one of two possible spots, making analysis difficult using the methods of this chapter.

Sunday	Monday	Tuesday	Wednesday	Thursday	Friday	Saturday
Start	1 1 Piece of Mail	2 Inheritance	3 3 Pieces of Mail	4 Deal	5 2 Pieces of Mail	6 Weekend Company
7 Sweet Sunday	8 Surprise Bonus	9 Buyer	10 High Roll	11 1 Piece of Mail	12 Deal	13 High School Dance
14 Sweet Sunday	15 Deal	16 3 Pieces of Mail	17 Buyer	18 Buy Groceries	19 1 Piece of Mail	20 Buyer
21 Sweet Sunday	22 1 Piece of Mail	23 Home Repairs	24 2 Pieces of Mail	25 Deal	26 Town Election	27 Daylight Savings
28 Sweet Sunday	29 Lottery Draw	30 Buyer	31 Pay Day			

FIGURE 6.8: Pay Day Board

aim is to use probability matrices to find the long-term steady-state of the game, we won't discuss too many of the rules; you can find those easily online for any of the versions of the game. We will, however, briefly talk about the premise of the game.

The intent of the game is to mirror real life; throughout the month, players accumulate mail that can consist of bills, junk mail, insurance opportunities, and other offers that impact the game. Starting the game with $325, players also receive their monthly income on "Pay Day," space 31 on the board (all months in Pay Day have 31 days). They also must pay all of the month's bills at that time. At the start of the game, the players collectively decide on how many months they will place, and on each turn, they roll the single die and move that many spaces. On spaces marked "Deal" the player draws a card from the Deal deck and has the opportunity to purchase the item at the listed price (for instance, the Diamond Ring costs $300) and, should the player land on a "Buyer" space, the player may sell one of their purchased items for the sell amount (which is $600 for the Diamond Ring). Most other spaces are self-explanatory, and besides the rule that forces a player to stop on Pay Day (a roll of 5 from the space 29 has you stop on 31), the other interesting space is "Daylight Savings" where *all* players move back one space. Of course, this also means that you can never *end* your turn on day 27.

Since, as part of the Pay Day space procedure, after receiving your $325, paying bills, and performing other tasks related to loans or savings, you place your playing piece back on "Start Here" before space 1, we can effectively think of the "Start Here" space and day 31 as the same location. The matrix PD presented in Figure 6.9 is the probability matrix that we'll be using, but you might ask how it was obtained!

The first column and row of PD is reserved for the "Start Here/31" space, so the month begins on day 1 which is in column/row number two. Almost all of the rows are very similar; from any one space, you can reach any of the next six spaces by rolling the die, and each has the same probability, 1/6, of being reached. For example, if you look at the first row highlighted in light gray (starting your turn on "Start Here/31"), there is a 1/6 chance of reaching

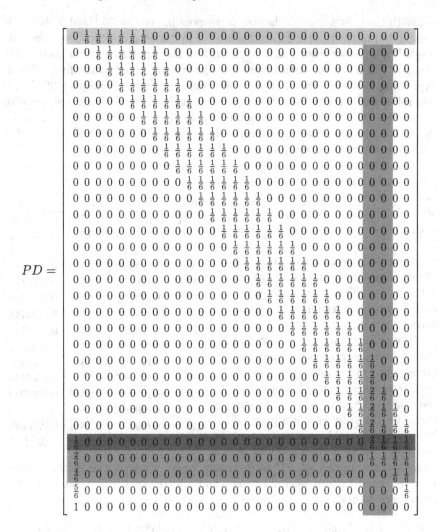

FIGURE 6.9: Pay Day Matrix

space 1 by rolling a one on the die, a 1/6 chance of reaching space 2 by rolling a two, and so on. Note that there is no possible way to remain on the "Start Here/31" space.

The interesting thing that happens with the "Daylight Savings" square (day 27); since we already remarked that it is not possible the end your turn on this square, it does not appear in the matrix PD. If it were present, we would have a row of all zeroes, violating the conditions of a probability matrix given in Definition 6.1; it would have appeared between the gray rows (and gray columns).

With that, let's discuss the row corresponding to day 25 then! This row is highlighted in the matrix PD of Figure 6.9 in darker gray (fifth from the bottom). Rolling a six results in landing on Pay Day exactly, so with probability 1/6, you will be on the "Start Here/31" space represented by the first column, explaining the entry there. If you roll a one, you will be on space 26, but if you roll a two, you land on "Daylight Savings" and move back one space to 26, so with probability 2/6 you will end your turn on day 26 (represented by the green column). Then, by rolling a three, four, or five, you will land on day 28, 29, or 30, respectively, so each of these columns gets a 1/6.

The blue row, representing day 28, is explained similarly. Rolling a one or two puts you on space 29 or 30, each with probability 1/6, but on a three, four, five, or six, you land on Pay Day and must stop, so the chance that you end your turn on "Start Here/31" from day 28 is then actually 4/6! The other rows in the matrix can be determined in similar ways.

If you have access to software that can take powers of matrices quite well, you might take a look at PD raised to various powers to see what happens as more and more turns are taken by you in a game of Pay Day. You could also try and find the steady-state vector this way by watching how powers of PD settle down; when doing so, the vector you obtain (rounded to four decimal places and splitting the thirty numbers into three lines to fit) is

TABLE 6.3: Pay Day Space Probabilities

Space	Prob	Space	Prob
SH/31	0.106111	15	0.030360
1	0.017685	16	0.030462
2	0.020633	17	0.030422
3	0.024072	18	0.030304
4	0.028083	19	0.030211
5	0.032764	20	0.030308
6	0.038225	21	0.030344
7	0.026910	22	0.030342
8	0.028448	23	0.030322
9	0.029750	24	0.030305
10	0.030697	25	0.030305
11	0.031132	26	0.066709
12	0.030860	28	0.031331
13	0.029633	29	0.031495
14	0.030087	30	0.031691

$$[0.1061, 0.0177, 0.0206, 0.0241, 0.0281, 0.0328, 0.0382, 0.0269, 0.0284, 0.0298,$$
$$0.0307, 0.0311, 0.0309, 0.0296, 0.0301, 0.0304, 0.0305, 0.0304, 0.0303, 0.0302,$$
$$0.0303, 0.0303, 0.0303, 0.0303, 0.0303, 0.0303, 0.0667, 0.0313, 0.0315, 0.0317]$$

which shows us that while most spaces are almost equally likely to be landed upon long term, there are some days that occur more frequently. Remember, the first entry of this vector represents the long-term probability of being on "Start Here/31," and the other days follow, in order, with "Daylight Savings" left out. The more accurate (rounded to six decimal places) can be found in Table 6.3.

Since you are forced to stop on "Start Here/31" at the end of the month, it should make sense that it has the highest long-term probability; of any other square, you spend most of your time here. The second highest is day 26, the space immediately before "Daylight Savings." This should also make sense, as when you are close to day 27, you virtually always have two chances to end your turn on day 26, about twice as much as any other space. You can also reason that space 1 should be a bit harder to get to, since the only way of landing there is by being on "Start Here/31" and rolling a one. With all of the other probabilities available, you might ask how can we use this information?

Example 6.3 There are some bills that you might receive on mail days while playing Pay Day that are actually quite expensive and hard to pay off without taking out a loan in the game. One natural question that you might ask, based on the steady-state we've obtained, is how many pieces of mail are you likely to get during a month of play? Examining the game's board of Figure 6.8, note that three pieces of mail are obtained on days 3 and 16, two pieces of mail are obtained on days 5 and 24, and one piece of mail is obtained on days 1, 11, 19, and 22. The probabilities associated with these quantities are

$$\text{Three Pieces} = \underbrace{0.024072 + 0.030462}_{\text{days 3, 16}} = 0.054534$$

$$\text{Two Pieces} = \underbrace{0.032764 + 0.030305}_{\text{days 5, 24}} = 0.063069$$

$$\text{One Piece} = \underbrace{0.017685 + 0.031132 + 0.030211 + 0.030342}_{\text{days 1, 11, 19, 22}} = 0.109370$$

where we have used the six-digit rounding of Table 6.3. We can combine this together to get the expected number of mail pieces per roll, as in

$$\text{Expected Mail/Roll} = 3 \cdot 0.054534 + 2 \cdot 0.063069 + 1 \cdot 0.109370$$
$$+ 0 \cdot 0.773027 = 0.399110,$$

where the number 0.773027 is the probability of landing on a space with no mail, found easily by subtracting the three probabilities found (for getting one, two, or three pieces of mail) from 1.

Of course, there are more rolls per month than just one, so we'll need to account for that. With 31 spaces available, and an average die roll value of

$$1 \cdot \left(\frac{1}{6}\right) + 2 \cdot \left(\frac{1}{6}\right) + 3 \cdot \left(\frac{1}{6}\right) + 4 \cdot \left(\frac{1}{6}\right) + 5 \cdot \left(\frac{1}{6}\right) + 6 \cdot \left(\frac{1}{6}\right) = \frac{21}{6} = 3.5$$

spaces rolled per roll, we would get about $31/3.5 \approx 8.86$ rolls per month. Bringing it all together, we have

$$\underbrace{0.399110}_{\text{mail/roll}} \cdot \underbrace{8.86}_{\text{rolls/month}} \approx 3.536$$

pieces of mail received per month.

We can, of course, look at a different question for a different type of space. The "Deal" spaces and "Buyer" spaces interact well during the game to make you a lot of money! The "Buyer" spaces will be dealt with as an exercise, including an exploration of whether it can make sense (or not) to buy Deal cards later in the game. As a small warm-up, though, let's explore the Deal spaces.

Example 6.4 The "Deal" spaces on the Pay Day board happen on days 4, 12, 15, and 25, which you can see in Figure 6.8. From Table 6.3, we can compute the probability of landing on a "Deal" space on any given roll (a fancy way of saying long-term). We obtain

$$\text{Deal Probability} = \underbrace{0.028083 + 0.030860 + 0.030360 + 0.030305}_{\text{days 4, 12, 15, 25}} = 0.119608$$

as that probability, which also implies that the probability of *not* landing on a "Deal" space is about 0.880392. How can this be useful? Suppose that you wanted to know the probability of landing on at least one "Deal" space in eight turns (just shy of an average month). One way to estimate this would be to say that landing on at least one space is the opposite of *not* landing on a "Deal" space eight turns in a row. This would lead to

$$\text{Deal in Eight} = 1 - (0.880392)^8 \approx 0.639082$$

for about a 63.9% chance of hitting a "Deal" space in eight turns. Note that we have very subtly used the Binomial Distribution of Definition 4.1 here, where a "success" is *not* landing on "Deal" and we want 8 trials with 8 successes.

This analysis of Pay Day works extremely well for *solitaire* games where your goal would be to accumulate as much money as you can over the course of the game. When other players are involved, *their* landing on "Daylight Savings" of course will move *all* players back one space, so the probabilities for each space would be adjusted slightly to account for this. We will not consider that complication here and, rather, move on to a more well-known game.

6.3 Monopoly

We are now ready to explore one of the most famous board games ever developed. Chances are that most of you reading this have played Monopoly at some point in your life. The "fast-dealing property trading" game has been around since Parker Brothers first sold the game in 1935 which features properties

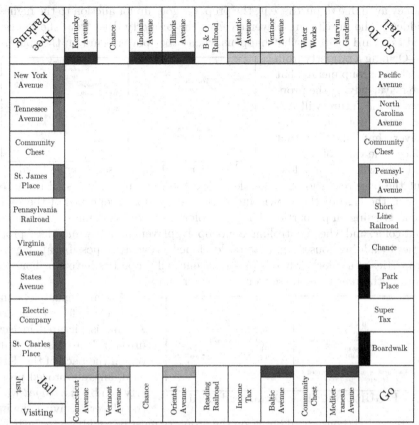

Note that, starting at "Go" and proceeding clockwise around the board, the eight color groups are brown, light blue, purple, orange, red, yellow, green, and dark blue.

FIGURE 6.10: Monopoly Board

named after locations in Atlantic City, New Jersey (not all of the locations listed on the board still exist, but most of them still do).

The premise of the game is fairly simple; players roll two dice on their turn and move around the board, pictured in Figure 6.10, the number of spaces indicated by the sum rolled. Some spaces, such as Income Tax and Luxury Tax, have amounts that players need to pay to the bank, while other spaces such as Free Parking and Just Visiting have no game effect. Landing on either Community Chest or Chance requires that you draw a card from the respective deck, performing the instructions indicated on the card (some cards have a monetary penalty or reward, some move you to another spot on the board, and others offer an opportunity to get out of jail for free). Most of the other squares, including the colored properties, railroads, and utilities, are available for players to purchase.

Landing on an unowned property gives that player the first opportunity to buy that space; should the player choose not to, or does not have the money available to purchase it, the property goes up for auction to the highest bidder.[5] The purchasing player receives the property deed from the banker, and this card dictates what happens when other players land on the space.

Owning a property from a colored group means that the player buying the property in the future will collect rent from any other player that lands on that space. The value of the rent to begin with is fairly low,

	Title Deed
Tennessee Avenue	Tennessee Avenue
	Rent $14
	With 1 House $70
	With 2 Houses $200
Mortgaged	With 3 Houses $550
for $90	With 4 Houses $750
	With HOTEL $950
Card must be turned this side up if	Mortgage Value $90
property is mortgaged	Houses cost $100 each
	Hotels, $100, plus 4 houses
	If a player owns ALL the Lots of any Color-Group, the rent is Doubled on Unimproved Lots in that group

FIGURE 6.11: Tennessee Avenue Deed

but the base rent shown is doubled when the same player owns all three (or two in the case of the brown and dark blue groups) properties of the same color. Owning all properties of a single color is known as having a *monopoly* of that color, and when controlling a monopoly, players may *improve* these properties by adding houses (up to four, developed as evenly as possible among the properties of a color) or hotels (available once all properties have four houses), causing the rent to increase to much higher values.

B & O Railroad	B & O Railroad	
	Rent	$25
	If 2 R.R.'s are owned	$50
Mortgaged	If 3　"　"　"	$100
for $100	If 4　"　"　"	$200
Card must be turned this side up if		
property is mortgaged	Mortgage Value	$100

FIGURE 6.12: B & O Railroad Deed

As an example, the deed card for Tennessee Avenue is pictured in Figure 6.11. Note that, as written on the back of the deed card, an unimproved property can be *mortgaged* with the player receiving the indicated value (houses and hotels on the property must be sold back to the bank for half of the purchase price before an improved property can be mortgaged). Consult your Monopoly rules for more details about mortgages.

Pictured in Figure 6.12 is the deed for one of the railroads. Each of these is the same except for the name, and certainly you can get a lot of money from your competitors the more railroads that you

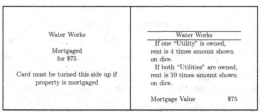

Water Works	Water Works
	If one "Utility" is owned,
Mortgaged	rent is 4 times amount shown
for $75	on dice.
	If both "Utilities" are owned,
Card must be turned this side up if	rent is 10 times amount shown
property is mortgaged	on dice.
	Mortgage Value $75

FIGURE 6.13: Water Works Deed

own. One strategy that we can discuss is that of making sure to get all of the railroads (either by landing on them and purchasing them, buying them from auction, or trading them from other players)! Of course, other players typically do not like someone to have a monopoly on railroads, so this can

---

[5]This rule is very commonly ignored as a house rule, and ignoring this rule is one of the biggest reasons that forces games to last longer than they ordinarily should.

sometimes be quite difficult. The utilities also provide a benefit to owning both, as you can see in the Water Works deed of Figure 6.13.

There are two rules regarding movement that we'll need to make use of rather early in our study of Monopoly. Given the relative ease of movement rules in Pay Day, things are somewhat more complicated for Monopoly, but we'll be able to deal with these complications fairly well. First, as you may know, if you happen to roll doubles (such as ⚁ ⚁ or ⚃ ⚃) on your turn, you then get to take an additional turn. If, on the additional turn, you again roll doubles, you then get to take yet another additional turn! If on this third roll in a row you roll doubles, instead of moving as normal, you get sent to jail. The second rule involves *leaving* jail. If you start your turn in jail, you can decide to play a "Get Out of Jail Free" card, or pay $50, to roll the dice and move out of jail as normal. If you choose not to, however, you still get to roll the dice and, if you roll doubles, move out of jail the indicated number of spaces (with no free turn, however, even though you rolled doubles). If you do not roll doubles on your third turn (third attempt), you *must* pay the $50 to leave and then move the indicated number of spaces.

Both of these rules will affect our probability matrix construction as we're about to see. It should be noted here that the *choice* of whether to leave jail early or stay in jail as long as possible (a possibly good strategy if you're waiting for some opponents to land on your properties and give you additional

TABLE 6.4: Two Dice Sum Probabilities

Sum	Prob	Sum	Prob
2	1/36	8	5/36
3	2/36	9	4/36
4	3/36	10	3/36
5	4/36	11	2/36
6	5/36	12	1/36
7	6/36		

money to support landing on someone else's well-developed Tennessee Avenue, for example) yields two different play strategies, "Short Jail" or "Long Jail," that will slightly alter your steady-state vector. We will, of course, look at both versions.

The development of the probability matrix for Monopoly is, as you might guess, more complicated than the Pay Day model or any of the "fake" games discussed before that. In fact, we will use a three-layered process to just develop one row of the matrix; this method will give you an insight into how I developed the matrix that we will see and will also show you how other games can be modeled by considering one complication at a time. The final model that we illustrate here does have one (small) flaw that I'll discuss afterwards; the full version would increase the matrix size from 1,600 entries to 144,000 entries. I've chosen to go with a model that is easier to understand and follow while still only introducing very little error.

Note that in our examples to follow, as well as the final probability matrix given, the order of the columns (and rows of the matrix) is based on the Monopoly board of Figure 6.10; the first column and row corresponds to "Go" and order proceeds clockwise around the board. We do *not* include "Go

To Jail" in our matrix at all, but rather list "In Jail" as a separate space immediately following "Just Visiting" in the matrix.

> **Example 6.5 (Go, Version 1)** Let's consider what happens when you start a turn on the "Go" space. Rolling two dice gives you eleven possible sums ranging from 2 to 12, with the probabilities listed in Table 6.4 (these values come from looking at the raw possible outcomes of Figure 1.2). Thus, for instance, the probability that you start on Go and land on Community Chest (South) would be 1/36, the chance that you land on Vermont Avenue is 5/36, and so on. Note that, here, there is no chance of landing on many of the board spaces due to just the *roll* itself, which makes this first version of our model quite simple and inaccurate, but it's a good start. This first row's entries, given here as a list of numbers, separated by commas, represents the probability of starting at Go and ending at Go, Mediterranean Avenue, Community Chest (South), Baltic Avenue, and so on, with "In Jail" listed immediately after "Just Visiting" and the "Go To Jail" space omitted since landing there sends you immediately to jail. We obtain
>
> $$\left(0, 0, \frac{1}{36}, \frac{2}{36}, \frac{3}{36}, \frac{4}{36}, \frac{5}{36}, \frac{6}{36}, \frac{5}{36}, \frac{4}{36}, \frac{3}{36}, 0, \frac{2}{36}, \frac{1}{36}, 0, 0, 0, \right.$$
>
> $$\left. 0,0,0,0,0,0,0,0,0,0,0,0,0,0,0,0,0,0,0,0,0,0,0,0 \right)$$
>
> as the numbers for this first row. All other rows for this model are very similar.

The first change that we need to make to Example 6.5 comes from the fact that if you roll doubles three times in a row, you get sent to jail. Considering a roll sum of five, where the only possibilities are ⚀⚃, ⚁⚂, ⚃⚀, and ⚂ ⚁, there are no changes. The chance of rolling a five and actually getting to advance your token around the board five spaces is still 4/36.

TABLE 6.5: Adjusted Two Dice Probabilities

Sum	Prob	Sum	Prob
2	35/1296	8	179/1296
3	2/36	9	4/36
4	107/1296	10	107/1296
5	4/36	11	2/36
6	179/1296	12	35/1296
7	6/36	Jail	1/216

However, consider a roll sum of four, where you can obtain this with ⚀⚂, ⚂⚀, and ⚁ ⚁. If you do roll ⚁⚁ *and* the previous two rolls were doubles, then instead of moving four spaces, you go to jail. This happens with probability

$$\left(\frac{1}{6}\right)^2 \cdot \frac{1}{36} = \frac{1}{1296}$$

where the 1/6 term represents the probability of rolling doubles, the two as an exponent forces this to happen twice in a row prior to this roll, and the 1/36 term is the chance of ⚁ ⚁ on this roll of the dice. The probability that we roll a pair of twos on the current roll and did *not* get doubles on *both* of the previous rolls is

$$\left(1 - \left(\frac{1}{6}\right)^2\right) \cdot \frac{1}{36} = \frac{35}{1296}$$

and this makes the probability of rolling a sum of four, and actually being allowed to move four spaces, equal

$$\frac{2}{36} + \frac{35}{1296}$$

since there is a 2/36 chance of rolling a non-double sum of four. For sums of two, six, eight, ten, and twelve, a similar method can be applied, giving the adjusted probabilities of Table 6.5 where the "Jail" entry is there to remind us that we do indeed have a chance of landing in jail each turn due to just the roll of the dice via this rule.

Of course, you should note that these probabilities do *not* apply to the first (or second) roll you make each turn, but since our goal is to determine the *long term* steady-state, we shall use these new probabilities in our model.

Example 6.6 (Go, Version 2) Again, let's consider what happens when you start a turn on the "Go" space. With the adjusted probabilities of Table 6.5, we can take the discussion of Example 6.5 and obtain the first row of a probability matrix (corresponding to starting on Go) for Monopoly movement as

$$\left(0, 0, \frac{35}{1296}, \frac{2}{36}, \frac{107}{1296}, \frac{4}{36}, \frac{179}{1296}, \frac{6}{36}, \frac{179}{1296}, \frac{4}{36}, \frac{107}{1296}, \frac{1}{216}, \frac{2}{36}, \frac{35}{1296},\right.$$
$$\left. 0, 0, 0, 0, 0, 0, 0, 0, 0, 0, 0, 0, 0, 0, 0, 0, 0, 0, 0, 0, 0, 0, 0, 0, 0, 0\right)$$

with other rows, again, developed similarly.

TABLE 6.6: Community Chest Destinations

Community Chest	
Destination	Prob
Jail	1/16
Go	1/16
Stay	14/16

The last (and biggest) complication that we'll need to deal with is the extra movement caused by the various "Chance" and "Community Chest" cards that are drawn when landing on those spaces. For the Community Chest cards at least, this is fairly simple, as of the 16 cards available, 14 of them do *not* involve moving to another square (they involve either gaining or losing money in some fashion). One card sends you directly to jail, and another sends you to Go. These probabilities are summarized in Table 6.6.

By comparison, the cards drawn upon landing on Chance have a much larger chance to send you to another location on the board. Table 6.7 shows all of the possibilities, and it's important to note that the "Next Railroad" and "Next Utility" always refer to the closest appropriate space located on the board moving in a clockwise (the same as the flow of play) direction.

In particular, note that the Chance space located on the east side of the board (as we see the board in Figure 6.10, with north at the top of the page) forces some interesting situations to occur. From this space, Reading Railroad is fairly likely since it is specifically mentioned by one Chance card, and is also the *nearest* railroad as well. In addition, from this space, you might draw the "Go Back 3 Spaces" card, sending you back to the Community Chest square where you have a non-zero probability of ending up somewhere else again! Let's look at Examples 6.5 and 6.6 again to arrive at our final version of the "Go" row for our probability matrix.

TABLE 6.7: Chance Destinations

Chance Destination	Prob
Jail	1/16
Go	1/16
Reading RR	1/16
St. Charles Place	1/16
Illinois Ave	1/16
Boardwalk	1/16
Next RR	2/16
Next Utility	1/16
Back 3 Spaces	1/16
Stay	6/16

Example 6.7 (Go, Version 3) To determine what changes we need to make, it's important to first look at which card-drawing squares are accessible from our starting square. If we start on "Go," a roll of 2 will send us to the southern Community Chest, and a roll of 7 will send us to the southern Chance, so both types of card-drawing squares are possible locations. Note that both types are not necessarily reachable from all starting squares for a turn, but at least one card-drawing spot is reachable no matter where you start. Using Table 6.6 and Table 6.7 is now required for us to determine a roll's true end location. What needs to be changed from our model of Example 6.6? It's time to find out.

The Community Chest situation is the easiest to look at; since there is a 14/16 chance that you will *stay* on Community Chest upon landing on it, we need to adjust the probability of ending our roll from the 35/1296 of Table 6.5 to the more accurate

$$\text{Probability Community Chest} = \underbrace{\left(\frac{35}{1296}\right)}_{\text{roll 2}} \cdot \underbrace{\left(\frac{14}{16}\right)}_{\text{stay}} = \frac{245}{10368}$$

which is the probability that we roll a two (and do not get sent to jail because it was our third double in a row), and then draw a Community Chest card that does *not* send us somewhere else. The two cards that do not leave us on that square send us to either Go or Jail, so one possible way of getting to Go when we've started on Go is to roll a two and then draw the card that sends us there, giving us

$$\text{Community Chest, then Go} = \underbrace{\left(\frac{35}{1296}\right)}_{\text{roll 2}} \cdot \underbrace{\left(\frac{1}{16}\right)}_{\text{Go}} = \frac{35}{20736}$$

as the chance that we roll a two and draw the card to end up at Go. Since there is only one Community Chest card that sends us to Jail as well, this is also the same probability that we end up in Jail via the Community Chest. At this point, there is definitely more to consider.

The Chance cards *also* also give us a chance to end up on Go, in Jail, and a plethora of other locations as well! I find it easier myself to look at each location and determine the probability of ending there, having started at Go. For instance, for starting on Go and then ending at Go, you can roll a two and draw the Community Chest card to send you there (as we saw above), or roll a seven and draw the Chance card that sends you there, giving us

$$\text{Prob Start Go, End Go} = \underbrace{\left(\frac{35}{1296}\right) \cdot \left(\frac{1}{16}\right)}_{\substack{\text{``roll'' 2, draw} \\ \text{go to Go}}} + \underbrace{\left(\frac{6}{36}\right) \cdot \left(\frac{1}{16}\right)}_{\substack{\text{roll 7, draw} \\ \text{go to Go}}} = \frac{251}{20736}$$

as the probability of starting on Go and ending at Go.

For the probability of starting on Go and ending up in Jail, however, we need to consider

$$\text{Prob Start Go, End Jail} = \underbrace{\left(\frac{35}{1296}\right) \cdot \left(\frac{1}{16}\right)}_{\substack{\text{``roll'' 2, draw} \\ \text{go to Jail}}} + \underbrace{\left(\frac{6}{36}\right) \cdot \left(\frac{1}{16}\right)}_{\substack{\text{roll 7, draw,} \\ \text{go to Jail}}} + \underbrace{\left(\frac{1}{216}\right)}_{\substack{\text{third} \\ \text{double}}}$$

$$= \frac{347}{20736}$$

which, while almost the same, adds in the probability that we roll our third double in a row from earlier. The Reading Railroad can be reached by either rolling a five or by rolling a seven and drawing the Chance card that moves you there, resulting in

$$\text{Prob Start Go, End Reading RR} = \underbrace{\left(\frac{4}{36}\right)}_{\substack{\text{roll} \\ \text{a 5}}} + \underbrace{\left(\frac{6}{36}\right) \cdot \left(\frac{1}{16}\right)}_{\substack{\text{roll 7, draw,} \\ \text{go Reading}}} = \frac{35}{288}.$$

As another example, consider Income Tax, where a roll of four lands you there naturally, but you could roll a seven, and draw the Chance card that sends you back three spaces (in a sneaky fashion)! The probability for this is

$$\text{Prob Start Go, End Income Tax} = \underbrace{\left(\frac{107}{1296}\right)}_{\substack{\text{``roll''} \\ \text{a 4}}} + \underbrace{\left(\frac{6}{36}\right) \cdot \left(\frac{1}{16}\right)}_{\substack{\text{roll 7, draw,} \\ \text{go back 3}}} = \frac{241}{2592}.$$

As we continue, note that for ending on Illinois Avenue, starting on Go, the probability is

$$\text{Prob Start Go, End Illinois Ave} = \underbrace{\left(\frac{6}{36}\right) \cdot \left(\frac{1}{16}\right)}_{\substack{\text{roll 7, draw,} \\ \text{go Ill. Ave}}} = \frac{1}{96}$$

since the only option is reaching Chance and drawing the specific card that sends you to Illinois Avenue. Of course, there are some spaces, such as Connecticut Avenue, for which the probability is easy to compute, since the only way to reach Connecticut Avenue from Go is to roll a nine, which happens with probability 4/36 as we have seen.

The first row of our probability matrix, corresponding to the probabilities of landing on each square, starting on Go, is then

$$\left(\frac{251}{20736}, 0, \frac{245}{10368}, \frac{1}{18}, \frac{241}{2592}, \frac{35}{288}, \frac{179}{1296}, \frac{1}{16}, \frac{179}{1296}, \frac{1}{9}, \frac{107}{1296}, \frac{347}{20736}, \frac{19}{288}, \right.$$
$$\left. \frac{97}{2592}, 0, 0, \frac{1}{48}, 0, 0, 0, 0, 0, 0, 0, 0, \frac{1}{96}, 0, 0, 0, 0, 0, 0, 0, 0, 0, 0, 0, 0, 0, \frac{1}{96}\right)$$

where the numbers not obtained earlier in this example can be computed similarly to the other locations we have already seen.

This, of course, gives us just one of the forty rows for the probability matrix associated to Monopoly; there will be plenty of opportunities for you to practice this yourself; there are, however, two more examples that I want to cover. The first involves what happens if you start your turn on the Water Works space. Note that from here on out, "roll" for us, when discussing an even value, means rolling that value *and* having it not be our third double in a row.

Example 6.8 (Water Works) Since we can reach both a Community Chest space and a Chance space, we will have several ways to reach Jail and Go, including by a natural roll of 2 or 12 on the dice. In addition, since drawing the Chance card that moves you back three spaces puts you on the Community Chest space here, ending up in Jail or on Go indeed happens in many scenarios. Looking specifically at Jail, we can reach that spot from Water Works by

- rolling a two on the dice,

- rolling a five on the dice and drawing the appropriate Community Chest card,

- rolling an eight on the dice and drawing the appropriate Chance card,

- rolling an eight on the dice, drawing the Chance card to move back three spaces, and then drawing the appropriate Community Chest card, or

- rolling our third set of doubles in a row.

This results in a probability of

$$\text{Start WW, End Jail} = \underbrace{\left(\frac{35}{1296}\right)}_{\text{roll 2}} + \underbrace{\left(\frac{4}{36}\right) \cdot \left(\frac{1}{16}\right)}_{\substack{\text{roll 5, draw} \\ \text{go to Jail}}} + \underbrace{\left(\frac{179}{1296}\right) \cdot \left(\frac{1}{16}\right)}_{\substack{\text{roll 8, draw} \\ \text{go to Jail}}}$$

$$+ \underbrace{\left(\frac{179}{1296}\right) \cdot \left(\frac{1}{16}\right) \cdot \left(\frac{1}{16}\right)}_{\substack{\text{roll 8, draw go back 3,} \\ \text{draw go to Jail}}} + \underbrace{\left(\frac{1}{216}\right)}_{\substack{\text{third} \\ \text{double}}} = \frac{5281}{110592}$$

which is just shy of about 5%. The probability of reaching Go is very similar and simply ignores the 1/216 term for rolling the third double in a row.

For reaching the Reading Railroad, note that not only is there the regular Chance card that moves you there, it also happens to be the *nearest* railroad, measured from the eastern Chance space! The probability of ending your move on the Reading Railroad is then

$$\text{WW to Reading RR} = \underbrace{\left(\frac{179}{1296}\right) \cdot \left(\frac{1}{16}\right)}_{\substack{\text{roll 8, draw go} \\ \text{to Reading RR}}} + \underbrace{\left(\frac{179}{1296}\right) \cdot \left(\frac{2}{16}\right)}_{\substack{\text{roll 8, draw go} \\ \text{to nearest RR}}} = \frac{179}{6912}.$$

Most of the other destinations are straightforward; before giving the matrix row, however, let's look at the probability of ending on Community Chest, which requires rolling a 5 and drawing a card that keeps you there, or rolling an 8, drawing the card to move back three spaces, and then drawing a Community Chest card that keeps you there. Remember from Table 6.6 that 14 of the 16 cards leave you on Community Chest. We end up with

$$\text{WW to CC} = \underbrace{\left(\frac{4}{36}\right) \cdot \left(\frac{14}{16}\right)}_{\substack{\text{roll 5, draw} \\ \text{and stay here}}} + \underbrace{\left(\frac{179}{1296}\right) \cdot \left(\frac{1}{16}\right) \cdot \left(\frac{14}{16}\right)}_{\substack{\text{roll 8, draw go back 3,} \\ \text{draw and stay here}}} = \frac{17381}{165888}$$

as that probability.

The final row for the probability matrix corresponding to Water Works is

$$\left(\frac{4769}{110592}, 0, 0, 0, 0, \frac{179}{6912}, 0, 0, 0, 0, 0, \frac{5281}{110592}, \frac{179}{20736}, \frac{179}{20736}, 0, 0, 0, 0, 0, 0, 0, 0, \right.$$

$$\left. 0, 0, 0, \frac{179}{20736}, 0, 0, 0, 0, 0, \frac{1}{18}, \frac{107}{1296}, \frac{17381}{165888}, \frac{179}{1296}, \frac{1}{6}, \frac{179}{3456}, \frac{1}{9}, \frac{107}{1296}, \frac{1331}{20736} \right)$$

where the order of the numbers is as usual.

As we've alluded to earlier, the model that we'll develop depends a bit on whether you choose to employ a "short jail" strategy or a "long jail" strategy. For the former, as soon as you end up in jail, for the next turn you will always pay the $50 to get out (or use a "get out of jail free" card), roll the dice, and move the indicated number of spaces. For the "long jail" strategy, you plan to stay in jail as long as possible, rolling the dice the first two turns in jail (successfully escaping by rolling doubles), and only on the third turn in jail will you roll the dice and leave (paying $50 if doubles are not rolled). The latter strategy can be especially beneficial late in the game when you can attempt to avoid landing on spaces with hotels by simply staying on the same space; since the rules still allow you to *collect* money from other players while in jail, this can be *extremely* nice). We will explore the "short jail" strategy here.

Example 6.9 (Jail, Short) The matrix row corresponding to the "short jail" strategy, as you can see from the matrix of Figure 6.14, is very similar to the row for Just Visiting. In fact, the two rows are identical except that the probabilities of rolling a 2, 4, 6, 8, 10, or 12 are taken from Table 6.4 instead of Table 6.5 since, when starting your turn In Jail, there is no possible chance that rolling doubles will result in your *third* double in a row. This means, for example, that the probability of reaching the Electric Company from Jail is 1/36, the probability of reaching States Avenue is 2/36, and so on, up to the probability of reaching the northern Chance square being 1/36 as well. See if you agree that the probability of reaching Go from Jail is given by the calculation

$$\text{Start Jail, End Go} = \underbrace{\left(\frac{6}{36}\right) \cdot \left(\frac{1}{16}\right)}_{\substack{\text{roll 7, draw} \\ \text{go to Go}}} + \underbrace{\left(\frac{1}{36}\right) \cdot \left(\frac{1}{16}\right)}_{\substack{\text{roll 12, draw} \\ \text{go to Go}}} = \frac{7}{576}$$

where, again, *roll 12* here means rolling a natural twelve via Table 6.4. The probability of ending at New York Avenue, due to it being three spaces before a Chance spot, is

$$\text{Start Jail, End NY} = \underbrace{\left(\frac{4}{36}\right)}_{\text{roll 9}} + \underbrace{\left(\frac{1}{36}\right) \cdot \left(\frac{1}{16}\right)}_{\substack{\text{roll 12, draw} \\ \text{go back 3}}} = \frac{65}{576}.$$

For our last example within the "short jail" strategy, note that you can still reach Boardwalk from Jail by drawing the corresponding Chance card; this happens with probability

$$\text{Start Jail, End Boardwalk} = \underbrace{\left(\frac{1}{36}\right) \cdot \left(\frac{1}{16}\right)}_{\substack{\text{roll 12, draw} \\ \text{Boardwalk}}} = \frac{1}{576}.$$

which is a probability slightly less than 0.2%. Putting all of these together, along with the other spaces possible, yields a matrix row of

$$\left(\frac{7}{576}, 0, 0, 0, 0, \frac{1}{576}, 0, 0, 0, 0, 0, \frac{7}{576}, \frac{1}{576}, \frac{1}{36}, \frac{1}{18}, \frac{1}{12}, \frac{1}{9}, \frac{5}{36}, \frac{7}{48}, \frac{5}{36}, \right.$$
$$\left. \frac{65}{576}, \frac{1}{12}, \frac{1}{18}, \frac{1}{96}, 0, \frac{1}{576}, \frac{1}{288}, 0, 0, \frac{1}{576}, 0, 0, 0, 0, 0, 0, 0, 0, 0, \frac{1}{576}\right)$$

where the order, again, is as usual.

Combining Examples 6.7, 6.8, and 6.9, along with the rows for the other 37 spaces of the board, we obtain the probability matrix for Monopoly, PM, shown in Figure 6.14. Remember that the order of the columns and rows starts with Go, proceeds clockwise around the Monopoly board from Figure 6.10, omits the "Go To Jail" space, and includes Jail immediately after the column (or row) corresponding to Just Visiting.

Even before we see the steady-state vector associated to the "short jail" strategy in Monopoly, there are some interesting patterns that are present in matrix PM. By looking at the columns, we can see that both Go and Jail can be reached anytime, regardless of which space you are currently on. In addition, both the Reading Railroad and Boardwalk are reachable from all but seven spaces of the board. Illinois Avenue can be landed on from all but five spaces, and St. Charles Place from all but three. These facts are primarily due to the effect of Chance cards and, for Go and Jail, Community Chest cards. Note that the "advance to the nearest utility" Chance cards also give both the Electric Company and the Water Works a larger amount of non-zero entries in their columns.

It is now time to see the steady-state vector; since it has 40 entries, and contains values that we will want to soon use, these probabilities are given in Table 6.8 rather in vector form so that we have easy access to them.

One thing that may surprise you if you haven't seen much of the mathematics behind Monopoly before is that, aside from Jail, the red property of Illinois Avenue has the highest long-term probability. This means that as the game continues, players are more likely to spend time there than on any other space (minus Jail), and it is the single most likely property to land on, for purchase, and for development with houses and hotels). We will consider the "long jail" option later, but for now, it's important for us to realize that, probabilities aside, not all of the properties are worth the same amount!

Question

What are the properties in Monopoly *really* worth?

To answer this question, we'll need some data from the game's deed cards, and some calculations that we are already equipped to do.

FIGURE 6.14: Monopoly Matrix *PM*, Short Jail

TABLE 6.8: Steady-State Vector Values for Monopoly, Short Jail

Space	Prob		Space	Prob
Go	0.0309071		New York Ave	0.0308864
Mediterranean Ave	0.0212834		Free Parking	0.0288584
Comm Chest (South)	0.0187994		Kentucky Ave	0.0283661
Baltic Ave	0.0215936		Chance (North)	0.0104828
Income Tax	0.0232170		Indiana Ave	0.0273401
Reading RR	0.0295896		Illinois Ave	0.0318586
Oriental Ave	0.0225527		B & O RR	0.0306493
Chance (South)	0.0086363		Atlantic Ave	0.0270773
Vermont Ave	0.0231383		Ventnor Ave	0.0267693
Connecticut Ave	0.0229617		Water Works	0.0280695
Just Visiting	0.0226176		Marvin Gardens	0.0258308
In Jail	0.0402954		Pacific Ave	0.0267389
St. Charles Place	0.0269635		North Carolina Ave	0.0262337
Electric Company	0.0259792		Comm Chest (East)	0.0236293
States Ave	0.0237101		Pennsylvania Ave	0.0249801
Virginia Ave	0.0246382		Short Line	0.0242937
Pennsylvania RR	0.0292213		Chance (East)	0.0086547
St. James Place	0.0279591		Park Place	0.0218313
Comm Chest (West)	0.0260042		Luxury Tax	0.0217571
Tennessee Ave	0.0294002		Boardwalk	0.0262247

The Properties' *Real* Values

As you might reason, a property that nets a little less rent for a hotel but is landed on more frequently than a more "rewarding" property might actually be worth more to you as the game continues. What we need to do is balance the long-term probabilities for each space against what that property can earn for you when players land upon it. We have some values for these properties, given in Table 6.9, and we also have the probabilities for these spaces listed in Table 6.8. What have we done so far in this book when probabilities and values were around? I hope by now the concept of expected value is coming back to you again.

In our analysis, we won't worry about the actual price for each property; the costs paid to purchase a space can be variable (through the auction rule), and these sunken costs should not factor into our long-term desire to have our opponents pay us large sums of (fake Monopoly) money. We will also not worry about the costs to improve each property, again citing sunk costs as the reason there. For the curious, however, note that houses and hotels cost $50 each for the southern locations, $100 each for the western ones, $150 each in the north, and $200 for the expensive properties of the east. Improvements can only be made by a player, again, when that player owns all properties of a color group, and only as evenly as possible within a color group; owning all the properties of a single color also doubles the base rent listed in Table 6.9 for unimproved properties. The house rent listed there is collected based on the number of houses on the particular property your opponent lands upon, and, like the hotel rent, is collected *instead* of double the base rent and *not* in addition to it.

TABLE 6.9: Property Information, Costs in Dollars ($)

Property	Price	Base Rent	House Rent				Hotel Rent
			1	2	3	4	
Mediterranean Ave	60	2	10	30	90	160	250
Baltic Ave	80	4	20	60	180	320	450
Oriental Ave	100	6	30	90	270	400	550
Vermont Ave							
Connecticut Ave	120	8	40	100	300	450	600
St. Charles Place	140	10	50	150	450	625	750
States Ave							
Virginia Ave	160	12	60	180	500	700	900
St. James Place	180	14	70	200	550	750	950
Tennessee Ave							
New York Ave	200	16	80	220	600	800	1000
Kentucky Ave	220	18	90	250	700	875	1050
Indiana Ave							
Illinois Ave	240	20	100	300	750	925	1100
Atlantic Ave	260	22	110	330	800	975	1150
Ventnor Ave							
Marvin Gardens	280	24	120	360	850	1025	1200
Pacific Ave	300	26	130	390	900	1100	1275
North Carolina Ave							
Pennsylvania Ave	320	28	150	450	1000	1200	1400
Park Place	350	35	175	500	1100	1300	1500
Boardwalk	400	50	200	600	1400	1700	2000

Using this information, let's take a look at how Illinois Avenue, the improvable property with the highest long-term probability, compares with its higher-valued neighbor Atlantic Avenue. For base rent, the expected income per turn for Illinois Avenue computes to

Illinois Avenue Base Rent EV $= 0.0318586 \cdot (\$20) \approx \$0.64$

whereas for Atlantic Avenue we obtain

Atlantic Avenue Base Rent EV $= 0.0270773 \cdot (\$22) \approx \$0.60$

per turn as the expected value. Despite the increased cost to purchase Atlantic Avenue and despite the (slightly) higher rent, the higher chance of landing on Illinois Avenue makes it a bit more valuable in the long run. This difference is magnified when we compute the expected income per turn with hotels, as we would obtain

$$\text{Illinois Avenue Hotel EV} = 0.0318586 \cdot (\$1100) \approx \$35.04$$

versus

$$\text{Atlantic Avenue Hotel EV} = 0.0270773 \cdot (\$1150) \approx \$31.14$$

and, as you can imagine, the same "superiority" for Illinois Avenue applies for any quantity of houses as well.

Since Monopoly is a real estate *trading* game, and owning all the properties of a color is required to build just one house, perhaps a better analysis is looking at the expected value, per turn, of the color groups themselves! We will do that here for three color groups for hotels only. More analysis will be left for you as exercises.

Example 6.10 The two dark blue properties of Boardwalk and Park Place are often sought after by beginning Monopoly players. Is this mathematically justified? The expected value of the color *group* per turn can be computed by adding together the individual property expected values to obtain

$$\text{Dark Blue EV} = \underbrace{0.0218313 \cdot (\$1500)}_{\text{Park Place}} + \underbrace{0.0262247 \cdot (\$2000)}_{\text{Boardwalk}} \approx \$85.19$$

which, as a reminder, means that as the game continues, you can reasonably expect the dark blue property group, with hotels, to net you about $85 per turn. For the yellow properties, we would obtain the very similar calculation, with three terms since three yellow properties exist, of

$$\text{Yellow EV} = \underbrace{0.0270773 \cdot (\$1150)}_{\text{Atlantic Ave}} + \underbrace{0.0267693 \cdot (\$1150)}_{\text{Ventnor Ave}} + \underbrace{0.0258308 \cdot (\$1200)}_{\text{Marvin Gardens}}$$
$$\approx \$92.92$$

which is a good bit better than the dark blue group, but also includes three properties instead of just two. Finally for this example, we look at the red group which contains Illinois Avenue. Here, we get

$$\text{Red EV} = \underbrace{0.0283661 \cdot (\$1050)}_{\text{Kentucky Ave}} + \underbrace{0.0273401 \cdot (\$1050)}_{\text{Indiana Ave}} + \underbrace{0.0318586 \cdot (\$1100)}_{\text{Illinois Ave}}$$
$$\approx \$93.54$$

which is even better than the yellow group!

We can see the relative value of each color group in Table 6.10. The ranking of the dark blue and brown groups is a bit unfair since each of those only includes two properties while the others all contain three. The slightly easier ability to acquire and improve these groups might be considered as you play in order to correctly rank these in your own strategy; you might divide the group expected values by the number of properties to get a quick comparison.

That said, the fact that the expected value of the red color group is higher than the yellow color group gives you a bit of an advantage against unaware players. It should be fairly easy to trade a yellow property you own for a red property (and perhaps some cash or something else) from another player!

Before we discuss the "long jail" strategy and how that game play affects our probabilities, it's worth taking a look at the railroads, especially since the combined probability of landing on a railroad is

TABLE 6.10: Color Group Hotel EV

Group	EV
Green	$102.51
Red	$93.54
Yellow	$92.92
Orange	$85.38
Dark Blue	$85.19
Light Purple	$60.18
Light Blue	$38.91
Brown	$15.04

$$\underbrace{0.0295896}_{\text{Reading}} + \underbrace{0.0292213}_{\text{Pennsylvania}} + \underbrace{0.0306493}_{\text{B \& O}} + \underbrace{0.0242937}_{\text{Short Line}} = 0.1137539,$$

which is the largest probability for the *type* of square. From Figure 6.12 we see that the "rent" for landing on a railroad depends on the number of railroads owned by that player. Owning all four gives an expected value of the collection as

$$\text{Four Railroad EV} = 0.1137539 \cdot (\$200) \approx \$22.75$$

per turn, which is fairly nice. It becomes even better when we realize that the Chance cards that send players to the nearest railroad instruct that player to pay *twice* the required amount.

Example 6.11 We might wonder what three-railroad subset is best to own. We could compute the expected values of all four possible three-railroad subsets, but since the rent paid depends only on the number of railroads owned and not on *which* are owned, the best three-railroad subset must therefore include the three railroads with the highest long-term probabilities. Since the Short Line has the lowest probability, the best three-railroad subset for the end game includes Reading Railroad, Pennsylvania Railroad, and the B & O Railroad. The combined probability of these three is 0.0894602 and, combining the "rent" of $100, gives an expected value

$$\text{Railroads, Minus Short Line, EV} = 0.0894602 \cdot (\$100) \approx \$8.95$$

which is a good bit higher than

$$\text{Railroads, Minus B \& O, EV} = 0.0831046 \cdot (\$100) \approx \$8.31,$$

the worst possible three-railroad subset.

More analysis using the "short jail" probabilities of Table 6.9 can certainly be done, but at this point, we will leave that for the exercises and your own imagination.

The "Long Jail" Strategy

Recall that the "long jail" strategy involves staying in jail as long as possible, perhaps to avoid landing on your opponent's developed red (or orange) properties in the near future. The rules of Monopoly state that when in jail, unless you pay \$50 or use a "Get Out of Jail Free" card before rolling, you will only leave jail if you roll doubles (with probability 6/36) except that on your third attempt, you will leave no matter what you roll (having to pay \$50 or use a card if doubles are not rolled).

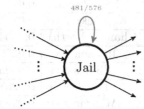

481/576

FIGURE 6.15: Long Jail, Attempt 1

Your first thought on modifying our previous work should be that only the In Jail space changes; all other probabilities for our probability matrix should remain the same. The modification for In Jail would appear similar to Figure 6.15 where we have a large chance of staying there. Only a double will let us escape our cell, and there's even a small chance (due to Chance) of landing back in jail again! The 481/576 that appears there is the 5/6 chance of staying due to not rolling doubles combined with the 1/576 chance that we roll a 12 and then draw the appropriate Chance card.

The problem with this modification is that it only works for the first two turns in jail! On the third roll, the probability for what spaces we might end on starting In Jail is identical to the "short jail" strategy since we *must* leave! Note that Figures 6.15 and 6.16 include arrows from and to nowhere; these indicate the locations that either send you *to* jail or to places that you can reach *from* jail; the exact names and the probabilities for these are not needed here and are omitted.

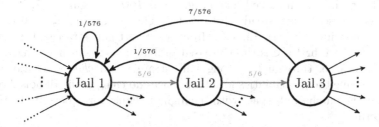

FIGURE 6.16: Long Jail, Fixed

For us to accurately keep track of how many turns we have been In Jail, we need to introduce two "fake" jail spaces to our model! Figure 6.16 shows three spaces, "In Jail (1)," "In Jail (2)," and "In Jail (3)" that are used not only to keep track of a player's location, but also how many turns they have remained there. Note that the 5/6 chance of remaining "In Jail" keeps us at the same physical location on the board, but does move us into a different virtual space; the third virtual jail space has arrows exiting as in the "short jail" version. The 1/576 probability of leaving jail, landing on Chance, and then ending up *back* in jail sends us to "In Jail (1)" since while we did end in jail, we successfully left.

TABLE 6.11: Steady-State Vector Values for Monopoly, Long Jail

Space	Prob	Space	Prob
Go	0.0290833	New York Ave	0.0280994
Mediterranean Ave	0.0200357	Free Parking	0.0282570
Comm Chest (South)	0.0176990	Kentucky Ave	0.0261104
Baltic Ave	0.0203323	Chance (North)	0.0104493
Income Tax	0.0218646	Indiana Ave	0.0256274
Reading RR	0.0279631	Illinois Ave	0.0299051
Oriental Ave	0.0212455	B & O RR	0.0288901
Chance (South)	0.0081359	Atlantic Ave	0.0253594
Vermont Ave	0.0217993	Ventnor Ave	0.0251460
Connecticut Ave	0.0216344	Water Works	0.0264968
Just Visiting	0.0213125	Marvin Gardens	0.0243307
In Jail	0.0959311	Pacific Ave	0.0251824
St. Charles Place	0.0255033	North Carolina Ave	0.0247052
Electric Company	0.0260934	Comm Chest (East)	0.0222292
States Ave	0.0217045	Pennsylvania Ave	0.0235078
Virginia Ave	0.0242348	Short Line	0.0228581
Pennsylvania RR	0.0263278	Chance (East)	0.0081460
St. James Place	0.0268036	Park Place	0.0205515
Comm Chest (West)	0.0229539	Luxury Tax	0.0204828
Tennessee Ave	0.0282195	Boardwalk	0.0247878

After seeing the modified matrix, given as $\widehat{PM}$ in Figure 6.17, it's time, of course, to see a new steady-state vector. This is given in Table 6.11, and when comparing this with Table 6.8 from the "short jail" strategy, the biggest change that we first see is the increased probability of a turn ending In Jail simply because, through the "long jail" strategy, we *choose* to stay there longer. While the probabilities for other spaces do change, the Illinois Avenue remains the purchasable space with the highest chance for a player to be there, long-term.

You might wonder if the relative worth of the color groups stays the same with this strategy. Certainly the expected dollar amounts per turn will change since the probabilities have changed, but are the *rankings* presented in Table 6.10 for the "short jail" strategy preserved? You will have the opportunity to

FIGURE 6.17: Monopoly Matrix $\widehat{PM}$, Long Jail

determine this for yourself later, but for now, let's see what happens for the red and yellow color groups by repeating what we did in Example 6.10.

Example 6.12 For the yellow properties, using the new probabilities given in Table 6.11 and data of Table 6.9, we would obtain

$$\text{Yellow EV} = \underbrace{0.0253594 \cdot (\$1150)}_{\text{Atlantic Ave}} + \underbrace{0.0251460 \cdot (\$1150)}_{\text{Ventnor Ave}} + \underbrace{0.0243307 \cdot (\$1200)}_{\text{Marvin Gardens}}$$
$$\approx \$87.28$$

which is about $5.50 less than for the "short jail" strategy. By comparison, for the red group we get

$$\text{Red EV} = \underbrace{0.0261104 \cdot (\$1050)}_{\text{Kentucky Ave}} + \underbrace{0.0256274 \cdot (\$1050)}_{\text{Indiana Ave}} + \underbrace{0.0299051 \cdot (\$1100)}_{\text{Illinois Ave}}$$
$$\approx \$87.22$$

which is now slightly worse than the yellow group.

As we see in Example 6.12 compared to Example 6.10 earlier, the red properties aren't quite as valuable compared to the yellow properties now; when players use the "short jail" strategy, they are going around the board faster, giving those red properties a slight advantage over the yellow. That advantage goes away when players choose to stay in jail longer. For the cost, though, a difference of $0.05 per turn better for the yellow properties is not much.

Fixing the Model

As I mentioned earlier, our models for "short jail" and "long jail" are not perfect. When starting a turn, there's no possible way to end up in jail as a result of having rolled our third consecutive doubles this turn; we need to actually have moved a few times! Our model accounts for this by appealing to the fact that, since we're studying long-term probabilities already, we can approximate this rule by using the adjusted probabilities for two dice of Table 6.5. A better (and correct) model uses the idea introduced for the "long jail" strategy. In that strategy, we introduced two additional virtual spaces to the board to keep track of the number of turns spent in jail. The full model for Monopoly can be found by introducing two additional virtual spaces for *each* square on the board that keep track of the number of doubles for a given turn that it took to reach that location. For instance, in addition to the standard "Park Place" square, we add "Park Place 1D" and "Park Place 2D" representing, respectively, having reached Park Place by one doubles or two doubles.

Figure 6.18 shows an example of how these spaces would work. Starting on

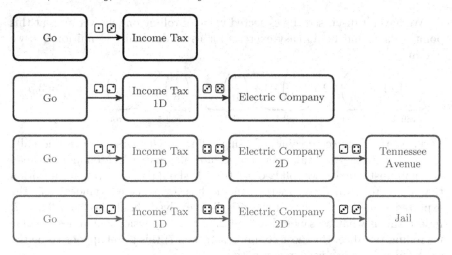

FIGURE 6.18: Fixed Model, Example

Go, a roll of ⚀⚄ sends us to Income Tax as normal, but a roll of ⚂⚀ sends us to the virtual "Income Tax 1D" space instead. So far this turn we have rolled doubles once, and the virtual space keeps track of that. From "Income Tax 1D," a roll of ⚃⚅ moves us to the regular Electric Company since our turn now ends (the count for doubles is reset to zero this way), while a roll of ⚄⚄ moves us to "Electric Company 2D" since, so far this turn, we have rolled doubles twice. Finally, a roll of ⚄⚄ starting at "Electric Company 2D" successfully sends us to the regular Tennessee Avenue (resetting our doubles count), but a roll of ⚃⚃ lands us In Jail (or on the virtual space "In Jail (1)" for the "long jail" strategy).

Since this model effectively adds two additional spaces per original space on the board, the matrix becomes way too big for this book (the versions we have already seen just barely fit on the page in a readable format) and, since the probabilities obtained do not differ much at all from the correct probabilities, the instructive nature of our models outweighed the slightly more correct version with 120 real or virtual spaces. In Tables T.4 and T.5 toward the end of the book, however, you will find the steady-state vectors for both the "short jail" and "long jail."

6.4   Spread, Revisited

Illinois Avenue does not owe its popularity entirely due to the Chance card that sends players there; after all, Boardwalk and St. Charles Place have that same advantage. The other thing that Illinois Avenue has going for it is its proximity to jail (where, of any other space, players spend the most time).

We haven't discussed the expected value of rolling one die yet and, at this point, this should be an easy exercise for us to do. Using Definition 2.4, we obtain

$$\underbrace{\left(\frac{1}{6}\right)\cdot 1}_{\text{roll 1}}+\underbrace{\left(\frac{1}{6}\right)\cdot 2}_{\text{roll 2}}+\underbrace{\left(\frac{1}{6}\right)\cdot 3}_{\text{roll 3}}+\underbrace{\left(\frac{1}{6}\right)\cdot 4}_{\text{roll 4}}+\underbrace{\left(\frac{1}{6}\right)\cdot 5}_{\text{roll 5}}+\underbrace{\left(\frac{1}{6}\right)\cdot 6}_{\text{roll 6}}=\frac{7}{2}=3.5$$

as the average (expected value) of a single die roll. Of course, we can't actually *roll* a 3.5, but expected values are not usually attainable. In Monopoly, instead of rolling just one die, we roll two per turn. It should agree with your intuition that if we roll two dice, where each die has an expected value of 3.5, the expected value for the sum of the two dice involved should be $3.5 + 3.5 = 7$. Indeed this is what we saw in Figure 1.2; of the 36 physical outcomes possible for rolling two dice, the most common sum was 7; this also happens to be the expected value since we can compute

$$\underbrace{\left(\frac{1}{36}\right)\cdot 2}_{\text{roll 2}}+\underbrace{\left(\frac{2}{36}\right)\cdot 3}_{\text{roll 3}}+\underbrace{\left(\frac{3}{36}\right)\cdot 4}_{\text{roll 4}}+\underbrace{\left(\frac{4}{36}\right)\cdot 5}_{\text{roll 5}}+\underbrace{\left(\frac{5}{36}\right)\cdot 6}_{\text{roll 6}}+\underbrace{\left(\frac{6}{36}\right)\cdot 7}_{\text{roll 7}}$$

$$+\underbrace{\left(\frac{5}{36}\right)\cdot 8}_{\text{roll 8}}+\underbrace{\left(\frac{4}{36}\right)\cdot 9}_{\text{roll 9}}+\underbrace{\left(\frac{3}{36}\right)\cdot 10}_{\text{roll 10}}+\underbrace{\left(\frac{2}{36}\right)\cdot 11}_{\text{roll 11}}+\underbrace{\left(\frac{1}{36}\right)\cdot 12}_{\text{roll 12}}=7$$

to match what we thought earlier.

It makes sense that on the roll of *four* dice (two rolls of the dice in Monopoly), the expected sum is 14 (given here without justification). By looking at the Monopoly board of Figure 6.10, you will notice that 14 spaces away from Jail is our favorite Illinois Avenue!

Getting a sum of 14 on four dice (or two iterations of rolling two dice) is also the most common (and highest probability) result. Of the $6^4 = 1,296$ different physical outcomes for rolling four dice, exactly 146 of them add to fourteen.[6] This probability of $p = 146/1296$ for getting a sum

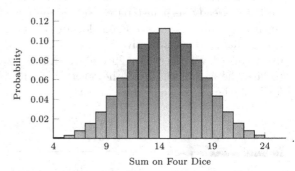

FIGURE 6.19: Distribution of Sums for Four Dice

of fourteen, along with the probabilities for the other possible sums, are shown in Figure 6.19. The standard deviation for the *sum* after rolling 4 dice hap-

---

[6]You will verify this as an exercise.

pens to be about 3.42,[7] so you might reasonably assume that a likely range for your number of spaces away from jail after two turns would be

$$14 \pm 3.42 = (14 - 3.42, 14 + 3.42) = (10.58, 17.42),$$

or anywhere from Kentucky Avenue (or Free Parking) to Ventnor Avenue (or Water Works), with Illinois Avenue the most likely.

FIGURE 6.20: Fourteen Distribution

Instead, we might just look at how often we actually do roll a sum of 14 with four dice. Since, on any given roll of the four dice (or twice rolling two dice) the probability is fixed at 146/1296, we can use the binomial distribution of Definition 4.1 to see that the probability of obtaining fourteen k times on n trials is given by the formula

$$\text{Probability of } k \text{ Fourteens in } n \text{ Trials} = \binom{n}{k} \cdot \left(\frac{146}{1296}\right)^k \cdot \left(\frac{1150}{1296}\right)^{n-k}$$

where the case when $n = 50$ is pictured in Figure 6.20 for various values of k that have probabilities high enough above zero to be shown.

Using Definitions 4.2 and 4.3 for the expected value and standard deviation for binomial distributions, here we get an expected number of fourteens in 50 trials of

$$\mu = n \cdot p = 50 \cdot \frac{146}{1296} \approx 5.634$$

with a standard deviation of about

$$\sigma = \sqrt{n \cdot p \cdot q} = \sqrt{50 \cdot \frac{146}{1296} \cdot \frac{1150}{1296}} \approx 2.236$$

meaning that, if we assume $n = 50$ trials (perhaps assuming that in our games of Monopoly players will end up in jail about 50 times), about 68% of the time players will experience a number of fourteens in the interval

$$5.634 \pm 2.236 = (5.634 - 2.238, 5.634 + 2.238) = (3.396, 7.872).$$

That is, we'll get a few extra stays in our Illinois Avenue hotel just because of its location. You are free, of course, to vary the value of n to see how this number changes.

---

[7]Consult an advanced probability book for more details; the probability distribution for rolling one die is a discrete uniform distribution for which the standard deviation is well-known, and the standard deviation for the sum of four independent such distributions is obtained by adding the variances (the square of standard deviation) together from those for distribution, and taking the square root.

6.5   Exercises

6.1 Imagine a game with two spaces, A and B, where the probability of moving from A to B is 0.40 and the probability of moving from B to A is 0.75 (the probability of staying on space A is therefore 0.60 and that for staying on space B is 0.25).

    (a) Draw a probability diagram illustrating this game.

    (b) Create the probability matrix, P, that describes this game.

    (c) Compute P^2 and use it to determine the probability of starting on space A and ending on space B after two moves.

    (d) What is the probability of starting on space A and ending on space B after *three* moves?

    (e) If you have access to a computer to compute matrix powers, compute higher and higher powers of P until your matrix stabilizes; use the resulting matrix to give the long-term probability of being on space A and being on space B.

6.2 Repeat Exercise 6.1 for a game where the probability of moving from A to B is 0.60 and the probability of moving from B to A is also 0.60; do you find these results surprising? Why or why not?

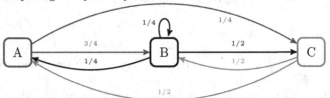

FIGURE 6.21: Probability Diagram for Exercise 6.3

6.3 Now imagine the game with three spaces, A, B, and C, with the probability diagram given in Figure 6.21.

    (a) Create the probability matrix, P, for this game.

    (b) Again, compute P^2 and use it to determine the probability of starting on space B and ending on space C after two moves.

    (c) What is the probability of starting on space C and ending on space A after *three* moves?

    (d) If you have access to a computer, compute higher and higher powers of P until you are able to determine the steady-state vector for this game. If this game were like Monopoly, which of spaces A, B, or C would you prefer to own?

6.4 Recall the game given in Figure 6.5; how would this diagram change if you were also allowed to reach "End" by rolling a 2 if you are currently on space D? Give the modified probability matrix for the game with this change.

6.5 Again, recall the game given in Figure 6.5. Now, suppose that you were forced to "wrap around" if you did not reach "End" by an exact count; that is, for example, rolling a 2 from space E sent you past "End" back to "Start" and rolling a 5 from space D would send you past "End," through "Start," ending on space B.

    (a) How would you modify the probability diagram for this version of the game?

(b) Give the modified probability matrix for this version of the game.

(c) If you have access to a computer, compute higher and higher powers of your probability matrix to determine the new steady-state vector; how does this vector compare to the vector for the original game?

6.6 The Pay Day board of Figure 6.8 pictures four spaces with "Buyer" where, as a player, you can sell one of the Deal cards you may have purchased earlier in the game. These spaces are days 9, 17, 20, and 30.

(a) As in Example 6.4, determine the probability of landing on a "Buyer" square on any given roll, and then determine the probability you will land on a "Buyer" square in any given month of the game.

(b) The Diamond Ring card costs $300 to purchase, but is worth $600 if you are able to sell it before the end of the game. Unsold Deal cards when the game ends are worth nothing. Determine the probability p you would need of landing on a "Buyer" square to make purchasing the Diamond Ring card have positive value.

(c) Assuming you currently have no other Deal cards, how many rolls of the game possibly remaining would you want to have, to make purchasing a Diamond Ring card viable?

6.7 The probability of rolling doubles three times in a row during a game of Monopoly is given by $\left(\frac{1}{6}\right)^3 = \frac{1}{216}$ since the chance of rolling doubles on a single throw of the dice is $\frac{1}{6}$. Imagine using a special pair of dice, where one die has sides given as 1, 2, 2, 3, 3, and 4 while the other die has sides given as 1, 3, 4, 5, 6, and 8 (these are called the Sicherman Dice, and a table of outcomes is given in Table 11.6 on page 360).

(a) Determine the probability of rolling doubles once on a single throw of the dice.

(b) Determine the probability of rolling doubles three times in a row using these dice and comment on how this is different from using standard dice.

6.8 Using the "short jail" strategy for Monopoly, use the probabilities of Table 6.8 and data in Table 6.9 to do the following:

(a) Verify the expected values (for hotels) of the color groups given in Table 6.10 for the colors we did not do in Example 6.10.

(b) Create a table like Table 6.10 that shows the expected values for base rent, one house, two houses, three houses, and four houses. How does the "advantage" for the red properties compared to the yellow properties change as the properties are improved?

6.9 In Monopoly, similar to Example 6.11, determine the best and worst-two railroad subsets using probabilities and then compute the expected values for these subsets.

6.10 Now, using the "long jail" strategy in Monopoly, use the probabilities of Table 6.11 (and data in Table 6.9) to determine the expected value of hotels for each color group as we started in Example 6.12.

6.11 Verify that there are 146 physical outcomes when rolling four dice that result in a sum of fourteen; you can do this via clever counting as opposed to explicitly listing all of them!

6.12 The following exercises use the *real* steady-state values of Table T.4 for the "short jail" strategy in Monopoly.

(a) Use the steady-state values, along with the data of Table 6.9, to find the expected values, for hotels, of each of the eight color groups.

(b) Determine the expected values for base rent, one house, two houses, three houses, and four houses.

(c) Similar to Example 6.11, find the best and worst three-railroad subsets and then find the best and worst two-railroad subsets.

(d) How do your results compare to using the slightly imperfect steady-state values of Table 6.8?

6.13 Repeat Exercise 6.12 for the "long jail" strategy in Monopoly by using the *real* values of Table T.5.

7

Sports Mathematics: Probability Meets Athletics

A discussion about mathematics and games should not leave out sports. Success in various sports can oftentimes be connected with mathematics, probability, and statistics. The popular book, later made into a film, *Moneyball* highlights the story of the general manager, Billy Beane, of the Oakland Athletics (of Major League Baseball) using data to revolutionize how decisions are made in the big leagues. This current book does not aim to replicate that and

FIGURE 7.1: Tennis in Action

will stay true to our focus of probability. If you are interested in learning more about the mathematics of sports, such as how the shape of a ball can determine its flight path or how it is not literally possible to "keep your eye on the ball" when batting in baseball, check out [31] by my colleague and friend Roland Minton, who discusses "sports math" in a friendly way that covers these two questions, as well as a short introduction to the popular area of sports analytics. For more on the latter, specifically, [47] provides an overview of those methods.

7.1 Sports Betting

From small office bets about the Super Bowl to official sports books in Las Vegas and around the world, sports betting has been around for a long time. Its popularity has soared recently with more and more access being granted to gamblers around the world by moving betting systems online. While you may have visions in your head, popularized by pop culture, of bookies in smoke-filled rooms answering phones to take bets and "enforcers" paying hurtful visits to bettors that have fallen behind on their payments, today's operations are much more streamlined with flashy websites and promotions driving the industry.

Being successful at betting on sports involves more skill and less probability than the rest of this book. Inside knowledge and an extreme familiarity with the sport, and its players, can help a bettor make more successful bets. However, even an amateur at sports betting can be successful and enjoy watching a game knowing that some of their money is on the line. With that, it's important to offer a few words of warning. First, know that the book maker's job is to have more inside knowledge and more extreme familiarity with the sport than virtually anyone else; their goal is to set the payouts for any wager so that the sports book is guaranteed to make money in the long term. Second, any form of gambling should always be thought of as entertainment and you should never spend more on entertainment than you can afford. What follows in this section is an exploration of how sports betting works, in terms of how to wager and how the book maker "guarantees" their share of the money, and the mathematics that comes into play.

Moneyline and Spread Betting

As I'm writing this chapter in August of 2020, Major League Baseball (MLB) is just starting its season in a very different way due to the COVID-19 pandemic. On Monday, August 3, the Philadelphia Phillies played the New York Yankees, and the moneyline odds on that game were +260 for the Phillies and −300 for the Yankees according to one Las Vegas casino.

What does this mean for placing bets? For the Phillies, the presence of the + sign means that they are the underdogs for the game and not expected to win. For betting, the +260 means that a bettor would wager $100 and, if the Phillies were to win, would walk away with $260 in winnings (in addition to receiving their original $100 back). On the Yankees side, the − sign means that they are favored to win the game, and the −300 means that a better would need to wager $300 in order to win $100 (again, also receiving their original wager back). Take a moment to note that the amount wagered and the amount won is *different* based on whether the sign is + or − on a bet.

Using the casino payouts we discussed in Definition 2.3, for the Phillies, we would list the casino odds as 260:100 (for every $100 wagered, you win $260) and, for the Yankees, we would list them as 100:300 (for every $300 wagered, you win $100).

Surely, the odds established by the book maker at the casino should encode some information about their guess about the probability that the Yankees will win the game. Let's explore this a bit. First, let p represent the probability that the Yankees win; then, the expected value of a bet on the Yankees is

$$\text{Yankees Bet Expected Value} = \underbrace{\$100 \cdot p}_{\text{win}} + \underbrace{-\$300 \cdot (1-p)}_{\text{lose}} = 400p - 300$$

where we ignore the possibility of a tie. Similarly, the probability that the Phillies win would then be $(1-p)$, so their expected value is

$$\text{Phillies Bet Expected Value} = \underbrace{\$260 \cdot (1-p)}_{\text{win}} + \underbrace{-\$100 \cdot p}_{\text{lose}} = -360p + 260.$$

Assuming that the casino wants to make money, each of these should be negative; for the Yankees, solving

$$400p - 300 < 0$$

gives $p < 0.750$ whereas for the Phillies, solving

$$-360p + 260 < 0$$

yields $p > 0.722$. Unlike what we saw in Chapter 2, rather than the probability of something happening driving the casino's choice for payout, it appears that the payouts are driving the probability that an event happens (such as the Yankees winning). However, this is not really the case. The two expected values combine to show that $0.722 < p < 0.750$ if the casino wants to make money on both of these wagers in the long run. What *is* happening is that the book maker has decided that the probability that the Yankees win (based on strength of the team and how the Yankees pitcher will do versus the Phillies in combination with how the pitcher for the Phillies will fare against the Yankees) is somewhere in the middle of that range, perhaps around $p = 0.736$. Using this value of p in the expected value formula for a bet on the Yankees gives $-\$5.60$ as the expected value. You will see a similar value if you check the Phillies expected value as well. This means that on each wager, the casino is taking about $5, give or take a little, where the give or take is due to rounding.

The "house advantage" here is the service charge, commonly called the "vigorish" or "vig" in sports betting parlance, which the book maker or casino takes on each wager. Note that the mathematical average of the 260 and 300 in this example, ignoring signs, is 280, and if the casino offered odds of 280:100 for the Phillies and 100:280 for the Yankees, the expected value calculations done earlier for the two teams would result in a single value for p and not a range. In fact, if we naively apply Theorem 2.2 to these odds, we would see an estimated value of p as $\frac{280}{100+280} \approx 0.736$, the same value that was in the center of the range when the actual sports bet moneyline odds were used. If the Phillies win, rather than getting $280, the casino only pays $260, and on the Yankees, rather than charging $280 to possibly win $100, they charge $300. The casino is making $20 on each bet! For the curious, the Yankees did win that game, by a score of 6 to 3.

Example 7.1 Consider the Pittsburgh Pirates playing the Minnesota Twins where the moneyline odds are $+200$ for the Pirates and -220 for the Twins. What are the payout odds, the service charge, and what probability does the book maker estimate of the Twins winning?

First, recall that, since the Twins are favored, the payout will only be $100 for each $220 that we wager, so the payout odds for the Twins are 100:220; similarly, since the situation is flipped for the Pirates, the payout odds would be 200:100. By looking at the midpoint between 200 and 220, we would assume that the book maker is charging $10 on each bet and really considers the probability of the Twins winning as $p = \frac{210}{100+210} = \frac{21}{31} \approx 0.677$.

For the curious, we can also compute the expected value of each wager, based on this value of p. For the Twins, we get

$$\text{Twins Bet Expected Value} = \$100 \cdot \frac{21}{31} + -\$220 \cdot \left(1 - \frac{21}{31}\right) \approx -\$3.23$$

and for the Pirates we get

$$\text{Pirates Bet Expected Value} = \$200 \cdot \left(1 - \frac{21}{31}\right) + -\$100 \cdot \frac{21}{31} \approx -\$3.23$$

as well.

Of course, we can generalize this to an arbitrary moneyline wager.

Theorem 7.1 (Moneyline Facts) Suppose that Team A is the underdog in a match up, with moneyline payout given as $+A$ and that Team B is the favored team to win the match, with moneyline payout of $-B$. Then:

1. The payout odds for Team A are $A{:}100$ and for Team B are $100{:}B$.

2. Setting $T = \frac{A+B}{2}$, the perceived true odds for Team A are $T{:}100$ and for Team B are $100{:}T$.

3. The perceived probability of Team B winning is $p = \frac{T}{100+T}$.

4. The service charge, or vigorish, is $S = \frac{|A-B|}{2}$.

5. The expected value of a wager on either Team A or Team B, assuming the value of T and S are as above, is

$$-\frac{100 \cdot S}{100 + T}.$$

We close out the moneyline discussion for now by using Theorem 7.1 on a different casino's payouts for the Phillies and Yankees game discussed earlier. For that same game, a different sports book had the Phillies at $+245$ and the Yankees at -280. Using Theorem 7.1, we see that the payout odds for the Phillies are $245{:}100$ and that for the Yankees is $100{:}280$, with perceived odds of $262.5{:}100$ and $100{:}262.5$, respectively. The book maker estimates the probability of the Yankees winning at $p = \frac{262.5}{362.5} \approx 0.724$, a bit lower than the other casino we looked at earlier. Similarly, the vigorish is smaller at just $\$17.50$ as opposed to the $\$20$ from the other casino, which gives a larger (less negative) expected value of about $-\$4.83$.

As a segue to our discussion of "the spread" it's worth noting that sometimes *neither* team in a moneyline bet is favored and the book makers are giving the teams an equal chance of winning, for a value of p at or very close to $p = 0.50$. In this case, the moneyline will be shown as -110 for both teams.[1] In this case, Theorem 7.1 does not apply. Here, the payout odds for both teams is 100:110, you'll need to wager \$110 to have a chance at winning \$100. We can then see that the expected value is

$$\text{Expected Value} = \underbrace{\$100 \cdot 0.50}_{\text{win}} + \underbrace{-\$110 \cdot 0.50}_{\text{lose}} = -\$5.00$$

for a service charge of \$10, as expected given that the true odds for two evenly matched teams should be 100:100, and by requiring players to bet odds of 100:110, the book maker is keeping that extra \$10.

Spread betting is not all that different from the moneyline wagers discussed previously since, in the most common form of spread betting, rather than the book maker setting a moneyline wager where the favored team requires more money wagered just to win \$100, the book maker lists a "point spread" that a team must cover in order to make the two sides of a wager as equal as possible. This is commonly used for sports with point totals that can get somewhat high, such as in the National Football League or the National Basketball Association.

For instance, consider the 2008 Super Bowl (Super Bowl XLII) between the New York Giants and the New England Patriots. The final spread for that game was "Patriots -12" which means that in order for a bet on the Patriots to win, not only would they have to beat the New York Giants, but they would have to beat them by 12 points; similarly, a bet on the Giants to "win" would mean that they trail the Patriots by less than 12 points at the end of the game (or, of course, win the game, which they ended up doing by a score of 17 for the Giants and 14 for the Patriots). In the 2005 Super Bowl, the final spread was "Patriots -7" but, for that event, even though the Patriots won the game 24 to 21, a bet on the Patriots would have lost since, by subtracting seven from the Patriots' score, they "lose" by the spread-adjusted score of 17-21.

Both of the bets (spread included) would have equivalent moneylines listed as -110, requiring players to use betting odds of 100:110 as we just saw at the end of the previous section. You might think that the goal of the book maker is to end up with p, the probability of a team winning, with the point spread included (called covering the spread), as close as possible to 0.50. However, the real goal of the book maker is to set the point amount to be covered in order to encourage bettors, as a whole, to wager equal amounts of money on each side. The end result of this would be that the money from the losing bets would be used to cover the money won by anyone with a winning wager, and the book maker or casino would keep the \$10 service charge levied on each

---

[1]Occasionally, some book makers will use -105 for one team and -115 for the other team, mainly to encourage betting on the team with the better odds in order to make sure the money wagered on each team does not get too far out of whack; other book makers may use -120 (or something close) for both teams in order to capture a higher service charge.

wager by using the moneyline odds of 100:110. The same is true of other bet types similar to point spreads, such as over/under wagers. Since we've already discussed betting when the odds are set at 100:110 for two "evenly matched" wagers, there is nothing much else to mention here.

Horse Racing

What we've looked at so far in terms of sports betting thus far offers the possibility that the book maker or casino loses money on a given wager if not enough losing money is collected in order to cover the winning wagers. This can happen when the person setting the odds does not keep up with betting trends and can especially happen if some large, late, bets come in. It can also happen when the outcome is surprising; the Giants' upset of the Patriots in 2008 mentioned earlier resulted in an estimated loss of over $2.5 million for Las Vegas Casinos!

In contrast, horse racing uses a system know as parimutuel betting; in this system, the actual winnings are taken from all of the money collected on bets, minus a take for the track or state. Related systems are used for other sports or games, such as the state lotteries we explore in Chapter 9.

While there are many exotic bets that are possible in horse racing, we only discuss the standard bets of win, place, and show here; a horse that wins, of course, finishes first in a given race. To place, a horse need only finish first or second, and to show a horse must finish in one of the top three spots. We proceed through a hypothetical "horse race" as presented in [33], mostly because I enjoy the horse names and methodology that appears there. Consider a race with five horses, Double Six, Bi-Nomial, Flip Ahead, Roll Four, and None Above, where the requirement for winning is given in Table 7.1

TABLE 7.1: Horses, Post Positions, and Rules for Winning

Post Position	Horse	Requirement for Winning	Odds
1	Double Six	Roll double six with a pair of dice	35:1
2	Bi-Nomial	Exactly 3 heads in 4 flips of a coin	109:35
3	Flip Ahead	Get a head in 1 flip of a coin	61:35
4	Roll Four	Roll a four on a single die	541:35
5	None Above	No other horse won the round	401:175

In horse racing, professionals at the track set what are known as the morning odds; these are typically based on the history of the horse, the jockey scheduled to ride the horse, and any track conditions that favor or provide a disadvantage for the jockey-horse combination. In our example, we find the morning odds in post order. By now, we know that the probability of rolling a pair of sixes is $\frac{1}{36}$, providing morning odds of 35:1 for Double Six. In order for

Bi-Nomial to win, we need Double Six to lose first, and then the requirement of Bi-Nomial to happen, so we have

$$\text{Bi-Nomial Win} = \underbrace{\left(\frac{35}{36}\right)}_{\text{Double Six loss}} \cdot \underbrace{\binom{4}{3}\left(\frac{1}{2}\right)^4}_{\text{3 heads, 4 flips}} = \frac{35}{144}$$

which gives morning odds of 109:35 as seen in Table 7.1. As an exercise, you will verify the remaining odds in that table.

It's worth noting that the morning line that appears at the track would likely be "prettier" than the true odds we can compute. For instance, the line for Roll Four might appear as 15:1. In parimutuel betting, since the winnings are paid out of the money collected on all bets, the odds displayed when you place a wager may be very different than the morning line, as when money is collected for a wager, the odds are recomputed by the *totalizer* which takes into account the amount wagered on each horse for each bet and displays the live odds for each horse. Bettors can look at the final odds (those displayed at the start of a race) to help figure out what a potential payout might be.

TABLE 7.2: Win, Place, and Show Pools with Totals

Post	Horse	Win $	Place $	Show $
1	Double Six	975	1000	600
2	Bi-Nomial	4000	3000	1300
3	Flip Ahead	4300	4000	4500
4	Roll Four	1070	400	100
5	None Above	2155	1600	1000
Totals		12500	10000	7500
−20% for Track/State		2500	2000	1500
Total for Payouts		10000	8000	6000

What appears in Table 7.2 is a list of hypothetical bet totals for each of win, place, and show for our five horses. Note that the 20% take for the track or state is deducted from the totals, showing the final pool of money to give out to winning bets. We use Roll Four as our favored horse (the one that we probably placed bets on) in order to illustrate how the actual payout is determined.

Example 7.2 To figure out how much a "win" bet would pay, should Roll Four actually win the race, we take the total amount wagered on "win" which, from Table 7.2, was $10000 and divide it by the amount wagered on Roll Four to win, getting $10000/1070 = 9.346$ as the initial dollar return on a $1 wager; since prices displayed at race tracks are typically based on a $2 ticket, we double this amount getting 18.692 and then round *down* to the nearest ten cents to get $18.60 as the payout (which includes the original wager) for a $2 win bet on Roll Four.

The computation for both place and show are more complicated, and we only look at how to compute the payout for a show bet. Suppose that Roll Four, None Above, and Bi-Nomial are the winning, placing, and show-ing horses for the race. First, from the $6,000 pot for show bets, in order to guarantee that successful show bets at least get their money back, we subtract $2,400 from the pot (this amount comes from pooling the $100 wagered on Roll Four to show, $1,000 wagered on None Above to show, and the $1,300 wagered on Bi-Nomial to show), giving us only $3,600 to work with. Splitting this three equal ways for each of the three showing horses gives us $1,200 for winnings (beyond the initial wager) for a show bet on Roll Four. Since $100 was the total amount wagered for this horse to show, we end up with a payout of $1200/100 = 12$ per dollar wagered, which, doubled, gives us $24 as the additional amount above the original bet. Our final payout for a show wager on Roll Four is $26 once that original bet is added back.

There are two observations from Example 7.2 that are worth mentioning. First, if you take the $18.60 paid out to the 535 successful $2 win bets on Roll Four, you only get $9,951; the track takes an extra $49 in as its take due to the way the final payout is rounded (this extra amount is called the *breakage*). Second, you might also notice that a $2 wager on Roll Four to *show* pays better than the $2 wager on Roll Four to *win*; as you might imagine, it is much easier for a horse to show than it is for a horse to win, so this juxtaposition of payouts is surprising and completely reinforces the notion that payouts for horse races are based entirely on how much is wagered on each horse and how those wagers are made. In our example, only 1% of the show bets were on Roll Four, while over 8% of the win bets were on Roll Four. If the idea of horse race betting is interesting to you, more details can be found in [33].

7.2 Game Theory and Sports

To see how game theory can apply to sports, consider soccer and a new, young, player who kicks with their right foot. For those of you that have played soccer or similar games, this means that kicks aimed to the left tend to be stronger and more accurate than those that are kicked to the right. Penalty kicks in soccer are a simple one-on-one play where the kicker tries to get the ball past the goalie.

For our simple model, adapted from [31], since the action moves so fast that the goalie does not have time to look and see what direction the ball has been kicked, think, and then move, the goalie is relying on instinct to guess the direction that the ball has been kicked. Suppose that a kick to the left is successful only 60% of the time when the goalie correctly predicts a left kick, but is 100% successful when the goalie guesses wrong, and if the kicker takes a risk and kicks a less accurate shot to the right, the kick is successful 20% of the time when the goalie guesses right, and only 40% successful if the goalie guesses left but the kick was made to the right.

Using the kicker as the row player and the goalie as the column player, we can summarize this information into the following zero sum game.

	GL	GR
KL	60	100
KR	40	20

Does it make sense that the column player's payoff amounts should be the negative of these? Well, as positive numbers, these represent percentages or probabilities, which certainly seems okay, but as negative numbers, as probabilities, something slightly off seems to be happening. What's important here is that the kicker, having 60 and 100 in the first row, would prefer that the goalie pick the second column (higher chance of making the goal) and, on the flip side, the goalie would prefer to pick the first column; we can reasonably assume that each player prefers a column by the same amount, and they just prefer the opposite columns. For the numbers in a game theory matrix to make sense, only relative differences and ratios are required to make a zero sum game's number consistent.[2] Further, as we saw in Exercise 5.12, all constant sum games are equivalent to zero sum games, and this could easily be thought of as a constant sum game (for the KL and GL entry, we could have written (60, 40) to represent the probability the kicker and goalie, respectively, prevail, and all ordered pairs would sum to 100).

By now, this game is easy to solve; the first row dominates the second row, so the kicker will always choose to kick left; given that, the goalie will always choose to go left, which results in a goal 60% of the time. A better model uses more realistic numbers.

Example 7.3 Consider the following soccer game matrix that uses more realistic probabilities, acknowledging that an unblocked kick to the right is likely much more successful than the 40% in the previous model, even for a young player.

	GL	GR
KL	64	94
KR	89	44

First, a quick check for dominance reveals that none exists, and since

---

[2] We did not cover this in Chapter 5, but coming up with the numbers in a game matrix uses the concept of *utility*, which you learn more about by consulting introductory economics books or by reading the chapter on utility in [49].

the maximin is 64 and the minimax is 89, there is no saddle point. The solution will be a mixed strategy courtesy of Theorem 5.4. For the kicker (row player), you should verify that the solution is ($\frac{3}{5}$KL, $\frac{2}{5}$KR) and for the goalie (column player), the solution is ($\frac{2}{3}$GL, $\frac{1}{3}$GR). Using the goalie column probabilities and the KL row, we get the value of the game as

$$\text{Value of the Game} = \underbrace{\left(\frac{2}{3}\right)\cdot 64}_{\text{GL}} + \underbrace{\left(\frac{1}{3}\right)\cdot 94}_{\text{GR}} = 74$$

which indicates that if each player does their optimal strategy, about 74% of the penalty kicks will be successful. As usual, recall that if either player chooses to deviate from their optimal strategy, the *other* player will gain an advantage.

How realistic are the numbers of Example 7.3? In 2008, researchers Palacios-Huerta and Volji used data from over two thousand penalty kicks taken from various professional soccer leagues in Europe to derive estimates for the first row of 60 and 95 (versus the 64 and 94 we used) and for the second row of 90 and 70 (versus the 89 and 44 we used), which can be found in [34]. In a laboratory experiment, they found that players selected their options with probabilities very close to the optimal strategy for their numbers, which you will discover in Exercise 5.6.

Tossing the ball into a different court, researchers Walker and Wooders, in [57], found that professional tennis players also play using strategies predicted by game theory. They analyzed videotapes of ten tennis matches that included highly ranked players from major tournaments (most of the data came from the final match of a championship match). The setup here is that the serving player can choose to aim their serve towards their opponent's left or their opponent's right, and the opponent needs to make a guess about where the serve might be heading (if their guess is wrong, they'll spend extra energy to maneuver to the correct side, leading to a lessened probability of winning the point).

Example 7.4 An example matrix, with entries representing the probability that the server wins the point, is given below:

	GL	GR
SL	0.58	0.79
SR	0.73	0.49

Note that the rows represent the server, choosing to serve left (SL) or right (SR). Again, we check for dominance first and notice that none exists among rows or columns. Similarly, upon checking for saddle points, none are to be found. Theorem 5.4 applies in order for us to find a mixed strategy, which yields ($\frac{8}{15}$SL, $\frac{7}{15}$SR) for the serving (row) player and ($\frac{2}{3}$GL, $\frac{1}{3}$GR) for the receiving (column) player, with a value of 0.65, meaning the server can expect to earn the point about 65% of the time, assuming both players use their game-theoretic solution.

7.3   Probability Matrices and Sports

In Chapter 6, we discussed using probability matrices to model board game movement. There are several sports where it can make sense to model the flow of a game using a probability matrix; baseball is a prime example of one of those sports. Spaces on an imaginary game board for baseball would indicate which bases have players on them as well as the number of outs for the current half-inning (top or bottom). The hitting team would start on the "zero outs and zero players on bases" space and move around the board as batters come to the plate. A double on the first play would move the team's pawn to "zero outs and a player on second" which means that a double on the second play would, in addition to scoring a run for the team, keep the pawn on to "zero outs and a player on second." If the team already had two outs and had a player on first base and third base, the pawn would be at "two outs and players on first and third" as another example.

For zero outs, there are eight possible configurations for players on bases.[3] Since a team could have either zero, one, or two outs, this results in a total of 24 spaces on our virtual game board for baseball so far; however, four more spaces need to be added which correspond to ending the half-inning. These spaces indicate getting a third out, with scoring either zero, one, two, or three runs with that play (while no run is scored if the half-inning is ended by a forced out or the batter never reaching first base, a third out can happen for other reasons, in which case up to three players might score).

Before we get into looking at what this looks like for the MLB, let's consider a simpler, but similar, game called One Base which appears in [31]. In this game, there is only one base (thus the name of the imaginary game) and a batter will either get a single, a home run, or an out. This miniature version of baseball ends a half-inning on only two outs, so we have fewer spaces to consider. The main spaces on our virtual board are "0 outs, 0 on," "0 outs, 1 on," "1 out, 0 on," and "1 out, 1 on" and we will consider a batter's probability of making a home run, a single, and an out, respectively, as 0.1, 0.2, and 0.7. We only need one space, called "2 outs," to end the half-inning, since our players will never be able to end the half-inning and also score a run given our batter's limited capabilities. Let's look at how the theory from Chapter 6 applies here.

---

[3]No players on, a player on first, on second, or on third, players on first and second, first and third, or second and third, or bases loaded.

Example 7.5 In order to determine the long-term steady state for this game, we need a transition matrix. Using the probabilities given, that ends up being

$$P = \begin{bmatrix} 0.1 & 0.2 & 0.7 & 0.0 & 0.0 \\ 0.1 & 0.2 & 0.7 & 0.0 & 0.0 \\ 0.0 & 0.0 & 0.1 & 0.2 & 0.7 \\ 0.0 & 0.0 & 0.1 & 0.2 & 0.7 \\ 0.0 & 0.0 & 0.0 & 0.0 & 1.0 \end{bmatrix}$$

where the row and column order is "0 outs, 0 on," "0 outs, 1 on," "1 out, 0 on," "1 out, 1 on," and "2 outs." For instance, the fourth row represents "1 out, 1 on" and the batter will hit a home run with probability 0.1, moving the state to "1 out, 0 on" (since both the runner on base and the batter will score), the batter will hit a single with probability 0.2, keeping the state at "1 out, 1 on" (since the batter will end up on base and the runner on base will score), and the batter will get an out with probability 0.7, moving the state to "2 outs" which ends the half-inning. Note that from the first two rows, it is not possible to reach "1 out, 1 on" or "2 outs" since with zero outs, the only possibility is that the batter will get an out (ending at "1 out, 0 on"), end up on base (ending at "0 outs, 1 on"), or hit a home run (staying at "0 outs, 0 on"). The last row represents "2 outs," a state that marks the end of the half-inning and cannot be moved from by batter actions.

The matrix P gives the probability of moving to any other state given a current state with just the current batter; using matrix multiplication, we can compute P^2 which represents starting in a state (given by a row) and ending in a state (given by a column) after *two* batters. For this, we get

$$P^2 = \begin{bmatrix} 0.03 & 0.06 & 0.28 & 0.14 & 0.49 \\ 0.03 & 0.06 & 0.28 & 0.14 & 0.49 \\ 0.00 & 0.00 & 0.03 & 0.06 & 0.91 \\ 0.00 & 0.00 & 0.03 & 0.06 & 0.91 \\ 0.00 & 0.00 & 0.00 & 0.00 & 1.00 \end{bmatrix}$$

where we notice that the probability of ending the inning, starting with zero outs (regardless of whether or not a player is on base), is about 49% as we can see in the right-most column of the first and second rows.

The steady-state vector should not be too surprising; baseball half-innings, and therefore One Base half-innings, do eventually end, and if you compute P^7, or any higher power, and round to two decimals, you'll see that the steady-state vector is

$$P_S = \begin{bmatrix} 0.00 & 0.00 & 0.00 & 0.00 & 1.00 \end{bmatrix}$$

indicating that, yes, half-innings do come to an end.

We will look at the MLB soon, but there are two natural questions that arise from our study of One Base.

Question

What is the expected number of runs scored for each state and how long is an expected half-inning for One Base?

To answer the first question, let a represent the expected number of runs for the state represented by the first row of the matrix P in Example 7.5 ("0 outs, 0 on"), with b, c, d, and e that for the other four rows, in order. The *current* batter who starts with no outs and no runners on base will only score a run by hitting a home run, with probability 0.1, so the expected number of runs for *just that batter* is $1 \cdot 0.1 = 0.1$. However, future runs may be scored by other batters in the round, so we have

$$a = 0.1 + 0.1a + 0.2b + 0.7c + 0.0d + 0.0e$$

where the coefficients come from P which, again, gives the probability of reaching a given state from a current state. This makes it easy to write down an expression for c since, again, a batter in that current situation only expects to score 0.1 runs on average, so we have

$$c = 0.1 + 0.0a + 0.0b + 0.1c + 0.2d + 0.7e$$

as the next piece in our puzzle.

From states two and four ("0 outs, 1 on" and "1 out, 1 on"), the *current* batter can achieve two runs for their team by hitting a home run (two runs with probability 0.1 gives 0.2 expected runs as the contribution) or one run by hitting a single (one run with probability 0.2 gives 0.2 expected runs as the contribution for this possibility). Adding the two means that 0.4 runs is the expected, average, contribution to runs by the *current* batter, so we end up with

$$b = 0.4 + 0.1a + 0.2b + 0.7c + 0.0d + 0.0e$$

and

$$d = 0.4 + 0.0a + 0.0b + 0.1c + 0.2d + 0.7e$$

as two additional pieces to our almost-solved puzzle. It should be no surprise that there are no runs to be scored once we reach the last row, so

$$e = 0.0 + 0.0a + 0.0b + 0.0c + 0.0d + 1.0e$$

which is a really fancy way of writing $e = 0$ but works well in keeping our equations parallel.

Solving for the values of a, b, c, d, and e is a nice linear algebra task, which you can read more about in [39]; the computed solution when solving this system of equations is $a = 0.52$, $b = 1.03$, $c = 0.23$, $d = 0.53$, and $e = 0.00$. The best state for getting the most runs is that for b, which is "0 outs, 1 on" and the worst state is that for c which is "1 out, 0 on" – this makes perfect sense!

To answer the second question about how long the inning may last, we appeal to methods previously studied. The expected number of batters is obtained via

$$\text{Average \# Batters} = \underbrace{2 \cdot (0.7)^2}_{\text{2 batters}} + \underbrace{3 \cdot \binom{2}{1}(0.3)(0.7)^2}_{\text{3 batters}} + \underbrace{4 \cdot \binom{3}{2}(0.3)^2(0.7)^2}_{\text{4 batters}} + \cdots$$

$$= \sum_{i=2}^{\infty} i \cdot \binom{i-1}{i-2}(0.3)^{i-2}(0.7)^2$$

$$= (0.7)^2 \sum_{i=2}^{\infty} i(i-1)(0.3)^{i-2} = \frac{20}{7} \approx 2.8571$$

where the term for having four batters in an half-inning comes from knowing that there will be eventually two outs (leading to the squared 0.7 term) and needing two of the first three batters to be non-outs (leading to the squared 0.3 term along with the choose notation). The multiplication by four finishes out the expected value calculation, combined with some calculus to evaluate the infinite sum. The end result is that just slightly less than three batters are expected to take the plate in any half-inning of One Base.

Turning our attention to Major League Baseball, Table 7.3 contains the data to create the probability matrix for all MLB teams for the 2018 regular season, with data courtesy of Retrosheet (see their website [35] for a lot more data). In that table, note that the shorthand "1O13B" means that there is one out and both first base and third base have players on them, while "3O2R" means that the third out has been reached, with two runs scoring on that play.[4] A few interesting (or perhaps not) observations are noted. First, note how similar the probability numbers are for the block containing zero outs (for both the rows and columns), the block containing one out, and the block containing two outs are; despite how all players may think to play differently based on the current number of outs, the data shows that this just does not happen. Second, if you look at the (0O0B,1O0B), (1O0B,2O0B), and (2O0B,3O0R) entries, all of them are equal to about 0.69, meaning that regardless of the number of current outs, the probability that the next batter will get an out (with no players on base) is about 69%. This is very close to the haphazard 70% chosen for an out for the One Base game we previously studied.

---

[4]In order to take the data and create Table 7.3, a few decisions had to be made; I chose only to include plays that resulted in the batter directly affecting the number of outs or players on bases. This means that things such as walks, intentional or not, are included, while stolen bases and wild throws are not considered; certainly, more rows and columns could be added to account for many more things in a baseball game!

Note: Table 7.3 is printed rotated (landscape). It is a 28 × 28 one–plate‑appearance transition (probability) matrix for MLB states, where each state is coded as [outs]O[baserunners]B (e.g. 0O12B = 0 outs, runners on 1st and 2nd) and the 3‑out end‑of‑inning states are coded 3O0R, 3O1R, 3O2R, 3O3R. Rows = current state, columns = next state. Blank cells are 0.

	0O0B	0O1B	0O2B	0O12B	0O3B	0O13B	0O23B	0O123B	1O0B	1O1B	1O2B	1O12B	1O3B	1O13B	1O23B	1O123B	2O0B	2O1B	2O2B	2O12B	2O3B	2O13B	2O23B	2O123B	3O0R	3O1R	3O2R	3O3R
0O0B	0.033	0.227	0.047	0	0.005	0.040	0.035	0	0.688	0.115	0.006	0	0.054	0	0	0	0.692	0.479	0.072	0	0.233	0.180	0.030	0.029	0.691	0.472	0.084	0
0O1B	0.032	0.000	0.013	0.200	0.006	0.083	0.004	0.159	0.001	0.006	0.072	0.396	0.007	0.261	0.324	0.357	0.001	0.002	0.462	0	0.076	0.165	0.025	0.025	0.642	0.001	0.423	0
0O2B	0.025	0.048	0.050	0.114	0.003	0.021	0.032	0.076	0.002	0	0.004	0	0.008	0.096	0	0	0.001	0.003	0.004	0	0.076	0.045	0.001	0.027	0.688	0.002	0.004	0
0O12B	0.027	0.139	0.017	0.028	0.007	0.128	0.025	0.120	0	0	0	0	0	0	0	0	0.214	0.172	0.066	0.085	0.027	0.119	0.043	0.027	0.644	0.192	0.046	0.387
0O3B	0.022	0	0.055	0	0.009	0.042	0.007	0	0.211	0.054	0	0	0	0.029	0	0	0.002	0.001	0.002	0	0.043	0.043	0.119	0.024	0.679	0.003	0.072	0
0O13B	0.041	0.049	0.006	0.139	0.010	0.082	0.055	0	0.001	0	0.001	0.085	0.004	0.046	0.023	0	0.003	0.001	0.025	0.085	0.025	0.066	0.054	0.023	0.634	0.054	0.050	0.131
0O23B	0.023	0	0.059	0.003	0.003	0.027	0.007	0.120	0.002	0	0	0.039	0.003	0.034	0.189	0	0.001	0	0.054	0.039	0.022	0.054	0.004	0.066	0.700	0.001	0.016	0.004
0O123B	0.029	0	0.008	0.027	0	0	0.055	0.184	0	0	0.002	0.068	0	0.038	0.116	0.357	0.002	0	0.002	0.039	0.036	0.022	0.003	0	0.700	0.045	0.016	0.045
1O0B									0.032								0.692	0.001	0.004	0	0.180				0.691	0.001	0.001	0
1O1B									0.033								0.001	0.479	0.072						0.642	0.227	0.016	0
1O2B									0.028								0.001	0.002	0.462						0.688	0.060	0.051	0.139
1O12B									0.031								0.214	0.172	0.066	0.396					0.644	0.139	0.015	0.131
1O3B									0.035								0.002	0.001	0.002		0.353				0.679	0.174	0.044	0.041
1O13B									0.021								0.003	0.001	0.025			0.261			0.634	0.054	0.050	0.079
1O23B									0.030								0.001	0	0.054				0.324		0.700	0.001	0.016	0.024
1O123B																								0.357	0.700	0.045	0.016	0.045
2O0B																	0.692								0.691	0.001	0.044	0.005
2O1B																	0.001	0.479	0.072						0.642	0.227	0.016	0.008
2O2B																	0.001	0.002	0.462						0.688	0.060	0.051	0.005
2O12B																	0.214	0.172	0.066	0.396					0.644	0.139	0.044	0.007
2O3B																	0.002	0.001	0.002		0.353				0.679	0.174	0.044	0.015
2O13B																	0.003	0.001	0.025	0.085		0.261			0.634	0.054	0.050	0.006
2O23B																	0.001	0	0.054	0.039	0.023		0.324		0.700	0.004	0.016	0.008
2O123B																	0.002	0	0.002	0.039	0.026	0.040	0.026	0.116	0.700	0.003	0.006	0.001
3O0R																									1.000	0	0	0
3O1R																									0	1.000	0	0
3O2R																									0	0	1.000	0
3O3R																									0	0	0	1.000

TABLE 7.3: Probability Matrix for MLB, 2018

You might wonder about the steady state for this matrix; this probability matrix is a bit different than the others we have seen in this book, since there are four "absorbing states" that end a half-inning (corresponding to zero through three scored runs when the third out is achieved). When computing very high powers of the matrix to find out what happens, we find that the steady state vector is

$$S = \begin{bmatrix} 0 & 0 & \cdots & 0 & 0 & 0.9897 & 0.0058 & 0.0040 & 0.0005 \end{bmatrix}$$

showing that most of the time a half-inning ends when the third out happens without any runs being scored. Feel free to visit the Retrosheet website to look at data and do some analyses for any season from 1916 to the present day!

The expected, or average, number of batters for a half-inning for MLB can also be computed, similar to what we did for One Base, using $p = 0.69$ as the probability of a given batter getting an out. We get

$$\text{Average \# Batters} = \underbrace{3 \cdot (0.69)^3}_{\text{3 batters}} + \underbrace{4 \cdot \binom{3}{1}(0.31)(0.69)^3 + \cdots}_{\text{4 batters}}$$

$$= \sum_{i=3}^{\infty} i \cdot \binom{i-1}{i-3}(0.31)^{i-3}(0.69)^3$$

$$= (0.69)^3 \sum_{i=3}^{\infty} i \left(\frac{(i-1)(i-2)}{2} \right)(0.31)^{i-3} = \frac{100}{23} \approx 4.3478$$

which means that we expect to see a little over four batters per half-inning during a Major League Baseball game.

7.4   Winning a Tennis Tournament

To commence the end of this chapter, we return to the game of tennis (see Figure 7.1). In Section 7.2 we looked at a very simple model of how tennis players choose the direction of their serve and, for the receiving player, how to make an initial guess on what direction the serve will be going. A different question that one can ask is how likely is a certain player to win a tennis *set* overall. For those that have not played tennis or do not know how a set works, a set is made up of several games; in order to win a set, a player needs to win six games, except that if their opponent has five wins, then the set needs to be won by a player winning seven games. This means that while a player can win a set by winning six games to their opponent's four, it is not possible to take the set by having six wins with their opponent having five; in these cases, the set win would come with the winner taking seven games to their opponent's five, or, if the set reaches a tie with six wins for each player, then the seventh win for either player successfully completes the set. Note that different tournaments do use different rules on how a set win happens, with several implementing a tie-breaker system and others requiring that someone wins a set by getting a two-game lead past a 6-6 tie; to begin, we omit breaking ties and the two-game lead requirement (called winning by *advantage set*).

In a *game*, players earn points, with 15, 30, and 40 being the names for the first, second, and third point. A current score of zero points is called "love" and in the case that each player has won three points, rather than announce the score as 40-all or 40-40, the score is called to be *deuce*. To win the game, a player needs to win a fourth point while also being ahead of their opponent by at least two points; in a deuce situation, play will continue until one player achieves that two point advantage. We do include this requirement in our model since virtually all tennis tournament systems keep this rule. In a single game, a specific player takes the role of server and will deliver their serve to their opponent for all points, and when moving between games in a set, the serving player alternates.

For a bit of history, the 2010 match at Wimbledon between players John Isner and Nicolas Mahut holds the record for the longest set in professional tennis, where Isner took the fifth set in this first-round match by 70-68; this set took over eight hours to play, but you can see that advantage set play could take an amazingly long time to complete!

Game Win Probability

Our model is based on work by my friend and colleague Adam Childers, who based his work on work done in [32]. When a game begins, the player who serves can score immediately by making a legal serve that is not touched by their opponent, which is called an ace; otherwise, the game will proceed into open play where the players hit the ball back and forth until the ball is missed or returned in an illegal way.[5] Let s_1 represent the probability that the first player wins a point on an ace and let s_2 be the same for the second player; note that the serve is often considered one of the most important parts of the game. For open play, let o_i, for i being 1 or 2, be the probability that the respective player wins the point during open play. Note in this case, since someone has to win, the relationship $o_2 = 1 - o_1$ applies. It should be noted here that the values of s_1, s_2, and o_1 (along with o_2 via the equation we just saw) will vary for each pair of players that are involved in a tournament, but these values can be estimated using historical data; it's also very likely that book makers will estimate these very well.

---

[5]In many cases, the server will hit a legal serve where the opponent touches the ball but is unable to return it back into play. This is sometimes called a "service winner" and our "ace" category will include both pure aces and service winners.

Using these parameters, we can see that

$$\text{Probability Player 1 Wins While Serving} = p_1 = \underbrace{s_1}_{\text{serve win}} + \underbrace{o_1 \cdot (1 - s_1)}_{\text{open play}}$$

since the way the first player can win the point when it is their game to serve is to either take the point outright on the serve, or proceed to open play, with probability $1 - s_1$, and capture the point then. Similarly, we have

$$\text{Probability Player 2 Wins While Serving} = p_2 = \underbrace{s_2}_{\text{serve win}} + \underbrace{o_2 \cdot (1 - s_2)}_{\text{open play}}$$

for the other player.[6]

Our initial goal is to determine the probability, g_1, that the first player wins a game when they are serving. There are several ways that this could happen. First, the player could win all four points in a row; that is, winning the point when the score is 40-0. We get

$$\text{Win 40-0} = p_1^4$$

as that probability, using an assumption, that for professional players, points are independent of each other. To get to a score of 40-15 and win, the opponent needs to win one of the first four points, while you still need four points, giving us

$$\text{Win 40-15} = \underbrace{\binom{4}{1} (1 - p_1)^1 \cdot p_1^4}_{\text{lose one point}}$$

and it then follows that

$$\text{Win 40-30} = \underbrace{\binom{5}{2} (1 - p_1)^2 \cdot p_1^4}_{\text{lose two points}}$$

is the probability for taking the point if our opponent wins two points of the first five to get the score to 40-30.

The case when the score reaches deuce (40-40) and we need to win by two points is a bit more complicated. Of the first six points, our opponent needs to win three and we need to win three; this is a nice binomial distribution with $n = 6$, $x = 3$, and using p_1 as the probability in Definition 4.1. In addition, we need to win the *last two* points. Between these possibilities is any number of pairs of points where we win a point and our opponent wins a point. To illustrate this, consider that this game looks like

$$\circ \circ \circ \circ \circ \circ \quad | \quad \cdots \quad | \quad \odot \odot$$

---

where, again, for those first six points, represented by ∘, each player wins exactly three of these; the binomial coefficient counts the number of distinct ways we could fill in those circles. The ⊙ symbols at the end mean that we must win the last two points if we are to win the game, while the $\cdots$ could be filled up with ∘●, ●∘, ●∘●∘, ●∘∘●, and many more, infinite, possibilities, but *not* anything such as ●●∘∘ since, in this case, the ● player won by two points after the first two points. Of course, the "any number of pairs of points" could also be zero. This give us

$$\text{Win on Deuce} = \underbrace{\binom{6}{3} p_1^3 (1 - p_1)^3}_{\text{win three of first six}} \cdot \underbrace{\left(\sum_{j=0}^{\infty} (2p_1 (1 - p_1))^j \right)}_{\text{pairs of alternating wins}} \cdot \underbrace{p_1^2}_{\text{win by two}}$$

as the last piece of the puzzle. The infinite sum is of a special form in mathematics that's called a geometric series, which has a nice formula to evaluate the sum without needing to actually add together an infinite number of terms.

Theorem 7.2 (Geometric Series Formula) For a geometric series of the form below, we have that

$$\sum_{j=0}^{\infty} a \cdot r^j = \frac{a}{1 - r}$$

provided that $|r| < 1$.

If you've had a good bit of calculus, you'll recognize this formula and may question whether or not the series actually converges; note that since p_1 is at most 1, either $p_1 \leq 1/2$ or $1 - p_1 \leq 1/2$, so multiplying by 2 and bringing those terms together results in a value of r in Theorem 7.2 less than 1, as needed.

Overall, by evaluating the binomial coefficients and using Theorem 7.2, we obtain

$$g_1 = p_1^4 + 4(1 - p_1) p_1^4 + 10(1 - p_1)^2 p_1^4 + \frac{20p_1^3 (1 - p_1)^3 \cdot p_1^2}{2p_1(1 - p_1)}.$$

We can, of course, calculate g_2, the probability that our opponent wins the game on their serve, very similarly to get

$$g_2 = p_2^4 + 4(1 - p_2) p_2^4 + 10(1 - p_2)^2 p_2^4 + \frac{20p_2^3 (1 - p_2)^3 \cdot p_2^2}{2p_2(1 - p_2)}$$

which will allow us to determine the probability of winning a set.

First, though, in looking at the formula for g_1, you might wonder what the probability of a player winning a game is if they have a certain probability of winning a point.

Example 7.6 Suppose that a player has a probability of winning a point of $p_1 = 0.56$. Keep in mind that this single probability value includes whatever values of s_1 and o_1 are appropriate for that player. In using the formula for g_1, we substitute $p_1 = 0.56$ to get $g_1 \approx 0.64682$ which means that, when serving, this player has a probability of winning the game of about 65%.

In Figure 7.2, the graph of how g_1 changes as p_1 ranges from 0 to 1 is displayed. You can see, as expected, that a probability of winning a point of 50% results in the probability of winning a game of 50% as well. Past a certain point, approximately when $p_1 \geq 0.75$, the probability of winning a game does not increase much at all; indeed, a value of $p_1 = 0.75$ gives a value of $g_1 \approx 0.95$, so improving your game so

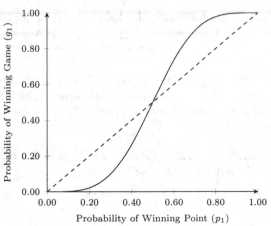

FIGURE 7.2: Probability of Winning Game

that your individual chance of winning a point increases beyond 75% does not yield a much higher chance of taking a game. Further, at $p_1 = 0.85$, the probability of winning the game increases to 99.3%, at which point it would be almost insane to put any further effort into improving your point play, on your serve, toward that opponent! The dashed line, which represents the line $y = x$, shows how a tennis game's structure magnifies the differences between p and g for each player.

Set Win Probability

To discuss the probability of winning a set, we'll assume that our first player, represented by p_1 in terms of winning a point on their turn to serve, serves in the first game of a set; to look at things from the point of view of the opponent, note that since *someone* must win the set, if t_1 is the probability that we win the set and t_2 is the probability our opponent takes the set, we must have $t_2 = 1 - t_1$.

The simplest case for winning a set is when we, as the first player, win all six games in a row; in this case, winning by a set score of 6-0, we will serve for three games and our opponent will serve for three games. The probability that we win a game when serving is given by g_1, and the probability we win a game when *not serving* is $1 - g_2$ (since g_2 is the probability our opponent wins when they serve). Then the probability of winning the set 6-0 is given by

$$\text{Win Set 6-0} = \underbrace{g_1^3}_{\substack{\text{win when} \\ \text{serving}}} \cdot \underbrace{(1 - g_2)^3}_{\substack{\text{win when} \\ \text{receiving}}}.$$

We won't walk through each of the other cases in long form; instead let us look at the case where we win the set by a score of 6-4. This will allow you to make direct connections to the other part of the probability of winning a set by any possibility and you can mirror the thoughts here.

A set with final score 6-4 means that 10 games were played, which means that each player was able to take the role of server for 5; it's important to note here that we must have won the last game, since it was the game that ended the set, and with the assumption that we serve first, this means that our opponent served the last game, and we won with probability $1 - g_2$. Otherwise, of the 5 games that *we* serve and the 4 games that *our opponent* serves, our opponent wins 4 (and, thus, we lose 4). It matters for our calculation whether the games we lost happened on our serve or not, so let b represent the number of games lost when *we* served (that is, the number of games won by our opponent on our serve).

With this notation, this means that we should choose b of the 5 games *we* serve to lose, and $4 - b$ of the 4 games our opponent serves to lose; this makes sense since if our opponent wins 4 games, and b of them happened on our serve, the remainder, $4 - b$, must happen on their serve. The pieces of the probability calculation become

$$\text{Win Set 6-4} = \sum_{b=0}^{4} \underbrace{\binom{5}{b} g_1^{5-b}(1 - g_1)^b}_{b \text{ losses on our serve}} \underbrace{\binom{4}{4-b} g_2^{4-b}(1 - g_2)^b}_{4-b \text{ losses on their serve}} \underbrace{(1 - g_2)}_{\substack{\text{win final} \\ \text{game}}}$$

under the brace: losses on our serve

which seems a bit messy, but illustrates where each piece possibility falls. The cases for 6-1, 6-2, and 6-3 can be reasoned similarly.

For the 7-5 case, note that it must be the case the first 10 games resulted in a score of 5-5 (otherwise, someone would have won 6 games at some point, prompting an end to the set, with 6-4 in the most-games case). Then, in order to win the set, we have to win the last two games, where one we serve and one we do not. This gives us

$$\text{Win Set 7-5} = \sum_{b=0}^{5} \underbrace{\binom{5}{b} g_1^{5-b}(1 - g_1)^b}_{b \text{ losses on our serve}} \underbrace{\binom{5}{5-b} g_2^{5-b}(1 - g_2)^b}_{5-b \text{ losses on their serve}} \underbrace{g_1(1 - g_2)}_{\substack{\text{win final} \\ \text{two games}}}$$

under the brace: losses on our serve

as that probability.

Finally, when we reach a situation where the set score is 6-6, we should first talk about how different tournaments address determining the set winner. In some tournaments, such as the French Open, "advantage set" scoring is used meaning that the set winner will be the player who wins at least six games with at least a two-game lead over their opponent; in this model, a set could continue for as long as is needed, such as the Wimbledon set score of 70-68 previously mentioned. Since 2019, Wimbledon has used a modified advantage set rule; while winning six games, but two games ahead of your opponent, is still the general requirement, if the set score becomes 12-12, a final tie-break game is played to seven points. By comparison, most tournaments use "tie-break set" rules whereby a set score of 6-6 results in a seven-point tie-breaking game immediately, where the player who served in the first game serves first, followed by two points with the other player serving, and going back and forth in two-point groups to determine the player that serves for a point; of course, in the tie-break game, players still need to win by a margin of two points or more once seven points are reached.

You might wonder why the thirteenth game is not just a last regular game that is played; the short answer is that it provides one player an advantage over the other in an unfair way based solely on whether or not they were selected to serve in the first game. To explore this further, and in order to keep our calculations similar and consistent, we consider that hypothetical case here. For our model, the "7-6" case for us happens when we win that thirteenth game to take the set. To begin, note that for the same reason as the 7-5 case, the first 10 games must have resulted in a 5-5 set score, and in order to get to 7-6, the next two games (games 11 and 12) had to result in a win by each player (in either order), and the last game is a win for us. The contribution to our probability for us to win, then lose, on games 11 and 12 means we win when we serve, and the opponent wins when they serve, which happens with probability $g_1 g_2$. The opposite order, when we lose on our serve and win on the opponent's serve for games 11 and 12, happens with probability $(1 - g_1)(1 - g_2)$. It's the messiest part of our overall calculation, but we get

$$\text{Win Set 7-6} = \underbrace{\sum_{b=0}^{5}}_{\substack{\text{losses on} \\ \text{our serve}}} \left(\underbrace{\binom{5}{b} g_1^{5-b}(1 - g_1)^b}_{b \text{ losses on our serve}} \underbrace{\binom{5}{5-b} g_2^{5-b}(1 - g_2)^b}_{5-b \text{ losses on their serve}} \right.$$

$$\left. \times \underbrace{(g_1 g_2 + (1 - g_1)(1 - g_2))}_{\text{games 11 and 12}} \underbrace{g_1}_{\substack{\text{win final} \\ \text{game}}} \right)$$

which means it's time to bring it all together to determine t_1, the probability that we win a set, which is based on the values of g_1 and g_2.

$$t_1 = \underbrace{g_1^3(1-g_2)^3}_{\text{win 6-0}} + \underbrace{\sum_{b=0}^{1} \binom{3}{b} g_1^{3-b}(1-g_1)^b \binom{3}{1-b} g_2^{1-b}(1-g_2)^{3-(1-b)} g_1}_{\text{win 6-1}}$$

$$+ \underbrace{\sum_{b=0}^{2} \binom{4}{b} g_1^{4-b}(1-g_1)^b \binom{3}{2-b} g_2^{2-b}(1-g_2)^{3-(2-b)}(1-g_2)}_{\text{win 6-2}}$$

$$+ \underbrace{\sum_{b=0}^{3} \binom{4}{b} g_1^{4-b}(1-g_1)^b \binom{4}{3-b} g_2^{3-b}(1-g_2)^{4-(3-b)} g_1}_{\text{win 6-3}}$$

$$+ \underbrace{\sum_{b=0}^{4} \binom{5}{b} g_1^{5-b}(1-g_1)^b \binom{4}{4-b} g_2^{4-b}(1-g_2)^b(1-g_2)}_{\text{win 6-4}}$$

$$+ \underbrace{\sum_{b=0}^{5} \binom{5}{b} g_1^{5-b}(1-g_1)^b \binom{5}{5-b} g_2^{5-b}(1-g_2)^b g_1(1-g_2)}_{\text{win 7-5}}$$

$$+ \underbrace{\sum_{b=0}^{5} \binom{5}{b} g_1^{5-b}(1-g_1)^b \binom{5}{5-b} g_2^{5-b}(1-g_2)^b (g_1 g_2 + (1-g_1)(1-g_2)) g_1}_{\text{win 7-6}}$$

Figure 7.3 shows the effect on t_1 as both g_1 and g_2 range from 0 to 1. There are some interesting takeaways here. First, note that when $g_1 = g_2 = 0$ (both players are terrible when they serve, since both players only win a game when they are *not* serving), we have $t_1 = 0$; the ability to win when not serving benefits the player that serves in the *second* game of a match (and all even numbered games).

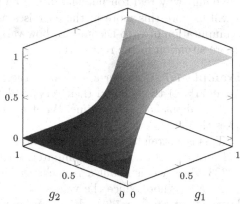

FIGURE 7.3: Probability of Winning Set

Similarly, when $g_1 = g_2 = 1$ (both players are so good that they always win the game if they are serving for that game), we have $t_1 = 1$ which benefits the player that serves first; this seems to make sense intuitively in both cases, as when both players are bad when serving, our opponent wins the first game and gets ahead in the set, and when both players are perfect when serving, we win the first game and start off with the lead.

If you analyze this further, if $g_1 = g_2 = 0$, our opponent wins the first game, third game, fifth game, seventh game, ninth game, which puts us into a 5-5 situation after game ten, and then the opponent takes game eleven as well as the final game thirteen. If $g_1 = g_2 = 1$, then we win the odd-numbered games, taking the set win on game thirteen.

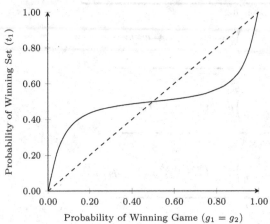

FIGURE 7.4: Chance of Winning Set, $g_1 = g_2$

When we look at the cross section of Figure 7.3 when $g_1 = g_2$ so that each player has an equal chance of winning the game when it's their turn to serve, something interesting happens; this restricted, two-dimensional, graph is shown in Figure 7.4. With evenly matched players, this graph would show a flat line at $g_1 = g_2 = 0.50$ if the probability of winning the tie break when the set score becomes 6-6 were 0.50. This curve shows the strong influence of letting the first player serve in the 6-6 game and explains why real tournaments use complicated tie breaking procedures! You will explore this more in the exercises as you consider modifications to our computation of t_1 to account for how various tournaments resolve the 6-6 set score situation in the real world.

Example 7.7 Consider a set where the first player serving wins a game about 71% of the time on their serve and the opposing player wins, when serving, about 55% of the time. What is the likelihood that the first player takes the set?

This is a straight-forward application of finding t_1 in the case that $p_1 = 0.71$ and $p_2 = 0.55$. Using a computer, the value of t_1 ends up being about 0.7383 which yields about a 74% chance that the first player takes the set.

As a side note, since the values of $p_1 = 0.71$ and $p_2 = 0.55$ are on the flat parts of Figures 7.3 and 7.4, our unrealistic assumption about what happens in a 6-6 set score likely does not affect our answer for this example much.

A tennis match usually consists of three or five sets, and continues until a player wins as soon as a player wins more than half of the sets; there is no need to win by two sets, so a match score of 3-2 is fine if the players are competing for best of five. We omit a discussion of match win probability here, mainly due to the complication that the person who serves first in the second set is controlled by whomever served first in the last game of the previous set (the privilege of serving first does not alternate between sets, only between games). For a rough estimate on a match win probability, though, the next section may help.

7.5   Repeated Play: Best of Seven

Many sports championships, such as the Stanley Cup games for the National Hockey League (NHL), use a "best of seven" model to determine the champion. That is, the finals involve playing up to seven games, and the first team to accumulate four wins is declared the victor.

Question

Does a best of seven model for sports championships promote a better or worse chance to win based upon a team's single-game win probability?

The question posed is very natural and can be studied using mathematics and probability. On the surface, this may seem similar to figuring out the probability of winning a tennis game based on the probability of winning a point, but that model is complicated by the need to win by two points; that does not apply here and our model will be simpler. There are some limitations, however, since some sports, such as baseball, may have different game win probabilities based on several factors, such as the starting pitcher. For our model, we consider one solely based on the probability of winning a game, which is held constant between games.[7]

Assuming that the probability of winning a game for the team of interest is denoted by g_1, we only have four cases to consider. First, the championship could be won in just four games when the team takes the victory by a championship score of 4-0. This is an easy case to do, since the probability is just g_1^4 to win 4-0. Let's consider the 4-2 case, where our opponent wins two games, as the case of interest, and the other cases will become clearer.

To win 4 games to 2 game, we need to lose 2 games; since a total of 6 games are being played, those losses must occur in the first 5 games and we

---

[7]While home team advantage could factor in, we will ignore that here and mention that the work we have done with tennis, between the narrative in this chapter as well as the exercises, would allow you to adjust the model to take home team advantage into account. As a side note to this footnote, the 2020 Stanley Cup championship match, due to the COVID-19 pandemic, has all of its games played on neutral territory for both teams.

win the last game. Otherwise, we would have won the championship earlier. We choose 2 of the first 5 games to lose, obtaining

$$\text{Win 4-2} = \binom{5}{2} g_1^4 (1 - g_1)^2$$

as the probability, since we need to win 4 games and lose 2 games of the series. This gives us

$$\text{Win Series} = \underbrace{g_1^4}_{4\text{-}0} + \underbrace{\binom{4}{1} g_1^4 (1 - g_1)}_{4\text{-}1} + \underbrace{\binom{5}{2} g_1^4 (1 - g_1)^2}_{4\text{-}2} + \underbrace{\binom{6}{3} g_1^4 (1 - g_1)^3}_{4\text{-}3}$$

as the probability of taking the championship. Note, though, that we could also have written the probability of winning 4-0 as

$$\text{Win 4-0} = \binom{3}{0} g_1^4 (1 - g_1)^0$$

which matches the format of the other terms; here, we choose none of the first 3 games to lose, and the power of 0 on the "lose" factor results in no contribution to the probability. We can simplify our formula to

$$\text{Win Championship} = \sum_{l=0}^{3} \binom{3 + l}{l} g_1^4 (1 - g_1)^l$$

which is quite nice!

Example 7.8 Suppose that the Pittsburgh Penguins have a 57% chance of winning any single game against their opponent in a best of seven series. Using the formula we developed, we can use $g_1 = 0.57$ to determine that the probability that the Penguins win the series is

$$\sum_{l=0}^{3} \binom{3 + l}{l} (0.57)^4 \cdot (0.43)^l \approx 0.65016$$

or about 65%.

As you may know, some championships, especially regional or divisional matches, are best of five. Modifying our probability calculation is fairly easy, but rather than simply do the best of five computation where the victor needs to win three games, let's consider a series where the tournament winner needs to win n games, where a total of $2n - 1$ games may be played (the case we looked at, best of seven, would be $n = 4$). In this model, if the victor needs n game wins, then the opponent could win anywhere from 0 to $n - 1$ of the games; if l is the number of losses we have (which is also the number of wins by the opponent), then any particular series has $n + l - 1$ games possible. By using the discussion above when $n = 4$, the following result should seem reasonable.

Theorem 7.3 (Best of Series Probability) If a team has an individual probability of winning a single game of g_1 and needs n game wins to capture a championship "best of" series, then

$$\text{Win Championship} = \sum_{l=0}^{n-1} \binom{n+l-1}{l} g_1^n (1 - g_1)^l$$

is the probability the team is victorious in the series.

Figure 7.5 shows how the probability of winning a series varies as g_1 ranges between 0 and 1 for n values of 3, 4, 5, and 6. Note that at $g_1 = 0.50$, the probability of winning the series is 50% regardless of series length; this should make sense. As g_1 grows above 0.50, the longer the series, the higher the probability that we win gets. This should also agree with your intuition, as if a team is truly stronger than their

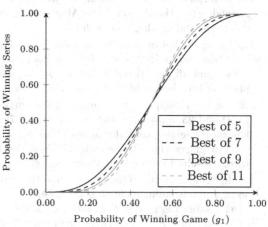

FIGURE 7.5: Probability of Winning Series

opponent (and, therefore, $g_1 > 0.50$), then repeated play should ferret this out. We would hope that "the better" team will win the championship, and increasing the number of games in the series reduces variability. With more games, the probability that the underdog wins is decreased since, as we remember from Theorem 4.3, the variance and standard deviation for a binomial distribution decreases as the number of games or events increases.

7.6   Exercises

7.1 For each of the follow pairs of teams and the given moneyline payouts, determine the perceived true odds for both teams, the service charge (vigorish), and the expected value of a value on either team.

  (a) Toronto Blue Jays −230, San Diego Padres +190

  (b) Tampa Bay Rays −160, Houston Astros +150

  (c) Philadelphia Phillies −145, San Francisco Giants +125

7.2 As alluded to in Section 7.1, when teams are evenly matched, the moneyline payouts for both teams are listed as −110. Explain why a casino or book maker would not set the odds as −110 for one team and +110 for the other.

7.3 Verify the odds in Table 7.1 for fictional horses Flip Ahead, Roll Four, and None Above.

7.4 Determine the payout on a $2 "win" wager for the horses Double Six, Bi-Nomial, Flip Ahead, and None Above as we did in Example 7.2. In addition, determine the breakage for each horse.

7.5 Using the bet information in Table 7.2 and assuming that the finishing order is Double Six, Flip Ahead, and Roll Four, determine the payout table for all horses and all bets (Double Six to win, to place, and to show, Flip Ahead to place and to show, and Roll Four to show). Assume a $2 wager as usual.

7.6 Repeat Example 7.3 by changing the entries in the first row of the matrix to 60 and 95 and the second row to 90 and 70. In [34], the authors found in a laboratory setting that the player chose KL about 0.3636 of the time and the column player chose GL about 0.4545 of the time. How close are your game theory results to those laboratory results?

7.7 Imagine that the "One Base" game that appears in Section 7.3 is modified so that a half-inning ends after *three* outs.

  (a) Modify the matrix P to account for this change.

  (b) Compute P^2 to determine the probability of starting in a given state and ending in another state after two batters have been to the plate.

  (c) Determine the expected number of runs scored for each state.

  (d) Determine the expected length (number of batters) in a half-inning for this modified version of One Base.

7.8 In a tennis game, suppose a player no longer needs to win by two points. Recompute g_1, the probability that the first player wins a game. Afterwards, repeat Example 7.6 and recreate Figure 7.2 using your formula for g_1, and then compare your results to the original.

7.9 Suppose that for a game of tennis, a player now needs to win a fifth point and still needs to be ahead of their opponent by two points. As in Exercise 7.8, recompute g_1, repeat Example 7.6, recreate Figure 7.2 and then compare your results to the original.

7.10 When computing t_1, the probability that the first player wins a *set* during a tennis tournament, we explored what would happen if the 6-6 scenario were handled by having a thirteenth game played with the player who served in the first game serving in this last game. For each of the following changes to how the 6-6 scenario is handled, modify the computation of t_1 to fit the scenario.

(a) French Open rules, where a player needs to win by two games

(b) tie-break set rules, where a last game is played to seven points, with a player needing to win by two points, *and* the serving order is the first player, two serves by the second player, two serves by the first player, two serves by the second player, and so on

(c) Wimbledon rules where a player needs to win by two games, but if the set score becomes 12-12, then a final tie-break game is played as in part (b)

8

Blackjack: Previous Methods Revisited

My favorite casino game probably has to be Blackjack. It seems to be a favorite among gamblers in Las Vegas (there are often more active Blackjack tables running than all of the other table games combined) and also among my students. My guess is that the reason for this is the excitement and decision-making, balanced with the relative ease of play and simple rules. Of course, we'll discuss all of those in time, while also taking a look at some of the mathematics of Blackjack (and some variants), but for now, a few short notes from the history of Blackjack are in order.

FIGURE 8.1: Blackjack

8.1 Blackjack

The origins of the game that we call "Blackjack" today dates back to the late 1500s, back in the time of Miguel de Cervantes. The work that you are most likely familiar with of his is *Don Quixote*, and from that work, what you probably remember most is the title character fighting windmills, imagining them to be giants. His tale of *Rinconete y Cortadillo* features cheats working in Seville, Spain, most often cheating at a game called *Ventiuna* (Spanish for twenty-one).

The game did not start being popular in the United States until the mid-1900s; early versions of the game before that time offered a special bonus prize for obtaining one of the two black Jacks in the deck. While some people still call the game "twenty-one," the name Blackjack became much more common and is what most people use these days. Variants of the game still exist, and we will discuss a few of them, but for the most part, what we'll discuss refers to the Blackjack game you will find in most US casinos.

Gameplay

A game of Blackjack is played at a table where you (and possibly several other players) play against the dealer. It's important to note that you win and lose completely based on your cards versus the dealer's cards, so you can more or less not care one bit about what the other players at the table might do. Of course, it's always exciting when anyone at the table wins because you do all have the dealer as the common opponent.

Any number of decks of cards can be used; six or eight deck games are common in casinos, and it's worth mentioning now that games with fewer decks are typically more favorable to you, the player, so finding games with fewer decks is a good idea. Game play begins with you receiving two cards and the dealer receiving two cards; typically, your cards will be face up at the table, and only one of the dealer's cards will be face up (the face down card is called the *hole* card). With the understanding that numeric cards are worth their face value, face cards (the Jack, Queen, and King) are worth 10, and an ace can be worth 1 or 11 (whichever is favorable to that hand), the goal is to have a hand that has value greater than the dealer's hand.

Of course, if that were the extent of the rules, the game would be fairly boring! Both you and the dealer have the opportunity to receive additional cards, subject to the following rules. Players always draw additional cards to resolve their hand entirely before the dealer reveals his or her hole card and resolves his or her hand. Your options are that

- you may *stand*, taking no additional cards and ending your turn, if you like your current total,

- you may *hit*, drawing one additional card, and then having the opportunity to hit or stand again,

- you may *double down*, doubling your wager, and drawing exactly one additional card, ending your turn,

- you may *split* your hand into two hands if you have a pair of the same-valued card (suit doesn't matter), offering an identical wager on the second hand and receiving an additional card for both hands; at this point, both hands are played separately as though they were new hands, or

- you may *surrender* half of your bet to the dealer to take back the other half and end the round.

The dealer's rule is rather simple; if the dealer has 17 or above, the dealer will stand, and if the dealer has less than 17, the dealer will hit until his or her total is 17 or above. Note that casinos vary on the rule when "soft 17" is involved (a hand such as A♦ 6♣ that can total either 7 or 17 is considered "soft").

For both you and the dealer, going over a total of 21 is called *busting*. The instant you bust, your bet is lost and the casino collects your wager, regardless of what happens with the dealer. Should the dealer bust, then all players remaining in the game (who haven't busted) will be paid at 1:1 odds.

Three special remarks are in order. First, receiving a total of 21 on your initial two cards, such as in Figure 8.1, is called receiving a Blackjack and is paid out immediately at a 3:2 payout, unless the dealer also has Blackjack, in which case your wager is a push (tie). Second, if the dealer has 21 with his or her two cards (the dealer will peek at the hole card if the face up dealer card is a 10-value card or an ace), his or her Blackjack immediately beats all player hands that are not Blackjack and the round will end without any players needing to make decisions. Third, in the event that the dealer has an ace showing as the face up card, the dealer will offer players an independent side bet called *insurance* (players may wager up to half of their original bet) that pays 2:1 if the dealer has Blackjack, and is a loss if the dealer does not. For this last remark, since almost one-third of the deck is ten-valued cards, it seems like an attractive bet, but (in general) it has a high house advantage; however, as we shall see, with knowledge of the cards remaining, an insurance bet could end up in your favor.

Before we discuss detailed strategy and ways to think about obtaining said strategy, let's answer a question that might be on your mind. What is the probability of obtaining a Blackjack on your initial two cards?

Example 8.1 We already have the mathematical framework needed to answer this question; much as we did in Chapter 3, we need to know the number of two-card hands that result in Blackjack, along with the total number of two-card hands. Of the 4 aces in the deck, we need to choose 1, and of the 16 ten-valued cards in the deck (four each of tens, Jacks, Queens, and Kings), we need to also choose 1. How about the total number of two-card hands? I think 52 choose 2 would be appropriate, giving us

$$\text{Probability of Blackjack} = \frac{\overbrace{\binom{4}{1}}^{\text{ace}} \cdot \overbrace{\binom{16}{1}}^{\text{``ten''}}}{\underbrace{\binom{52}{2}}_{\text{total}}} = \frac{4 \cdot 16}{1,326} = 0.0483,$$

or about 5%, a percentage that many people would say "happens a lot" even though 5% is only a "one in twenty" chance (or odds of 19:1).

Assuming that the dealer's hole card is the first of his or her cards dealt, and we're using only one deck of cards with no other players at the table, this is the exact probability of you receiving Blackjack. Once any cards other than your own are visible, though, the probability of you receiving Blackjack changes slightly, since the chance of getting the second card you need to complete your hand changes as you see other cards dealt (typically, at a Blackjack table, everyone gets their first card before anyone receives their second). What happens as more decks are used? The probability of getting Blackjack on your first two cards decreases; let's see exactly how.

For two decks, you can see that there will be 8 aces available and 32 ten-valued cards, along with 104 cards total; for four decks, that increases to 16 aces, 64 ten-valued cards, and 208 cards total. Surely we can write down a formula for the chance of Blackjack on your first two cards for n decks. With 4 aces in each deck, and n decks, there will be $4n$ aces; similarly, we will have $16n$ ten-valued cards and $52n$ total cards. Following Example 8.1, we get

$$\text{Probability of Blackjack} = \frac{\overbrace{\binom{4n}{1}}^{\text{ace}} \cdot \overbrace{\binom{16n}{1}}^{\text{"ten"}}}{\underbrace{\binom{52n}{2}}_{\text{total}}} = \frac{(4n) \cdot (16n)}{\binom{52n}{2}}$$

where there are some nice cancelations when using the formula for binomial coefficients (Theorem 3.2) to help us out, since

$$\binom{52n}{2} = \frac{(52n)!}{2!(52n-2)!} = \frac{(52n)(52n-1)\cancel{(52n-2)}\cancel{(52n-3)}\cdots\cancel{3}\cdot\cancel{2}\cdot\cancel{1}}{2\cdot\cancel{(52n-2)}\cancel{(52n-3)}\cdots\cancel{3}\cdot\cancel{2}\cdot\cancel{1}}$$
$$= \frac{(52n)(52n-1)}{2}.$$

At this point, our probability of Blackjack, using some algebra, simplifies, for n decks, to

$$\text{Probability of Blackjack} = \frac{128n^2}{(52n)(52n-1)}$$

which, using $n = 1$, you can verify gives the same result as we found in Example 8.1. The probabilities for 2, 4, 6, and 8 decks are also given in Table 8.1 where you can observe that the probabilities indeed shrink slightly as n, the number of decks, increases.

Decks	Probability
1	0.048266
2	0.047797
4	0.047566
6	0.047490
8	0.047451
∞	0.047337

TABLE 8.1: Probabilities for 21

You might wonder about the last line of that table; as n gets larger and larger, the probability of Blackjack keeps decreasing, but it never falls below 0.047337. If you've had (or will be taking) calculus, you may be (or will become) familiar with the concept of a limit, where an analysis of a function's behavior can be studied as the inputs become very large. In the denominator, $(52n)(52n-1) = 2704n^2 - 52n$, the term with n^2 grows *so* much faster than the term with n, that in the *limit* as n approaches infinity, we can drop the negligible term with n and treat this fraction as

$$\frac{128n^2}{(52n)(52n-1)} = \frac{128n^2}{2,704n^2 - 52n} \sim \frac{128n^2}{2,704n^2} = \frac{128}{2,704} \approx 0.047337.$$

One small takeaway from this right now is that *assuming* an infinite deck in order to determine probabilities, while of course introducing *some* error, only introduces a relatively small error.

Card Counting

My guess is that most of you reading this book have heard of card counting. You might also have an intuitive feel of why it can be a good idea. Here we aim to explore some of the ideas involved in card counting and hopefully give you concrete evidence that *any* idea of what cards are left in a deck of cards is an advantage. Before we start, though, I should mention mathematician Edward Thorp who, in the 1960s, was the first person to prove that card counting can give the player an advantage over the house at Blackjack; several of the examples here are borrowed from his work, and all of them are at least inspired by him. If you have time, it's worth reading his *Beat the Dealer* book (found as [53] in the Bibliography).

Question

How does knowledge of which cards remain in a partially used deck of cards help at Blackjack?

Suppose, for some reason, you know that the next six cards of the deck consist only of 2 sevens and 4 eights as you prepare to play a round of Blackjack against the dealer. Perhaps you have been counting cards, and there aren't many cards left in the deck. Also suppose that there are no other players at the table so these 6 cards are the only cards that will be involved in the next hand. How much should you bet, and what should your strategy be? Take a moment to think about your response before you move on to the next paragraph.

In this situation, you should bet as much as the table limit will allow; you are guaranteed to win! Whatever two cards you receive, simply *stand* and the dealer will bust. Why is this? If the dealer has a pair of eights, for a total of 16, the dealer will hit and either a seven or eight will force a total greater than 21. The same is true if the dealer receives an eight and a seven, for a total of 15, since a seven or eight brings the total to 22 or 23 respectively. If the dealer gets a pair of sevens, for a total of 14, another seven would be okay, but at this point, *there are only eights left in the deck*! The dealer must bust, and you will win. From this example you can see that knowledge of the cards left in the deck is important! The advantage, of course, is usually not this obvious.

Let's take some time now to examine the insurance side bet in Blackjack. Recall that this is only offered by the dealer when the dealer's face up card is an ace; you are placing the side bet that the dealer indeed has a ten-valued card as his or her hole. For these two examples, we assume that the A♦ is showing as the face up card, and we're playing with another friend at the table so that we have a little more information present. We assume one deck of cards for play.

Example 8.2 Suppose that, in addition to the dealer's face up ace, we have 4♥ Q♦ and our friend has 4♠ 8♠. While insurance has a maximum bet of half of our original bet, here we are only interested in the house advantage on insurance itself, so we assume a $1.00 bet. Should we take insurance? Of the 47 unknown cards (we have seen five cards, but the hole card is still unknown), there are 15 ten-valued cards left (we have seen one in our hand). The probability of the hole card being ten-valued is 15/47 (and thus we win the insurance bet at a 2:1 payout), so we have

$$\text{Expected Winnings} = \underbrace{\left(\frac{15}{47}\right) \cdot \$2.00}_{\text{win}} + \underbrace{\left(\frac{32}{47}\right) \cdot (-\$1.00)}_{\text{lose}} = -0.04255,$$

not too surprising since the wagers we have talked about in the book so far have all been house advantages. We should *not* take insurance here.

As you can imagine, if, between us and our friend, you have *more* ten-valued cards, the expected winnings calculation becomes more negative, for an even greater house advantage! We need more ten-valued cards among the *unknown* cards for the dealer's hole card. What happens if between us and our friend there are *no* ten-valued cards present?

Example 8.3 Now suppose that we have 2♠ 6♦ and our friend has 3♠ 8♥ (and the dealer still has a face up ace). Should we take insurance? Of the 47 unknown cards, there are now 16 ten-valued cards left, and the probability of the hole card being ten-valued is now 16/47. We have

$$\text{Expected Winnings} = \underbrace{\left(\frac{16}{47}\right) \cdot \$2.00}_{\text{win}} + \underbrace{\left(\frac{31}{47}\right) \cdot (-\$1.00)}_{\text{lose}} = 0.02128,$$

giving *us* the advantage! We should *definitely* take insurance!

By now, you might guess that knowing the number of ten-valued cards left in the deck relative to the number of cards remaining is key; in fact, what we should do is look to find the *proportion* of ten-valued cards remaining that tips the scale in favor of taking insurance versus not taking insurance. That is, we'll be employing the method of "counting tens" that Thorp studies in depth.

Let the ratio of non-ten-valued cards to ten-valued cards be represented by R; that is, if there are b ten-valued cards left and a other cards remaining, $R = a/b$.

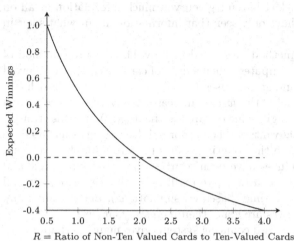

You might wonder why we don't just remember the actual number of tens. In practice, it's a lot easier to remember a single deck starting as the count $(36, 16)$ where a would equal 36 and b would equal 16 (or a multiple of this pair for more than one deck), and as cards are dealt and seen, *one* of these numbers will change with every card. It's a much easier habit to get into than needing to count all of the used cards separately, since we do only care about the proportion.

R = Ratio of Non-Ten Valued Cards to Ten-Valued Cards

FIGURE 8.2: Effect of R on Expected Winnings

Example 8.4 There are b ten-valued cards remaining out of a total of $a+b$ cards, so the probability of winning an insurance bet is $b/(a+b)$. Similarly, the probability of losing the bet is $a/(a+b)$. The expected winnings becomes

$$\text{Expected Winnings} = \underbrace{\left(\frac{b}{a+b}\right) \cdot \$2.00}_{\text{win}} + \underbrace{\left(\frac{a}{a+b}\right) \cdot (-\$1.00)}_{\text{lose}} = \frac{2b-a}{a+b},$$

which, for us to want a favorable bet, needs to be greater than 0. That is, we want

$$\frac{2b-a}{a+b} > 0$$

which happens when $2b - a > 0$ since the denominator is already a positive quantity. This happens when $2b > a$.

From the above, it should now become clear that the insurance bet is an *even-money* wager when there are twice as many ten-valued cards as other cards unknown, and it becomes a *player-favorable* wager when there are even more unseen ten-valued cards! That is, when $R = 2.00$, it doesn't matter if we take insurance or not, but when $R < 2.00$, we should absolutely take the insurance side bet, as we can see in Figure 8.2. If you know the number of ten-valued cards left, multiply by two, and get something equal to or larger than the number of other cards left, take insurance.

Without getting into the details of the expected winnings or house advantage of Blackjack quite yet, I do want to spend a little more time on the subject of card counting. At the very least, if you want to count cards at Blackjack and try your luck, I haven't given you much information at all on *how* to do it; rather, you have only seen that information about what's left in the deck can indeed help.

Perhaps the simplest method, described first by Thorp, is the method of counting only fives. As he computes, when a deck of cards is completely devoid of fives, the player's advantage increases by 3.07% over the standard house advantage in Blackjack, and is the *largest* increase in advantage obtained by removing all of any other single value of cards in the deck. While this strategy only gives you that extra advantage when a normal deck of cards has used up all of the fives (this is where the counting comes in; watch for the fives, and bet higher when all of the fives have been seen), with one deck of cards, and only you against the dealer at a table, you will be in this situation, with all fives seen, about 9.8% of the time (based on simulations). Bet the minimum amount allowed until the fives are gone, then increase your bet!

A quick search online for card counting systems gives many; perhaps one of the most complicated is the method of counting tens provided by Thorp that uses the ratio R we used earlier combined with a basic strategy chart that tells you, for each possible scenario, what value of R is the tipping point between hitting and standing (or making other decisions as well). As you might guess, this is not the best method used by beginning card counters, as it takes a *lot* of practice both keeping the ratio in your head as well as memorizing tables of numbers.

Instead, I will leave you with a card counting system that does give the player an advantage and is the same system that I introduce when giving talks on Blackjack; the system here is called Hi-Lo. As with any card counting system, though, if you want to put this into use, you should spend a great deal of time practicing by cycling through decks of cards (somewhat quickly) and keeping a count, and verifying the count at certain times to make sure that you are correct.

For the Hi-Lo system, any time the dealer shuffles the cards, the count is set to 0; for each two, three, four, five, or six that you see as part of any player's hand (including the dealer), increase the count by 1. For each ten-valued card or ace that you see, decrease the count by 1. For betting, increase your bet by a factor of two when the count reaches 4 times the number of decks in use, by a factor of four when the count reaches 8 times the number of decks in use, and so on and so forth. Your standard-sized bet before increasing is likely to be the table minimum. Throughout play, this system offers you, the player, an advantage of about 1% over the house, and has a *betting correlation* of about 0.97 (this is a measure of how close the card count mirrors what actually remains in the deck in terms of strategy).

For a wonderful account that shows exactly how knowledge of just a single card in a Blackjack shoe of cards can tilt the advantage from the house to the player, see *Busting Vega$* by Ben Mezrich (in the Bibliography as [30]). As a word of warning, modern casinos have adapted their techniques to counterbalance the effect of card counting and other techniques that attempt to "beat the system" by inventions such as the continuous shuffling machine. My advice, if you choose to play, is to use basic strategy and treat any form gambling, including Blackjack, as entertainment where the loss of money is expected.

Decision Making, Probability Trees, and Basic Strategy

At this point, it should be clear that you can determine the probability of getting any specific hand for the initial hand at Blackjack (at least for a freshly shuffled deck), or even any specific *total* for your hand. This information is certainly interesting mathematically and perhaps even fun to determine, but it doesn't help us at all once we need to make a decision about hitting, standing, splitting, doubling down, or surrendering.

Question

Does there exist a strategy that can tell us what the best decision is at Blackjack?

To really create a strategy that will work for any possible situation, and not even including changes in strategy that card counting might introduce, we'll need to do an in-depth study of each situation and determine the best possible outcome for the player, given the limited information that we have when a decision needs to be made. Your starting hand could have a point total of 2 through 21, and the dealer could have any card showing, so there are *a lot* of different possibilities. Add to that changes due to the number of decks and variations in rules being used, and it should make sense that the full strategy we'll present below comes from an in-depth computer analysis of each of these possibilities for a given rule set.

My aim here is not to spend time looking at each of these, of course. Rather, we'll spend time looking at the mathematics behind a particular situation. You could, perhaps, call this a small glimpse into what calculations would be necessary, and what programming might be behind a computer program, to create a "basic strategy" chart; charts, by the way, that casinos are generally more than happy for you to have sitting at a Blackjack table (even with best play, the house still has an advantage, and if a chart provides you comfort and safety in wagering, they'll take your bets happily).

Suppose that your hand is 8♦ 10♣ and the dealer has Q♥ showing as his or her face up card. With your point total of 18 against a dealer 10 showing, what should you do? If you are even just somewhat familiar with Blackjack, you're probably screaming that you should definitely stand, but *why*? Let's figure out the probability that we'll lose by hitting (that is, we want to determine the probability that we will bust). Of course, since there are fewer cards that will have us *not* bust, we will work with that first.

With one deck of cards, there are 49 unknown cards, and only an ace, a two, or a three will keep us safe if we decide to hit. The probability that we hit and draw one of those cards, since all four cards for each of the aces, twos, or threes are available, is $12/49 \approx 0.24489796$. For four decks of cards, of the 205 unknown cards (there are 208 cards total, and three have been seen), all sixteen cards for aces, twos, or threes in the deck give us a probability of not busting as $48/205 \approx 0.23414634$, and for eight decks of cards the probability is $96/413 \approx 0.23244552$.

What happens as the number of decks continues to increase? With an *infinite* number of decks, the probability of any specific non-ten-valued card is set to $4/52$, and the probability of a ten-valued card is $16/52$. The *infinite deck assumption* in Blackjack is a convenient way to work with probabilities, as it doesn't require us to keep track of what cards have been seen, and with a larger number of decks, it makes sense that seeing any one particular card matters less and less. The probability of not busting with this assumption, given the proportion of aces, twos, and threes, fixed at $4/52$ for each, is $12/52 \approx 0.23076923$, very close to the eight deck probability. Let's formalize this idea.

> **Theorem 8.1 (Infinite Deck Assumption)** The *infinite deck assumption* in Blackjack fixes the probability of getting any non-ten-valued card as $4/52$ and getting a ten-valued card as $16/52$. This assumption introduces very little error in calculation, and can be used to determine strategies for Blackjack games with a large number of decks.

From this point on, we will use this assumption, and if you are worried about this (and it's reasonable to be worried), of the 290 decision boxes for a basic strategy chart, in the eight deck version, using the Infinite Deck Assumption of Theorem 8.1 only results in the wrong decision for *one* entry compared to perfect play; this small error only affects the overall expected value at Blackjack by an almost negligible amount.

So, where does that leave us? The probability of not busting is about 0.23076923, so the probability of actually busting, and therefore immediately losing against the dealer, is the opposite probability of 0.76923077. Now, we need to determine the probability that we lose by *standing*, taking into account what the dealer may have and the play that proceeds for the dealer. If the probability of losing by hitting is greater than standing, then our decision shall be to stand; otherwise, hitting would be best.

In our situation, we'll lose if the dealer manages to obtain a total of 19, 20, or 21 after he or she turns their hole card face up and receives additional cards if needed (to reach 17 or higher). Remember, since we are at the point of making *our* decision, the dealer has already checked for and has not gotten Blackjack, so the dealer does not have an ace face down. Let's take a look at this point-by-point for the dealer. How could he or she end up with a point total of 19? The possibilities are that

- the dealer has a 9 as the hole card,

- the dealer has a 6 as the hole card and draws a 3 when hitting,

- the dealer has a 5 as the hole card, and either draws a 4 when hitting, or draws an ace, followed by a 3 when hitting,

- the dealer has a 4 as the hole card, and either draws a 5 when hitting, draws a 2 followed by a 3, or draws an ace and follows any of the possibilities as though he or she follows the "5 as the hole card" situation,

- the dealer has a 3 as the hole card, and either draws a 6 when hitting, draws a 3 followed any another 3, draws a 2 and follows the "5 as the hole card" situation, or draws an ace and follows the "4 as the hole card" situation, or

- the dealer has a 2 as the hole card, either draws a 7 when hitting, draws a 4 followed by a 3, draws a 3 and follows "5 as the hole card," draws a 2 and follows "4 as the hole card," or draws a 2 and follows "3 as the hole card."

Note, again, that we haven't included "7 as the hole card" as a possibility since if the dealer has a ten-valued card and a 7, the point total of 17, along with the dealer's strict rules on hitting and standing, require that the dealer *stand* and allow us to win with our total of 18. As you might guess, we could explicitly list all of the (or-

TABLE 8.2: Dealer 19 with 10 Showing

9	63	54	5A3
45	423	4A4	4AA3
36	333	324	32A3
3A5	3A23	3AA4	3AAA3
27	243	234	23A3
25	223	22A4	22AA3
2A6	2A33	2A24	2A2A3
2AA5	2AA23	2AAA4	2AAAA3

dered) possibilities for cards that the dealer can get to beat us, as I have done in Table 8.2 where the hole card is in bold. We could also display this information in a *probability tree* as seen further in this chapter (in Figure 8.4). Let's take a closer look at the idea of probability trees first, however, since they can be useful in general and we will most definitely be using them later in this book.

Definition 8.2 A *probability tree* is a diagram that illustrates the flow of an experiment or procedure. Physical outcomes at any particular time are the *nodes* of the tree and are often either denoted by text representing the outcomes or as dots, with the outcomes listed as labels for those dots. Between dots are the *edges* of the tree, represented as arrows with a number attached to them. This number indicates the probability of that edge occurring (by the outcome it is pointing to actually happening).

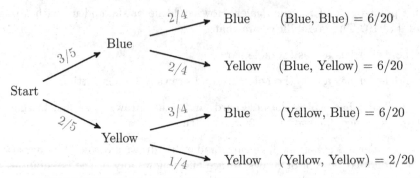

FIGURE 8.3: Probability Tree for Blue and Yellow Marbles

Of course, a probability tree could have extra features added that help highlight certain points about the experiment. For an example, consider the following classic problem that appears in most probability books.

Suppose we have an opaque jar that contains five marbles. Since it would be rather boring if all of the marbles were the same color, imagine that three of the marbles are blue, and two of the marbles are yellow. We will randomly draw one marble, and then, without putting that marble back in the jar, randomly draw a second marble. What does the probability tree look like for this situation? In Figure 8.3 you can see the nodes represented by what color marble is drawn for the first marble based on the edges leading out of the "Start" node. These edges are labeled with the probabilities for obtaining that color marble. From each of the two nodes after the start, we again have two edges leading out, one for "Blue" and one for "Yellow," and the probabilities are listed along those arrows. Note, for example, that after drawing a yellow marble for the first draw, the probability of getting a blue marble is 3/4 since three blue marbles remain out of a possible four.

Note that I've added some probabilities on the right side of Figure 8.3; for example, the 6/20 attached to (Blue, Blue) indicates the probability of starting this experience and getting a blue marble on both draws. The probability of drawing one of each color is 12/20, given by adding together the probabilities for the (ordered) situations (Blue, Yellow) and (Yellow, Blue). The relatively low probability for both marbles being yellow should be expected, since we're asking to draw the only two yellow marbles from a bag of five at the onset.

The probabilities for the second draw, as you can see from the probability tree, are *not* independent of the first draw. The chance of getting a yellow marble at the start is 2/5, but that *increases* to 2/4 if you draw a blue marble first, and it *decreases* to 1/4 if you draw a yellow marble first. These probabilities are known as *conditional probabilities* because, unlike for independent outcomes, the occurrence of something during the first draw *does* affect the probabilities of the second! Let's formalize this.

Definition 8.3 The *conditional probability* that an outcome B happens given that outcome A has already occurred is denoted by $P(B|A)$ and read as "the probability of B given A." It can be computed using probability trees, direct consideration, or using the formula

$$P(B|A) = \frac{P(B \text{ and } A)}{P(A)}$$

where you may recall from Chapter 1 that $P(B \text{ and } A)$ is the probability of both A and B occurring.

While we reasoned through the probabilities of Figure 8.3, it's worth seeing how Definition 8.3 can also give us some of these numbers.

Example 8.5 To determine $P(\text{blue second}|\text{yellow first})$, we can use Definition 8.3, using $P(\text{yellow first}) = 2/5$ and $P(\text{blue second and yellow first}) = 6/20$. Then we have

$$P(\text{blue second}|\text{yellow first}) = \frac{P(\text{blue second and yellow first})}{P(\text{yellow first})} = \frac{6/20}{2/5} = \frac{3}{4}$$

as we also found earlier, indicated in the probability tree.

Of course, we can use Definition 8.3 any time we know two of the values and desire to know the third. Let's see how that works.

Example 8.6 If we want to determine $P(\text{blue second and blue first})$, note that we already know the value of $P(\text{blue first})$, given as $3/5$ since three of the five balls are colored blue, and $P(\text{blue second}|\text{blue first})$ from the tree. The formula now looks like

$$\frac{2}{4} = \frac{P(\text{blue second and blue first})}{3/5}$$

so cross multiplication gives us the probability of both occurring as

$$P(\text{blue second and blue first}) = \frac{6}{20}$$

which is what appears in Figure 8.3.

Note that, as usual with formulas, they may appear unnecessary in simple situations; it is always nice to see that in situations we can work by hand that the formulas do indeed work, because when more difficult situations arise, the formula may be the only way out.

You can see, now, turning our attention back to the scenario where we

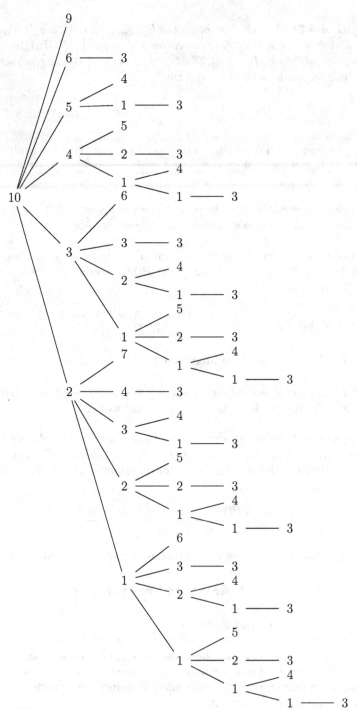

FIGURE 8.4: Probability Tree for Dealer 19 with 10 Showing

have a total of 18 (8♦ 10♣) and the dealer has a face up Q♥, probability trees can give a useful sense of the entire picture. I think, at least for me, it is easier to see how situations can change for the different point totals that the dealer can have when using a probability tree such as Figure 8.4 than it is from a list of possibilities as in Table 8.2. Following a path through the tree gives us the probability of that branch happening by multiplying as we go. For instance, the probability that the dealer's hole card is a 4, and he gets 19 by getting a 2 followed by a 3, using the infinite deck assumption, is about

$$\text{Probability of 4, 2, 3} = \underbrace{\frac{4}{52}}_{\text{four}} \cdot \underbrace{\frac{4}{52}}_{\text{two}} \cdot \underbrace{\frac{4}{52}}_{\text{three}} = \frac{1}{2,197} \approx 0.000455.$$

It should be clear that this same probability applies to *any* of the situations in Table 8.2 that involve three cards (including the hole card). For a situation that, say, involves five cards, you would have five 4/52 terms instead of three.

Example 8.7 Using Table 8.2 as a guide, since there is 1 situation involving one card, 5 situations involving two cards, 10 with three, 10 with four, 5 with five, and 1 with six, we can compute the probability that the dealer obtains 19 with a ten-valued card as the face up card as

$$\text{Probability of Dealer 19} = \underbrace{\left(\frac{4}{52}\right)}_{\text{1 card}} + \underbrace{5 \cdot \left(\frac{4}{52}\right)^2}_{\text{2 cards}} + \underbrace{10 \cdot \left(\frac{4}{52}\right)^3}_{\text{3 cards}}$$

$$+ \underbrace{10 \cdot \left(\frac{4}{52}\right)^4}_{\text{4 cards}} + \underbrace{5 \cdot \left(\frac{4}{52}\right)^5}_{\text{5 cards}} + \underbrace{\left(\frac{4}{52}\right)^6}_{\text{6 cards}} \approx 0.111424.$$

You will have a chance to repeat these calculations yourself for the probability that the dealer obtains a point total of 20 or 21 soon; there are subtle differences in the answers, since for a point total of 20, a hole card valued ten has probability different than the usual 4/52 that we used for the hole card of 9 above, and for a point total of 21, we already know the hole card is *not* an ace (otherwise the round would end immediately), so that possibility is eliminated. The probability that the dealer obtains 20 ends up equaling 0.342194, and the probability that the dealer obtains 21 is 0.034501. Note that, in the language of conditional probability, these two numbers represent $P(\text{dealer 20}|\text{up card 10})$ and $P(\text{dealer 21}|\text{up card 10})$.

Adding these three probabilities[1] together gives us 0.488119, the probability that we lose (by standing) because the dealer beat our point total of 18.

---

[1]If you are worried about the Infinite Deck Assumption messing with things, the actual probabilities (using eight decks) of the dealer obtaining a point total of 19, 20, and 21, respectively, are 0.11181, 0.34056, and 0.03474; the value of 0.488119 you see above is off by less than one one-thousandth.

We can now make our decision. On the one hand, the probability that we lose (by busting) if we hit, from earlier, is about 0.769231, and on the other hand, the probability that we lose if we stand is about 0.488119. I don't know about you, but I would rather have a 48.8% chance of losing than a 76.9% chance of losing! In this case, we should definitely *stand* and our decision is made.

If you're curious, the probability that we tie the dealer (he or she ends up with a point total of 18) is also about 0.111424, since we could replace the last card in each of the 32 possibilities of Table 8.2 with a card worth one point smaller without affecting the calculation. With a 48.8% chance of losing, and a 11.1% chance of pushing, that leaves a 40.1% of winning if we decide to stand. On a $1.00 wager, this gives an expected winnings calculation of

$$\text{Expected Winnings (Stand)} = \underbrace{0.401 \cdot \$1.00}_{\text{win}} + \underbrace{0.488 \cdot (-\$1.00)}_{\text{lose}} = -\$0.087,$$

ignoring the term for pushing since the amount won (or lost) is $0.00. This gives the advantage to the house, but the house has an even bigger advantage if we hit.

As a quick note, if you're a number fanatic, you may have recognized the numbers 1 5 10 10 5 1 that appeared in Example 8.7 as coefficients as coming from the fifth row of Pascal's Triangle, or perhaps even as the binomial coefficients that appear in the expansion of $(x+y)^5$. Indeed, these are also the six possibilities you can obtain when computing "5 choose r" for r ranging from 0 to 5, as with

$$\binom{5}{0} = 1 \quad \binom{5}{1} = 5 \quad \binom{5}{2} = 10 \quad \binom{5}{3} = 10 \quad \binom{5}{4} = 5 \quad \binom{5}{5} = 1.$$

Everyone, and especially fanatics, is encouraged to explore Section 11.4 that shows some neat connections between computing these probabilities and the straightforward mathematical theories involving partitions and components.

At this point it's time to display the basic strategy tables, noting that I've included the tables for a common set of specific Blackjack rules[2] that you will find using eight decks. Various online basic strategy table generators will allow you to create similar tables for the specific rules at your table; we'll discuss some of those rules in the next section.

After a bit of explanation, the tables are fairly straightforward to use; the "hard total" table is most commonly used and provides you guidance on most initial hands, and all hands if additional cards are obtained by hitting (or you obtain new hands through splitting). Look at the column corresponding to the dealer's face up card and the row corresponding to your point total and the color and letter of the box tells you what to do. While **S** and **H** are clear

---

[2]Perhaps the most important "rule" option that I've used that may differentiate this set of tables from others that you might see is the rule that the dealer will *hit* on soft 17; that is, the dealer hits any hand he or she has that could count as 7 or 17. This variation is very common in Atlantic City and Las Vegas among eight-deck Blackjack games.

TABLE 8.3: Basic Strategy for Hard Totals

Hard Total	Dealer's Up Card									
	2	3	4	5	6	7	8	9	10	A
18+	S	S	S	S	S	S	S	S	S	S
17	S	S	S	S	S	S	S	S	S	RS
16	S	S	S	S	S	H	H	RH	RH	RH
15	S	S	S	S	S	H	H	H	RH	RH
14	S	S	S	S	S	H	H	H	H	H
13	S	S	S	S	S	H	H	H	H	H
12	H	H	S	S	S	H	H	H	H	H
11	DH	DH	DH	DH	DH	DH	DH	DH	DH	DH
10	DH	DH	DH	DH	DH	DH	DH	DH	H	H
9	H	DH	DH	DH	DH	H	H	H	H	H
5-8	H	H	H	H	H	H	H	H	H	H

TABLE 8.4: Basic Strategy for Soft Totals

Soft Total	Dealer's Up Card									
	2	3	4	5	6	7	8	9	10	A
20+	S	S	S	S	S	S	S	S	S	S
19	S	S	S	S	DS	S	S	S	S	S
18	DS	DS	DS	DS	DS	S	S	H	H	H
17	H	DH	DH	DH	DH	H	H	H	H	H
16	H	H	DH	DH	DH	H	H	H	H	H
15	H	H	DH	DH	DH	H	H	H	H	H
14	H	H	H	DH	DH	H	H	H	H	H
13	H	H	H	DH	DH	H	H	H	H	H

for standing and hitting, along with **P** for splitting a pair, the other symbols require a bit of explanation. Both **DH** and **DS** represent using the double down option if available, otherwise for **DH**, you should hit, and similarly for **DS**, you should stand. In the same vein, **RH** and **RS** mean that you should surrender, and if that option is not available, you should hit or stand as appropriate. As I mentioned earlier, most of the time, you'll use Table 8.3. If your hand is *soft* in that it contains an ace and could be considered two different point totals (such as 7 or 17), use Table 8.4, and if your hand is a pair (including, for example, J♦ K♠), use Table 8.5 to decide if you should split.

Example 8.8 In the situation where we have 8♦ 10♣ for a point total of 18 and the dealer has Q♥, consulting the basic strategy table Table 8.3 at the 18+ row and 10 column tells us to stand. Note that this is the same decision we arrived at earlier.

At this point, I should mention that you are completely equipped with

TABLE 8.5: Basic Strategy for Paired Hands

Pair					Dealer's Up Card					
Hand	2	3	4	5	6	7	8	9	10	A
A-A	P	P	P	P	P	P	P	P	P	P
10-10	S	S	S	S	S	S	S	S	S	S
9-9	P	P	P	P	P	S	P	P	S	S
8-8	P	P	P	P	P	P	P	P	P	RP
7-7	P	P	P	P	P	P	H	H	H	H
6-6	P	P	P	P	P	H	H	H	H	H
5-5	DH	DH	DH	DH	DH	DH	DH	DH	H	H
4-4	H	H	H	P	P	H	H	H	H	H
3-3	P	P	P	P	P	P	H	H	H	H
2-2	P	P	P	P	P	P	H	H	H	H

enough knowledge to replicate the basic strategy chart by considering each situation as we did above for a player 18 versus a dealer 10. Using the Infinite Deck Assumption will yield the same exact table, with only one difference (as noted earlier). In particular, if you decide to try your hand at some of these entries, the entry that changes is in Table 8.4, where you should hit a soft total of 13 versus a dealer 5 rather than doubling down. I think a very small difference in strategy that, over a long run of Blackjack, would have a relatively small effect on the house advantage is worth the savings in computation if working by hand. Of course, the basic strategy presented here was generated by my computer program that enumerates all possibilities for what happens after the player and dealer receive their initial hands and essentially "plays out" all of these to determine the best outcome for you.

Before we move on, let's take one last look at the decision-making process; when does it make sense to surrender? Why do Tables 8.3 and 8.5 suggest surrendering at particular times? It all has to do with the probability of actually winning the given hand assuming best decision making.

Example 8.9 In order to make the decision between surrendering or not surrendering, it's important to know the probability that you will win the hand you currently have *assuming you hit or stand correctly*. Let's denote by p the probability that you will win the hand you currently have (making the highest probability of winning decision available to you). With a bet of $1.00, we can compute the expected value of not surrendering as

$$\text{Expected Winnings No Surrender} = \underbrace{p \cdot \$1.00}_{\text{win}} + \underbrace{(1-p) \cdot (-\$1.00)}_{\text{lose}} = \$(2p-1)$$

and, if we surrender, we give up half of our bet to the house, for an expected value of

$$\text{Expected Winnings (Surrender)} = -\$0.50.$$

The tipping point is when these two quantities are equal; that is, we want to know when is $2p - 1 = -0.50$. A little bit of algebra gives a value of $p = 0.25$. Whenever your probability of winning is less than 25%, even with best play, it makes sense to surrender since your expected winnings are *lower* than simply giving up half of your bet.

Note that in Example 8.9 above, we assume that you and the dealer will not push; considering the push option is left for you as an exercise later.

8.2 Blackjack Variants

In this section, I want to take some time to discuss some variants of Blackjack. First, we'll look at different, sometimes subtle, rules that are in effect at casino tables that are labeled as "Blackjack" and then we will look at some recent versions that change the game in more fundamental ways.

Rule Variations for Normal Blackjack

The rule set that I've used to generate the basic strategy given in Tables 8.3, 8.4, 8.5 results in a house advantage of about 0.48362%. That means for each dollar you wager, you're expected, in the long run, to lose about half of a cent. This makes Blackjack, along with the game of Craps discussed in Chapter 2, two of the games with the lowest house advantage in a casino; counting cards makes Blackjack even better.

Here are some of the rule "options" that different casinos (and even different tables within the same casino) use:

- The *number of decks* can vary widely, typically between 1 and 8 decks. A lower number of decks is more favorable to the player, but is less common, and the difference in house advantage for 1 deck is about 0.48% lower than for 8 decks. Our basic strategy assumes *eight decks*.

- The dealer's rule for *soft 17* can be to either stand (stands on all 17s) or hit (hits on soft 17s). The house gets an extra 0.2% advantage on hitting soft 17, and our basic strategy assumes *hitting soft 17*.

- Sometimes the *surrender* option is not available at all; when it is offered, it can be a late or early surrender (early meaning that the player can surrender *before* the dealer checks for Blackjack if he or she has an ace or ten-valued card). As you might guess, no option to surrender at all is the most favorable to the house, and early surrender is most favorable to the player (and is also very rare). Our basic strategy assumes *late surrender* as it is the most common.

- *Splitting* can also vary from table to table. It is somewhat common for casinos to not allow resplitting (splitting a pair that was formed on a split itself), not allow resplitting aces (and, sometimes, to not allow hitting on split aces), or to not allow splitting of aces at all. These different options increase or decrease the house advantage by as little as 0.03% and as much as 0.13%. Our strategy assumes that you can *always split, resplit aces, and hit after splitting aces*.

- Sometimes different options are presented to restrict the player's ability to *double down*; in particular, some casinos have moved to allowing doubling on hard totals of 9, 10, or 11 (called the Reno Rule), and also some casinos disallow doubling after a split; these rules increase the house advantage by, respectively, 0.01% and 0.12%. Even rarer is allowing players to double on any amount of cards (giving the player a decrease in disadvantage of 0.23%). Our strategy allows for *doubling on any two cards*, the most common option.

- More recently, some casinos have been changing the *payout on Blackjack* from the standard 3:2 payout to 7:5, 6:5, and even 1:1 in some cases; stay away from these games if at all possible, as these give the house an additional advantage of 0.45%, 1.39%, and 2.27%, respectively. We assume the standard *3:2 payout*.

Of course, you may see many different rules available at a table, but these are at least the most common and the ones you should be aware of; ask your dealer if you want to know the specific rules in play at the table where you are (the rule about hitting soft 17 or not is usually printed on the table itself).

Blackjack Switch

The game of Blackjack Switch[3] is an invention of Geoff Hall from 2001. The premise of this variant is that each player is required to play and bet equal wagers on two hands of Blackjack at the same time. Once the dealer checks for Blackjack (if he or she has a face up ace or ten-valued card), the player has the option to switch the second cards dealt to the player's hands. As an example, consider the scenario where the player is dealt J♣ 4♥ to one hand (with the four being the second card) and 7♠ 9♦ to the other (with the nine being the second card). The player could choose to leave the hands the same, or swap the 4♥ and 9♦ to form the new hands J♣ 9♦ and 7♠ 4♥.

After the switch decision is made, Blackjack is played per traditional rules (more or less) with each of the hands separately, and each hand wins or loses against the dealer on its own. Since it should be clear to you that having the opportunity to switch cards gives you the chance to drastically improve some hands, the house loses its advantage; to compensate for this, a Blackjack is paid at 1:1 odds (instead of the usual 3:2 odds), and *a dealer point total of 22*

---

[3]Blackjack Switch is a registered trademark of Customized Casino Games, Limited.

TABLE 8.6: Blackjack Switch Basic Strategy for Hard Totals

Hard Total	Dealer's Up Card									
	2	3	4	5	6	7	8	9	10	A
18+	S	S	S	S	S	S	S	S	S	S
17	S	S	S	S	S	S	S	S	S	RS
16	S	S	S	S	S	H	H	RH	RH	RH
15	S	S	S	S	S	H	H	H	RH	RH
14	S	S	S	S	S	H	H	H	H	H
13	S	S	S	S	S	H	H	H	H	H
12	H	H	H	S⁴	S	H	H	H	H	H
11	DH	DH	DH	DH	DH	DH	DH	DH	H	H
10	DH	DH	DH	DH	DH	DH	DH	H	H	H
9	H	H	H	DH	DH	H	H	H	H	H
5-8	H	H	H	H	H	H	H	H	H	H

pushes against all non-busted player hands except a natural Blackjack. This swings the advantage back to the house. Note here that a point total of 21 with two cards formed *as the result of a switch* is not considered a natural Blackjack. Traditional choices of Blackjack variations (as just discussed in the last section) for Blackjack Switch use six decks of cards, the rule where the dealer hits a soft 17, standard doubling rules, and only one round of splitting is possible.

As you might imagine, once the decision to switch or not to switch happens, a version of basic strategy can be determined that dictates how best to play each hand. The major reason that a new set of basic strategy tables is required is the fact that the dealer no longer busts when reaching a point total of 22. The change to using six decks and restricting splitting has a small effect on strategy, but it is the "twenty-two push" addition that makes the changes to basic strategy that you will find in Tables 8.6, 8.7, and 8.8. Note that, like the normal basic strategy you saw earlier, Table 8.6 is used for hard point totals, Table 8.7 is used when your hand involves an ace that can be counted as either 1 or 11 without busting, and pairs are treated in Table 8.8. You should take a few minutes now to compare these tables with those found earlier for standard Blackjack and hypothesize about possible reasons for particular changes. I should comment again that these tables may appear slightly different than others you find online or at some casinos due to the specific Blackjack options used. I'd like to take a moment here to thank Roanoke College student Heather Cook who, for an independent study project, worked with me on studying this game, creating these basic strategy tables, and working with expected values of the game.

By now, you should be able to predict what comes next. Now that basic strategy has been created, there is only one question that remains.

---

⁴For 10-2 only, you should hit rather than stay for an increase in the expected value of 0.00132.

TABLE 8.7: Blackjack Switch Basic Strategy for Soft Totals

Soft Total	Dealer's Up Card									
	2	3	4	5	6	7	8	9	10	A
19+	S	S	S	S	S	S	S	S	S	S
18	S	S	S	DS	DS	S	S	H	H	H
17	H	H	H	DH	DH	H	H	H	H	H
16	H	H	H	H	DH	H	H	H	H	H
13-15	H	H	H	H	H	H	H	H	H	H

TABLE 8.8: Blackjack Switch Basic Strategy for Paired Hands

Pair Hand	Dealer's Up Card									
	2	3	4	5	6	7	8	9	10	A
A-A	P	P	P	P	P	P	P	P	P	P
10-10	S	S	S	S	S	S	S	S	S	S
9-9	S	S	P	P	P	S	P	P	S	S
8-8	P	P	P	P	P	P	P	RP	RH	RH
7-7	S	P	P	P	P	P	H	H	H	RH
6-6	H	H	P	P	P	H	H	H	H	H
5-5	DH	DH	DH	DH	DH	DH	DH	H	H	H
4-4	H	H	H	H	H	H	H	H	H	H
3-3	H	H	H	P	P	P	H	H	H	H
2-2	H	H	H	P	P	P	H	H	H	H

Question

When should the second cards be switched and when should they not be switched?

If you've been paying attention throughout this book, you can probably guess that the concept of expected value will come into play. The short answer to this question is that, given the expected values of the two hands that you currently have and the expected values of the two hands that you can form by switching, choose the option that gives you the higher *sum* of expected values for those two hands. This, of course, means that having the expected values handy for each possible hand dealt to the player versus the dealer's different face up cards.

These expected values, thanks to Heather and the computer program that generated them, can be found at the end of the book in the Tables section as Tables T.8 and T.9. Note that these tables do *not* tell you what to do after you've made the switch decision; that part of the game is handled by the new basic strategy. Instead, these numbers represent the expected winnings (on a $1.00 wager per individual hand), assuming that you follow the basic strategy chart once you have switched or not. Let's consider an example.

Example 8.10 Suppose, as in the opening paragraph about Blackjack Switch, you have the two hands **J♣** 4♥ and **7♠** 9♦ (second card dealt for each hand will be listed second). Also, for this example, we know that the dealer's face up card is 6♦. What should we do? Consulting Table T.8, the 10-4 hand versus a dealer 6 gives us an expected value of -0.21042, and the 9-7 hand gives -0.21400; our reasoning then yields

$$\text{Expected Winnings (No Switch)} = \underbrace{-\$0.21042}_{\text{10-4}} + \underbrace{-\$0.21400}_{\text{9-7}} = -\$0.42442,$$

but how does that compare to switching? If we *do switch*, we have the two hands **J♣** 9♦ and **7♠** 4♥. The 10-9 hand versus a dealer 6 gives an expected value of $+\$0.36114$ while the 7-4 hand gives an expected value of $+\$0.49691$, combining to get

$$\text{Expected Winnings (Switch)} = \underbrace{\$0.36114}_{\text{10-9}} + \underbrace{\$0.49691}_{\text{7-4}} = \$0.85805.$$

In this case, we should *definitely* switch! In fact we are very much in a favorable position with these two hands. Common sense, though, after playing a good bit of Blackjack should give you the intuition that the point totals of 19 and 11 are great; you should never complain about having a total of 19, and a total of 11 versus a 6 is a great double down opportunity!

There are, of course, situations where switching is not the ideal thing to do. As a quick example that you can think of, being dealt **J♣** 9♦ and **7♠** 4♥ initially, as we saw in Example 8.10, is very favorable to you as a player and switching makes things worse. Let's consider one more example here that is less obvious.

Example 8.11 Suppose your initial two hands at Blackjack Switch are A♦ **7♠** and 10♥ **5♠**, and the dealer has a three showing. Consulting Tables T.8 and T.9 as appropriate, the A-7 hand gives an expected value of $+\$0.04355$, while the 10-5 hand gives $-\$0.34404$ for a combined result of

$$\text{Expected Winnings (No Switch)} = \underbrace{\$0.04355}_{\text{A-7}} + \underbrace{-\$0.34404}_{\text{10-5}} = -\$0.30049,$$

while switching to A♦ **5♠** and 10♥ **7♠** gives expected values of $-\$0.08745$ and $-\$0.21830$, respectively. Combining these gives

$$\text{Expected Winnings (Switch)} = \underbrace{-\$0.08745}_{\text{A-5}} + \underbrace{-\$0.21830}_{\text{10-7}} = -\$0.30575$$

for the switched version. Since the non-switched hands are slightly better, we should not switch. Notice that standing on A-7 (from the Blackjack Switch basic strategy chart in Table 8.7) sounds relatively safe and provides the only positive expectation result from among the four hands potentially involved in this scenario.

Gaming mathematician Cindy Liu and professional gambler Arnold Snyder have developed easier systems to use when deciding to switch or not, as has professional actuary and author Michael Shackleford. For those systems, I encourage you to search online (Shackleford runs the popular gaming website WizardOfOdds.com, to which Liu has contributed, and Snyder published *The Big Book of Blackjack*, found as [48] in the Bibliography).

8.3   Exercises

8.1 Determine the probability of getting a hand totaling 20 on the initial two cards in Blackjack. Remember that either a pair of ten-valued cards, or an ace and a nine will give you such a hand.

8.2 Repeat Exercise 8.1 for using the following number of decks at Blackjack and comment on the effect of more decks on the chance of getting a hand total of 20.

    (a) two decks   **(b)** four decks   **(c)** six decks   **(d)** eight decks
    (e) infinite decks

8.3 Suppose the dealer has an ace showing face up, and you are the only player at the table. Determine the expected value of the insurance side bet for the following hands. Assume the casino is only using one deck of cards.

    (a) 5♠ 8♦   **(b)** 9♣ J♠   **(c)** J♦ Q♥

8.4 Repeat Exercise 8.3 assuming that you know two other players at the table, combined, have cards 2♦ 3♠ 6♦ 7♥. How does this extra information help or hurt us?

8.5 Suppose an urn, famous for holding marbles of different colors, has 5 blue marbles and 3 red marbles. Draw a probability tree similar to that of Figure 8.3 for the process of drawing two marbles from the urn without replacement. Then, use your probability tree to give the following probabilities.

    (a) $P(\text{red second}|\text{blue first})$ **(b)** $P(\text{blue second}|\text{blue first})$
    (c) $P(\text{blue second}|\text{red first})$ **(d)** $P(\text{red second and blue first})$
    (e) $P(\text{blue second and blue first})$

8.6 Imagine that a special new urn holds marbles of *three* different colors. This one contains 3 red marbles, 4 blue marbles, and 5 yellow marbles. Draw the probability tree as in Figure 8.3 for drawing two marbles without replacement.

8.7 Using Figure 8.4 and Example 8.7 as guides, draw the probability trees for both the dealer ending with a point total of 20 and ending with a point total of 21. Verify the probabilities of 0.342194 and 0.034501, respectively, given in the text.

8.8 Consider Example 8.9 and, instead of having the probability of losing represented by $1 - p$, let the probability of losing be $1 - p - r$ where r is the probability of a tie; determine the relationship between p and r that tells you when surrendering in the best option.

8.9 This exercise uses Table T.6, appearing at the end of this book, that lists the expected values for all player hard point totals versus all dealer face up cards (assuming basic strategy is used); use that table to answer the following questions.

(a) Suppose that you currently have a (hard) point total less than 18; would you rather see the dealer having a face up 6 or face up 7? Discuss how you made your decision and how the rules of Blackjack encourage and support this decision.

(b) If you currently have a (hard) point total of 18, would you rather a face up 6 or face up 7 for the dealer? How did you make your decision?

(c) Observe that no column in Table T.6 has *all* positive entries; what does this say about the relative safety of player hands versus the dealer?

8.10 Looking at Table T.7 which also appears at the end of this book, what face up card(s) would you like to see for the dealer's hand when your hand has a pair or an ace? Would you prefer to see that the dealer has a face up 3 or face up 4? Why?

8.11 This exercise (partially) examines how powerful the technique "Busting the Dealer" from the MIT blackjack team of the early 1990s (see [30]) can be. Use the Infinite Deck Assumption of Theorem 8.1 in your probability calculations.

(a) List all 54 possible unordered non-blackjack two-card hands that the dealer can receive at Blackjack. Treat all ten-valued cards as one type of card.

(b) For each hand listed in part (a), determine the probability of that hand (the total sum of your probabilities should be near 100% and would be 100% if you add the probability the dealer has blackjack).

(c) Add up the probabilities from part (b) that correspond to point totals for which the dealer would bust if the dealer had 16 points or less and had to draw, assuming that the next card is a ten-valued card.

(d) Comment on how useful it would be if you were able to predict exactly when a ten-valued card would be the next card of the deck (players could easily all stand in order to force the dealer to receive that card).

9

A Mix of Other Games

As you might imagine, it would be quite impossible to discuss every game that has ever existed from a mathematical point of view. It is, however, true that there are interesting mathematical questions about most games that have been played throughout time. I apologize if your favorite game was not included in this book; in selecting the games discussed, I've tried to choose some of my own personal fa-

FIGURE 9.1: Old Backgammon

vorites while balancing the desire to introduce mathematical topics in a cohesive way. This chapter is the home to some other games that either just didn't fit into one of the categories of the initial eight chapters or were similar enough in style to an earlier-presented game that, while having very interesting questions to explore, did not illustrate the mathematical toolbox items as well. Of course, I have used the "author's prerogative" in completing this chapter, so games are here simply because I have found the game to be fun, or found the questions about the game fun. I sincerely hope that you have enjoyed this book and the information contained within, but since you are not quite through the main portion of the entire book, please read on and enjoy!

9.1 The Lottery

As of the writing of this book, 45 of the 50 states in the USA currently have some form of state lottery. Chances are that you have seen some of these games, and perhaps have given a few a try. The so-called "Instant Win" tickets are usually given in the form of a card with several areas that can be removed by scratching a coin against the surface. The win conditions themselves usually take several forms, from finding three or more of the same dollar figure, anywhere or possibly in a row or column, and playing miniature versions of games such as Bingo and Monopoly, to being given a list of numbers or symbols called "winning numbers," matching "your numbers" to any of those to win a prize, such as the game pictured in Figure 9.2 courtesy of the Virginia Lottery.

The probabilities associated with these scratch-off tickets, at least to me, are not terribly interesting. Once the lottery officials decide exactly how much, either as a percentage or fixed dollar amount, they wish to pay out for the game and the amount and distribution of possible payout figures (usually ranging from prizes such as a free ticket, small dollar amounts roughly equal to the cost of the ticket, to huge dollar amounts), the tickets are printed in some semi-randomized fashion and given out to retailers to sell. Only luck and perhaps knowledge of what prizes have not yet been claimed can be used as possible winning "strategies" so, for us, there isn't much to study. It is nice, though, to know

FIGURE 9.2: Scratch-Off Ticket   that, at least in Virginia, the officials stop selling tickets for a particular game once the last top prize has been claimed.

Pick 3

Instead, the type of game that I want us to look at together is what lotteries call a "draw" game; a certain number of balls are placed in a machine that uses air to blow them around in a random manner. At particular times, an opening is created at the top of the machine so that one ball can come up to the top, and, if these balls are marked with numbers, this gives a random method to the selection of a subset of numbers. By purchasing a ticket and selecting your own set of numbers, payouts can be made based on how many of your numbers

FIGURE 9.3: Pick 3 Slip

match the lottery drawing's numbers, perhaps in a specific order or not. Two examples of such games are the multi-state lotteries Powerball and Mega Millions; we will discuss Mega Millions later but, for now, let's consider a simpler version of a draw game offered by the Virginia Lottery called Pick 3.

Pick 3 is played twice a day in Virginia; drawings happen both during the day and also later in the evening, and for each of these, players select a three-digit number between 000 and 999, and optionally choose different ways to play their number. A slip used to select your number, given to the lottery official or machine along with your cash, is pictured in Figure 9.3. One option is to select a number, such as 556, and mark *exact* which means that you win if the number drawn is precisely 556, in that order. By comparison, the *any* option will win if your number comes up in any order. By now, you should be able to quickly determine that there are 3 different orderings of the number 556, so this gives you three times the probability of winning. Of course, with a higher chance of winning, the reward is three times lower.

TABLE 9.1: Pick 3 Winnings for a $1 Bet

Bet	Winnings
Exact	$500
Any (6 Way)	$80
Any (3 Way)	$160
Front Pair	$50
Split Pair	$50
Back Pair	$50

The winnings for essentially each type of wager on a $1 bet is given in Table 9.1 (for a $0.50 wager, divide the winnings in half). Note that *any* has two possible payouts, depending on whether your chosen number has six different orderings, such as with 246, or only three different orderings, as with our 556. A *50/50* bet takes your $1 wager and puts half of it on exact for your number and the other half on any, with the payout determined normally from there. The *combo* bet, for a number such as 246, is a shorthand for paying $6, with $1 wagered on each possible ordering (and, thus, for a number like 556, costs $3 with $1 wagered on each of the three orderings). The *front pair*, *back pair*, and *split pair* wagers are essentially a "Pick 2" bet where you win if your number matches the first two, last two, or first and last digits, respectively, of the drawn number.

Question

Where does the money wagered on Pick 3 go?

The first question we can ask, of course, is how much does the lottery give back, on average, for a $1 exact wager for Pick 3. We can compute this as we've done in the past.

Example 9.1 With exactly 1,000 different possible three-digit numbers to choose from, the probability that your selected number matches precisely is $1/1,000$. In the words of Chapter 2 via Definition 2.1, we could also say that the true mathematical odds are 999:1 for winning, since there are 999 ways to lose and only one way to win. The *winnings* on the $1 wager can be computed, using Table 9.1 for the payout, as

$$\text{Expected Winnings} = \underbrace{\left(\frac{1}{1,000}\right) \cdot \$500}_{\text{win}} + \underbrace{\left(\frac{999}{1,000}\right) \cdot \$0}_{\text{lose}} = \$0.50$$

but, unlike casino games where your bet is returned to you, in all cases here we give up the $1 to the lottery, resulting in an expected value of −$0.50. We could also redo the calculation, subtracting the $1 from both the payout and loss to obtain

$$\text{Expected Value} = \underbrace{\left(\frac{1}{1,000}\right) \cdot \$499}_{\text{win}} + \underbrace{\left(\frac{999}{1,000}\right) \cdot (-\$1)}_{\text{lose}} = -\$0.50$$

for a 50% house advantage to the lottery.

So where *does* the money go? From Example 9.1 we see that $0.50 of every wagered dollar is returned, on average, to players of the game. Sure, maybe you did not win, but for those other players that did, the lottery must pay out. Of the remaining half dollar, anywhere from $0.05 to $0.10 typically goes to cover lottery operations (employees) and advertising, and the rest goes to support some organization or cause dictated by state law. For example, in Virginia, the proceeds go to support public education, and in Pennsylvania, the proceeds support senior citizens. There can be some controversy regarding how these funds are actually earmarked for these causes, and since our goal in this book is to study the mathematics and probabilities of these games, I will not argue for or against using a lottery as a way to pay for a state's expenses.

Before we move on, let's take a look at one other Pick 3 wager and compute the expected value of that bet.

Example 9.2 As mentioned earlier, the *50/50* bet for Pick 3 is a bet that takes half of your $1 wager and places it on *exact* and the other half and places it on *any*. Since the particular number we choose does play a role here, let's work with a "3 Way" number such as 556.

With probability 1/1, 000 we will win *both* the exact wager and the any wager (if 556 is the drawing), so on a $0.50 bet for each, we win $250 and $80 on these, respectively, for a total of $330 (note that these dollar amounts are half of what's listed in Table 9.1 since our wager is half as much). Subtracting the dollar we spent purchasing the ticket, our net profit for this win would be $329. Note, though, that with probability 2/1, 000 we still win the any wager (if 565 or 655 were the drawn number), so we win a net profit of $79 here (the $80 from a $0.50 wager on "Any (3 Way)" of Table 9.1, minus $1 spent). Combining this together, we get an expected value of the *50/50* wager of

$$\text{EV} = \underbrace{\left(\frac{1}{1,000}\right) \cdot \$329}_{\text{win, exact}} + \underbrace{\left(\frac{2}{1,000}\right) \cdot \$79}_{\substack{\text{win, any,} \\ \text{not exact}}} + \underbrace{\left(\frac{997}{1,000}\right) \cdot (-\$1)}_{\text{lose}} = -\$0.51,$$

giving a slightly larger lottery advantage than a standard exact bet.

At this point, and as you will realize in the exercises later, the "house" advantage for Pick 3 is more or less about 50%, giving the lottery officials room to pay their employees and move funds over to support public education, senior citizens, or some other group. Now, chances are that you have not played a Pick 3 or similar type of game that has a low payout relative to more famous multi-state drawings such as the Mega Millions.

Mega Millions

As of this writing, all states in the USA that have a lottery system participate in the Mega Millions multi-state lottery; drawings are twice a week, with one drawing picking a single winning list of numbers for all 45 states (and two other jurisdictions that participate). Tickets cost $2 and players select five numbers from 1 to 70, along with selecting 1 more number, called the "Mega Ball," from 1 to 25; this Mega Ball selection is independent of the first five se-

FIGURE 9.4: Mega Millions Play Slip

lected numbers, so even if you pick 12 as one of your first five numbers, you could also select 12 as your Mega Ball number. You win the jackpot if you match all five numbers in addition to matching the Mega Ball as well. A play slip is pictured in Figure 9.4.[1] In most jurisdictions, players can also wager an additional $1 using the "Megaplier" option which increases the prize for all non-jackpot amounts by a random amount chosen before each drawing that can range from twice to five times the normal prize.

Question

How likely am I to win the Mega Millions jackpot, and does it ever make sense to play?

At this point, we should be experts in answering the first part of the question. Since there is exactly one possible way to win, by matching all five of your numbers and the Mega Ball as well, we need only count the number of possible Mega Millions tickets. From the first 70 numbers, we must choose 5 of them, and from the 25 possible Mega Ball numbers, we need to choose 1.

---

[1]The rules for Mega Millions are correct as of July, 2020 (the format changed in October of 2017), but one should note that occasionally the range of numbers, for both the pick five portion and the Mega Ball portion, tend to change over time; the methods we use here, though, should be easily adaptable to future versions.

This gives us

$$\text{Number of Tickets} = \underbrace{\binom{70}{5}}_{\text{pick five}} \cdot \underbrace{\binom{25}{1}}_{\text{Mega Ball}} = 12,103,014 \cdot 25 = 302,575,350$$

which is a boat-load (or two) of possible tickets! This means, of course, that the probability of winning the jackpot is 1 in about 302.5 million, or true mathematical odds of 302,575,349:1. Other prizes are given in Table 9.2, where the Pick 5 column indicates the quantity of the pick five portion that you match, the Ball column indicates whether you matched the Mega Ball or not, and the Probability column gives the probability of winning that exact prize. We will see how to compute some of these later.

TABLE 9.2: Mega Millions Prizes

Pick 5	Ball	Prize	Probability[2]
5	Yes	Jackpot	$1/302,575,350$
5	No	$1,000,000	$1/12,607,306$
4	Yes	$10,000	$1/931,001$
4	No	$500	$1/38,792$
3	Yes	$200	$1/14,547$
3	No	$10	$1/606$
2	Yes	$10	$1/693$
1	Yes	$4	$1/89$
0	Yes	$2	$1/37$

The question about ever playing is more interesting; we can use expected value computations to help answer that. Because the amount of the jackpot varies (each $2 wager increases the jackpot by about $1 since roughly half of each wager is retained by the state lottery that sold the ticket), the expected value becomes a function of the jackpot amount. Since the advertised jackpots are in the form of an *annuity* that is paid out over several years based on prevailing interest rates, we will use the "cash option" (or lump sum) amount in our computations to come. This number is typically also announced, albeit not as prominently, and is typically about 60% to 70% of the advertised annuity amount.

We can definitely use the information in Table 9.2 to develop the expected value formula, in terms of J, however, since, as a footnote alluded to, the probabilities given in that table are *rounded*, we should take some time to develop the unrounded values. First, note that the probability for the jackpot is correct, since we have already computed the number of possible Mega Millions tickets. For matching all five balls of the first draw and *not* matching

---

[2]Note that the probabilities given here are the official values listed on the Mega Millions website; these numbers have been slightly rounded, as we shall see.

the Mega Ball, the probability can be computed as

$$\text{Probability } 5+0 = \frac{\overbrace{\binom{5}{5}}^{\text{match five}}}{\underbrace{\binom{70}{5}}_{\substack{\text{all possible}}}} \cdot \underbrace{\binom{24}{25}}_{\substack{\text{don't match}\\\text{Mega Ball}}} = \frac{4}{50,429,225} \approx \frac{1}{12,607,306}$$

where the last number is the approximation given by the Mega Millions website. To match three of the first five and also the Mega Ball, winning \$200, we would get

$$\text{Probability } 3+1 = \frac{\overbrace{\binom{5}{3} \cdot \binom{65}{2}}^{\text{match three, two others}}}{\underbrace{\binom{75}{5}}_{\substack{\text{all possible}}}} \cdot \underbrace{\binom{1}{25}}_{\substack{\text{match}\\\text{Mega Ball}}} = \frac{416}{6,051,507} \approx \frac{1}{14,547}.$$

The remainder of the probabilities, along with the two we have computed, are given in Table 9.3. You'll be asked to verify some of the other probabilities later, but now that we have the true values, we can proceed with the task of computing the expected value. Note that the probability for "Else" is obtained easily by adding up all of the probabilities for winning something and subtracting from 100%, and the shorthand notation used, $x + y$, in the result means matching x of the first five numbers for the first part of the draw, and y is either 0 or 1, with 1 indicating a matched Mega Ball and 0 indicating otherwise.

TABLE 9.3: Mega Millions Probabilities

Result	Probability
5+1	$1/302,575,350$
5+0	$4/50,429,225$
4+1	$13/12,103,014$
4+0	$52/2,017,169$
3+1	$416/6,051,507$
3+0	$3,328/2,017,169$
2+1	$416/288,167$
1+1	$3,224/288,167$
0+1	$196,664/7,204,175$
Else	$6,903,936/7,204,175$

In our expected value computation, we use J to represent the unknown value of the lump sum jackpot. Note that while we typically only list each prize (or payout) once while determining expected values, because we have two different Mega Millions results that offer the same prize with different probabilities, we will have two entries with the same payout in Example 9.3. No worries, since remember that when using expected value, we take the sum of all possible *results* and not necessarily all possible payouts.

Example 9.3 Since the probabilities are already given in Table 9.3, our job is fairly easy. Note that in our computation, we again subtract \$2 from each of the prizes given, since any ticket, win or lose, costs that much. We obtain

$$\text{Expected Value} = \underbrace{\left(\frac{1}{302,575,350}\right) \cdot (J-2)}_{5+1} + \underbrace{\left(\frac{4}{50,429,225}\right) \cdot 999,998}_{5+0}$$

$$+ \underbrace{\left(\frac{13}{12,103,014}\right) \cdot 9,998}_{4+1} + \underbrace{\left(\frac{52}{2,017,169}\right) \cdot 498}_{4+0}$$

$$+ \underbrace{\left(\frac{416}{6,051,507}\right) \cdot 198}_{3+1} + \underbrace{\left(\frac{3,328}{2,017,169}\right) \cdot 8}_{3+0} + \underbrace{\left(\frac{416}{288,167}\right) \cdot 8}_{2+1}$$

$$+ \underbrace{\left(\frac{3,224}{288,167}\right) \cdot 2}_{1+1} + \underbrace{\left(\frac{196,664}{7,204,175}\right) \cdot 0}_{0+1} + \underbrace{\left(\frac{6,903,936}{7,204,175}\right) \cdot (-2)}_{\text{completely lose}}$$

$$= \frac{J-2}{302575350} - \frac{265,210,061}{151,287,675}.$$

Since a game is in *our* favor if the expected value is positive, we should consider Mega Millions to be a game in our favor, using the result from Example 9.3, whenever

$$\frac{J-2}{302575350} - \frac{265,210,061}{151,287,675} > 0$$

which happens when $J > 530,420,124$. That is, if the lump sum jackpot is greater than \$530,420,124 (or annuity jackpot greater than around \$816,030,144), using expected value *only* as a guide, it is worth our time and money to play. Of course, that assumes that only one ticket matches the jackpot! Should two people both have winning jackpot tickets, each person receives half of the jackpot, so you would need to replace J in Example 9.3 by $0.50J$, leading to a values of J (yielding positive expected value) greater than \$1,060,840,248. Or, supposing that three or four tickets match all five numbers and the Mega Ball, values greater than \$1,591,260,372 and \$2,121,680,496, respectively.

As with any game of chance, I should remind you that any wager should be done as a form of entertainment only. The odds are definitely stacked against the player, and even for the lottery, the odds are definitely not favorable. Sure, you may have a positive expectation when the jackpot is large enough, but that positive expectation is based entirely on the prize for the single most unlikely outcome. No matter what the jackpot is, you are expected to lose almost 96% of the time!

9.2 Bingo

While Bingo could very well be considered a lottery itself, I think the popularity of the game along with its more in-person play style earns it a separate section in this book. The game is very simple to play. Each player gets one (or more) Bingo cards; each card consists of five columns of numbers, a column associated to each of the letters in the name Bingo. The "B" column will

B	I	N	G	O
12	18	41	47	61
7	26	39	54	70
4	27	·	49	63
5	23	35	58	73
3	30	32	52	75

FIGURE 9.5: Bingo Card

consist of five numbers selected from the range 1 through 15. Column "I" has five numbers chosen from 16 to 30. As a built-in added bonus, the "N" column only has four numbers selected from 31 to 45; the middle space in this column is always designated as a Free Space and if this space is included in the target pattern, then all players start with one space filled in already! The "G" and "O" columns have five numbers each, with those numbers chosen from the range 46 to 60 and 61 to 75, respectively. A sample Bingo card is shown in Figure 9.5. Once players have their Bingo cards, the caller selects a pattern and then selects from the 75 numbers at random, calling each number with its associated letter one at a time. The first player to match the desired pattern on their card with the Bingo numbers that have been called stands up, yells "Bingo," and after verification, wins the round. There may be some cost to play based on the number of cards you desire, and the winner of a round typically wins some sort of prize.

Question

How many essentially different Bingo cards are possible?

Counting Cards

This seems like a very simple question, and given the experiences we've had throughout this book with various counting techniques, it should be straightforward to answer. However, the answer *very much* depends on the pattern being used. Why is this true? The standard Bingo pattern that most people learn is the regular "five in a row" game where a player wins by matching five numbers in a row either horizontally across a row, vertically down a column, or diagonally down one of the two possible diagonals. Looking at the card of Figure 9.5, switching B12 and B7 does not matter as far as the "B" column goes (order doesn't matter within columns), but it *does* matter for horizontal matchings. The row B7 I26 N39 O54 G70 is different with the swap of B12 and B7 (order within columns matters for purposes of matching rows). While counting Bingo cards still isn't hard, per se, the changing patterns for various rounds of Bingo do make counting Bingo cards interesting. Note that by "essentially different" we are defining two cards to be different if there exists a win condition (set of called numbers) for one card that is *not* also a win for the other.

> **Example 9.4** To count the number of essentially different Bingo cards for the standard "five in a row" game, note that since order within columns matters for purposes of lining up the rows, we need to select five of the possible 15 numbers for columns B, I, G, and O, and four of the possible 15 numbers for column N, in such a way that order matters; we'll use the permutations of Definition 4.4 to obtain
>
> $$\text{Bingo Cards} = \underbrace{{}_{15}P_5}_{\text{B}} \cdot \underbrace{{}_{15}P_5}_{\text{I}} \cdot \underbrace{{}_{15}P_4}_{\text{N}} \cdot \underbrace{{}_{15}P_5}_{\text{G}} \cdot \underbrace{{}_{15}P_5}_{\text{O}}$$
> $$= \underbrace{(15 \cdot 14 \cdot 13 \cdot 12 \cdot 11)^4}_{\text{B, I, G, and O}} \cdot \underbrace{15 \cdot 14 \cdot 13 \cdot 12}_{\text{N}}$$

which, equalling 552,446,474,061,128,648,601,600,000, is a fairly large number! We have slightly overcounted, however, because rows 1 and 2 can be swapped for rows 4 and 3, respectively, without any change in game play, as you can see in Figure 9.6, the card obtained from Figure 9.5 by exchanging rows as we mentioned (anytime the card in Figure 9.5 wins, the card in Figure 9.6 also wins, including in diagonal cases, and vice-versa). We need to divide our total by two, getting

Essentially Different Bingo Cards = 276, 223, 237, 030, 564, 324, 300, 800, 000.

Note, however, that one "problem" with the five in a row game is that even with two different cards, they could still win at the exact same time. Two cards whose only difference is a variation in ordering of the B numbers will both win in the B column at the same time (order doesn't matter within columns). Two cards with the exact same row 2 will win at the same time even if every other number is different or shifted around!

B	I	N	G	O
3	30	32	52	75
5	23	35	58	73
4	27	·	49	63
7	26	39	54	70
12	18	41	47	61

FIGURE 9.6: Bingo Card Number 2

Another pattern that is very popular in Bingo is the "blackout" (or "coverall") game where the winner is the first player to get every single space on their Bingo card called. In this case, counting is a *lot* easier and produces fewer essentially different Bingo cards.

Example 9.5 To count the number of different cards for blackout, first note that order within columns does not matter at all. From any of the 15 numbers possible for the B column, we simply need to choose a subset of 5 numbers. The same is true for the I, G, and O columns, and for the N column we choose 4 of the possible 15. Using the combinations of Definition 3.2, we get

$$\text{Blackout Cards} = \underbrace{\binom{15}{5}}_{B} \cdot \underbrace{\binom{15}{5}}_{I} \cdot \underbrace{\binom{15}{4}}_{N} \cdot \underbrace{\binom{15}{5}}_{G} \cdot \underbrace{\binom{15}{5}}_{O}$$
$$= 111,007,923,832,370,565$$

essentially different Bingo cards as it applies to the blackout pattern.

A very similar pattern that comes up playing Bingo (especially in the theater or pool-side on a cruise) is the X. This pattern requires the four corners (itself a popular game as well) along with the four spaces diagonally adjacent to the free space in the middle. It also requires the free space, but everyone gets this space, not surprisingly, for free.

Example 9.6 Of the five locations for the B column, we only care about the top and bottom entries; this also applies to the O column. Similarly, we only care about the entries in rows 2 and 4 for columns I and G. For each specific column, the order of the numbers for the two spaces is irrelevant, so we can count the number of essentially different Bingo cards for X as

$$\text{X Cards} = \underbrace{\binom{15}{2}}_{B} \cdot \underbrace{\binom{15}{2}}_{I} \cdot \underbrace{\binom{15}{2}}_{G} \cdot \underbrace{\binom{15}{2}}_{O} = 121,550,625$$

which is much less than we've seen for other patterns. Note that the number of essentially different cards for the "four corners" game is computed very similarly as

$$\text{Four Corners Cards} = \underbrace{\binom{15}{2}}_{B} \cdot \underbrace{\binom{15}{2}}_{O} = 11,025$$

which is a very manageable number compared to the patterns we have discussed earlier.

Note that for both of these cases, there are many different ways to fill in the spaces not used by the X (or the four corners), and even among the spaces included, the two numbers in a specific column could be swapped. Each of these variants for a set choice of numbers for B, I, G, and O results in the same exact win condition, so even though your neighbor's card looks very different, it could very well be the same as yours for these patterns.

Examples 9.4, 9.5, and 9.6 illustrate ways to count the number of relevant Bingo cards possible for given patterns, but computing the probabilities involved in Bingo is much harder. In particular, as you can imagine based on how the counts depended on the patterns, the way to compute probabilities *also* depends on the pattern. The

	# of Different Cards in Play			
Game	500	1,000	1,500	2,000
Standard	12.10	10.84	10.22	9.85
Blackout	60.17	58.84	58.08	57.55
Blackout (2nd)	62.46	61.56	61.06	60.72
Blackout (3rd)	63.86	63.11	62.70	62.43
Blackout (4th)	65.01	64.31	63.95	63.71
X^3	28.63	26.63	25.54	24.81
N	38.52	36.92	36.04	35.43
Big H	38.55	36.97	36.09	35.49

TABLE 9.4: Expected Number of Calls Until Bingo

expected number of calls for the first Bingo to be called for different games (including a second, third, and fourth chance at blackout), based on simulations, are given in Table 9.4.

Let's take a look at the blackout game some more. In particular, our goal will be to determine the probability that someone will be able to shout out Bingo after the caller has called n numbers. We look at the specific case of $n = 24$ in this example. As you can probably guess at this point, the chance that your specific 24 numbers are called as the very first 24 numbers might be rather small.

Example 9.7 Since there are 24 distinct numbers on a Bingo card, the rarest case of getting Bingo early cannot happen until $n = 24$ and each of the numbers on a card have been announced. Of the 75 numbers that can be called, we need to choose 24 of them, and among the ways that this can be done, count the number of valid Bingo cards with those numbers. Note that the numerator below was seen in Example 9.5 and ensures that of the 24 numbers called, 5 are from the B column, 5 from the I column, and so on. The probability that someone has Bingo for the pattern blackout on call 24 is

$$\text{Probability Blackout } (n = 24) = \frac{\overbrace{\binom{15}{5} \cdot \binom{15}{5} \cdot \binom{15}{4} \cdot \binom{15}{5} \cdot \binom{15}{5}}^{\text{different Bingo cards}}}{\underbrace{\binom{75}{24}}_{\text{possible calls}}}$$

$$= \frac{111007923832370565}{25778699578994555700} \approx 0.00430619.$$

---

[3]These values do not include calling numbers from the N column; if for some reason those numbers are still called, the expected number of calls becomes 35.62, 33.13, 31.77, and 30.84 for the indicated number of different cards in play.

Of course, this is the probability that someone calls Bingo on turn 24 assuming that *all* of the possible blackout Bingo cards are in play! The probability of any specific card winning on turn 24 would be much, much smaller, with

$$\text{Probability Specific Card Blackout } (n = 24) = \frac{1}{25778699578994555700}$$
$$\approx 0.00000000000000000003879172$$

as that probability. Don't expect that to happen any time soon; you have a much higher chance of winning the Mega Millions jackpot *twice* before you have your specific Bingo card win at blackout on turn 24.

Bingo Probabilities

A better analysis is to determine the probability that your bingo card wins at blackout on an arbitrary number, n, of picks. That is, we might compute the chance that your bingo card wins on turn 25, on turn 26, or on any other turn. It turns out that this is relatively easy to do.

Example 9.8 To win on turn 25 *exactly*, one of the twenty-four numbers on your card needs to *not* be called, and this can be chosen in 1 of 24 ways; in addition, of the first 24 numbers called, exactly one of them needs to not be on your card. With 75 numbers total and 24 used on your card, we can select this "bad" number by choosing 1 of those 51 numbers not on the card. By considering the sample space of all ways to select these 24 numbers, we effectively compute the probability that after 24 calls, you're "one away" as the bingo players would say. Multiplying by 1/51 to get our last number on that twenty-fifth call gives us a computation of

$$\text{Specific Card Blackout } (n = 25) = \frac{\overbrace{\binom{24}{1} \cdot \binom{51}{1}}^{\text{missing number, "bad" number}}}{\underbrace{\binom{75}{24}}_{\text{possible calls}}} \cdot \underbrace{\left(\frac{1}{51}\right)}_{\substack{\text{get last} \\ \text{number}}}$$

$$= \frac{2}{2148224964916212975}$$
$$\approx 0.000000000000000000931001$$

which is still a small number but about 10 times as likely as we found in Example 9.7 when n was 24.

Following in the footsteps of Example 9.8 for $n = 26$, the changes that need to be made are minimal. We still need one of the numbers on our card to be missing, but there will be two "bad" numbers to choose from the 51 "bad" numbers available. The sample space will expand to choosing 25 of the possible 75 numbers, and that last call will match the missing number on our card with probability 1/50 now since 25 numbers have already been used. This gives us

$$\text{Specific Card Blackout } (n = 26) = \underbrace{\frac{\overbrace{\binom{24}{1} \cdot \binom{51}{2}}^{\text{missing number, "bad" ball}}}{\underbrace{\binom{75}{25}}_{\text{possible calls}}}} \cdot \underbrace{\left(\frac{1}{50}\right)}_{\text{get last ball}}$$

$$\approx 0.000000000000000116375$$

which is, as expected, an even higher chance.

Theorem 9.1 For $24 \leq n \leq 75$, the probability that your specific card will win at blackout bingo exactly on pull n is given by

$$PBB(n) = \frac{\binom{24}{1} \cdot \binom{51}{n-24}}{\binom{75}{n-1}} \cdot \left(\frac{1}{76-n}\right)$$

where the format for the formula follows as in Example 9.8.

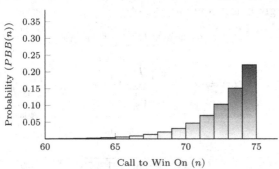

FIGURE 9.7: Blackout Bingo Probabilities

Figure 9.7 shows the probability of your specific bingo card winning during a blackout game for values of n ranging from 60 to 75. Other values are not shown as the probability of winning on any of the values of n below 60 is below 0.04%, and the *cumulative* probability that you win with your card on turns 24 through 59 is barely 0.084% which may seem crazy, but how many times have you personally won at blackout early in the game?

When might you *expect* to have your bingo card covered? We can use the expected value formula that we've used in the past to compute the average number of calls needed for *your* card to win. This is

$$\text{Expect Calls to Win} = \sum_{i=24}^{75} i \cdot PBB(i) = 72.96$$

or almost 73 calls! While this may seem high at first, simulations done by bingo system manufacturer Arrow International show that with 500 distinct bingo cards in play, it *still* takes about 60.17 calls for at least one card to win, decreasing slightly to 55.58 calls needed for 6,000 distinct bingo cards in play. While your individual chances are not that high, at the onset, at least, they are the same as everyone else's.

9.3 Uno

Another card game that is popular among families is Uno.[4] In this game, players try to be the first to have played all of the cards in their hand; each player starts with seven cards randomly dealt from the deck. A standard deck in Uno has 108 cards; of these cards are two each of the numbers 1 through 9 from each of the four colors (red, yellow, green, and blue). There is also a single number card of 0 for each

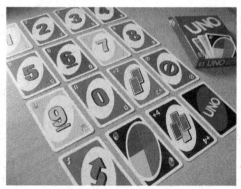

FIGURE 9.8: Uno Cards

of the colors. Special action cards exist as well, with two "skip" cards, two "reverse" cards, and two "draw two" cards in all four of the colors, as well as four "wild" cards and four "draw four plus wild" cards (there are just eight of the various wild cards as they do not correspond to any specific color). Several of these Uno cards are shown in Figure 9.8.

Our work in Chapter 3 mostly prepared us to answer questions about Uno. The main difference here, which will be even more present with Scrabble in Section 9.6, is that some cards are not unique. With a regular deck of playing cards (jokers not included), there are 52 distinct cards; here, however, there are a total of four "draw four plus wild" cards. We proceed to explore a few questions about Uno.

---

[4]Uno is a registered trademark of Mattel, Inc.

Question

What's the probability of getting at least one "wild" or "draw four plus wild" cards in our opening hand?

Example 9.9 To determine the probability of getting at least one type of wild card in our opening hand, we make use of the fact that it's much easier to figure out the chance that an opening hand has *no* wild cards at all. As you may have already been thinking, the total number of possible initial Uno hands is gotten by choosing 7 cards from the 108; to get none of the wild cards at all, we need to choose 7 cards from the 100 cards that are not wild in any form. This give us

$$\text{Probability 1+ Wild Cards} = 1 - \frac{\binom{100}{7}}{\binom{108}{7}} = 1 - \frac{16007560800}{27883218168} \approx 0.4259$$

which means that, even though there are only eight "wild" or "draw four plus wild" cards in the deck, we still have about a 43% chance of getting one in our opening hand.

We can modify this example slightly to figure out the probability of getting *any* special card in your opening hand. Besides those eight cards that serve as a type of wild, there are six cards in each of the colors that are "special" for a total of 32 special cards. The probability of getting at least one of them is

$$\text{Probability 1+ Special Cards} = 1 - \frac{\binom{76}{7}}{\binom{108}{7}} = 1 - \frac{2186189400}{27883218168} \approx 0.9216$$

so there's about a 92% chance that your opening hand will have at least one non-numbered card. If you play a game of Uno and get only numbered cards in your hand, you're very likely in the minority in that game!

Both of these calculations were relatively easy; the next one requires a bit more work.

Question

What is the chance that an opening hand has at least one number card from blue and at least one number card from red?

Example 9.10 There is some subtly in this question. Since there are 19 numbered cards in a given color (two each of the numbers 1 through 9, along with a single 0), we would initially consider choosing 1 of the 19 blue cards and choosing 1 of the 19 red cards; the other five cards in your hand could come from any the remaining cards. The issue that some of those other five cards might *also* be blue or red, and we'd rather not add together the probabilities of getting exactly one blue and one red numbered card, two blue and one red numbered cards, one blue and two red numbered cards, three blue and one red numbered cards, and so on and so forth.

Instead, let us start with all possible hands and then subtract both those hands that have no blue numbered cards and those hands that have no red numbered cards. This, however, removes a bit *too much* since any hand that has neither blue numbered cards nor red numbered cards is removed by both subtractions, so we need to add them back to get a final count. This gives us

$$\text{Number of Hands} = \underbrace{\binom{108}{7}}_{\text{all hands}} - \underbrace{\binom{89}{7}}_{\text{no blue}} - \underbrace{\binom{89}{7}}_{\text{no red}} + \underbrace{\binom{70}{7}}_{\substack{\text{no blue} \\ \text{and} \\ \text{no red}}} = 15,301,455,744$$

as the number of hands that have at least one numbered card from each of blue and red. Note that the term that chooses 7 out of 89 for the "no blue" entry comes from seeing there are 89 cards in the deck that are non-blue or blue, but not numbered (the same is also true for red), and the term of "no blue and no red" comes from only having 70 cards that fit the criteria.

The final probability, then, we get by dividing by the total number of hands. We get

$$\text{Desired Probability} = \frac{15301455744}{27883218168} \approx 0.5488$$

as our answer.

What we did in Example 9.10 is an instance of using a process called inclusion-exclusion; we will cover that method more formally in Section 9.6.

9.4 Baccarat

The game of baccarat has recently become a popular casino game, especially among players that wish to wager a large sum of money. Traditionally, the game was not available for usual table play in casinos and was instead *only* played in special rooms designed for the game and under increased security due to the large sums of money being exchanged (in private rooms, it is not uncommon for tens or hundreds of thousands of dollars to be wagered on a single hand). While the game has been popular in Europe for many centuries, its rise in North America may be due to its appearance in popular media; baccarat is James Bond's favorite game and can be seen in many of the 007 movies!

The version that can be found in American casinos these days is the "punto banco" variant of baccarat and is the one that we will discuss here; you can read about other versions of the game in other sources or online. Our version will also use the standard 8 decks of cards, though 4-deck and 6-deck tables can occasionally be found.

For game play, two cards are dealt to the player hand (even if there are multiple players at a table, only one player hand is dealt) and two cards are dealt to the banker hand. Depending on the point total for each hand, either the player hand, banker hand, or both hands will receive a third (and final) card. Note that all cards are dealt face up. Point totals are compared, and the hand with the higher point total wins.

Note that gamblers wagering on this game can bet (before any cards are dealt) that the player hand will win *or* the banker hand will win (the names player and banker are based on tradition). Bets on the player hand are paid even money (1:1 payout) if it wins, and bets on the banker hand are paid at 95% (19:20 payout due to a 5% commission on betting on the banker). Ties are either returned to the bettor as a push (which we will assume here) or left on the table for the next hand.

Cards two through nine have a point value equal to the number on the card; tens and face cards (Jacks, Queens, and Kings) are worth zero points, while aces are worth one point. All point totals are computed modulo ten, a fancy way of saying that *only* the right-most digit in the total (the ones place) matters! The exact rules for when each of the hands receives a third card are strictly based on the point total of the hand; there is no strategy in determining whether or not a hand should draw because the following rules completely specify how game play is to proceed. First, if either the player hand or banker hand has a point total of 8 or 9, no one receives a third card and the result is determined. Otherwise,

- the player hand receives a third card if its point total is between 0 and 5, inclusive, and

- the player hand *stands* with its total if the point total is 6 or 7 (remember, if the total were eight or nine, neither hand would receive an additional card).

Then, for the banker's hand,

- if the player hand *did not* receive a third card, the banker hand only gets a third card if the banker's total is between 0 and 5, inclusive,

- otherwise, if the player drew a 2 or 3, the banker draws on a total of between 0 and 4 inclusive, and

- if the player drew a 4 or 5, the banker draws on a total of between 0 and 5 inclusive, and

- if the player drew a 6 or 7, the banker draws on a total of between 0 and 6, inclusive, and

- if the player drew an 8, the banker draws on a total of between 0 and 2 inclusive, and

- otherwise, if the player drew a 9, 10, face card, or ace, the banker draws on a total of between 0 and 3 inclusive.

Note that players may also bet that the two hands will *tie* with a payout of 8:1.

The chance that either the player hand wins or the dealer hand wins (or a tie is obtained) are very easy to compute with a program; if you have experience with computer programming, I encourage you to give this a shot by iterating over all possible *six card* sequences, determining the probability for each of those, and

TABLE 9.5: Probability of Baccarat Outcomes

Hand	Probability		
Total	Player Win	Banker Win	Tie
0	0.000000000	0.000000000	0.005797838
1	0.004929418	0.004859781	0.004101157
2	0.008868546	0.008939180	0.004002603
3	0.012593319	0.014590230	0.004451528
4	0.017238677	0.032682428	0.007261153
5	0.024575528	0.043357075	0.007939353
6	0.062551817	0.053863716	0.019240164
7	0.081634947	0.076880493	0.020350027
8	0.111200262	0.106016854	0.010979401
9	0.122654096	0.117407667	0.011032745
Total	0.446246609	0.458597423	0.095155968

determining the outcome of that six card sequence (for many of these hands, the fifth and sixth cards will be irrelevant for which hand wins, but since the first four (or five) cards for each of these sequences will result in the *same* outcome, considering them separately is okay). The results are given in Table 9.5 where the overall probability for a player win, dealer win, or tie is in the last row (the other rows are the probability of a win for each player or a tie with the given (winning or tying) total).

With probabilities in hand, we can now compute some expected values as we have done in the past. Let's do the expected values for betting on the player hand and for betting on the banker's hand; the tie bet will be left for you later, along with some other variants commonly seen in casinos.

Example 9.11 Recall that a bet on the player's hand has a payout of 1:1 casino odds. We compute the expected value of this bet using the probabilities from Table 9.5, remembering that for a $1 bet, a win nets us $1, a loss results in −$1 for us, and a tie results in no profit ($0) for us. The result is an expected value of

$$\underbrace{0.446246609 \cdot \$1}_{\text{win}} + \underbrace{0.458597423 \cdot (-\$1)}_{\text{lose}} + \underbrace{0.095155968 \cdot \$0}_{\text{tie}} \approx -\$0.0124$$

or about a 1.24% house advantage for betting on the player's hand.

For the banker's hand, recall that the bet is *almost* an even-money wager except that there's a 5% commission charged on a win; the casino odds of 19:20 amount to the equivalent casino odds of 0.95:1, so our $1 wager will only net us $0.95 on a win. The expected value we compute as

$$\underbrace{0.458597423 \cdot \$0.95}_{\text{win}} + \underbrace{0.446246609 \cdot (-\$1)}_{\text{lose}} + \underbrace{0.095155968 \cdot \$0}_{\text{tie}} \approx -\$0.0106$$

for a 1.06% house advantage (including the commission) for betting on the banker's hand.

After the card counting discussion in Chapter 8, you might imagine that card counting can be advantageous in baccarat. Surprisingly, Edward Thorp (the mathematician who first proved that card counting can overcome the house advantage in blackjack) found, using mathematical and statistical analysis, that this is *not* the case. He wrote that "advantages in baccarat are very small, they are very rare and the few that occur are nearly always in the last five to twenty cards" left in the shoe [54]. Because of this, even though the probabilities in Table 9.5 assume a full shoe of eight decks, they are essentially correct no matter how many cards have been seen. Despite this, keeping track of the cards seen has been part of the tradition for baccarat for years; casinos have recently even installed video monitors displaying what cards have been used from the deck. If knowledge of the cards seen could even remotely give bettors an advantage, surely they wouldn't want this information known, would they?

9.5 Farkle

Farkle is a folk game that has been around for awhile that has recently made a comeback; it has been popular at parties for years and more recently on Facebook, and due to this rise in popularity many commercial packagings of the game are now available. On the surface, Farkle may share a lot of similarity with Yahtzee, but at its heart the game is quite different in how it's played and strategies that can be used.

Players use 6 six-sided dice to play; on their turn, the dice are rolled. At least some of the dice that score points must be saved and then the player has the option to roll again. This process continues until the player

- has no scoring dice among the most recently rolled dice, called a "Farkle," upon which any points accumulated this turn are lost,

- scores with all six of his or her dice, upon which the player temporarily records the turn's score and may reroll *all* six dice and continue, or

- decides to stop risking the accumulated points and decides to stop rolling and score.

Since Farkle has its roots as a folk game, the exact rules for scoring vary. Our coverage will assume a particular set of scoring *categories* but we will not worry about the actual points scored; you can do your own research into the game and find a scoring system that works best for you and the folks you play with.[5] A fairly common set of scoring categories is given in Table 9.6; note that fives and ones are always scoring dice by themselves but can also

TABLE 9.6: Farkle Scoring Categories

Category	Example
Six of a Kind	⚃ ⚃ ⚃ ⚃ ⚃ ⚃
Five of a Kind	⚃ ⚃ ⚃ ⚃ ⚃
Four of a Kind	⚃ ⚃ ⚃ ⚃
Three of a Kind	⚃ ⚃ ⚃
Straight	⚀ ⚁ ⚂ ⚃ ⚄ ⚅
Three Pairs	⚃ ⚃ ⚃ ⚃ ⚅ ⚅
A Single Five	⚄
A Single One	⚀

be a part of another scoring category, such as three of a kind, as well. Scoring categories on all rerolls must be achieved with the rerolled dice *only*; a saved five and two new fives from a reroll cannot combine to count as a three of a kind with fives.

Question

What is the probability of a Farkle?

Since farkling ends the round with you forfeiting any points accumulated so far that round, it makes sense to determine the probability of this happening given the number of dice you have yet to roll. That is, when rolling n dice, where $1 \leq n \leq 6$, what is the probability that those n dice will result in zero points. Determining these probabilities now is made especially nice since ones and fives are always scoring dice, so we need only consider twos, threes, fours, and sixes as possible outcomes for dice in the patterns that produce zero points. We'll do our analysis using the multinomial coefficients found in Definition 4.5.

Example 9.12 To farkle out when rolling all six dice ($n = 6$), recall that three of a kind (or more of a kind) will score, so we're looking for all possible rolls that do not have three of a kind or better; since three pairs is also a scoring category, this means from among twos, threes, fours, and sixes, we must roll two pair and two singletons such as in the roll ⚃ ⚃ ⚃ ⚅ ⚁ ⚁ (there cannot be only one pair as the other four dice, chosen from the other three dice types, would also need to form a pair). We're making two multi-chooses here, as first, from the 4 dice faces available, we need to choose 2

---

[5]Different scoring systems have various ways of scoring the "of a kind" categories and how these categories interact with ones and fives.

for the pairs and 2 for the singletons. Then we need to choose, from the 6 dice, which dice form the first pair, second pair, first singleton, and second singleton. This probability is then

$$\text{Farkle with 6 Dice} = \frac{\overbrace{\binom{4}{2,2}}^{\text{numbers}} \cdot \overbrace{\binom{6}{2,2,1,1}}^{\text{choose dice}}}{\underbrace{6 \cdot 6 \cdot 6 \cdot 6 \cdot 6 \cdot 6}_{\text{six dice possibilities}}} = \frac{1080}{46656} \approx 0.023148$$

since with six dice, there are $6^6 = 46,656$ possible outcomes.

Example 9.12 shows that there's about a 2.3% chance that you farkle out on the first roll. The probability, of course, will change when you have fewer dice, and you can reasonably expect that probability to increase. For $n = 5$ dice, there are two patterns that must be considered, and this scenario is left for you in the exercises. For now, let's take a look at $n = 4$.

Example 9.13 With $n = 4$ dice, it's possible that we get two pairs, a pair and two singletons, or four singletons for the non-scoring patterns. For two pairs, we again, from the 4 numbers, choose 2 of them to represent our pairs, and then from the 4 dice, choose 2 to represent the first pair and 2 to represent the second pair. This probability is

$$\text{Farkle with Two Pair, 4 Dice} = \frac{\overbrace{\binom{4}{2}}^{\text{numbers}} \cdot \overbrace{\binom{4}{2,2}}^{\text{choose dice}}}{\underbrace{6 \cdot 6 \cdot 6 \cdot 6}_{\text{four dice possibilities}}} = \frac{36}{1296} \approx 0.027778.$$

For the pattern with a pair and two singletons, of the 4 numbers (remember, we are not including ones or fives), we choose 1 for the pair, 2 for the singletons, and 1 for the unused number. Then, from the 4 dice we are rolling, choose 2 to give us the pair, 1 for the first singleton, and 1 for the last singleton, resulting in a probability of

$$\text{Farkle with One Pair, 4 Dice} = \frac{\overbrace{\binom{4}{1,2,1}}^{\text{numbers}} \cdot \overbrace{\binom{4}{2,1,1}}^{\text{choose dice}}}{\underbrace{6 \cdot 6 \cdot 6 \cdot 6}_{\text{four dice possibilities}}} = \frac{144}{1296} \approx 0.111111.$$

Finally, for four unmatched dice, we choose 4 numbers from the 4 possible numbers, and then for the 4 dice, we choose 1 for the first singleton, 1 for the second singleton, and so on, resulting in a probability of

$$\text{Farkle with No Pair, 4 Dice} = \frac{\overbrace{\binom{4}{4}}^{\text{numbers}} \cdot \overbrace{\binom{4}{1,1,1,1}}^{\text{choose dice}}}{\underbrace{6 \cdot 6 \cdot 6 \cdot 6}_{\text{four dice possibilities}}} = \frac{24}{1296} \approx 0.018519$$

and, after combining all three probabilities together, we get a total probability of a farkle out result with four dice of approximately 0.157407.

The case when $n = 3$ is very similar, but there is an alternate way to think about it as well; we discuss both here in Example 9.14.

Example 9.14 When only three dice are being rolled, the only ways to not score is to obtain a pair and a singleton or three singletons (using twos, threes, fours, and sixes as we have been). The first occurs with probability

$$\text{Farkle with One Pair, 3 Dice} = \frac{\overbrace{\binom{4}{1,1,2}}^{\text{numbers}} \cdot \overbrace{\binom{3}{2,1}}^{\text{choose dice}}}{\underbrace{6 \cdot 6 \cdot 6}_{\text{three dice possibilities}}} = \frac{36}{216} \approx 0.166667$$

where the first multinomial coefficient chooses, from the 4 dice numbers available, 1 number for the pair, 1 number for the singleton, and 2 unused numbers. Note that the denominator is 6^3 when only $n = 3$ dice are involved. For three singletons, we obtain

$$\text{Farkle with No Pair, 3 Dice} = \frac{\overbrace{\binom{4}{3}}^{\text{numbers}} \cdot \overbrace{\binom{3}{1,1,1}}^{\text{choose dice}}}{\underbrace{6 \cdot 6 \cdot 6}_{\text{three dice possibilities}}} = \frac{24}{216} \approx 0.111111$$

for a total probability with three dice of about 0.277778. Note that we could have used a multinomial coefficient instead of a binomial coefficient, but it so happens that

$$\binom{4}{3} = \binom{4}{3,1} = 4.$$

This probability could have been obtained with a different line of thought. Among all possible three-dice rolls involving only twos, threes, fours, and sixes (there are $4^3 = 64$ of these), four of these result in three of a kind and will be scoring dice. Thus, only 60 combinations are non-scoring, out of the total sample space for rolling three six-sided dice with $6^3 = 216$ elements result in no score, and $60/216 \approx 0.277778$ as we found earlier.

TABLE 9.7: Farkle Out Probabilities

Number of Dice	Probability
6	0.023148
5	0.077160
4	0.157407
3	0.277778
2	0.444444
1	0.666667

For rolling $n = 2$ or $n = 1$ dice, our job is much easier. When rolling two dice, the only way to farkle out is to roll two non-scoring dice. There are $4^2 = 16$ ways to do that out of the possible $6^2 = 36$ outcomes, so we get a probability of $4/9 \approx 0.444444$. For one die, we farkle out if we roll a two, three, four, or six, so that probability is $4/6 \approx 0.666667$. A summary of the probabilities for these various numbers of dice and farkling out is given in Table 9.7.

We should take a look at some probabilities associated with *scoring* rolls before we move on to other games. First, let's consider the probability of getting a three of a kind on the opening roll (or a reroll when all six dice are scored).

Example 9.15 Unlike the previous examples, we now consider all six dice again; we are trying to count the number of ways to get a three of a kind with any of the dice, including ones and fives. Note that along with the three of a kind, the other three dice could also form a three of a kind, form a pair and a singleton, or form three singletons. In the first case, from the 6 possible numbers, we choose 2 for the three of a kinds; from the 6 dice, we choose 3 for the first triple and 3 for the second triple, resulting in

$$\text{Two 3 of a Kinds} = \underbrace{\binom{6}{2}}_{\text{numbers}} \cdot \underbrace{\binom{6}{3,3}}_{\text{dice}} = 300$$

different double three of a kind rolls. For a triple, a pair, and a singleton, from the 6 numbers we choose 1 value for the triple, 1 value for the pair, 1 value for the singleton, and 3 values for the unused numbers, followed by choosing 3 dice for the triple, 2 for the pair, and 1 for the singleton out of the 6 dice, giving us

$$\text{3 of a Kind and a Pair} = \underbrace{\binom{6}{1,1,1,3}}_{\text{numbers}} \cdot \underbrace{\binom{6}{3,2,1}}_{\text{dice}} = 7,200$$

for that count. Finally, for a triple and three singletons, we choose, from the 6 numbers, 1 for the triple, 3 for the singletons, and 2 for the unused faces. From the 6 dice then, we choose 3 dice for the triple and 1 for each of the singletons, yielding a quantity of

$$\text{3 of a Kind, 3 Singles} = \underbrace{\binom{6}{1,3,2}}_{\text{numbers}} \cdot \underbrace{\binom{6}{3,1,1,1}}_{\text{dice}} = 7,200$$

for a *total* number of three of a kinds equalling the sum of these three values, or 14,700. The probability is obtained by dividing by the total number of outcomes for rolling 6 dice of 6 sides, $6^6 = 46,656$ to obtain a probability of 0.315072.

A typical scoring system awards 200 points for a triple of twos, 300 points for a triple of threes, 400, 500, and 600 for a triple of fours, fives, or sixes, respectively. A triple of ones is treated very specially and differs among Farkle players, but we'll assume a typical 1,000 points for a triple of ones.

Using this information, we can determine the expected contribution from three of a kinds to the score on your first roll. Note, though, that *double* three of a kinds in one roll is often treated differently, so we will consider that a separate category and ignore those cases here. From the remaining 14,400 cases of Example 9.15, exactly one-sixth will correspond to triple ones, one-sixth will correspond to triple twos, and so on. The probability of a *specific* three of a kind is then $2400/46656 = 25/486$, so the expected contribution from the three of a kind category on the first six dice is

$$\underbrace{1000 \cdot \left(\frac{25}{486}\right)}_{\text{ones}} + \underbrace{200 \cdot \left(\frac{25}{486}\right)}_{\text{twos}} + \underbrace{300 \cdot \left(\frac{25}{486}\right)}_{\text{threes}} + \underbrace{400 \cdot \left(\frac{25}{486}\right)}_{\text{fours}}$$

$$+ \underbrace{500 \cdot \left(\frac{25}{486}\right)}_{\text{fives}} + \underbrace{600 \cdot \left(\frac{25}{486}\right)}_{\text{sixes}} \approx 154.32.$$

Farkle can be a lot of fun, especially in groups where the excitement of scoring (or watching your opponent farkle out) can be heard all over the house. The exercises give you the opportunity to explore these probabilities and expected values more. Note that a "best strategy" according to expected value, as in Yahtzee, is possible by working backwards as we did with Mini-Yahtzee of Figure 4.4 but, again, it would be quite huge. Rather than present that here, due to the plethora of different scoring systems, I will leave you with something to keep in mind as you roll. As we saw in Table 9.7, the more dice you can reroll, the lower your chance of getting a farkle is. Thus, for a roll such as ⊡ ⊠ ⊠ ⊞ ⊞ ⊞, it makes sense to *only* keep the one and reroll all other dice, even though the fives are scoring dice. Don't be afraid to reroll some scoring dice in the hopes of getting a higher score with the reroll!

9.6 Scrabble

The creator of Scrabble,[6] Alfred Butts, gathered many, many documents in English and analyzed the frequency of letters that occurred in words of differ-

---

[6]Scrabble is a registered trademark of Mattel, Inc.

ent lengths; the distribution of letter tiles in a game of Scrabble, where players work to create words on the board in a crossword-like fashion, is based on his analysis (see Table 9.8). In a game of Scrabble, players start by drawing seven tiles from the bag to create their "opening" hand. The first player to play will get to have a blank board and is able to construct any word that they can using their seven tiles (future players must play off of at least one existing letter on the board).

The questions that we want to ask and answer about Scrabble involve figuring out the number of opening hands that contain the appropriate letters so that certain words could be made; of course, by dividing by the total number of possible opening hands (there are 100 tiles total of which we choose 7), we could also obtain the probability of getting a hand that can form a particular word.

TABLE 9.8: Letter Tile Distribution

Letter	Qty	Letter	Qty	Letter	Qty
A	9	J	1	S	4
B	2	K	1	T	6
C	2	L	4	U	4
D	4	M	2	V	2
E	12	N	6	W	2
F	2	O	8	X	1
G	3	P	2	Y	2
H	2	Q	1	Z	1
I	9	R	6	Blank	2

Question

How many opening "hands" in Scrabble can form the word "IN" without using blank tiles? How about the word "DOG" with no blanks? What about a natural "PEEL"?

As alluded to in Example 9.10 when we were discussing Uno, our main discussion here will be a technique known as inclusion-exclusion. Let's begin by considering the number of hands that allow us to form the word "in" where we need at least one I and at least one N. We take the total number of possible hands, *subtract* the number that do not have any Is, *subtract* the number that do not have any Ns, and then *add back* those that do not have any Is and also do not have any Ns. This gives us

$$\text{Hands for IN} = \binom{100}{7} - \binom{91}{7} - \binom{94}{7} + \binom{85}{7} = 2,613,550,002$$

as the number of starting hands that can form the word "in." With 16,007,560,800 possible starting hands (the $\binom{100}{7}$ part of it), the probability of being able to form "in" with your opening hand is just a little over 16%. With longer words, though, the process gets a bit more involved.

Example 9.16 To figure out the number of starting hands that can form the word "dog" we need to be a bit more careful. First, we start with all possible hands as we did before, with

$$\text{All Hands} = \binom{100}{7} = 16,007,560,800$$

after which we subtract those hands that do not contain a D, subtract hands without an O, and subtract hands without a G; those contributions are

$$\text{Subtract No Ds} = -\binom{96}{7} = -11,919,192,480$$

$$\text{Subtract No Os} = -\binom{92}{7} = -8,760,554,088$$

$$\text{Subtract No Gs} = -\binom{97}{7} = -12,846,240,784$$

but now we've gotten rid of too many! Any opening hand that contains neither a D nor a G, for instance, was subtracted *twice* from our running total, so we need to add these hands back once since, by subtracting twice and adding once, we have a net total of being subtracted of just once, as desired. The same is true for the pairs D and O, as well as for O and G; our new contributions of

$$\text{Add No Ds or Os} = +\binom{88}{7} = +6,348,337,336$$

$$\text{Add No Ds or Gs} = +\binom{93}{7} = +9,473,622,444$$

$$\text{Add No Os or Gs} = +\binom{89}{7} = +6,890,268,572$$

are needed. The final problem for us here is with any starting hand that does not contain any Ds, Os, or Gs at all! These hands were in our initial total count, subtracted three times, and added back three times; they need to be subtracted off one last time, so an adjustment of

$$\text{Subtract No Ds, Os, or Gs} = -\binom{85}{7} = -4,935,847,320$$

is needed.

If we take all of these contributions together, we get a final total of $257,954,480$ which, after dividing by $16,007,560,800$, gives a probability of about 1.6% that my opening hand will be able to form the word "dog." As much as I love my dog Lilly, there's not too high of a chance that I can even put "dog" on the board with my opening hand in Scrabble.

For a more general and detailed coverage of the inclusion-
exclusion principle, consult more
advanced probability textbooks.
Here, it's enough for us to work
with the alternating subtractions and additions. If we had
a word with four distinct letters
to study, we would end up having four subtractions at the first
step, six additions as the second
step (there are $\binom{4}{2} = 6$ pairs that
would need to be added back),
and then four more subtractions

FIGURE 9.9: Scrabble Tournament in
France

before we would do the final addition. Why does this work? Looking back at
Example 9.16, consider an opening hand that has a D, but no Os or Gs. This
hand is included in the overall total, is subtracted off for "no Os" and for "no
Gs," and is added back for "no Os or Gs" – after the initial inclusion, it's
subtracted off twice and added back once, for a net total of being subtracted
exactly once, just as it should be (we cannot form the word "dog" with a
hand that only has the D and no other needed letters). When in doubt, go
ahead and consider specific hands and follow when that hand is added and
subtracted.

We end our coverage of Scrabble by answering the last of the questions;
the double letter in "peel" will cause a small complication, but it's nothing
that we can't handle.

Example 9.17 The word "peel" only has three distinct letters, so we can
mostly proceed the same way that we did in Example 9.16. The complication
is that when we talk about the letter E, we'll be talking about hands that
don't have *two* Es, meaning we'll be adding and subtracting hands that
have 1 or 0 Es. We'll include the total and the first round of subtractions
in our first contributions to the calculation, starting with

$$\text{All Hands} = \binom{100}{7} = 16,007,560,800$$

$$\text{Subtract No Ps} = -\binom{98}{7} = -13,834,413,152$$

$$\text{Subtract 1 or 0 Es} = -\left(\binom{88}{7} + \binom{12}{1}\binom{88}{6}\right) = -12,851,512,168$$

$$\text{Subtract No Ls} = -\binom{96}{7} = -11,919,192,480$$

where the term for 1 or 0 Es comes from considering 0 Es first (as we did for any other letter) and also including hands where there is exactly 1 E (choosing one of the twelve possible) and 6 non-E tiles.

With this, the rest of the contributions should be clear and follow the template in Example 9.16. We have

$$\text{Add No Ps and 1 or 0 Es} = + \left(\binom{86}{7} + \binom{12}{1}\binom{86}{6} \right)$$
$$= +11,015,061,804$$

$$\text{Add No Ps or Ls} = + \binom{94}{7} = +10,235,867,928$$

$$\text{Add No Ls and 1 or 0 Es} = + \left(\binom{84}{7} + \binom{12}{1}\binom{84}{6} \right)$$
$$= +9,407,144,304$$

$$\text{Subtract No Ps, Ls, and 1 or 0 Es} = - \left(\binom{82}{7} + \binom{12}{1}\binom{82}{6} \right)$$
$$= -8,003,698,560$$

as the final contributions. Bringing it all together leaves $56,818,476$ hands as the final number, or a probability of 0.35%, due to the low number of Ps and Ls relative to Es in the mix.

Scrabble continues to be an extremely popular game, with several tournaments (see Figure 9.9) held each year!

9.7 Backgammon

Backgammon is a fun game of chance and strategy that has its roots in Mesopotamia and has been around since 3000 BC. It was popular all throughout Europe during medieval times and continues to be a game played by many. An old backgammon board was featured as Figure 9.1 at this chapter's start, recovered from the Swedish warship *Vasa* from the 1620s. Our study of this game that involves a pair of six-sided

FIGURE 9.10: Backgammon Board

dice and movement of many pieces around a board will be short and will barely scratch the surface as far as the rules go and also as far as the mathematics go. We study it here only to see how the concepts we have seen throughout this book can be applied to a game that, on the surface, might appear to not involve much mathematics at all. Readers more interested in the game are encouraged to consult any of the numerous books about the game; *Backgammon for Winners* by Bill Robertie, [40] in the Bibliography, is highly recommended. The two topics we will explore, though, are "hitting" your opponents pieces off of the board and use of the "doubling cube" that enters the picture when backgammon is played for points or money.

The goal of backgammon is to move all of your pieces around the board into the six-pointed section designated as your home, and then win by removing all of your pieces from the board (called *bearing off*) before your opponent does. In the backgammon board shown in Figure 9.10, black's home is in the lower right, with black moving counter-clockwise, and red's home is in the upper right, moving clockwise around the board. Upon rolling the two six-sided dice, a player can either move a single piece a number of spaces equal to the *sum* of the dice *or* for each die, move a single piece the number of spaces indicated. If doubles are rolled, rather than "two" moves, the play gets four moves as though he had rolled four dice of the indicated number instead of just two. A pointed-space may not be landed on if your opponent has two or more pieces on it, but if your opponent only has a single piece on a point, you may land there, removing the opponent's piece from the board (it will have a chance to reenter the game later); this is called *hitting* your opponent's piece. Note that the sides of the board are connected in that the left-most point on the top side of Figure 9.10 is connected to the left-most point on the bottom of the board. In essence, backgammon can be considered a "race" around the board, where the opportunity to hit your opponent's pieces or build up a safe zone of two or more pieces at the same point present a conflict between the two players who move around the board in opposite directions.

Probabilities for Hitting a Piece

Question

What is the probability that my roll will allow me to hit off my opponent's piece that is n spaces away?

The ability to move one *or* two pieces (or possibly four when doubles are rolled) per turn makes this study a little different than Monopoly where we simply total the numbers on both dice to determine the move. If you have a piece that is n spaces away from a single piece of your opponent, we need to find all possible rolls for which the rules of the game allow that piece to move n spaces somehow. We look at a few cases here.

Example 9.18 Suppose your opponent has a single piece located one point away from a piece of yours. How many ways can the dice be rolled that give us the opportunity to advance one piece one space? Using Table 1.2 as we have done many times now, *any* roll where the white die shows a 1 or the black die shows a 1 will give us the opportunity we desire. By inspection of that table, there are 11 rolls that will make us happy, out of a possible 36 rolls total, giving

$$\text{Probability We Can Move One Piece One Space} = \frac{11}{36} \approx 0.305556$$

as that probability.

Observe how Example 9.18 shows that the ability to move a single piece a number of spaces shown on just one of the dice completely changes the probability of getting to move a specific number! When taking the sum of two dice and moving a single piece that number of spaces, it is completely impossible to move just one space.

Example 9.19 If your opponent's piece is three spaces away from one of your own, how many ways can we get to move that number of spaces? Again, any of the 11 rolls with a three from Table 1.2 will work, but we also have a few more possibilities here. First, we could roll ⊡ ▨ or ⊡ ▨ which gives us two more ways to hit the opponent's piece and knock it off of the board, but we could *also* roll double ones, giving us the ability to move one piece three spaces (with the ability to move a piece one space thereafter)! This gives us

$$\text{Probability We Can Move One Piece Three Spaces} = \frac{14}{36} \approx 0.388889$$

as the probability desired.

With Monopoly and other games of simple movement rules using two dice, 7 is the most often rolled number; does that remain the case here? Let's take a look.

Example 9.20 To move a piece *six* spaces and hit your opponent's piece, we could roll any of the 11 results that contain a six and be happy. We could also roll any of the 5 results from Table 1.2 that result in a *sum* of six. Finally, we could also roll ⊡ ▨ and use three of the four movements allowed to move our piece six spaces (note that ⊡ ▨, while allowing four moves, two of which would be moving a single piece six spaces, was already included in our list of results that add to six). This gives us

$$\text{Probability We Can Move One Piece Six Spaces} = \frac{17}{36} \approx 0.472222$$

as a fairly high probability!

However, for moving a piece *seven* spaces, we no longer have the benefit of an outcome that has a single die with a face showing seven pips. Since seven is also not a multiple of any of the numbers 1 through 6, we cannot benefit from rolling doubles either. The only way to hit an opponent's piece seven spaces away is to roll a *sum* of 7, which we can do in any of the 6 usual ways, resulting in

$$\text{Probability We Can Move One Piece Seven Spaces} = \frac{6}{36} \approx 0.166667,$$

much lower than the probability of moving six spaces.

Indeed, you will see later that 6 is the single highest probability for reaching your opponent's single piece with one of yours!

Because of the rule regarding doubles, other distances away from your pieces not usually present are also possible; for instance, a roll of 🎲 🎲 would allow a single piece to move 16 spaces; other numbers such as 15, 18, 20, and 24 are also possible (with 18 reached by rolling 🎲 🎲 and using three of the four moves on a single piece). While you will complete the other cases as an exercise later, my general advice, used by good backgammon players, is to never leave a single piece of yours on a point that your opponent can reach with a roll of six or fewer. As common sense would also dictate, generally speaking, the farther away from your opponent's pieces you can leave a point with a single piece of yours, the better.

The Doubling Cube

Backgammon games played for points or money often feature a special die used to keep track of the stakes involved for the current game; this die is called the *doubling cube* and is pictured in the far left of Figure 9.10 as the die showing 64. At the start of a game, the cube is not used but, at the start of a player's turn, before rolling, that player may put the doubling cube into play, showing a side of 2. The other player must immediately either agree to play the current game at *double* the stakes or concede. Assuming the player accepts, the accepting player now controls the doubling cube and, before he or she rolls on his or her turn, may then change the face to 4, doubling the stakes yet again. Since your opponent redoubled, you now have to choose to continue the game at four times the original stakes or concede; if you continue, control of the doubling cube is now yours. Control of the cube passes back and forth using these same rules. While 64 is the highest number shown on the die, there is no maximum value and games can be played for 128 times the original stake or higher. Note that there are several optional rules that modify conditions on and use of the doubling cube; if you find yourself playing backgammon for points or money, be sure to agree on any optional rules used.

Question

When should I accept a doubling and when should the doubling cube be used?

While the question is simple, the answer is complex and depends on your ability to assess the probability that you win the game given the current board conditions. For our discussion, we assume that the players involved in the game are able to determine with relative accuracy the value of p, their chance at winning the game. While this is hard to determine in general, backgammon strategy books often include useful guides on determining good estimates of p. First, we look at what happens when your opponent offers to double the stakes.

Example 9.21 Your opponent decides to use the doubling cube to double the current stakes in a game worth s points or dollars. You estimate that the chance you win is about p, and now that we have a value and a probability, it should make sense that our concept of expected value should come out of hiding. If you concede, you immediately lose s, so we have

Expected Value, Concede $= -s$

but, if we accept the new stakes of $2s$, the expected value is

$$\text{Expected Value, Accept} = \underbrace{p \cdot (2s)}_{\text{win}} + \underbrace{(1-p) \cdot (-2s)}_{\text{lose}} = s \cdot (4p - 2)$$

where our decision point comes when the expected value for accepting is higher than the expected value for conceding. This happens when

$$s \cdot (4p - 2) > -s$$

which, by algebra, means that $p > 1/4$. You should accept if you estimate your chance of winning is greater than 25% and concede otherwise.

The flip side of Example 9.21 is about the offensive side of the game; when should you offer to double the current game in order to benefit your long-term coffers of money or points?

Example 9.22 At the current money, you estimate your chance of winning as p and the current stakes are again denoted by s. We proceed as in Example 9.21 where, now, if your opponent concedes, you *win* the current stakes. This give us

Expected Value, Opponent Concedes $= s$

and if the opponent accepts the new stakes of $2s$ we have

$$\text{Expected Value, Opponent Accepts} = \underbrace{p\,(2s)}_{\text{win}} + \underbrace{(1-p)\,(-2s)}_{\text{lose}} = s\,(4p - 2).$$

Setting these two values equal to each other, we find that $p = 3/4$ is the value at which you do not care if your opponent accepts or not. When $p > 3/4$ you definitely want your opponent to accept doubling!

Of course, this doesn't have anything to do with *not* doubling, so we compute the expected value of the current game with the current stakes as

$$\text{Expected Value, No Double} = \underbrace{p \cdot s}_{\text{win}} + \underbrace{(1-p)\,(-s)}_{\text{lose}} = s\,(2p - 1)\,.$$

The expected value of not doubling equals the expected value of your opponent accepting when $p = 1/2$, so if your chance of winning is greater than 50%, doubling makes sense, and you should hope that your opponent definitely accepts when $p > 3/4$.

The fact that doubling gives up control of the cube itself is very hard to factor into the mathematical analysis, and, of course, use of the cube is subject to some intimidation and bluffing (though with near-perfect play by both players, everyone should be able to determine the same estimates on p for each player). The doubling cube does, however, make for this interesting mathematical analysis and can turn backgammon into a dynamic game full of action and suspense on top of the strategy already present.

9.8 Memory

One of the first games that children learn to play is known as *Concentration* or *Memory* where the goal is to turn over face-down cards with the hope of finding two matching cards. While commercial versions are available, it's quite easy to play this game with a standard deck of playing cards; with this version, cards are considered "matching" if the two cards have the same rank

FIGURE 9.11: Memory Game

and same color (so that, for instance, J♥ and J♦ match, while 5♠ and 5♥ do not). As a competitive game, players take turns turning over two cards, one at a time, and if those cards match, that player takes another turn; otherwise, the cards are flipped back over and the next player takes their turn. The winner for this version of the game is the player with the most matches once all of the cards have been taken. Figure 9.11 shows a slightly less competitive version of the game being played.

You may find it somewhat surprising at first that the competitive form of Memory *does* have interesting strategies; in particular, note that each unknown card flipped over gives both players new information, so as the second flip of your turn it may not be in your best interest to reveal a new card, especially late in the game.

What we wish to examine here, though, is the solitaire version of Memory where the goal is to finish the game in the shortest amount of moves. Since the exact placement of cards on the table while playing is irrelevant, we can easily assume that all cards are laid out in a single row. As a strategy, quick thought should convince you that, assuming perfect memory of the player, the strategy for play is

- turn over both cards of a known pair and remove them from the board; if the positions of a pair is not known,

- flip over any unknown card, and

  - if the other card of that pair's position is known, flip it over to complete the pair,

  - otherwise, flip over any other unknown card and hope for a lucky match.

For our purposes, we will assume that the player *does* have perfect memory, and since the order of the initial layout is random, we may also assume that cards are always flipped over, according to this strategy, from left to right.

Question

How many essentially different Memory games are there?

That this question is the first we'll ask about Memory should not be a surprise; any study we've done about probability or expected value has always considered the complete sample space, and our coverage here is no different. First, rather than playing cards, we imagine the numbers 1 through n written on cards, giving us a total of $2n$ cards in play. How many possible games are there when $n = 2$? We can list these very explicitly, as seen in Table 9.9. However, as far as game play is concerned, 1122 plays exactly the same as 2211, as does 1212 and 2121, with 1221 and 2112 playing exactly the same as well. Of the 6 layouts shown in Table 9.9, only 3 are essentially different. Since explicit lists will become too long as n grows, we can also consider a computational approach to obtaining this figure.

TABLE 9.9: Layouts for Memory, $n = 2$

1122	2211
1212	2121
1221	2112

Example 9.23 First, note that when $n = 2$, we have 4 cards in play. From these four cards, we need to choose 2 spots for the ones and 2 spots for the twos, so the multinomial coefficients of Definition 4.5 will be useful.

However, there are also permutations (of Definition 4.4) in play; the two face values for the cards can be permuted in $2! = 2$ ways, so we'll divide our total by 2 in order to determine the essentially different layouts. This gives us

$$\text{Essentially Different Layouts, } n = 2 = \frac{\overbrace{\binom{4}{2,2}}^{\text{layout of cards}}}{\underbrace{2!}_{\text{permutations of values}}} = \frac{6}{2} = 3$$

as this quantity.

For $n = 3$, the number of possible layouts, before permutations are considered, grows from the 6 of Example 9.23 to 90, so the computational method will indeed be better as we move forward.

Example 9.24 The case when $n = 3$ involves a total of 6 cards; from these, we choose 2 spots for the ones, 2 spots for the twos, and 2 spots for the threes. We'll have $3! = 6$ permutations of the *values* 1, 2, and 3, so our final count of essentially different layouts will be

$$\text{Essentially Different Layouts, } n = 3 = \frac{\overbrace{\binom{6}{2,2,2}}^{\text{layout of cards}}}{\underbrace{3!}_{\text{permutations of values}}} = \frac{90}{6} = 15.$$

The general formula should now be clear; when $2n$ cards are involved, we again choose 2 spots for the ones, 2 spots for the twos, continuing until we choose 2 spots for the pair of cards of value n; there will be $n!$ permutations of the values, giving us the following general result.

Theorem 9.2 The number of essentially different layouts for Memory, using $2n$ cards with values 1 through n, is

$$\text{Essentially Different Layouts} = \frac{\binom{2n}{2,2,\dots,2}}{n!} = \frac{\frac{(2n)!}{2^n}}{n!} = (2n-1)!!$$

where $(2n-1)!! = (2n-1) \cdot (2n-3) \cdots 5 \cdot 3 \cdot 1$ is the *double factorial* function.

The last two equalities of Theorem 9.2 can be verified through some algebra and the definition of the double factorial. Note that this theorem follows the reasoning behind Examples 9.23 and 9.24 exactly.

Question

What is the expected number of moves needed to finish a game of Memory?

While this question is easy to ask, the answer is much harder to find. In fact, we will not consider the general case in this book; mathematicians Velleman and Warrington in 2013 found a very good estimate on this value for arbitrary values of n (see [56] in the Bibliography) and that result will be given later in this section.

First, however, we can consider easily the case of having the *shortest* game. In order for us to have the quickest game of Memory possible, each time we turn over two cards, they must match. That means, for a game with $2n$ cards with n values, the layout

$$1\ 1\ 2\ 2\ 3\ 3 \cdots (n-1)\ (n-1)\ n\ n$$

will give us a game that takes exactly n moves. Using our strategy, it should be clear that each card will be flipped over either once or twice (and never more than twice), there will be between $2n$ and $4n$ flips; with two flips per turn, this gives us a total number of turns from n to $2n$. We have already exhibited a case when the low end, n, can be achieved; is it possible that a game takes $2n$ moves? The answer is no. Take a moment to verify that our strategy forces the *last* card to only ever be turned face up once in any game. This new upper bound of $2n - 1$ is actually achievable, as in the layout

$$1\ 2\ 3\ 1\ 4\ 2 \cdots k\ k-2 \cdots n-1\ n-3\ n-2\ n-1\ n$$

where you can verify that it will take $2n - 1$ moves to finish this game (a quick verification can be done by writing this layout down for a specific value of n).

Since we have relatively small quantities of possible layouts for $n = 2$ and $n = 3$, we can compute the expected length for these games fairly easily.

Example 9.25 With 4 cards in the case of $n = 2$, recall that the 3 possible essentially different layouts is given as the first column of Table 9.9. Each of these is equally likely, so with $1/3$ probability we end up in the case where the game takes 2 moves (with the layout 1122) and with probability $2/3$ we end up in the case where it takes 3 moves to end the game (with layouts 1212 or 1221). This gives us

$$\text{Expected Length, } n = 2 = \underbrace{\left(\frac{1}{3}\right) \cdot 2}_{\text{length 2}} + \underbrace{\left(\frac{2}{3}\right) \cdot 3}_{\text{length 3}} = \frac{8}{3} \approx 2.67$$

as the expected number of moves to end the game.

TABLE 9.10: Layouts and Game Lengths for Memory, $n = 3$

Layout	Length	Layout	Length	Layout	Length
112233	3	121233	4	122133	4
112323	4	121323	5	122313	5
113223	4	123123	5	123213	5
131223	4	132123	5	132213	4
311223	4	312123	5	312213	4

We can do the same for the case when $n = 3$. First, in Table 9.10, you can find the 15 essentially different layouts. An easy way to make this list is to take each of the 3 essentially different layouts for $n = 2$ in the first column of Table 9.9 and append "33" to the end. For each of these, take that first 3 and slide it to the left, getting 5 new layouts as that 3 goes from position 5, to position 4, then position 3, continuing until it reaches position 1. With a 1/15 chance of having a game of length 3, a 8/15 chance for a game of length 4, and a 6/15 chance for a worst-case game length of 5, we get

$$\text{Expected Length, } n = 3 = \underbrace{\left(\frac{1}{15}\right) \cdot 3}_{\text{length 3}} + \underbrace{\left(\frac{8}{15}\right) \cdot 4}_{\text{length 4}} + \underbrace{\left(\frac{6}{15}\right) \cdot 5}_{\text{length 5}} = \frac{65}{15} \approx 4.33$$

as the expected length when 6 cards (3 pairs) are involved.

At this point, courtesy of Velleman and Warrington [56], here is the general result for the expected length of a game of Memory.

Theorem 9.3 The expected length of a game of Memory with $2n$ cards is given by

$$\text{Expected Length} = EL(n) = (3 - 2\ln 2) \cdot n + 7/8 - 2\ln 2 + E(n)$$

where $E(n)$ is an error-term that approaches zero as n gets large.

While the exact format of the error-term, $E(n)$, involved in Theorem 9.3 is not precisely known, for $n = 3$ the estimate provided by this theorem, $EL(3) \approx 4.3298$, is off by less than half of one-hundredth of a move; since this error-term goes to zero as n increases, the estimates of Theorem 9.3 only *improve*. It's worth noting here that, except for this negligible error-term, the function $EL(n)$ is *linear*. That is, the expected length of a game of Memory increases by the same number of moves whether we go from 4 cards to 6 cards or from 40 cards to 42 cards; adding 2 more cards (1 more pair) increases the number of moves required to end the game by about

$$\text{Increase in Moves per Added Pair} \approx EL(n) - EL(n-1) \approx 1.61371$$

which is not a huge increase in the complexity of the game.

9.9   Zombie Dice

The game of Zombie Dice[7] is published by
Steve Jackson Games, creator and distributor of
the various games in the Munchkin series. The
premise of the game is that you are a zombie
trying to collect as many brains as you can (perhaps for eating) while avoiding being shot. Dice
are rolled to obtain results, with brains being
collected and stored, footprints indicating that
your target ran away but you can perhaps give
chase, and shotguns indicating that you've been
hit.

FIGURE   9.12:   Zombie
Dice

At the start of your turn, three dice are randomly selected from the dice
cup and then rolled. After seeing the result, you choose whether to score your
current total of brains or continue rolling. If you do choose to roll again, any
footprints that were rolled are marked to be rolled again, and you randomly
select enough dice from the cup to bring your total to three dice to roll. You
can keep rolling and rolling in this manner until you either decide to stop and
score the number of brains you have obtained this round or you reach three
shotguns. Unfortunately, these three shotgun blasts mean that your round is
over, and you *do not score* the brains you currently have. Why? Because, as
the zombie who was trying to be too greedy with brains, you have died.

There are three different colored six-sided dice involved in the game, each
representing a different level of threat (see Figure 9.12). The quantity of brains,
footprints, and shotguns as follows:

- Each *green* die has three brains, two footprints, and one shotgun. The game
  begins with 6 of these dice in the cup.

- Each *yellow* die has two brains, two footprints, and two shotguns, and there
  are 4 of these in the cup at the start.

- Each *red* die has one brain, two footprints, and three shotguns; at the start,
  there are 3 of these in the cup for you to hopefully not get.

As you may imagine, to play the game well, you must have awareness of how
many of each color remains in the cup, and, as you play, you must balance the
need to obtain more brains with the threat of shotgun blasts looming. If you
currently have one shotgun, does it make sense to risk, say, five brains that
you have stored? How about if you had two shotguns where a single roll of a
shotgun will make you lose those five brains? The game continues until one
player has at least 13 brains saved up from his or her turns of rolling, at which

---

[7]Zombie Dice is a registered trademark of Steve Jackson Games.

time any player who hasn't had the same number of turns as that player can complete one last turn. The winner is then the player with the highest number of brains (continue rolling among tied players to break the tie).

If you'd like to give this game a try, Steve Jackson Games publishes a free version of the game for the iPhone, iPad, and Windows Phone devices; the retail version is also fairly inexpensive with a suggested retail price of only $13.13. At this point I would like to again thank Roanoke College student Heather Cook, who also did an independent study project with me regarding the mathematics behind Zombie Dice and also went on to complete a Ph.D. at the University of Virginia. She would very much like it if I also pointed out that the stock number for Zombie Dice is 131313; apparently, the game designers really like the number 13.

Question

How can we make the optimal decision to stop rolling and score, or continue rolling, at Zombie Dice?

The question is simple, but the answer is not. You might be thinking at this point that the binomial distribution we looked at for Liar's Dice would come into play, and you would be somewhat correct. The change is that we *do* care about the three different results for each die. The concept of "success" or "failure" turns into "brain," "footprint," or "shotgun." Much like the ones, twos, threes, fours, fives, and sixes of Yahtzee where we were interested in particular amounts of each, the multinomial distribution of Definition 4.6 will be helpful. For now, though, we turn to some relatively simple questions about Zombie Dice that serve as inspiration for the development of the full theory.

If your goal is to get as many brains as possible, it shouldn't take you too long to realize that the green dice are your friends and the red dice are your enemies. Let's compare the effect of the different colored dice on rolling three brains.

Example 9.26 Suppose we pull three green dice from the cup; what is the probability that we'll get three brains? Thinking of "brain" as a success and everything else as a failure, the binomial distribution can help us out. With $n = 3$ dice and a desire to have three successes, using Definition 4.1 with a probability of success of $p = 1/2$ (since half of the faces of a green die are brains), we get

$$3 \text{ Brains, 3 Green Dice} = \binom{3}{3}\left(\frac{1}{2}\right)^3 \cdot \left(\frac{1}{2}\right)^0 = \frac{1}{8}$$

so a respectable 12.5% of the time, three green dice will give us three brains. How about for yellow and red dice? Using Definition 4.1 and changing the value of p as appropriate for the colors (recall that yellow dice have brains on 1/3 of the faces and red dice only have brains on one side) we get

$$3 \text{ Brains, 3 Yellow Dice} = \binom{3}{3}\left(\frac{1}{3}\right)^3 \cdot \left(\frac{2}{3}\right)^0 = \frac{1}{27}$$

$$3 \text{ Brains, 3 Red Dice} = \binom{3}{3}\left(\frac{1}{6}\right)^3 \cdot \left(\frac{5}{6}\right)^0 = \frac{1}{216}.$$

While Example 9.26 is fine and dandy on its own, note that these probabilities reflect your chances at getting three brains given that you already know you have already pulled particular dice from the cup! Things change, of course, if our initial dice selection consists of, say, two yellow dice and one green die. More on that will be coming in a bit.

Since we looked at the probabilities of getting a particular result (three brains) given the specific dice combinations of three green, three yellow, and three red, it makes sense now to determine the probabilities of actually *getting* those combinations from the cup of dice!

Example 9.27 To determine the chance that we will pull three green dice from the cup at the start of the game, we have to think about the process involved in much the same way as we determined the number of different named hands at poker in Chapter 3.

When a round begins for you, there are 13 dice in the cup, with 6 green, 4 yellow, and 3 red dice among them; if we need to select three dice at random, it would seem that the total number of different ways to pull three dice from the cup is given through the binomial coefficients (Definition 3.2). From 13 dice, we choose 3, and this becomes our entire sample space of possible dice pulled. Then, for three green dice, we need to choose 3 of the available 6, and for both yellow and red we need to choose 0 of the available amounts (4 and 3 respectively). This will end up giving us

$$\text{Probability of GGG} = \frac{\overbrace{\binom{6}{3}}^{\substack{\text{green} \\ \text{dice}}} \cdot \overbrace{\binom{4}{0}}^{\substack{\text{yellow} \\ \text{dice}}} \cdot \overbrace{\binom{3}{0}}^{\substack{\text{red} \\ \text{dice}}}}{\underbrace{\binom{13}{3}}_{\substack{\text{all dice} \\ \text{combinations}}}} = \frac{20 \cdot 1 \cdot 1}{286} \approx 0.06993.$$

For getting three yellow dice, not much needs to be changed. You should get

$$\text{Probability of YYY} = \frac{\overbrace{\binom{6}{0}}^{\substack{\text{green}\\\text{dice}}} \cdot \overbrace{\binom{4}{3}}^{\substack{\text{yellow}\\\text{dice}}} \cdot \overbrace{\binom{3}{0}}^{\substack{\text{red}\\\text{dice}}}}{\underbrace{\binom{13}{3}}_{\substack{\text{all dice}\\\text{combinations}}}} = \frac{1 \cdot 4 \cdot 1}{286} \approx 0.01399$$

where the result should agree with your intuition that having more green dice in the cup to start (six as opposed to four) drives up the probability that you can get three of them upon random selection.

If the two computations in Example 9.27 look very similar, as they should, you can imagine that we can develop a formula that gives the probability of getting various quantities of green, yellow, and red dice. See if the following agrees with your ideas on generalizing the formula.

$$\text{Probability of } g \text{ Green, } y \text{ Yellow, } r \text{ Red Dice} = \frac{\overbrace{\binom{6}{g}}^{\substack{\text{green}\\\text{dice}}} \cdot \overbrace{\binom{4}{y}}^{\substack{\text{yellow}\\\text{dice}}} \cdot \overbrace{\binom{3}{r}}^{\substack{\text{red}\\\text{dice}}}}{\underbrace{\binom{13}{3}}_{\substack{\text{all dice}\\\text{combinations}}}}$$

From this, we can compute the probability of any specific collection of g green, y yellow, and r red dice from a full cup of dice! This idea, and even further generalizations, will help us on the way to understanding the game of Zombie Dice and proper, mathematically influenced play.

At this point, there's a very special generalization that we need to develop before we can really answer our question about Zombie Dice. However, since "best play" for most of the games so far in this book has revolved around the idea of expected value, if you're thinking along those lines, bonus points for you! We seem to be marching down the avenue of trying to determine the probabilities of specific scenarios for Zombie Dice, and we'll step it up rather soon.

Example 9.26 assumed that we were starting with dice from the cup of the same color and wanted to get three brains; what would happen if, instead, we wanted to get 2 brains and 1 shotgun? How about 1 brain, 1 shotgun, and 1 footprint? As we alluded to earlier, we are outside the realm of just "success" and "failure." How does this work? Consider the next example.

Example 9.28 If we pull three green dice from the cup, how can we determine the probability of getting 2 brains and 1 shotgun? If we *do care* about *order*, this is a very easy question. The probability that the first green die we roll is a brain, the second green die is also a brain, and the third ends up being a shotgun is given by

$$\text{Probability Brain, Brain, Shotgun} = \underbrace{\frac{1}{2}}_{\text{brain}} \cdot \underbrace{\frac{1}{2}}_{\text{brain}} \cdot \underbrace{\frac{1}{6}}_{\text{shotgun}} = \frac{1}{24} \approx 0.04167$$

but, as we've learned from poker hands in Chapter 3, when order doesn't matter, we need to figure out how many ways there are to *order* these dice. Of the three dice available to use, we need to choose 2 of them to be brains, and the other is forced to be a shotgun (since we are only considering the 2 brain and 1 shotgun possibility here). This happens in

$$\binom{3}{2,1} = 3$$

ways, which shouldn't seem too surprising. Choosing the *two* dice to be brains is equivalent to choosing *one* die to be the shotgun, and of course there are three choices for which die is the shotgun! Combining this together, we get

$$\text{GGG Probability 2 Brains 1 Shotgun} = \underbrace{\binom{3}{2,1}}_{\text{orderings}} \cdot \underbrace{\frac{1}{2}}_{\text{brain}} \cdot \underbrace{\frac{1}{2}}_{\text{brain}} \cdot \underbrace{\frac{1}{6}}_{\text{shotgun}} = \frac{3}{24}$$

or, in other words, 12.5% of the time.

What changes in Example 9.28 if we want one each of brains, footprints, and shotguns? We can determine the probability of getting, say, a brain, followed by a shotgun, followed by a footprint by

$$\text{Probability Brain, Shotgun, then Footprint} = \underbrace{\frac{1}{2}}_{\text{brain}} \cdot \underbrace{\frac{1}{6}}_{\text{shotgun}} \cdot \underbrace{\frac{1}{3}}_{\text{footprint}} = \frac{1}{36},$$

which is about 0.02778, but how many ways can these be ordered to determine *all* possible ways to get one of each?

Example 9.29 Since we have three green dice in this example, using the idea of permutations given in Definition 4.4, you should reason that there are $3! = 6$ orderings for the set {brain, footprint, shotgun}. As I remarked before, and will mention again, this is the same number of orderings given by the multinomial coefficient $\binom{3}{1,1,1}$ since we have three different results for a die, and among the three dice, we are counting the number of ways to get one of each type. Combining this together with the specific probabilities of obtaining these results on the green dice, we get

$$\text{GGG Probability One of Each} = \underbrace{\binom{3}{1,1,1}}_{\text{orderings}} \cdot \underbrace{\frac{1}{2}}_{\text{brain}} \cdot \underbrace{\frac{1}{3}}_{\text{footprint}} \cdot \underbrace{\frac{1}{6}}_{\text{shotgun}} = \frac{6}{36}$$

which is approximately 0.16667.

What we're really doing here is applying the multinomial distribution of Definition 4.6 with probabilities that are *not all the same* as they were for Yahtzee.

Example 9.30 Repeating Example 9.29 but with three red dice, for the purposes of practicing more with Definition 4.6, we have $n = 3$ red dice with which to work, possibilities of R_1, R_2, and R_3 corresponding to brains, footprints, and shotguns respectively, and probabilities of $p_1 = 1/6$, $p_2 = 1/3$, and $p_3 = 1/2$, we are able to determine the probability of getting exactly one of each ($x_1 = x_2 = x_3 = 1$ in the definition) as

$$\text{RRR Probability 1 of Each} = \underbrace{\binom{3}{1,1,1}}_{\text{orderings}} \cdot \underbrace{\left(\frac{1}{6}\right)^1}_{\text{brain}} \cdot \underbrace{\left(\frac{1}{3}\right)^1}_{\text{footprint}} \cdot \underbrace{\left(\frac{1}{2}\right)^1}_{\text{shotgun}} = 6 \cdot \left(\frac{1}{36}\right)$$

$$= \frac{1}{6} \approx 0.16667$$

which is identical to our answer of Example 9.29. This should make sense since getting one *shotgun* on a red die is equivalent to getting one *brain* on a green die.

We can also repeat Example 9.28 using the multinomial distribution of Definition 4.6. There, changing the green dice again to red dice just to be a little bit different, recall that for brains, footprints, and shotguns, respectively, for outcomes R_1, R_2, and R_3, we have the same probabilities as in Example 9.30.

Example 9.31 For three red dice, to determine the probability of 2 brains and 1 shotgun using Definition 4.6, we will get

$$\text{RRR Probability 2 Brains, 1 Shotgun} = \underbrace{\binom{3}{2,0,1}}_{\text{orderings}} \cdot \underbrace{\left(\frac{1}{6}\right)^2}_{\text{brain}} \cdot \underbrace{\left(\frac{1}{3}\right)^0}_{\text{footprint}} \cdot \underbrace{\left(\frac{1}{2}\right)^1}_{\text{shotgun}}$$

$$= 3 \cdot \left(\frac{1}{72}\right) = \frac{1}{24} \approx 0.04167.$$

Note here that computation of 0! is involved; we remind you, again, that 0! is defined to have the value 1 as a matter of convenience.[8]

As your intuition should tell you, since brains are much less likely on red dice and shotguns are much *more* likely, the probability of getting 2 brains decreases from Example 9.28 to Example 9.30. This is why the chance of getting 2 brains and 1 shotgun decreases from one example to the other. Of course, the probability of a shotgun goes up, but, hey, we're only looking for 1 of those.

We can generalize this quite nicely, and obtain the following result that we will be using soon.

Theorem 9.4 (Zombie Dice) For Zombie Dice, suppose that we are rolling n green dice. Then the probability that we get x_b brains, x_f footprints, and x_s shotguns is given by

$$P_{\text{green}}^{(x_b, x_f, x_s)} = \underbrace{\binom{n}{x_b, x_f, x_s}}_{\text{orderings}} \cdot \underbrace{\left(\frac{1}{2}\right)^{x_b}}_{\text{brains}} \cdot \underbrace{\left(\frac{1}{3}\right)^{x_f}}_{\text{footprints}} \cdot \underbrace{\left(\frac{1}{6}\right)^{x_s}}_{\text{shotguns}}$$

where we have the restriction that $x_b + x_f + x_s = n$. Similarly, the same formula for yellow dice and red dice is given by, without labeled braces as I put above,

$$P_{\text{yellow}}^{(x_b, x_f, x_s)} = \binom{n}{x_b, x_f, x_s} \cdot \left(\frac{1}{3}\right)^{x_b} \cdot \left(\frac{1}{3}\right)^{x_f} \cdot \left(\frac{1}{3}\right)^{x_s}$$

$$P_{\text{red}}^{(x_b, x_f, x_s)} = \binom{n}{x_b, x_f, x_s} \cdot \left(\frac{1}{6}\right)^{x_b} \cdot \left(\frac{1}{3}\right)^{x_f} \cdot \left(\frac{1}{2}\right)^{x_s}$$

where the same restriction $x_b + x_f + x_s = n$.

The astute reader will note that the formula given for yellow dice given in Theorem 9.4 can be simplified nicely since the probability of any particular face is always 1/3; I gave the full version in Theorem 9.4 because it parallels the other two colors, but you could certainly write

$$P_{\text{yellow}} = \binom{n}{x_b, x_f, x_s} \cdot \left(\frac{1}{3}\right)^{x_b} \cdot \left(\frac{1}{3}\right)^{x_f} \cdot \left(\frac{1}{3}\right)^{x_s} = \binom{n}{x_b, x_f, x_s} \cdot \left(\frac{1}{3}\right)^{n}$$

instead.

Our goal from earlier, you might remember, is to determine the best play for Zombie Dice. Given that we have some brains set aside, perhaps some

---

[8]Again, there are many good reasons that 0! = 1, but for us it's enough to say that it's both convenient and, defining it to have value 1, it simply works.

footprint dice to reroll, and a bunch of dice left in the cup. Given the footprints we may need to reroll and what dice are left in the cup that we might randomly select, is it better to stop rolling and score the brains (maybe we're very likely to get a bunch of shotguns, or already have 2 shotguns and one more will force us to lose our brains) or continue rolling? The answer, you may imagine, has to do with striking a balance between that risk of losing all of our brains if we continue and the reward to continuing and seeing a large possibility of increasing the quantity of our brains.

If we let b represent the amount of brains we currently have for this round, along with PE equal to the probability that on our next roll (randomly selecting dice from the cup to bring our footprint plus cup dice total to three) we will have 3 or more shotguns (ending the round and causing us to lose our b stored brains), we can consider what might happen in the future in terms of brains. Denote by EB the expected number of brains that the next roll, combined will all subsequent possible rolls. Then we should continue rolling if the decision equation

$$\text{Expected Value of Continuing} = \underbrace{-b \cdot PE}_{\text{round ends}} + \underbrace{EB \cdot (1 - PE)}_{\text{round continues}} > 0.$$

That is, we can think of the next roll as a "wager" where, instead of dollars, the brains we've accumulated this round serve as the bet. In this manner of thinking, we can essentially play the game until our expected value becomes negative (or zero).

Of course, determining how to compute PE and EB will be the main part of our work and is made extremely nice through the use of the summations we saw in Section 2.2. For example, first let us consider determining the probability of getting three shotguns on the very first roll of a round.

Example 9.32 We will be using the additive rule for probability here. In order to determine the chance that your Zombie Dice round will end on the very first three dice you pull from the cup, we need to consider all possibilities for the colors of dice pulled, and, for each of those, the probability of getting three shotguns. To do this, let S be the set of all triples of the form $\{g, y, r\}$ where we restrict each of g, y, and r to integers from 0 to 3 (that is, for example, $0 \le g \le 3$). Since g, y, and r will represent the number of green, yellow, and red dice, respectively, that we pull from the cup, we also require that $g + y + r = 3$. Thus S contains the elements $\{0, 2, 1\}$, $\{0, 0, 3\}$, and $\{1, 1, 1\}$ but not the elements $\{4, 0, 2\}$ or $\{1, 1, 0\}$. With this setup, the probability that the green dice pulled will result in shotguns is given by Theorem 9.4 as

$$\text{Prob. Green Shotguns} = \underbrace{\binom{g}{0, 0, g}}_{\text{all shotguns}} \cdot \underbrace{\left(\frac{1}{2}\right)^0}_{\text{brains}} \cdot \underbrace{\left(\frac{1}{3}\right)^0}_{\text{footprints}} \cdot \underbrace{\left(\frac{1}{6}\right)^g}_{\text{shotguns}} = P_{\text{green}}^{(0,0,g)}$$

since we require, for this example, that all of the green dice have shotguns.

Similarly, using the notation from Theorem 9.4, we must then have

$$\text{Prob. Yellow Shotguns} = P^{(0,0,y)}_{\text{yellow}} \qquad \text{Prob. Red Shotguns} = P^{(0,0,r)}_{\text{red}}$$

so that, because the different colored dice are completely independent of each other, indicating that we can multiply probabilities, the chance that we roll all shotguns given the dice combination of $\{g, y, r\}$ is

$$\text{Probability 3 Shotguns with } \{g, y, r\} = P^{(0,0,g)}_{\text{green}} \cdot P^{(0,0,y)}_{\text{yellow}} \cdot P^{(0,0,r)}_{\text{red}}.$$

Now, the actual probability that we pull $\{g, y, r\}$ from the cup, starting with the 13 dice given in the game with 6 greens, 4 yellow, and 3 reds, we recall is

$$\text{Probability Get } \{g, y, r\} = \frac{\binom{6}{g} \cdot \binom{4}{y} \cdot \binom{3}{r}}{\binom{13}{3}}$$

so that, combining everything together, the probability that our first round of rolling results in 3 shotguns is given by

$$\sum_{\{g,y,r\} \in S} \underbrace{\frac{\binom{6}{g} \cdot \binom{4}{y} \cdot \binom{3}{r}}{\binom{13}{3}}}_{\substack{\text{probability of} \\ \text{this combo}}} \cdot \underbrace{P^{(0,0,g)}_{\text{green}} \cdot P^{(0,0,y)}_{\text{yellow}} \cdot P^{(0,0,r)}_{\text{red}}}_{\text{probability 3 shotguns this combo}}$$

iterate over dice combos

which, while it may not be pretty, is almost as concise as we can be and at the very least mathematically correct!

In Table 9.11 I have listed the 10 different possible elements of the set S used in Example 9.32 (in the form of $\{g, y, r\}$ used there) paired with the actual dice colors

TABLE 9.11: Color Combinations

Combo	Element of S	Combo	Element of S
GGG	$\{3,0,0\}$	YYY	$\{0,3,0\}$
RRR	$\{0,0,3\}$	GGY	$\{2,1,0\}$
GGR	$\{2,0,1\}$	YYG	$\{1,2,0\}$
YYR	$\{0,2,1\}$	RRG	$\{1,0,2\}$
RRY	$\{0,1,2\}$	GYR	$\{1,1,1\}$

involved. Using these, we can compute the summands involved in the summation over S of that example. For instance, here we compute that for the element $\{0, 2, 1\}$, corresponding to YYR, we would obtain

$$P^{(0,0,0)}_{\text{green}} = \binom{0}{0,0,0} \cdot \left(\frac{1}{2}\right)^0 \cdot \left(\frac{1}{3}\right)^0 \cdot \left(\frac{1}{6}\right)^0 = 1$$

for the green dice,

$$P_{\text{yellow}}^{(0,0,2)} = \binom{2}{0,2,0} \cdot \left(\frac{1}{3}\right)^0 \cdot \left(\frac{1}{3}\right)^2 \cdot \left(\frac{1}{3}\right)^0 = \frac{1}{9} \approx 0.11111$$

for the yellow dice,
and finally

$$P_{\text{red}}^{(0,0,1)} = \binom{1}{0,0,1} \cdot \left(\frac{1}{6}\right)^0 \cdot \left(\frac{1}{3}\right)^0 \cdot \left(\frac{1}{2}\right)^1 = \frac{1}{2} = 0.50000$$

for the red dice. Note that it may seem silly to list the green case when we are only worrying about YYR, but it is definitely true that the probability of getting no shotguns with no green dice is 100%! Combining this together, using Example 9.32 as the guide, the probability that we select the dice combination YYR and then roll three shotguns is given by

$$\underbrace{\frac{\binom{6}{0} \cdot \binom{4}{2} \cdot \binom{3}{1}}{\binom{13}{3}}}_{\substack{\text{probability of} \\ \text{the YYR combo}}} \cdot \underbrace{P_{\text{green}}^{(0,0,0)} \cdot P_{\text{yellow}}^{(0,0,2)} \cdot P_{\text{red}}^{(0,0,1)}}_{\text{probability 3 shotguns with YYR}} = \frac{1 \cdot 6 \cdot 3}{286} \cdot 1 \cdot \frac{1}{9} \cdot \frac{1}{2} \approx 0.00350.$$

For a different dice configuration, such as RRR, we could also compute the probability that we get this dice combination and then roll three shotguns. We would get

$$\underbrace{\frac{\binom{6}{0} \cdot \binom{4}{0} \cdot \binom{3}{3}}{\binom{13}{3}}}_{\substack{\text{probability of} \\ \text{the RRR combo}}} \cdot \underbrace{P_{\text{green}}^{(0,0,0)} \cdot P_{\text{yellow}}^{(0,0,0)} \cdot P_{\text{red}}^{(0,0,3)}}_{\text{probability 3 shotguns with RRR}} = \frac{1 \cdot 1 \cdot 1}{286} \cdot 1 \cdot 1 \cdot \frac{1}{8} \approx 0.00044$$

where the computation of $P_{\text{red}}^{(0,0,3)}$ (and the others) is left for you to verify. Similarly, for the case where we get one of each die, GYR, we would get

$$\underbrace{\frac{\binom{6}{1} \cdot \binom{4}{1} \cdot \binom{3}{1}}{\binom{13}{3}}}_{\substack{\text{probability of} \\ \text{the GYR combo}}} \cdot \underbrace{P_{\text{green}}^{(0,0,1)} \cdot P_{\text{yellow}}^{(0,0,1)} \cdot P_{\text{red}}^{(0,0,1)}}_{\text{probability 3 shotguns with GYR}} = \frac{6 \cdot 4 \cdot 3}{286} \cdot \frac{1}{6} \cdot \frac{1}{3} \cdot \frac{1}{2} \approx 0.00699$$

for the probability of drawing the combination from the cup and then getting a shotgun on each die.

Instead of computing the probabilities for the other seven combinations given in Table 9.11 here, I have left these for you as an exercise later. The method above will work fine for the other seven cases, and when you finish those and add your results to the above, you should find that the total probability of rolling three shotguns on your initial roll for Zombie Dice is, courtesy of Example 9.32 and your probabilities,

$$
\underbrace{\sum_{\{g,y,r\}\in S}}_{\substack{\text{iterate over}\\\text{dice combos}}} \underbrace{\frac{\binom{6}{g}\cdot\binom{4}{y}\cdot\binom{3}{r}}{\binom{13}{3}}}_{\substack{\text{probability of}\\\text{this combo}}} \cdot \underbrace{P^{(0,0,g)}_{\text{green}}\cdot P^{(0,0,y)}_{\text{yellow}}\cdot P^{(0,0,r)}_{\text{red}}}_{\text{probability 3 shotguns this combo}} = \frac{94}{3,861} \approx 0.02435
$$

for only about a 2.4% chance of ending the round on your first roll of the dice.

At some point you might consider the use of summations and multinomial coefficients overkill here, but for us to actually develop a formula for the expected value of the next roll, we will definitely need them. First, we need to think about what happens later in the game when some dice have been used up, and we're considering what to do on the next roll. We might very well have some footprints that must be rerolled (if we choose to roll again), but otherwise we need to randomly draw dice from whatever is left in the cup. For example, if there are 3 green dice in the cup, and we already have 1 green footprint to reroll, when considering the combination GGR, we need to only choose 1 of the 3 green dice in the cup since we already have the green footprint to reroll. This idea gives us the following generalization.

Theorem 9.5 Suppose that there are currently g_c green dice in the cup, y_c yellow dice in the cup, and r_c red dice in the cup. Further suppose that we currently have g_f green footprints to reroll, y_f yellow footprints to reroll, and r_f red footprints to reroll. The probability that we end up rolling the dice combination $\{g, y, r\}$, keeping the same format for combinations as in Table 9.11, is given by the function

$$
C^{g_c,y_c,r_c}_{g_f,y_f,r_f}\{g,y,r\} = \mathbf{C}\{g,y,r\} = \frac{\binom{g_c}{g-g_f}\cdot\binom{y_c}{y-y_f}\cdot\binom{r_c}{r-r_f}}{\binom{g_c+y_c+r_c}{3-(g_f+y_f+r_f)}}
$$

where we also take the convention that $\mathbf{C}\{g,y,r\} = 0$ for impossible situations, such as the combination $\{1,1,1\}$ when $r_f = 2$ (since rerolling 2 red dice means we cannot possibly also have a green *and* a yellow).

There is one important, but small, point to make here. Virtually all of the functions we deal with from now on will depend on the six variables g_c, y_c, r_c, g_f, y_f, and r_f, and rather than overly complicate our work, we will use the shorthand notation introduced in Theorem 9.5. Any function that has these six variables as at least some of its parameters will be represented by accompanying notation using a boldface function name. Of course, a function may have more parameters, and we will list these at all times, but the boldface notation lets us get around having things just *look* more complicated than they really are.

You will want to take a few moments to digest Theorem 9.5. Note that when we have g_c choose $g - g_f$, we're saying that, from the g_c green dice left in the cup, while normally we would just choose g of them, with rerolling g_f green dice already, we need only to select $g - g_f$ dice from the cup. The other terms are similar, and the denominator considers all possible dice (with $g_c + y_c + r_c$) and from among all of those, chooses enough dice to bring us up to 3 to roll.

Putting our shotguns aside for now (they will, of course, be coming back), let's turn our attention to getting some brains. If we want to determine the expected number of brains for the next roll, we need to determine the probability of actually getting three, two, one, or zero brains on that roll. Why? If we let $PB(x)$ represent the probability of getting x brains on the next roll, then the expected number of brains can be computed, recalling Definition 2.4, as

$$\text{Expected Brains} = 0 \cdot PB(0) + 1 \cdot PB(1) + 2 \cdot PB(2) + 3 \cdot PB(3)$$

where, as you might imagine, the values of $PB(x)$ depend heavily on the number (and type) of dice in the cup and the dice to reroll. For better notation, denote the probability of getting x brains, given $g_c, y_c, r_c, g_f, y_f, r_f$ as in Theorem 9.5, as

$$\text{Probability of } x \text{ Brains} = B^{g_c,y_c,r_c}_{g_f,y_f,r_f}(x) = \mathbf{B}(x)$$

so that we can now adapt some of our previous work to determine the probability of getting three brains.

Using Example 9.32 as a guide, where we considered the probability of getting three *shotguns* on the first roll, we can adapt the summation given there for the probability of getting three brains on any roll, given the current values of green dice g_c and g_f, along with the values for yellow and red dice as well. We would get

$$\mathbf{B}(3) = \sum_{\{g,y,r\} \in S} \underbrace{\mathbf{C}\{g,y,r\}}_{\substack{\text{iterate over} \\ \text{dice combos}}} \cdot \underbrace{P^{(g,0,0)}_{\text{green}}}_{\substack{\text{probability of} \\ \text{this combo}}} \cdot \underbrace{P^{(y,0,0)}_{\text{yellow}} \cdot P^{(r,0,0)}_{\text{red}}}_{\text{probability 3 brains this combo}}$$

where the function $\mathbf{C}$ comes from Theorem 9.5.

Note the subtle difference between the notation $\{g, y, r\}$, which represents a specific selection of *dice* wherein here we have g green dice to roll, y yellow dice to roll, and r red dice to roll, and the notation (i, j, k), which represents, for a specific *color* of dice, getting i brains, j footprints, and k shotguns. Each element of S yields a specific set of (i, j, k) possibilities for green, a specific set of (i, j, k) possibilities for yellow, and another specific set of (i, j, k) possibilities for red. What do I mean? Perhaps an example is in order.

Example 9.33 Consider the dice combination $\{1, 2, 0\}$ which represents having 1 green die, 2 yellow dice, and no red dice. Suppose we are interested in determining the probability of getting only *two* brains with this combination. We could have both brains be on the yellow dice, giving $(2, 0, 0)$ for the brains, footprints, and shotguns possibility for yellow, but then the green die could be either a footprint, giving $(0, 1, 0)$ as the possibility, or a shotgun, giving $(0, 0, 1)$ as the possibility. On the other hand, the green die could yield one of the brains, meaning that one of the yellow dice would be a brain and either a footprint or a shotgun. This scenario gives $(1, 1, 0)$ or $(1, 0, 1)$ for the yellow dice, and $(1, 0, 0)$ for the green die.

What we need is a list of (i, j, k) possibilities for each color, based on the number of brains we are looking for, denoted by x, and the element of S (dice combination, say $\{g, y, r\}$) we currently have. Denote by $T(\{g, y, r\}; x)$ this set of possibil-

TABLE 9.12: $T(\{1, 2, 0\}; 2)$ Elements

Element #	Green	Yellow	Red
1	(0,1,0)	(2,0,0)	(0,0,0)
2	(0,0,1)	(2,0,0)	(0,0,0)
3	(1,0,0)	(1,1,0)	(0,0,0)
4	(1,0,0)	(1,0,1)	(0,0,0)

ities so that, based on Example 9.33, we have the elements of $T(\{1, 2, 0\}; 2)$ given as the rows of Table 9.12. All other possibilities for the function T are given in Table T.10 at the end of the book for your reference rather than including them all here. To be clear each element of $T(\{1, 2, 0\}; 2)$ (and all other values of T) is a set of three (i, j, k) triples, one for the green dice involved, one for the yellow dice involved, and one for the red dice involved. Below we will consider an element of T to consist of a (g_i, g_j, g_k), a (y_i, y_j, y_k), and a (r_i, r_j, r_k).

With this, we can now develop formulas for $\mathbf{B}(x)$ for the other values of x. The most general form will be

$$\mathbf{B}(x) = \underbrace{\sum_{\{g,y,r\}}}_{\substack{\text{dice} \\ \text{combos}}} \underbrace{\mathbf{C}\{g, y, r\}}_{\substack{\text{probability of} \\ \text{this combo}}} \underbrace{\sum_{T(\{g,y,r\};x)}}_{\substack{\text{iterate over} \\ T \text{ elements}}} \underbrace{P_{\text{green}}^{(g_i,g_j,g_k)} P_{\text{yellow}}^{(y_i,y_j,y_k)} P_{\text{red}}^{(r_i,r_j,r_k)}}_{\text{probability } x \text{ brains this combo}}$$

which, again, while not pretty, definitely is mathematically correct and virtually as simple as we can make it.

It's worth taking a moment to see what happens with the inside summation, based on the work we did in Example 9.33 and Table 9.12. We consider just a piece of $\mathbf{B}(2)$.

Example 9.34 Looking at $\mathbf{B}(2)$ and considering just one of the elements of S, $\{1, 2, 0\}$, for this example, the iteration over elements of T in the formula would proceed as follows. There are four elements of T that we will need to work with, given in Table 9.12 as well as later in the book as Table T.10. The first, as the table shows, is when we get both brains on the yellow dice and a footprint on the green die, indicated by the triple $(0, 1, 0)$ for green, $(2, 0, 0)$ for yellow, and $(0, 0, 0)$ for red. You should be able to follow the other three elements when they show up in

$$\underbrace{\sum_{T(\{g,y,r\};x)}}_{\substack{\text{iterate over} \\ T \text{ elements}}} \underbrace{P^{(g_i,g_j,g_k)}_{\text{green}} P^{(y_i,y_j,y_k)}_{\text{yellow}} P^{(r_i,r_j,r_k)}_{\text{red}}}_{\text{probability } x \text{ brains this combo}} = \underbrace{P^{(0,1,0)}_{\text{green}} P^{(2,0,0)}_{\text{yellow}} P^{(0,0,0)}_{\text{red}}}_{\text{probability for } T \text{ element 1}}$$

$$+ \underbrace{P^{(0,0,1)}_{\text{green}} P^{(2,0,0)}_{\text{yellow}} P^{(0,0,0)}_{\text{red}}}_{\text{probability for } T \text{ element 2}}$$

$$+ \underbrace{P^{(1,0,0)}_{\text{green}} P^{(1,1,0)}_{\text{yellow}} P^{(0,0,0)}_{\text{red}}}_{\text{probability for } T \text{ element 3}}$$

$$+ \underbrace{P^{(1,0,0)}_{\text{green}} P^{(1,0,1)}_{\text{yellow}} P^{(0,0,0)}_{\text{red}}}_{\text{probability for } T \text{ element 4}}$$

which, as a reminder, is just the inside summation of $\mathbf{B}(2)$ for *just one* specific element of S.

Finally, we can put together an expression for the expected number of brains for the next roll based on whatever dice might be left in the cup and any footprints you might have to reroll. This expression, in summation notation, is

$$\text{Expected Brains} = EB^{g_c, y_c, r_c}_{g_f, y_f, r_f} = \mathbf{EB} = \sum_{i=0}^{3} i \cdot \mathbf{B}(i)$$

and would certainly be long to compute here by hand. As you might expect, in the interest of saving space and not driving you any crazier, we omit this computation. As we shall see, though, there are two other reasons that we shouldn't compute this number exactly; at least, some small modifications should be in order.

Before that, though, we need to develop an analog for the probability of getting zero, one, two, or three shotguns. Although the triple for green dice (g_i, g_j, g_k) that appears in Table T.10 as values for $T(\{g, y, r\}; x)$ was developed for brains, there is quite honestly nothing special about brains here. We can use this for shotguns as well! Of course, the triple (g_i, g_j, g_k) would now be *backwards* as far as our formulas go, but if we simply switch the g_i and g_k before inserting them into our probability functions, we will be perfectly fine. If we denote by $S^{g_c, y_c, r_c}_{g_f, y_f, r_f}(x) = \mathbf{S}(x)$ the probability of getting x shotguns given the usual variables for the starting condition, we have

$$\mathbf{S}(x) = \underbrace{\sum_{\{g,y,r\}}}_{\substack{\text{dice} \\ \text{combos}}} \underbrace{\mathbf{C}\{g,y,r\}}_{\substack{\text{probability of} \\ \text{this combo}}} \underbrace{\sum_{T(\{g,y,r\};x)}}_{\substack{\text{iterate over} \\ T \text{ elements}}} \underbrace{P_{\text{green}}^{(g_k,g_j,g_i)} P_{\text{yellow}}^{(y_k,y_j,y_i)} P_{\text{red}}^{(r_k,r_j,r_i)}}_{\text{probability } x \text{ shotguns this combo}}$$

which looks more or less the same as for the brains. The only, very subtle, difference that we alluded to is the use of (g_k, g_j, g_i), with the role of g_i and g_k reversed, in $P_{\text{green}}^{(g_k,g_j,g_i)}$ since there, through the definitions of Theorem 9.4, the order (g_k, g_j, g_i) specifically means g_k brains, g_j footprints, and g_i shotguns. That order is the *reverse* of the elements of T when dealing with shotguns. Of course the same is true for the yellow and red dice and their functions.

The probability that our round will end on the next roll, as you might imagine, depends on how many shotguns we currently have. If you assume that we currently have s shotguns (where s is zero, one, or two), then if we roll $3 - s$ (or more) shotguns with the next single throw of the dice, we must discard our brains and end our round. Since we can only roll a maximum of 3 shotguns, the probability that the round ends due to shotguns on the next roll, given our usual variables and having s current shotguns, can be defined as

$$\text{Probability Round Ends} = PE_{g_f,y_f,r_f}^{g_c,y_c,r_c}(s) = \mathbf{PE}(s) = \sum_{i=3-s}^{3} \mathbf{S}(i)$$

where, for example, if we currently have 1 shotgun, we add together the probabilities of getting 2 or 3 shotguns on the next roll; that is, for example,

$$\mathbf{PE}(1) = \mathbf{S}(2) + \mathbf{S}(3).$$

Recall from earlier that our decision equation for whether or not we continue rolling was based on the expected value of the next roll, balancing the probability of the round ending (and losing all of the brains we currently have) with the reward of eating (rolling) more brains. There, we said we should roll again if

$$\text{Expected Value of Continuing} = \underbrace{-b \cdot PE}_{\text{round ends}} + \underbrace{EB \cdot (1 - PE)}_{\text{round continues}} > 0$$

which is still true, but now we have more proper functions with which to work. Specifically, we modify this slightly, so that given we currently have s shotguns, b brains, and the cup has g_c, y_c, and r_c green, yellow, and red dice, respectively, along with us needing to reroll (if we choose to reroll) g_f green dice, y_f yellow dice, and r_f red dice, we should roll again if

$$\text{Expected Value of Continuing} = \underbrace{-b \cdot \mathbf{PE}(s)}_{\text{round ends}} + \underbrace{EB \cdot (1 - \mathbf{PE}(s))}_{\text{round continues}} > 0.$$

While this is nice, and certainly computable given the development of our functions, it does have at least two limitations that should be corrected before I give you a good way to play Zombie Dice. The first of these problems is that currently the function **EB** allows for some brains that are generally not allowed. For instance, if the element of S we are currently iterating over is $\{1, 1, 1\}$ and we also currently have 2 shotguns, the $T(\{1, 1, 1\}; 2)$ element having $(1, 0, 0)$ for the green die, $(1, 0, 0)$ for the yellow die, and $(0, 0, 1)$ for the red die would result in 2 additional brains for us, but also end the round since we picked up a third shotgun. That is, some elements of T for specific elements of S would need to be ignored. The second limitation is that, should you choose to reroll and survive, you also have an opportunity to roll *again* and eat more brains! We need to consider not just the expected number of brains on the *next* roll, but rather *all future* rolls. This can be done recursively very easily and quickly with the use of computers.

If we solve the equation

$$-b \cdot \mathbf{PE}(s) + \mathbf{EB} \cdot (1 - \mathbf{PE}(s)) > 0$$

for the value of b, we can, for any set value of s, along with values for the quantity of dice remaining in the cup and current footprints, give a value b that can be considered the decision point. Actually solving for b, we obtain

$$\frac{\mathbf{EB} \cdot (1 - \mathbf{PE}(s))}{\mathbf{PE}(s)} > b$$

which means that we should continue rolling as long as the number of brains you currently have is less than the value on the left-hand side. Remember, though, since many of the variables change value after one roll, each time you do roll, you need to recompute the left-hand side!

Thankfully, with a bit of fun programming, I did take the time to remove the limitations discussed above so that I can present to you an optimal strategy for Zombie Dice play. In Table T.11 of the Tables section at the end of the book you will find the decision point b for each possible setup for dice in the cup, dice to reroll, and current amount of shotguns. There, you can look up the row corresponding to the current dice situation (cup and footprints), and then look at the "Decision" column corresponding to the number of shotguns you have (where the (SG 0), (SG 1), and (SG 2) columns, as you might guess, represent $s = 0$, $s = 1$, and $s = 2$ respectively). Table 9.13 below is a one-row excerpt from Table T.11.

TABLE 9.13: Sample Row of Table T.11

| Dice in Cup | | | Current FPs | | | Decision | Decision | Decision |
R	Y	G	R	Y	G	(SG 0)	(SG 1)	(SG 2)
2	3	1	1	0	1	78.338580	4.043669	0.180008

Example 9.35 Suppose that there are currently 1 green die, 3 yellow dice, and 2 red dice in the cup, and you have a red die and green die indicating a footprint that you would need to roll should you choose to continue. Using Table 9.13 as a guide, if we have no shotguns presently, we find $b = 78.338580$, meaning that you should continue rolling if you have 78 or fewer brains, and you should stop rolling if you have 79 or more brains. Unless you've been extremely lucky rolling brains and more brains, putting brains back into the cup to continue rolling, and haven't gotten a single shotgun, this value of b essentially means "roll again." Of course, if you had, say, 50 brains, you might stop anyway since you're already at 13 or above, and I doubt anyone will match or beat you. However, note that if you have 1 shotgun, with a value of $b = 4.043669$, the strategy says to stop rolling if you have 5 or more brains, and continue if you have 4 or less. With 2 shotguns already set aside, $b = 0.180008$, so if you already have at least 1 brain, stop rolling. The chance that you will get the third shotgun is too great to risk the brains (or brain) you currently have!

Example 9.35, as you think about it, makes great sense. We know that we'll be rerolling a red die and a green die, and the other randomly selected die to bring us up to three dice to roll is likely to be yellow or red; knowing that we already will have a 50% chance of getting a shotgun on the red die we would have to reroll, we can see why it makes sense to score a single brain if we already have 2 shotguns! Chances are we may be rolling two red dice, or, most likely, a red die and a yellow die, where the chance for that last shotgun, making us unhappy, is pretty high!

With this strategy, you will be a formidable opponent at Zombie Dice! While this strategy definitely works well as a solitaire effort, it also serves as your best play for the early and middle parts of play against real opponents. When the game is nearing the end (with someone in the current round who comes *before* you in the turn order reaching 13 or more brains, or someone in the current round coming *after* you having 10 or more brains), you will want to at least reach 13 brains or more, at whatever cost. Why the magic number 10 for players after you in the turn order? The expected number of brains on your turn, starting with 3 red dice, 4 yellow dice, and 6 green dice in the cup, no footprints, and no shotguns, happens to be 3.315559,[9] so if someone has 10 dice and has yet to go, you can reasonably expect them to get at least 13 and end the round without you having a chance to win!

If you would prefer a slightly less optimal but easier to remember strategy, consider Table 9.14 which summarizes the generalizations obtained from Table T.11, courtesy of Heather Cook. Recall that g_c and g_f refer to the number of green dice left in the cup and the number of green footprints that must be

---

[9]The number here is not computed in this book, but is rather part of the computation used to create Table T.11.

rerolled if you choose to continue, respectively, with similar notation for the yellow and red dice.

TABLE 9.14: Simpler Zombie Dice Strategy

Current Shotguns	Rule
0	Always keep rolling.

1	If we must roll 3 red dice, stop at 1 brain.
	If ($r_f = 2$ and $y_f = 1$) or $y_c > g_c$, stop at 2 brains.
	If ($r_f = 2$ and $g_f = 1$) or $g_c > y_c$, stop at 3 brains.
	If we must roll 3 green dice, roll again.
	If $g_f = 2$, roll again.

2	With $g_f = 3$, stop at 2 brains. Otherwise, stop at 1 brain.

While not 100% the same as following Table T.11, unless you are really good at memorizing many pages of numbers, it should treat you well.

Before we end this chapter, there's an important takeaway from this section. Yes, the development of strategy for Zombie Dice was more complicated than Yahtzee (while the actual solving of optimal strategy for Yahtzee uses computers, the setup done identically to Mini-Yahtzee, which was quite simple), but the lesson is that we can work within the rules of a given game to model a turn or round, and extract information about ways to make the best decisions. With the recent popularity of dice games with custom faces, colors, and other neat novelties, the ability to *model* these games is the most important thing. Feel free to pick up a copy of Zombie Dice to give it a shot, or think about strategies for other new dice games such as Martian Dice or Dungeon Roll.

9.10   Exercises

9.1 Research a "Pick 4" game that your state lottery offers (or find the Virginia Lottery online) and then compute the expected value of an exact match.

9.2 Referring to the "Pick 4" game you researched for Exercise 1, compute the expected value of at least two other bet types offered by that lottery.

9.3 Research the rules, cost, and payouts for the multi-state lottery Powerball and then repeat Example 9.3 to find the value, J, for the current jackpot size that yields an expected value of zero (note that you may need to compute the probabilities yourself rather than relying on the published numbers since they are likely to be rounded).

9.4 Count the number of essentially different Bingo cards for the "big H" pattern, comprised of all five spaces for the B and O columns, the middle spaces in the I and G columns, along with the free space.

9.5 Count the number of essentially different Bingo cards for the "big I" pattern, comprised of the first row, bottom row, and all remaining spaces in the N column.

9.6 Count the number of essentially different Bingo cards for the "skinny H" pattern, comprised of all five spaces for the I and G columns (along with the free space).

9.7 Bingo games on cruise ships often offer a special prize for obtaining bingo on a specific draw (using the blackout pattern); use Theorem 9.1 to find the probability of a specific card winning on draw $n = 50$.

9.8 Suppose that Bingo were changed so that each of the columns only had 10 possible numbers as opposed to 15. With this change, find the number of essentially different Bingo cards for the following patterns.

    (a) the X   **(b)** blackout   **(c)** the "big H"   **(d)** the "big I"

    (e) the "skinny H"   **(f)** five in a row, column, or diagonal

9.9 Imagine a new game played the same as Bingo except that there are only *four* columns and *four* rows, each with at most 10 numbers. There is no free space in this version of the game, and the columns, for lack of better names, use the letters G, A, M and E.

    (a) Determine the probability that a specific card of 4-by-4 Bingo wins on the sixteenth call under blackout rules.

    (b) Determine a formula for the probability that a specific card of 4-by-4 Bingo wins on call n under blackout rules, where $16 \leq n \leq 40$.

    (c) Determine the expected number of calls needed to completely cover your specific 4-by-4 Bingo card.

9.10 Repeat Exercise 9.9 for a game of Bingo with *three* columns and rows, where there are 8 possible numbers per column *and* a free space is included as the middle square in the middle column.

9.11 For a game of Uno, find the probability that your starting hand of seven cards has:

    (a) only special cards (no numbered cards at all).

    (b) only red cards.

    (c) no red cards.

    (d) no special cards (hand only has numbered cards).

    (e) only "wild" or "draw four plus wild" cards.

    (f) at least one card from each color.

9.12 The rules for Uno stipulate that the only time a player may play a "draw four plus wild" is when they do not have a card in their hand that matches the color on the discard pile; if a player does play one of these cards, another player may challenge its use (forcing them to draw the four cards instead if they played it illegally). Imagine your starting hand has three red cards, two blue cards, and two green cards; you play one of your red cards. The player to your left plays a "draw four plus wild" card; determine the probability that the card was played illegally.

9.13 Using Table 9.5, compute the house advantage for a tie bet (paid at 8:1 casino odds) at standard punto banco baccarat; would you recommend betting on a tie, or would you prefer to bet on the player's hand or dealer's hand?

9.14 Compute, as in Example 9.11, the expected value for betting on the banker's hand at standard punto banco baccarat *without* the 5% commission on wins. Why do you think that the casino charges this commission on wins?

9.15 A modern version of baccarat that can be found in casinos is called "EZ

Baccarat" and eliminates the commission charged on banker hand wins by offering a regular 1:1 payout on the banker's hand bet *except* when the banker hand wins with a point total of 7 with *three cards*; this results in a push instead. For this exercise, most of the probabilities of Table 9.11 still apply (see below) since the game play itself is not changed.

(a) Determine the house advantage for a bet on the banker's hand for this variant of baccarat assuming that the probability of a *three card* win with a point total of 7 for the dealer is 0.022535.

(b) This version offers a side bet for one round at baccarat that pays casino odds of 40:1 if the banker's hand wins on a point total of exactly 7 with three cards (called the "Dragon 7" bet); determine the house advantage for this wager.

(c) Also offered during "EZ Baccarat" is the "Panda 8" side bet that pays odds of 25:1 if the player's hand wins with a point total of exactly 8 (with three cards, using a probability of 0.034543); determine the house advantage on this bet.

9.16 Another version of baccarat offers a payout of 1:1 on the banker's hand *except* it pays 1:2 (or, equivalently, 0.50:1) on the banker's hand if the winning point total is 6; determine the house advantage on the banker's hand with this rule (use Table 9.5 as usual).

9.17 Verify that the probability that you farkle out with $n = 5$ dice during a game of Farkle is about 0.077160. Note that there are two patterns for you to consider as opposed to one in Example 9.12 and three in Example 9.13, so you will need to determine the patterns first.

9.18 In Farkle, as we saw in Example 9.15, there are 300 ways to get a double three of a kind with 6 dice. Determine the 15 different pairings of these triples, the scores (using the scoring system presented after that example) for these pairings, and use this information to determine the expected contribution to your score from the category "double triples" on the first roll.

9.19 Recall that for single three of a kinds in Farkle when rolling 6 dice, the two possibilities can be described using the patterns AAABBC and AAABCD where each letter represents a different value for the rolled dice. Determine the expected contribution for the *entire roll* when the initial roll is a single three of a kind by considering the possible assignments of 1, 5, and other numbers to the letters of the patterns (for instance, the pattern AAABBC could have A as 1, B as 5 and C as 4, which results in a different score than A as 4, B as 6 and C as 1). For pairs or singletons, treat ones as having value 100 each and fives as having value 50 each.

9.20 Find a scoring system for Farkle by researching the game online, and use the system you found to do the following for the four of a kind category.

(a) Determine the probability of rolling a four of a kind with the initial six dice.

(b) Determine the expected contribution for *only* the four of a kind to your initial roll.

(c) By considering all possible patterns, determine the expected score for an initial roll that includes a four of a kind.

9.21 Repeat Exercise 9.20 for the following scoring categories (note that part (c) is unnecessary for straights, six of a kinds, and three pairs, and can be omitted for rolls with only 1s or 5s) based on a scoring system for Farkle found by researching online.

(a) five of a kind   (b) six of a kind   (c) straight   (d) three pairs
(e) rolls with only ones and fives not listed in any other category

9.22 Using your information from Exercises 9.18 through 9.21, compute the expected value of the initial roll (or any roll of six dice) in Farkle.

9.23 In Scrabble, determine the number of starting "hands" of tiles, and the probability of getting those starting hands, that contain the appropriate letters to make each of the following words without using a blank tile.

(a) IS   (b) CAT   (c) THE   (d) GEE   (e) MOON
(f) GOAL

9.24 Suppose that some tiles go missing in your set of Scrabble, but you still decide to play a game with your friend. Determine the number of starting hands, and the probability of getting a starting hand, to form the word DOG, without using a blank tile, under the following circumstances.

(a) one of the D tiles is missing
(b) one of the E tiles is missing
(c) two of the G tiles are missing

9.25 Compare the results of Exercise 9.24 with Example 9.16.

9.26 The game of *Super Scrabble*[10] introduces a larger game board and twice as many tiles, for a total of 200 letter tiles. Players still draw seven tiles to start a game.

(a) Assuming there are 8 Ds, 15 Os, and 5 Gs, determine the number of starting hands, and the probability of getting a starting hand, with letters to form the word DOG without using blank tiles; compare your answer to Example 9.16.
(b) Assuming there are 6 Cs, 16 As, and 15 Ts, determine the number of starting hands, and the probability of getting a starting hand, with letters to form the word CAT without using blank tiles; compare your answer to Exercise 9.23 part (b).

9.27 For backgammon, as we did in Examples 9.18, 9.19, and 9.20, compute the probability of being able to hit an opponent's piece n spaces away for the values of n equal to 2, 4, 5, 8, 9, 10, 11, 12, 15, 16, 18, 20, and 24. Argue that any other number between 1 and 24 is not possible.

9.28 Repeat Examples 9.21 and 9.22 if the doubling cube in backgammon were replaced with a *tripling* cube to find an optimal strategy on accepting a triple or offering a triple. Note that rather than the 2, 4, 8, 16, and so on, of the doubling cube, the tripling cube has sides 3, 9, 27, 81, and so on.

9.29 Determine the expected length of a game of Memory when $n = 4$ pairs of cards are involved by following the steps below.

(a) For each of the layouts for $n = 3$ listed in Table 9.10, append "44" to the end and then move the first 4 into each of the other possible 6 positions for the layout (there should be 7 layouts obtained from each of the 15 layouts when $n = 3$).
(b) For each of your 105 essentially different layouts for $n = 4$, determine the number of moves required to end the game, using the strategy discussed in this chapter.

---

[10]Super Scrabble is a registered trademark of Winning Moves Game and licensed to Hasbro, Inc.

(c) Use the information you've accumulated so far to find the expected length of the game as we did in Example 9.25.

(d) Without the error-term included, how close is the estimate of Theorem 9.3 for $n = 4$ to the value you found? Comment on the usefulness of Theorem 9.3 in computing expected lengths.

9.30 Consider an alternate version of Memory where *triples* of cards are used instead of pairs; on each turn, you may turn over three cards, and if the three cards match, they are removed.

(a) Determine the number of essentially different layouts for $n = 2$ and $n = 3$ for this version as we did in Examples 9.23 and 9.24. Note that $n = 2$ involves 6 cards and $n = 3$ involves 9.

(b) Find a generalized formula giving the number of essentially different layouts for "triple Memory" when n triples are involved.

(c) By writing down a few different layouts when $n = 3$, devise a strategy for playing triple Memory; is it the case that one should only turn over a known card when the position of the other two cards of that triple are known?

(d) Determine the expected length of a game of triple Memory for $n = 2$.

9.31 For each of the other seven dice combinations given in Table 9.11, compute, for Zombie Dice, the probability of getting this combination from a fresh cup of dice, and then rolling three shotguns with those dice.

9.32 For this exercise, refer to Example 9.32, Table 9.11, and the discussion that involves computing the probability of getting three shotguns on your initial roll in Zombie Dice.

(a) Keeping the number of yellow and red dice in the cup the same, compute the probability of getting three shotguns on the initial roll if an additional green die is added to the cup. Do the same assuming that a green die is *removed* from the cup instead.

(b) Keeping the number of yellow and green dice in the cup the same, compute the probability of getting three shotguns on the initial roll if an additional red die is added to the cup. Do the same assuming that a red die is *removed* from the cup instead.

(c) Keeping the number of green and red dice in the cup the same, compute the probability of getting three shotguns on the initial roll if an additional yellow die is added to the cup. Do the same assuming that a yellow die is *removed* from the cup instead.

(d) Is it better to add an additional yellow die to the cup (negating the effect of red dice by reducing their proportion) or to remove a yellow die from the cup (enhancing the effect of green dice by increasing their proportion)?

9.33 For this exercise, use Table T.11 that appears in the Tables section of this book that gives optimal decisions when playing Zombie Dice.

(a) Suppose that you currently have 1 shotgun and 1 red footprint on the table. If there are 1 red, 2 yellow, and 1 green dice left in the cup, at how many brains should you stop rolling?

(b) With 1 red and 2 yellow dice left in the cup, suppose that you have 2 shotguns; if you currently have 3 green footprints to reroll, how many brains should you have to make rerolling a good option?

(c) Now, again with 1 red and 2 yellow dice left in the cup and 2 shotguns, if you currently only have 2 green footprints to reroll, how many brains do you need to make rerolling a good option?

(d) If the cup has 1 red and 4 green dice left in the cup and your friend currently has 1 shotgun, what advice would you give your friend if he or she doesn't have any footprints to reroll?

(e) Find all entries of Table T.11 where the number of brains to stop at when having 0 shotguns is below 13; what general advice would you give players if they currently have no shotguns?

9.34 Repeat parts (a) through (d) of Exercise 4.33 except this time use the Simple Strategy given in Table 9.14; do your answers change much, if at all?

10

Betting Systems: Can You Beat the System?

There's a bit of warning that applies for this chapter. There aren't really any games, *per se*, included here. I do, however, feel that this chapter is one of the most important in this book. The ideas contained here are lessons that I strongly feel are appropriate alongside any study of the mathematics behind games of chance. Questions that appear here, while not about any particular game, are most definitely game-

FIGURE 10.1: Casino Chips

related and have some interesting new mathematics we can explore together.

10.1 Betting Systems

When games of chance involving wagers are the topic, surely someone will point out a betting system that has generated profits for them. Luck, however, is almost always the driving force behind someone's winnings at a casino game. Ignoring card counting for now, there is no way, long term, for a player to profit from continuously playing Blackjack. Sitting at a roulette table and betting your money on red all day long might lead to a big profit for you, but it is more likely that you will walk away from the table having lost all of the money you've allotted for gambling. Let's explore this further.

Martingale System

Perhaps one of the most famous systems that people have tried (and studied) is the Martingale System, popular as early as the 1700s in France. The method is actually very easy to describe. Take $10 and bet it on any even-money wager (remember, this means a 1:1 payout) in the casino. If you win, take the $20 you now have and smile that you are now $10 richer. Otherwise, if you lose, bet $20 on that same even-money wager.

If, perhaps, on the second bet you win, you now have $40 in your pocket, but combined with the $10 loss from the first bet and the $20 you wagered on the second bet, your net winnings are only $10, the same as if you had won the first bet. On a loss of the second bet, double your wager and bet $40 on the next round, where a win will cover your $10 loss from the first bet, $20 loss from the second bet, and leave you $10 in the black. Can you see where this is going?

The Martingale system dictates that you should always double your bet after a loss; the instant that you win your bet you will have covered any losses thus far and made a modest profit equal to your very first bet. Yes, you can quite "easily" walk into a casino, take $100 and turn it into a guaranteed $200. Almost.

Recall that, at roulette, the probability of losing any of the even-money wagers (including red) is $20/38 = 10/19$. The probability that you lose, say, five times in a row is given by $\left(\frac{10}{19}\right)^5 \approx 0.0403$. That means there is more than a 95% chance that your $100 will become $200, starting with $100 and doubling your bet on a loss. It also means that a little more than 4% of the time your first five bets will *each* lose. How much are you out? Here is the amount of each bet for bets one through five, and the total, for this scenario.

Bet	1	2	3	4	5	Total
Amount	$100	$200	$400	$800	$1,600	$3,100

Note here that your last bet, trying to recoup your money and make a lousy $100 profit to make up for your string of four losses in a row, is a whopping $1,600! If you have heard anyone talk about how fast exponential growth is, you now know why!

Of course, as experienced users of the concept of expected value, we ought to see, given the idea that we'll only bet at most 5 rounds trying to win the coveted $100, what the expected value of this is. We see that

$$\text{Expected Value (5)} = \underbrace{\left(1 - \left(\frac{10}{19}\right)^5\right) \cdot \$100}_{\text{win once in first five}} + \underbrace{\left(\frac{10}{19}\right)^5 \cdot (-\$3100)}_{\text{lose all five}} \approx -\$29.24$$

for an expected loss. The loss doesn't happen often, but, when it does, it hurts! If money is not a concern, you might try a sixth or even seventh bet to try and win back your money, but if we walk into the casino, expecting to play at most six or seven times with the Martingale System, note that

$$\text{Expected Value (6)} = \underbrace{\left(1 - \left(\frac{10}{19}\right)^6\right) \cdot \$100}_{\text{win once in first six}} + \underbrace{\left(\frac{10}{19}\right)^6 \cdot (-\$6300)}_{\text{lose all six}} \approx -\$36.04$$

and

$$\text{Expected Value (7)} = \underbrace{\left(1 - \left(\frac{10}{19}\right)^7\right) \cdot \$100}_{\text{win once in first seven}} + \underbrace{\left(\frac{10}{19}\right)^7 \cdot (-\$12700)}_{\text{lose all seven}} \approx -\$43.20$$

where our situation is getting *worse*, not better. Wagering $12,700 to recoup losses and end up with a net profit of $100 doesn't sound like fun either.

Question

How bad, mathematically, can the Martingale System get?

We can generalize this to determine the expected value given any base bet amount, fixed number of rounds, and probability of winning, but first we need a small tool to quickly add up the values of bets made.

Theorem 10.1 For a positive integer n and value of r not equal to 1 (that is, $r \neq 1$), we have the shortcut formula

$$\sum_{i=1}^{n} a \cdot r^{i-1} = a \cdot \left(\frac{1-r^n}{1-r}\right) = a \cdot \left(\frac{r^n-1}{r-1}\right)$$

or, by shifting the index for i, the equivalent formula

$$\sum_{i=0}^{n-1} a \cdot r^{i} = a \cdot \left(\frac{1-r^n}{1-r}\right) = a \cdot \left(\frac{r^n-1}{r-1}\right).$$

Note that the reason for this formula is the neat algebra "trick" where

$$(1 + r + r^2 + \cdots + r^{n-1})(1 - r) = 1 - r^n$$

which makes things look quite nice.

With this, we consider the next example to illustrate how the expected value changes as we vary the parameters.

Example 10.1 Suppose that we make a base bet of B dollars, doubling the amount until we either win a round or have reached n losses in a row. To be as general as possible, we also denote the probability that we *lose* a bet by q (we use q here to be consistent, as p usually represents the chance of winning; for games where ties are not possible, remember that $q = 1 - p$).

If we *do* lose each of the n rounds, the total amount that we have bet is given by, using Theorem 10.1 and doubling on a loss,

$$\sum_{i=1}^{n} B \cdot 2^{i-1} = B \cdot \left(\frac{2^n - 1}{2 - 1}\right) = B \cdot (2^n - 1).$$

The expected value of these n rounds can then be calculated as we did earlier as

$$\text{Expected Value} = \underbrace{(1 - q^n) \cdot B}_{\text{win}} + \underbrace{q^n \cdot (-B(2^n - 1))}_{\text{lose}} = B(1 - (2q)^n).$$

In Figure 10.2, I have graphed the expected value function given in Example 10.1 as q ranges from 0.40 to 0.60. Since changing the value of B only affects the scaling of the graph, I chose $B = 100$, and the lines given refer to values of n equal to 4, 6, 8, and 10 for the solid, dashed, gray, and dashed gray lines, respectively. There are a few takeaways from Figure 10.2 we should discuss before we move on to some more bad news about the Martingale system and other systems in general.

One immediate observation that you can see is that each of these lines passes through an expected value of \$0.00 when $q = 0.50$. When playing a true even-money game (such as flipping a fair coin and betting on heads), using the Martingale strategy is just as fair to you as it is to the person or entity taking your money. It's also clear that the more rounds you decide to play, the *worse* your expected value

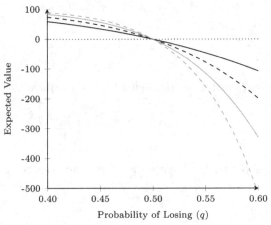

FIGURE 10.2: Martingale Expected Values

is when $q > 0.50$ (where you are more likely to lose than win). What you may find surprising, though, is that the positive expected winnings when $q < 0.50$ (perhaps for a good card counter at Blackjack) are not all that high compared to the losses possible for casino-favorable wagers. The exponential behavior of the Martingale system only applies to the losses and not the wins, unfortunately for us. Another way to put that is that the *penalty* for losing many times in a row grows exponentially, but the *reward* for a win is fixed at the value B.

(Loss) Streaks

Perhaps the most important question when considering the Martingale system is taking a look at how often streaks of losses actually occur. Suppose that you have $6,300 lying around that you feel comfortable using to gamble away at roulette, betting on even-money wagers such as red, black, even, and odd. Why that specific amount? It makes things fairly simple for us since $6,300 is exactly the amount of money you would be out if you were unlucky enough to get six losses in a row. How often could a six-loss streak happen? If you are only wagering a maximum of six times, it's fairly unlikely (with a probability of about 2.13%), but chances are, if you sit down at a roulette table and start wagering, you might want to stick around a little longer than six rounds. We need to determine the probability that a streak of six losses occurs in, say, a round of 20 bets. Or 40 bets, or 70 bets (the average number of betting rounds at a roulette table in two hours of play). While we develop the formula, let's use a round of 10 bets to keep the numbers somewhat small.

> **Example 10.2** Determining the probability of having a streak of at least 6 losses in 10 rounds of betting on 1:1 payout wagers at roulette is actually not too rough. One way of getting a streak of at least 6 losses, though, is by having those 6 losses in the first 6 bets. That is, we get a string of length ten that consists of the letters W and L where the format is
>
> $$\text{Six Losses at Start} = \underbrace{\text{LLLLLL}}_{\text{losses}} \underbrace{????}_{\text{rest}}$$
>
> where the question marks indicate that we don't care about whether those bets are wins or losses. Since a loss happens with probability $q = 10/19$, we obtain the probability of
>
> $$\text{Probability First Six Losses in Ten Rounds} = \left(\frac{10}{19}\right)^6 \approx 0.021256$$
>
> we alluded to earlier. If the streak of six losses does not happen right away, then we know that a win has to happen somewhere in the first six rounds. For example, perhaps the win happens first, obtaining a string of the form
>
> $$\text{Win First} = \text{W} \underbrace{********}_{\text{other nine}}$$
>
> where we could compute the probability by determining the chance that we have a win first, followed by the probability of obtaining a streak of six or more losses in *nine* rounds. That is, since the probability of winning is $p = 1 - 10/19 = 9/19$, we would compute
>
> $$\left(\frac{9}{19}\right) \cdot (\text{probability of streak of six or more in nine rounds})$$
>
> and add that to our running probability tally.

Generalizing a bit, if the streak of six losses does not happen immediately, then our string would start with any number of losses up to five, a win, and then obtaining a streak of six with the remaining rounds. What we need here is a little *recursion*, a very useful mathematical tool that takes a problem and expresses the solution in terms of "simpler" problems. That's exactly what we're doing here by taking a question about *ten* rounds and reducing it to questions about *less than* ten rounds. This process must eventually end, since recursion stops when the number of losses in our desired streak is the same as the number of rounds remaining!

To formalize this, let $P(n, k)$ represent the probability of getting a streak of at least k losses in n rounds of betting where $k > 0$. Then defining $P(n, k) = 0$ for $k > n$ since we cannot possibly get a streak of k losses when less than k rounds exist, we can express a recursive formula for $P(n, k)$ as

$$P(n, k) = \underbrace{\left(\frac{10}{19}\right)^k}_{\text{streak at start}} + \underbrace{\sum_{j=1}^{k} \left(\frac{10}{19}\right)^{j-1} \cdot \left(\frac{9}{19}\right)}_{j-1 \text{ losses, then win}} \cdot \underbrace{P(n-j, k)}_{\text{recurse}}$$

where we recall the summation notation from Definition 2.5. This will make our final answer $P(10, 6)$, but let's explore that outside of this example box.

Let's take a look at the first few "base cases" of the formula given in Example 10.2. Since computing $P(5, 6)$ or anything of the form $P(n, 6)$ where $n < 6$ would result in a probability of zero, the first case to consider is when $n = 6$ as well, where we obtain

$$P(6, 6) = \left(\frac{10}{19}\right)^6 + \sum_{j=1}^{6} \left(\frac{10}{19}\right)^{j-1} \cdot \left(\frac{9}{19}\right) \cdot P(6-j, 6)$$

$$= \left(\frac{10}{19}\right)^6 \approx 0.021256$$

as we saw before. Note that the summation term completely disappears since even when $j = 1$, the recursion term ends up being $P(5, 6)$ which is zero. Similarly, we can obtain

$$P(7, 6) = \left(\frac{10}{19}\right)^6 + \sum_{j=1}^{6} \left(\frac{10}{19}\right)^{j-1} \cdot \left(\frac{9}{19}\right) \cdot P(7-j, 6)$$

$$= \left(\frac{10}{19}\right)^6 + \underbrace{\left(\frac{10}{19}\right)^0 \cdot \left(\frac{9}{19}\right) \cdot P(6, 6)}_{\text{the } j = 1 \text{ term}} \approx 0.031324$$

where we make use of our previous computation of $P(6, 6)$. As a side note here, it is often very nice when working with recursive formulas to work on the smaller-valued cases first, as these are likely to be needed to answer the "larger" questions. As one last example, consider

$$P(8,6) = \left(\frac{10}{19}\right)^6 + \sum_{j=1}^{6} \left(\frac{10}{19}\right)^{j-1} \cdot \left(\frac{9}{19}\right) \cdot P(8-j,6)$$

$$= \left(\frac{10}{19}\right)^6 + \underbrace{\left(\frac{10}{19}\right)^0 \cdot \left(\frac{9}{19}\right) \cdot P(7,6)}_{\text{the } j=1 \text{ term}} + \underbrace{\left(\frac{10}{19}\right)^1 \cdot \left(\frac{9}{19}\right) \cdot P(6,6)}_{\text{the } j=2 \text{ term}}$$

$$\approx 0.041393$$

which shows us that the probability of getting that streak of six losses increases as the number of betting rounds increases. That should agree with your intuition.

You should take a moment to verify the values of $P(9,6)$ and $P(10,6)$ that are given in Table 10.1. In doing so, making use of the other values we have already computed (also listed in the table) is probably a super idea. From this, though, you can see that we expect a little more than 6% of the time a streak of at least six losses will happen when wagering at roulette's bets paying 1:1 odds. What happens with more betting rounds? That chance of six straight losses, costing you the $6,300, grows and grows, as seen in Figure 10.3!

TABLE 10.1: Recursion Values

$P(6,6)$	0.021256
$P(7,6)$	0.031324
$P(8,6)$	0.041393
$P(9,6)$	0.051462
$P(10,6)$	0.061530

While Figure 10.3 does show the increasing nature of the probability as the number of rounds increases, the amount at which it grows per one unit change in rounds shrinks. With 20 bets, the exact probability of finding a streak of at least six is about 15.7667% and for 40 bets, that probability grows to 32.1313%.

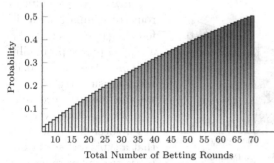

FIGURE 10.3: Streaks versus Probability

Our scenario where we sit down at a moderately busy roulette table and play for two hours (expecting, on average, about 70 rounds of roulette to be played) results in a 50.9151% chance that we'll end up with 6 losses in a row. Of course, this is roulette, and for only 6 losses. We can generalize Example 10.2 a bit more to reach a general theory on the probability of streaks.

Theorem 10.2 (Streaks Theorem, Part 1) The probability of having a streak of at least k successes in n trials, assuming each trial is independent as in the standard binomial distribution, is given recursively by

$$P(n,k) = p^k + \sum_{j=1}^{k} p^{j-1} \cdot q \cdot P(n-j,k)$$

where p is the probability of success, q is the probability of failure (equal to $1-p$), and we define $P(n,k) = 0$ when $n < k$.

If you have any background in programming, I should mention that it is very easy to create a function to compute the probability of streaks using Theorem 10.2. As a suggestion, since when n gets somewhat large, the amount of recursion needed goes up a great deal, you may wish to store the value of $P(n,k)$ for various values of n in an array or vector and refer to that first when needed. This will save much time, and is what we did by hand when computing $P(8,6)$. A wonderful study of this problem by Abraham de Moivre in the 1700s resulted in

$$P(n,k) = 1 + p^k \sum_{l=0}^{\lfloor \frac{n-r}{r+1} \rfloor} (-1)^l \binom{n-r-lr}{l} (qp^r)^l - \sum_{l=0}^{\lfloor \frac{n}{r+1} \rfloor} (-1)^l \binom{n-lr}{l} (qp^r)^l$$

as a nice closed formula for the probability for a streak of length at least k in n trials (the notation $z = \lfloor x \rfloor$ indicates taking the *floor* of x, meaning the integer $z \leq x$ that is closest to x); we did not list this in Theorem 10.2, as its development would require a lengthly study of difference equations and generating functions which are not included in this book.

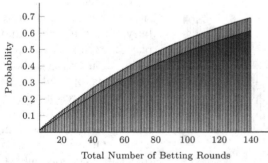

FIGURE 10.4: Streaks versus Probability, Blackjack

How do our results for roulette compare to a different game? Recall that a typical player probability of winning is about 0.495 using basic strategy,[1] so we can simply recompute $P(n,k)$ for whatever values we desire. Studies show that a typical Blackjack table with 5 players can run through 70 hands of Blackjack in *one* hour, so if we

---

[1]This is a special "weighted" probability that takes into account changes in betting induced by splitting, doubling down, and an extra special payout for a natural Blackjack.

want to compare that to roulette, we should look at $P(140, 6)$ with $p = 0.495$ and $q = 0.505$. On the other hand, if we happen to be good at counting cards and increase our (weighted) probability of winning to a respectable $p = 0.515$ with $q = 0.485$, that could be exciting! Figure 10.4 shows both, with the card counting example in dark gray and the regular basic strategy version behind in light gray. Going broke using the Martingale system seems very likely in just two hours, even if playing a player-favorable version of Blackjack in counting cards.

Anti-Martingale System

Assuming that the previous section scared you away from trying to use the Martingale system, perhaps a system called the anti-Martingale system would appeal to you? Unfortunately, if the name sounds good to you, you would still be wrong. The short version of a disclaimer about *any* betting system is that they simply do not work! Do not spend $29.95 online to purchase a betting system; if the system actually worked, why would they need to sell a betting system to make money?

The anti-Martingale system takes the huge negative component of the ordinary Martingale system and flips it upside down. A bettor who is playing an anti-Martingale strategy will double his or her bets upon *winning*, and on a loss will revert back to some standard base bet. The exponential growth for losses no longer applies here, but, on the other hand, it certainly does apply to the winnings. Sounds great, doesn't it?

Question

How likely is the anti-Martingale system to make money?

By now, you should expect that what I want to do is to look at this mathematically; if the anti-Martingale system sounds good on the surface (exponential growth for winnings and minimal losses), what makes it, again, a system that is likely to work differently in practice? Let's suppose we walk into a casino, sit down to play some basic strategy Blackjack or some craps, starting with a bankroll of $1,000, intending to make $100 bets as our base bet (we want to impress our date, perhaps). Since the probability of winning at basic strategy Blackjack or craps is fairly close to 50%, I grabbed a quarter from my desk and flipped it twenty times to generate a somewhat probable sequence of wins and losses; the result was

WWLWL WWWLL
WLWLW WWWLW

which appears somewhat lucky for us with so many wins!

TABLE 10.2: Anti-Martingale Flow

Bet	Start	Bet	Result	Gain
1	$1,000	$100	Win	+$100
2	$1,100	$200	Win	+$200
3	$1,300	$400	Loss	−$400
4	$900	$100	Win	+$100
5	$1,000	$200	Loss	−$200
6	$800	$100	Win	+$100
7	$900	$200	Win	+$200
8	$1,100	$400	Win	+$400
9	$1,500	$800	Loss	−$800
10	$700	$100	Loss	−$100
11	$600	$100	Win	+$100
12	$700	$100	Loss	−$100
13	$600	$100	Win	+$100
14	$700	$100	Loss	−$100
15	$600	$100	Win	+$100
16	$700	$200	Win	+$200
17	$900	$400	Win	+$400
18	$1,300	$800	Win	+$800
19	$2,100	$1,600	Loss	−$1,600
20	$500	$100	Win	+$100

In Table 10.2, you can see what happens for the anti-Martingale bettor for this sequence of wins and losses. Large streaks are fairly unlikely as we have seen before, but the anti-Martingale player only ends up with a lot of money if they know when to stop. In our scenario, using Table 10.2, we could easily have stopped when we were up to $1,500 and walked away from the table with a $500 profit, but we would have missed out on the opportunity to walk away with $2,100. Note, though, that each loss in this system costs us our base bet. Starting with $1,000 as we did, we can only afford 10 losses. You will never know when to stop.

It's worth taking a moment for you to try recreating a table similar to Table 10.2 using a different sequence of wins and losses. In particular, give it a try with the sequence

LLWLW LLLWW LWLWL LLLWL

which is the opposite of the sequence we used earlier (when I created the original sequence, I arbitrarily chose heads to represent a win, but this version would have virtually the same chance). You might notice that, unfortunately, you run out of money.

Our discussion of the anti-Martingale strategy so far has left off any discussion of the probabilities involved; that time has now come. Note that when we bet on the same thing (or at least bet with the same exact probabilities), we stay in the realm of independent, identical trials from a binomial distribution. Let's consider an example.

Example 10.3 Desiring some quick money, we walk into a casino with $300 trying to make $1,500 with the anti-Martingale strategy by making $100 bets. This means that we're looking for a streak of 4 wins, turning the initial bet of $100 into $200, then $400, then $800, and finally $1,600, for a profit of $1,500, *before* obtaining 3 losses, depleting our money coffers. One way that this could happen is for us to simply win with our first four bets; this sequence of wins and losses we can denote by WWWW. Another way is that we could have two wins, a loss, three wins, a loss, and then four wins via the sequence WWLWWWLWWWW. We could also start with a loss, pickup two wins, another loss, and then four wins as in LWWLWWWW, or have a single win, a single loss, and then four wins with WLWWWW. All 21 possibilities are

WWWW	LWWWW	LWLWWWW
LWWLWWWW	LWWWLWWWW	LLWWWW
WLWWWW	WLLWWWW	WLWLWWWW
WLWWLWWWW	WLWWWLWWWW	WWLWWWW
WWLLWWWW	WWLWLWWWW	WWLWWLWWWW
WWLWWWLWWWW	WWWLWWWW	WWWLWWLWWWW
WWWLWLWWWW	WWWLLWWWW	WWWLWWWLWWWW

where you should take a moment to verify that these are the *only* strings of wins and losses that give us a streak of 4 wins before 3 losses.

To determine the overall probability, we can take each string of wins and losses, count the number of wins and losses, and use the binomial distribution of Theorem 4.1; note that we leave off the binomial coefficients because we have already done the *counting* by hand. For this scenario, let's assume we're playing only red, black, even, or odd while at an American roulette table, with the probability of a win fixed at $p = 18/38$. For example, for the string WLWLWWWW, we obtain a probability of

$$\text{Probability of WLWLWWWW} = \underbrace{\left(\frac{18}{38}\right)^6}_{6 \text{ wins}} \cdot \underbrace{\left(\frac{20}{38}\right)^2}_{2 \text{ losses}} \approx 0.003129$$

and for the string WWWLWWWW we get a probability of

$$\text{Probability of WWWLWWWW} = \underbrace{\left(\frac{18}{38}\right)^7}_{7 \text{ wins}} \cdot \underbrace{\left(\frac{20}{38}\right)}_{1 \text{ loss}} \approx 0.002816.$$

Adding the probabilities together for all 21 strings gives a probability of making our \$1,500 before getting three losses of 0.143559.

Hopefully the format of these strings is not lost on you; for our example where we wanted to have a streak of 4 wins before 3 losses occurred, each string ends with exactly 4 wins. Once we have reached the fourth win in a row, we have accomplished our goal, so the betting (and string) ends precisely then. What happens *before* that streak of 4 wins is also very predictable! Except in the case where we win with the first 4 bets we make, each of those strings has the 4 wins preceded by one or more losses (up to two), and those losses have exactly 0 to 3 consecutive wins in a row before them; there cannot be 4 consecutive wins before a loss, as 4 wins in a row always ends the string.

In general, if we want to find the probability of a streak of k wins before l

losses happen, we would be looking at strings of the form

$$\underbrace{\rule{5cm}{0pt}}_{\text{pre-string}}\underbrace{\text{WW}\cdots\text{W}}_{k \text{ wins}}$$

where the pre-string contains anywhere from 0 to $l - 1$ "loss blocks" that consist of a number of wins between 0 and $k - 1$, inclusive, followed by a loss.

Compositions of positive integers are covered in Section 11.4; we do not need their full power here, but we do note that the format of the pre-string can be exactly described by a set of compositions. In short, a composition of a positive integer n is a list of integers $(\lambda_1, \lambda_2, \ldots, \lambda_m)$ such that $\lambda_1 + \lambda_2 + \cdots + \lambda_m = n$, where m is the number of *parts* of the composition and each part is a positive integer ($\lambda_i > 0$ for each i). For instance, $(4, 3)$, $(3, 4)$, $(5, 1, 1)$, and $(1, 5, 1)$ are each possible compositions of 7 (there are definitely more).

For our purposes here, we can assign to each pre-string a composition (and each composition describes a pre-string). For instance, the string WWLWL-WWWW from Example 10.3, with pre-string WWLWL, would be represented by the composition $(3, 2)$ of 5; the parts of the composition represent the *length* of a "loss block" (which includes the loss itself). When we're looking for a streak of k wins, the largest number of wins in a "loss block" is $k - 1$, so by counting the loss as well, the largest length for a block is k; we will only be considering compositions for which each part is no bigger than k (that is, $\lambda_i \leq k$ for all i). In addition, since we can only have at most $l - 1$ "loss blocks," we will be restricting ourselves further to compositions of at most $l - 1$ parts. Finally, the largest possible integer that we can actually create using compositions for pre-strings involves using the maximum number of blocks ($l - 1$), with each having the maximum number of wins allowed ($k - 1$), for a largest integer to use of $k \cdot (l - 1)$ since we *include* the loss as well as the wins in the length of the block. If we denote by $\bar{c}(m, n, k)$ the number of compositions of n with m parts, each part no larger than k, then we can write the probability of a streak of k wins before l losses as

$$\text{Streak of } k \text{ Wins Before } l \text{ Losses} = \underbrace{p^k}_{①} + \underbrace{\sum_{m=1}^{l-1}}_{②} \underbrace{\sum_{n=1}^{k\cdot(l-1)}}_{③} \underbrace{\bar{c}(m, n, k)}_{④} \cdot \underbrace{q^m \cdot p^{n-m+k}}_{⑤}$$

where the pieces are explained as follows. First, since compositions require that there be at least 1 part, and that part be of size at least 1, the case where we have no losses (and therefore have our streak of k wins right from the beginning) is not covered in our count of compositions, so we list this case separately as ①. The summations of ② and ③ signify having a number of parts between 1 and $l-1$, inclusive, and total integer we are using compositions of as $k \cdot (l - 1)$, respectively. Since, at the point we reach ④, n and m are fixed, and the probability term found in ⑤ only depends on the number of losses, m, the cumulative total of wins and losses for the block measured by integer n, and the desired streak length k at the end, we can simply multiply the probability by the number of compositions ④ that fit the current n and m criteria; you can think of this as replacing the binomial coefficients with a new count of terms fitting the same probability. For the probability term ⑤ itself, we note that with m blocks, there are m losses, resulting in the q^m term, while with total block length n and m losses, there must be $n - m$ wins in the block, and the streak of k at the end gives p^{n-m+k}.

While that was indeed a lot, it gives us a nice formula with pieces that are easily explained. Even better, with compositions being a topic of mathematical study in the middle and late 1900s, Morton Abramson (and others) has a nice formula for $\bar{c}(m, n, k)$ that gives us a second part of the Streaks Theorem; see [1] in the Bibliography for more on (restricted) composition theory.

> **Theorem 10.3 (Streaks Theorem, Part 2)** The probability of obtaining a streak of k successes before l failures are observed, assuming each trial is independent as in the standard binomial distribution, is given by
>
> $$\overline{P}(k,l) = p^k + \sum_{m=1}^{l-1} \sum_{n=1}^{k \cdot (l-1)} \sum_{j=0}^{m} (-1)^j \binom{m}{j} \cdot \binom{n - jk - 1}{m - 1} \cdot q^m \cdot p^{n-m+k}$$
>
> where p is the probability of success, q is the probability of failure (equal to $1 - p$), and we take $\binom{n}{k} = 0$ when $n < 0$ or $k < 0$.

Using Theorem 10.3 with $p = 18/38$, $q = 20/38$, $k = 4$ and $l = 3$ to replicate Example 10.3 gives the same probability of $\overline{P}(4,3) \approx 0.143559$ as before. It's always nice to see that the developed theory matches an example computed by hand!

Figure 10.5 shows the effect of changing l, emulating what would happen if we still desired to win \$1,500 by making \$100 bets using the anti-Martingale strategy with different starting amounts. For example, if we started with \$800, the probability of us getting the \$1,500 is $\overline{P}(4,8) \approx 0.338504$ and starting with \$1,500 in the

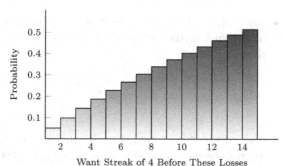

FIGURE 10.5: Loss Threshold versus Probability, $k = 4$

first place yields $\overline{P}(4, 15) \approx 0.539225$, just a bit better than laying all of our starting money down on an even-money wager in the first place.

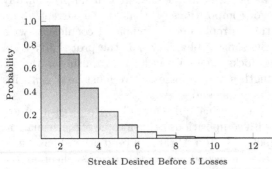

FIGURE 10.6: Streak Amount Probability, with
$l = 5$

If we instead change the value of k, trying to get a larger profit, while keeping the number of losses fixed (say at $l = 5$, indicating a starting bankroll of $500 in our scenario with roulette), we get the graph in Figure 10.6. Note that, as you might expect, the probability of getting more and more money (through doubling your bets with the anti-Martingale strategy) decreases very rapidly when the maximum amount of losses is fixed; given a finite amount of money, you are always putting it at risk when gambling.

As Edward Thorp argues nicely in his then-controversial book of 1985, *The Mathematics of Gambling* (see [54] in the Bibliography), under the assumptions that

- each individual bet in a game has a negative expected value,

- there is a maximum limit to the size of any round,

- the results of one round of play do not influence any other rounds, and

- there is a minimum bet forced upon the player,

no betting system can work! In fact, mathematics shows that under these conditions, any gambling "system" that exists is "worthless" (in Thorp's words) as

- any *series* of bets has a negative expected value, with expected value equal to the sum of the expected values for the individual bets (and this sum is therefore negative as well),

- with continued betting, a player's actual loss divided by total amount bet will get closer and closer to the expected loss divided by total amount bet, and

- with continued betting, it is virtually certain that the player will become a loser, eventually always stay a loser and never break even, and eventually lose all of his or her starting money, no matter how much it was to begin with.

With this, it is certainly worth repeating a lesson from earlier; do not wager any more money at a game of chance unless you can comfortably afford to do without that money.

The Kelly Criterion

No discussion of betting strategies would be anywhere close to complete without talking about the Kelly Criterion. This "system" for betting was published in 1956 by John Kelly, Jr., a scientist at Bell Labs, as an article in the area of information theory. Mathematician Ed Thorp, who we remember from earlier as the father of card counting in Blackjack, used this system to make large sums of money betting in Las Vegas, and this system and formula has become part of current investment theory; indeed, it has been reported that billionaire and investor Warren Buffet has been one of the more prominent users of the Kelly Criterion.

The bet amounts given by the Kelly system only make sense when playing a game with positive expectation. For us, that means we will assume that the probability of a win is $p > 0.50$ (and thus $q < 0.50$) making the quantity $p - q > 0$. We will see what this restriction is necessary later. In order to develop the system here, we will also assume that our wager is on an even-money game (with payout of 1:1), and we use the notation

- $X(k)$ is the expected amount of money we have after bet number k; $X(0)$ is our starting bankroll, and

- $B(k)$ is the amount we will wager for bet number k.

With this setup, we can see that

$$X(1) = \underbrace{X(0)}_{\text{current}} + \underbrace{p \cdot B(1)}_{\text{win}} - \underbrace{q \cdot B(1)}_{\text{lose}} = X(0) + (p - q) \cdot B(1)$$

and without much extra thought, it should be clear that

$$X(2) = \underbrace{X(1)}_{\text{current}} + \underbrace{p \cdot B(2)}_{\text{win}} - \underbrace{q \cdot B(2)}_{\text{lose}} = X(0) + (p - q) \cdot B(1) + (p - q) \cdot B(2)$$

$$= X(0) + (p - q) \cdot (B(1) + B(2))$$

or, in general for an arbitrary amount of bets n, we have

$$X(n) = X(0) + (p - q) \cdot \sum_{k=1}^{n} B(k).$$

If our goal is to *maximize* $X(n)$, our expected holdings of money after n bets, then it makes sense that since $(p - q)$ is a positive value by assumption, the calculation shows that we need to maximize $B(k)$ for each of the k bets. It follows from this that $B(1)$ should be as large as possible, and this makes $B(1) = X(0)$, our starting amount. Then if we win on that first bet, we have $X(1) = 2 \cdot X(0)$, and so the largest bet we can now make is $B(2) = 2 \cdot X(0)$ and the process continues. Of course, this means we are doubling our bet each time, hoping to get a streak of wins as in the anti-Martingale strategy. The difference here is that since we're maximizing our bet each time, *any* loss results in the loss of all of our money. The probability that we start with a streak of n wins is given by p^n, so the probability of "ruin" is $1 - p^n$ which, even though we are playing a favorable game with $0.50 < p < 1.00$, as n gets large, since p^n tends to 0, the probability of "ruin" of $1 - p^n$ tends to 1; that is, the chance of financial loss for us approaches 100% as n grows. This is definitely not desirable.

On the other hand, if we want to *minimize* the probability of ruin while still making bets and taking advantage of our positive expectation situation ($p > 0.50$), your intuition should tell you that we should *minimize* $B(k)$ at each bet; that is, we should minimize our bets and therefore bet the minimum allowed for whatever game you might be playing. Think about this concept as we explore the Gambler's Ruin concept soon and as Theorem 10.5 is introduced; the probability of ruin is minimized when we set the value of a betting unit as small as possible and consider a large total T involved there. For now, though, it's enough for us to agree that minimizing the chance of ruin means minimizing each bet, and this has the side effect from our calculation of $X(n)$ of also minimizing $X(n)$ as well. This is also not desirable.

What Kelly proposed was a *fixed fractional* system whereby each bet is set to be a fixed percentage of the *current* amount of money held by a player. That is, in our notation, we set $B(k) = f \cdot X(k-1)$; to determine how much I should bet right now, I look at the money I had after the last round, and take a percentage of that amount. Note that the bet sizes *do* change per round, as while I always use the same percentage, the amount of money I currently have to work with *does* change. Note that when $0 < f < 1$, total ruin can technically *never* happen, as the amount of money wagered is *always* less than the amount I currently have.

When a wager is won, note that our current money increased by a predictable amount, as

$$X(k + 1) = X(k) + f \cdot X(k) = (1 + f) \cdot X(k)$$

and, similarly, our current holdings change on a loss via

$$X(k + 1) = X(k) - f \cdot X(k) = (1 - f) \cdot X(k),$$

so it makes sense that after n wagers, if there happen to be w wins and l losses, the current amount of money we have is

$$X(n) = \underbrace{X(0)}_{\text{start}} \cdot \underbrace{(1 + f)^w}_{\text{wins}} \cdot \underbrace{(1 - f)^l}_{\text{losses}}$$

since each win multiplies our holdings by $(1 + f)$ and each loss does so by $(1 - f)$; the order of wins and losses is irrelevant.

Instead of maximizing (or minimizing) the expected value of $X(n)$ itself, the Kelly system maximizes the expected value of $\log X(n)$ as a function of f. While this may seem out of the blue, many economists and mathematicians, rather than working with raw dollar amounts, work with a notion of "utility." It is very common for utility functions based on measuring the "desirability" of money to be logarithmic. What follows is a simplified, less technical exploration of how the ideal value of f is found; it is still mathematically correct. The results that we shall see almost speak for themselves, and the technical details can be found in Kelly's original paper and Rotando and Thorp's more recent exposition of the system (found as [24] and [43] in the Bibliography).

Since any root function is increasing (for example, if m is larger than n, the square root of m is also larger than the square root of n) and the value of n is fixed while we maximize expected value with respect to the value of f, we can instead maximize the expected value of

$$g(f) = \log\left(X(n)\right)^{1/n} = \log\left(X(0) \cdot (1+f)^w \cdot (1-f)^l\right)^{1/n}$$

which, using convenient rules of logarithms, is the same as

$$g(f) = \frac{1}{n} \cdot X(0) + \frac{w}{n} \cdot \log(1+f) + \frac{l}{n} \cdot \log(1-f)$$

and since the expected value of w/n is p (we expect our ratio of wins to total bets to be the probability of a win) with l/n expected to be q, this simplifies to

$$g(f) = \frac{1}{n} \cdot X(0) + p \cdot \log(1+f) + q \cdot \log(1-f).$$

A bit of calculus (solving the derivative function $g'(f)$ for zero and using a Second Derivative Test argument) shows that the maximum of this function occurs when $f = p - q$; this is the percentage we should apply to our *current* funds when deciding how much to bet on the *next* round. When the game involved is *not* an even money game but rather a game with a b-to-1 payout, replacing $\log(1 + f)$ with $\log(1 + bf)$ and using calculus gives a modified bet amount, given in the following result.

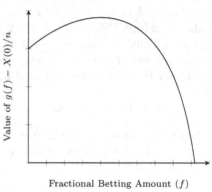

FIGURE 10.7: Behavior of $g(f)$

Note that a graph of $g(f)$, with the constant term removed to more clearly show its behavior, is given in Figure 10.7.

Theorem 10.4 (Kelly Criterion) For a wager with probability p of winning, probability q of losing where $q = 1 - p$, and payout of $b : 1$, the Kelly betting system dictates that you should wager the fraction of your current funds equal to

$$f = \frac{bp - q}{b}$$

which, in an even money situation where $b = 1$, simplifies to

$$f = p - q.$$

This bet maximizes the expected logarithm of funds.

A few notes about Theorem 10.4 are in order. First, note that when $b = 1$, if $p < q$ then the value of f is negative; for instance, betting on red for American roulette where $p = 18/38$ yields a fraction of your bankroll to bet of

$$f = p - q = \frac{18}{38} - \frac{20}{38} = -\frac{1}{19} \approx -0.0526,$$

meaning that you should actually bet about 5.26% of your bankroll that red will *not* come up. Unfortunately, casinos do not allow you to bet *against* red (they *do* allow you to bet on black, but betting against red means getting a black *or* green number). If $p = q = 0.50$ as in flipping a fair coin, then $f = p - q = p - p = 0$ so no wager should be made. The Kelly system only advises betting when the player has *some* advantage. Note that when $b \neq 1$, in order for $f > 0$, Theorem 10.4 requires that $bp - q > 0$. The payout needs to be high enough to overcome a perhaps increased chance of losing. If you wager on rolling a 5 or 6 when rolling a fair six-sided die, the true mathematical odds, recall from Definition 2.1, are $2 : 1$, with a probability of winning of $2/6$ and probability of losing of $4/6$. If $b = 2$ for payout, note that $f = 0$ and we should not bet. To make this situation favorable, the payout would need to be greater than 2, or the die would have to be slightly unfair and favor fives and sixes.

Let's consider a short example.

Example 10.4 For the case of moderate card counting when $p = 0.515$ and $q = 0.485$, the Kelly Criterion of Theorem 10.4 says that we should bet the fraction

$$f = p - q = 0.515 - 0.485 = 0.03$$

of our current funds each time. Starting with $1000, this means that our first bet would be $30. If we win this bet, our new funds would be $1030, for which a 3% fractional bet would have the next wager equal $30.90.

If instead we assume almost perfect card counting with $p = 0.53$ and $p = 0.47$, the Kelly Criterion gives us a betting fraction of

$$f = p - q = 0.53 - 0.47 = 0.06$$

so starting with $1000 again, we would bet $60 (and $63.60 on the next bet should we win that first bet).

A simulation of a few thousand bets for $p = 0.53$ are shown in Figures 10.8 and 10.9. The results for the Martingale strategy we saw earlier show a fairly steady linear growth except that in *both* simulations the current amount of money available drops below zero! Unless you had a wealthy friend or bank loan to help out, the Martingale results should effectively stop immediately once your cash available reaches zero (do *not* use a loan to cover your losses, as more losses are definitely around the corner; the graphs show multiple dips below zero). The anti-Martingale strategy, relying on streaks of wins, does not do well at all for long-term play; this makes sense since we only really talked about this strategy as attempting to reach a predetermined target.

Included alongside the Kelly system itself are slight tweaks to the fractional betting system. The line shown for Kelly (Over) indicates a bettor who wishes to make a slightly larger bet than the Kelly Criteria dictates; this can work well with a large streak of wins, but is punishing when many losses happen. The "over" Kelly system shown in the figures adds an additional 2% of the bankroll for each bet (betting 8% of current reserves rather than 6%); it performs almost the same as the regular Kelly bet, but falls short in the long-run. Similarly, the Kelly (Under) line is going for a bit less risk while betting; the variation in current dollar amounts is less, but so are the long-run cash amounts (the simulations used a bet of 5% instead of the prescribed 6%). These changes to the Kelly system offer many of the same positive effects, but do not perform as well; indeed, for a system to maximize the expected logarithmic bankroll, the Kelly bets given in Theorem 10.4 will beat all others in the long-run.

There are several neat properties that the system developed by Kelly has. If we write $G(f)$ for the function $g(f)$ with the term $X(0)/n$ removed, then the following are true.

- Whenever $G(f) > 0$, as n gets large, so does $X(n)$; in fact, for each M, the probability that $X(n) > M$ eventually is 100%. This means that even a slight "under bet" or "over bet" for the Kelly system will succeed.

- Whenever $G(f) < 0$, as n gets large, $X(n)$ tends to zero; for each $m > 0$, the probability that $0 < X(n) < m$ eventually is also 100%.

- If $G(f) = 0$, then the value of $X(n)$ oscillates randomly between (close to) zero and infinity.

- Given any different (not necessarily a fixed fractional system) strategy for betting with cash flow values of $X^*(n)$, as n gets large, $X(n)/X^*(n)$ gets arbitrarily large; this means that the Kelly Criterion (or any variation that also maximizes the logarithm of $X(n)$) dominates (does better than) any others.

- The expected amount of time for $X(n)$ to reach a specific value is at least as *small* for a strategy, such as the Kelly Criterion, that maximizes $\log X(n)$ as it is for any other.

While the last two bullets are especially positive notes for the Kelly Criterion, there are of course limitations and pitfalls of the system. Perhaps the most notable of those is the fact that sometimes a bet amount may not be possible. In the world of gambling, minimum bets exist, and even for slot machines (or video poker) where occasionally payouts of over 100% are available (especially when combined with promotions) and a penny is accepted as a wager, having only $0.25 available and trying to make a 2% fractional bet is impossible. Since we do not win any particular rounds or even win once in a particular length of rounds, reaching infeasible bet sizes is possible; after all, chance is involved and we should not wager what we cannot afford to lose.

As this section comes to an end, I should mention that this limitation can be overcome in the world of stocks where partial ownership is possible. If one can find a way to be in a positive weighted expectation scenario ($bp - q > 0$), then money can very well be made. William Poundstone has written a very good non-mathematical non-fiction account of how the Kelly Criterion was actually used to win large sums of money in his book *Fortune's Formula* (see [37] in the Bibliography); I highly recommend this book as a great read. As a teaser, Edward Thorp (whose name appears frequently throughout Poundstone's book) offers the following example from his own book referenced earlier in this chapter (see [54]).

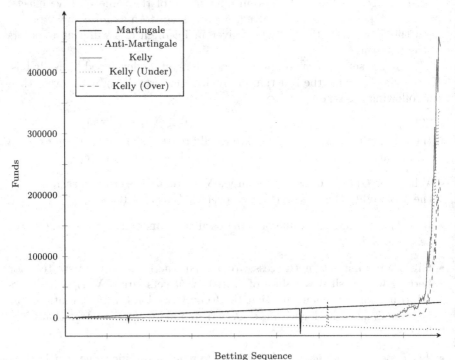

FIGURE 10.8: Betting Systems Simulation, 1

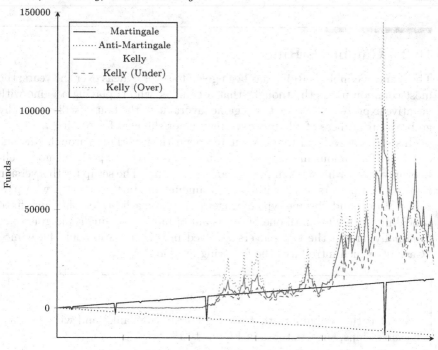

Betting Sequence

FIGURE 10.9: Betting Systems Simulation, 2

Example 10.5 Using an electronic device to gain an advantage at roulette, Throp was able to achieve an advantage of about 44% on a single number, amounting to a probability of $p = 0.04$ for winning with the standard 35:1 payout offered. Using Theorem 10.4 then with $b = 35$ and $q = 0.96$, the fixed fraction of our bankroll that we should bet each round is

$$f = \frac{bp - q}{b} = \frac{35 \cdot 0.04 - 0.96}{35} \approx 0.0126,$$

or about 1.26% of what we currently have. Note the relatively small percentage we bet, since the probability of winning is not too high. We don't want to lose a lot of money early. Using the Kelly system with this sort of advantage, after 1,000 bets, gives an expected current funds amount of about 11.47 times the starting amount (using growth rate projections provided by Kelly). Of course, using an electronic device in this way, especially today, is highly illegal and considered cheating, so don't do it!

10.2   Gambler's Ruin

The name "Gambler's Ruin" has been given to several ideas over the years; the most common use is the thought that a gambler who is playing a game with negative expected value (that is, a game favorable to the house) will eventually go broke, regardless of whatever strategy he or she uses for betting.

Instead, the version that I want to spend time working through was one developed by famous mathematician Christiaan Huygens, perhaps more well-known to you for his work in physics and astronomy. The setup for this version involves two players, each with a fixed amount of starting money, who play round after round of some type of wager (with, for each player, the same fixed probability of a win) until one player is out of funds. You might imagine that, eventually, one of the two players involved in this scenario could be named "Casino," which introduces the following question.

Question

What is the probability that we can "break the bank" and win all of a casino's chips, such as those in Figure 10.1?

In order to get a grasp on the question at hand, let's consider a small thought exercise; note that the answer for the exercise is given as a footnote so that you do have time to think about your response before seeing the solution.

Example 10.6 Suppose you and a friend each have $50 and are taking turns betting on the outcome of a fair coin. Each round, you and your friend put $1 into a pot and then flip the coin. If the coin happens to result in a head, you win, netting a profit of $1 and a loss of $1 for the friend. Otherwise, on a tail, your friend takes the pot, netting a $1 loss for you. Play continues until one of you has the $100 total.

What is the probability that you walk away with all of the money? Think about this for a few minutes, then take a look at the footnote[2] for the answer.

At this point, you should realize that this question is a bit more meaningful when that probability that you win a given round is something other than $p = 0.50$, but the method we can use to solve this problem doesn't need that assumption (though, as we will see, something strange does happen with our

---

[2]With a fair coin, you and your friend each have a 50% chance of winning any particular round; it should make sense that if you both start with the same amount of money as well, neither of you has any advantage over the other at this game, so the probability that you walk away with $100 must be 50% (with your friend also having a 50% chance of winning).

probability formulas when p *does* equal 0.50). We shall also follow our tradition of using q to represent the probability of a loss, which for us means $q = 1 - p$. Also, let T represent the total amount of money between you and your friend, and use P_i for the probability that we end up with all of the money given that we currently have \$$i$. Summarizing these ingredients, we have

p = probability of winning a round,

q = probability of not winning a round,

T = total amount of money between the two players,

P_i = probability we take it all if we currently have i dollars.

Using Example 10.6 as a reference, there we had $p = 0.50$, $q = 0.50$, $T = 100$ and $P_{50} = 0.50$, where the last value was reasoned through logic.

There are some interesting values of P_i that we can already compute. The values

$$P_0 = 0 \qquad\qquad P_T = 1$$

should make sense, since if we currently have no money, we can't possibly take everything, so P_0 must equal zero. On the other hand, if we already have all T dollars, then we've already won, so P_T must be 100%. Otherwise, note that we can easily define P_1 in terms of another probability. If we currently have exactly 1 dollar (or betting unit, or, simply, unit), then taking all of the money starting with 1 unit means that we need to win the next round, with probability p, and then win with 2 units! That is,

$$P_1 = p \cdot P_2$$

which seems very nice and perhaps even simple. How can we take all of the money if we currently have 2 units? Well, either we can win the next round, and then win with 3 units, *or* we can lose the next round and then win with 1 unit. The former of these happens with probability $p \cdot P_3$ and the latter happens with probability $q \cdot P_1$, giving us the relationship

$$P_2 = q \cdot P_1 + p \cdot P_3$$

and, continuing on, it should not be too hard to reason that the relationship

$$P_k = q \cdot P_{k-1} + p \cdot P_{k+1}$$

for any value of k between 1 and $T - 1$, inclusive.

These relationships form what mathematicians like to call a recurrence relation (or difference equation) and, depending on the exact nature of relationships such as these that define probabilities in concert with other probabilities, methods have been developed to answer the questions. While we won't go into the general methods of solving these (check out a discrete mathematics course or differential equations course for more details), there is an algebraic method available we can use. Since you may wish to make small tweaks to this system that would likely result in an algebraic way to solve things as well, let's do some algebra.

We can rewrite P_k to make our method clearer. Note that $P_k = p \cdot P_k + (1-p) \cdot P_k = p \cdot P_k + q \cdot P_k$, so we can use our previous formula for P_k to write

$$p \cdot P_k + q \cdot P_k = p \cdot P_{k+1} + q \cdot P_{k-1}$$

which, by moving all terms with a p to the left and those with a q on the right, and factoring we have

$$p \cdot (P_{k+1} - P_k) = q \cdot (P_k - P_{k-1})$$

or, dividing both sides by the non-zero probability p,

$$P_{k+1} - P_k = \left(\frac{q}{p}\right) \cdot (P_k - P_{k-1}),$$

which gives us a nice way to express differences of probabilities in terms of already previously computed probabilities (which should feel similar to the recursion of Example 10.2).

Being explicit and writing this out for values of k going from 1 to $T - 1$ (recalling that $P_0 = 0$), we get

$$P_2 - P_1 = \left(\frac{q}{p}\right) \cdot P_1$$

$$P_3 - P_2 = \left(\frac{q}{p}\right) \cdot (P_2 - P_1) = \left(\frac{q}{p}\right)^2 P_1$$

$$P_4 - P_3 = \left(\frac{q}{p}\right) \cdot (P_3 - P_2) = \left(\frac{q}{p}\right)^3 P_1$$

$$\vdots$$

$$P_k - P_{k-1} = \left(\frac{q}{p}\right) \cdot (P_{k-1} - P_{k-2}) = \left(\frac{q}{p}\right)^{k-1} P_1$$

$$\vdots$$

$$P_{T-1} - P_{T-2} = \left(\frac{q}{p}\right) \cdot (P_{T-2} - P_{T-3}) = \left(\frac{q}{p}\right)^{T-2} P_1$$

$$1 - P_{T-1} = \left(\frac{q}{p}\right) \cdot (P_{T-1} - P_{T-2}) = \left(\frac{q}{p}\right)^{T-1} P_1$$

where, again, the last line is slightly simpler because the term P_T (which you may have expected to see) was replaced by 1.

For amazing fun, we're about to add *all* of these equations together! Adding together all of the left-hand sides of the equations is wonderful because of the cancelation that happens; the P_3, for example, that appears in the third line is subtracted on the fourth line. The only terms that survive are the P_1 from the first and the 1 from the last, resulting in $1 - P_1$ for the left. The terms on the right look very similar, and factoring the P_1 out results in

$$1 - P_1 = P_1 \cdot \left(\frac{q}{p} + \left(\frac{q}{p}\right)^2 + \cdots + \left(\frac{q}{p}\right)^{T-1} \right) = P_1 \cdot \sum_{i=1}^{T-1} \left(\frac{q}{p}\right)^i,$$

and after solving for P_1, gives

$$P_1 = \frac{1}{1 + \dfrac{q}{p} + \left(\dfrac{q}{p}\right)^2 + \cdots + \left(\dfrac{q}{p}\right)^{T-1}}.$$

While this is useful, we could also choose only to add together the equations, starting with $P_2 - P_1$ and ending with $P_k - P_{k-1}$ in order to get an expression for P_k; adding these together yields

$$P_k - P_1 = P_1 \left(\frac{q}{p} + \left(\frac{q}{p}\right)^2 + \cdots + \left(\frac{q}{p}\right)^{k-1} \right) = P_1 \cdot \sum_{i=1}^{k-1} \left(\frac{q}{p}\right)^i$$

which, after solving for P_k and using the value for P_1 we found, results in

$$P_k = \frac{1 + \dfrac{q}{p} + \left(\dfrac{q}{p}\right)^2 + \cdots + \left(\dfrac{q}{p}\right)^{k-1}}{1 + \dfrac{q}{p} + \left(\dfrac{q}{p}\right)^2 + \cdots + \left(\dfrac{q}{p}\right)^{T-1}},$$

bringing us to the Gambler's Ruin Theorem.

Theorem 10.5 (Gambler's Ruin Theorem) For a total amount of betting units T, probability of success p, and probability of failure q, the probability that you will end up with all of the money, assuming you start with k units, is given by

$$P_k = \begin{cases} \dfrac{k}{T} & p = 0.50 \\[2ex] \dfrac{1 - \left(\dfrac{q}{p}\right)^k}{1 - \left(\dfrac{q}{p}\right)^T} & p \neq 0.50 \end{cases}$$

where the case when $p = 0.50$ comes easy, since q also equals 0.50 there, so q/p reduces to 1 in our equation for P_k. When $p \neq 0.50$, we can use the shortcut of Theorem 10.1 which gives the relationship

$$1 + \frac{q}{p} + \left(\frac{q}{p}\right)^2 + \cdots + \left(\frac{q}{p}\right)^{n-1} = \frac{1 - \left(\frac{q}{p}\right)^n}{1 - \frac{q}{p}}$$

which we can use with $n = k$ and $n = T$ to simplify the expression for P_k.

Using Theorem 10.5 is fairly easy now. First, note that, referring back to Example 10.6 where you and your friend start with \$50 each and flip a coin, we see that $P_{50} = 0.50$ where, in this situation, $T = 100$. It also would say that, if you started with \$25 and your friend had \$75, the chance that you walk away with all of the dough is $P_{25} = 25/100 = 0.25$ which may not be as obvious or able to be reasoned as logically. When flipping a fair coin, your probability of walking away with the cash is exactly equal to your proportion of the total pot.

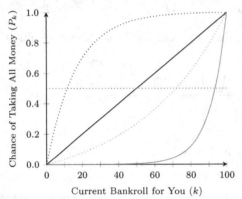

FIGURE 10.10: Gambler's Ruin for Popular Win Probabilities

By changing the value of p (and therefore q) used in Theorem 10.5, we can look at the change in P_k for various values of k. Assuming that $T = 100$ as before, meaning that a total of \$100 (or 100 betting units) are available between you and your opponent, Figure 10.10 shows the value of P_k for values of k ranging between 0 and 100 for four different "games." Remember, P_k is the probability that you will walk away with all of the betting units; in solid black the scenario where $p = 0.50$ is displayed, again showing that the probability of walking away with everything is equal to your proportion of the total. The dotted gray, solid gray, and dotted black curves show the behavior of P_k for roulette ($p \approx 0.4737$), Blackjack with basic strategy ($p = 0.495$), and Blackjack with moderately good card counting ($p = 0.515$), respectively. It should not be lost on you that the look of the curve changes drastically when the probability of a win shifts from below $1/2$ to above $1/2$, as it does for the card counting version dotted in black.

Since the card counting line in Figure 10.10 is *above* the neutral line given when $p = 0.50$, you might wonder exactly when your chance of "breaking the bank" exceeds 50%. That is, what starting bankroll k do you need to hold in order to have a greater than half chance of walking away with all of the dough? This question can be answered by an easy application of Theorem 10.5.

Example 10.7 When $p = 0.50$, using Theorem 10.5 is simple. We desire to solve the equation

$$P_k = \frac{k}{T} = 0.50$$

for k, and this results in $k = 0.50 \cdot T$. In the neutral situation where neither player has an advantage, you need to start with more than half of the money to have a more than 50% chance of walking away with your opponent's money. Using values other than 0.50 for P_k also shows that this relationship between your bankroll and chance of obtaining all of the money is strictly linear, exactly what Figure 10.10 shows.

For all other values of p, the probability of winning each bet, we are still interested in solving

$$P_k = \frac{1 - \left(\frac{q}{p}\right)^k}{1 - \left(\frac{q}{p}\right)^T} = 0.50$$

which can be done using standard algebra techniques to obtain

$$k = \frac{\ln\left(1 - \frac{1}{2}\left(1 - \left(\frac{q}{p}\right)^T\right)\right)}{\ln\left(\frac{q}{p}\right)}$$

for the value of k that gives you a 50% chance of breaking the bank. Note that the ln function is the standard natural logarithm, found on almost all calculators these days.

For the card counting scenario at Blackjack where our particular skill gives a net probability of winning as $p = 0.515$, using the formula of Example 10.7 gives $k = 11.51$ (using, still, $T = 100$). Since the 100 refers more to the "betting units" available, this means, rounding up, that if we start with 12 betting units, we have a more than half chance of taking the casino's 88 betting units home with us as well! In concrete money figures, this translates to us starting with $12, making $1 wagers, and getting $88 from the casino. It *also* translates to us starting with $1,200, making $100 wagers, and getting $8,800 from the casino.

On the flip side, when $p \approx 0.4737$ as it is for even-money wagers at American roulette, using the formula from Example 10.7 yields $k = 93.42$. We need to start with about 94 of the 100 total betting units available just to have a more than half chance of walking away with all 100 of them! Certainly this is not very favorable since the translation means that we should start with $94 just to get the casino's $6, and that certainly isn't worth the risk!

The fact that Blackjack with card counting looks so favorable here leads to the following question.

Question

How *long* does it take to break the bank?

The method to answer this question is very similar to the way we developed P_k in the first place. If we let E_k denote the expected number of bets needed to reach reward (break the bank) or ruin (no money left for us), starting with k betting units among a total number of betting units T, we have the relationship

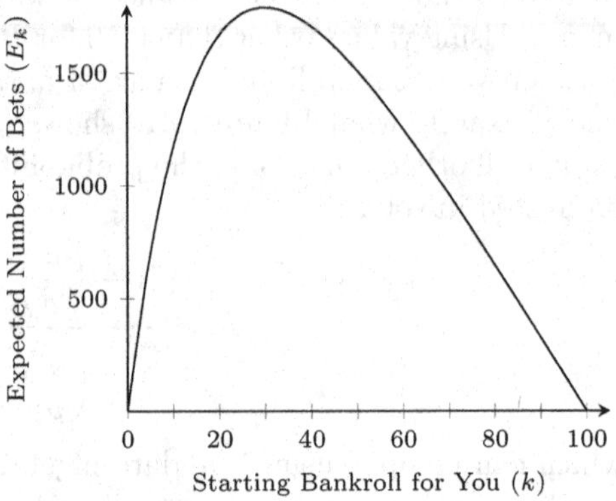

FIGURE 10.11: Card Counting Expected Number of Bets

$$E_k = 1 + p \cdot E_{k+1} + q \cdot E_{k-1}$$

since if we currently have k units, then we will settle for attaining reward or ruin by making the current bet (giving us the 1 term) and, if winning with probability p, attaining our outcome with the expected number of bets if we had $k + 1$ betting units at the start. The same reasoning explains the $q \cdot E_{k-1}$ term, for if the current bet is a loss, then we add on the expected number of bets needed starting with $k - 1$ units. As you should guess, both E_0 and E_T are defined to be 0, since we have reached ruin or reward, respectively.

Theorem 10.6 For a total amount of betting units T, probability of success p, and probability of failure q, the expected number of bets needed to reach reward or ruin, assuming you start with k units, is given by

$$E_k = \begin{cases} k \cdot (T - k) & p = 0.50 \\ \dfrac{k}{q - p} - \dfrac{T}{q - p} \cdot \left(\dfrac{\left(\frac{q}{p}\right)^k - 1}{\left(\frac{q}{p}\right)^T - 1} \right) & p \neq 0.50. \end{cases}$$

The full derivation of the previous result is a good bit messier than we saw for P_k so it has been omitted from this text, but you are more than encouraged to give it a try on your own. Difference equations and recurrence relations techniques quickly give us the following expected number of bets.

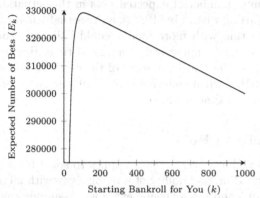

FIGURE 10.12: Card Counting Expected Number of Bets

With $p = 0.515$ and $T = 100$ as in our card counting scenario from earlier, the expected number of bets is shown in Figure 10.11. In particular, recall that a starting bankroll of $k = 12$ units was required to have a higher than 50% chance of reaching reward; the expected number of bets required to reach that reward (or ruin) is $E_{12} \approx 1315$ in this case.

The maximum number of expected bets is about 1790, occurring when $k = 30$ (for those of you who have had calculus, this is an easy exercise in optimization). For a more realistic scenario in really trying to break the bank, however, we need to increase the total amount of money available.

Example 10.8 Here, we assume that we walk into a casino with \$10,000, intending to make \$100 wagers while counting cards at Blackjack (using $p = 0.515$ as before). This gives us a starting bankroll of $k = 100$ since a betting unit here is \$100. Assuming we want to take the casino for a million dollars, we will work with 10,100 total betting units (a total dollar figure of \$1,010,000). Note that using Theorem 10.5 gives a value of $P_{100} \approx 0.9975$ so it's pretty likely that we'll end up taking the money.

The expected number of betting rounds, given by Theorem 10.6, is $E_{100} = 329,175$. How long would it take to actually play that number of rounds? If we assume 3 bets per minute, a fairly reasonable number that factors in other players and shuffling time, this would take

$$329,175 \text{ bets} \cdot \frac{1 \text{ minutes}}{3 \text{ bets}} \cdot \frac{1 \text{ hour}}{60 \text{ minutes}} \approx 1829 \text{ hours}$$

which is over 76 straight days of playing Blackjack and counting cards. Since eating and sleeping would be desirable to keep us well-rested for not making any mistakes, taking 8 hours per day to care for ourselves, this would be just over 114 days of playing, or about 4 months.

Imagine trying to play Blackjack, count cards, and not make mistakes for that length of time. You might also imagine how many casinos might ask you to leave on suspicion of counting cards.

Figure 10.12 shows the behavior of E_k for the situation presented in Example 10.8. Note that I have only graphed this for k ranging from 0 to 1,000 where the more interesting results are present. For values of k larger than 1,000, the graph resembles a line.

For the curious, the maximum number of expected bets in this situation happens when $k = 107$, so by starting with only \$10,000, we're actually making things tough for ourselves. Starting with more money could help, but you would still be playing for a *long* (and impractical) amount of time. Beating the casino isn't easy, even when we have at least *some* of the odds stacked in our favor courtesy of Theorem 10.5 when counting cards, but our hopes are ruined through the sheer amount of time required.

Minimizing the Probability of Ruin

If we want to minimize the probability of ruin as it applies to some betting strategy, consider the Gambler's Ruin probability of walking away with all of the money given in Theorem 10.5 and the following example. Assuming that $p = 0.515$ and $q = 0.485$ as in moderately good card counting, suppose that we currently have \$1,000 available to us, and our opponent (a small casino or simply a target amount to win from the casino) has \$100,000. Here, note that $q/p = 0.485/0.515 = 97/103$, which appears in our calculations to come. If we want a bet size of \$100, this means that we have a total of $k = 10$ betting units compared to a total of $T = 1,010$ betting units, and Theorem 10.5 says that the probability we succeed in walking away with the casino's money is

$$P_{10} = \frac{1 - \left(\dfrac{97}{103}\right)^{10}}{1 - \left(\dfrac{97}{103}\right)^{1010}} \approx 0.4513$$

which also means that the *probability of ruin* is $1 - 0.4513 = 0.5487$. On the other hand, if we instead want a bet size of \$10, our starting number of betting units is now $k = 100$ with a total of $T = 10,100$ available; the same theorem gives a probability of taking the casino's cash as

$$P_{100} = \frac{1 - \left(\dfrac{97}{103}\right)^{100}}{1 - \left(\dfrac{97}{103}\right)^{10100}} \approx 0.9975$$

for a *probability of ruin* of 0.0025. Note that the *smaller* betting size results in a much lower chance that we'll suffer ruin.

We can, of course, be more general. Suppose that we start with a bankroll of $X(0)$ and a desired amount to take from the casino as $C(0)$. We let the variable f represent the fraction of the bankroll that we will use for our bets.

This means that our bet size will be $b = f \cdot X(0)$, the amount of betting units we have available at the start is

$$k = \frac{X(0)}{f \cdot X(0)} = \frac{1}{f},$$

and the total number of betting units available is

$$T = \frac{X(0) + C(0)}{b} = \frac{X(0) + C(0)}{f \cdot X(0)}.$$

Subtracting from 1 the probability of taking all of the money given by Theorem 10.5 and using r to represent the ratio $r = q/p$, we obtain the following probability of ruin $R(f)$ given as

$$R(f) = 1 - \frac{1 - r^{\left(\frac{1}{f}\right)}}{1 - r^{\left(\frac{X(0)+C(0)}{f \cdot X(0)}\right)}} = \frac{r^{\left(\frac{1}{f}\right)} - r^{\left(\frac{X(0)+C(0)}{f \cdot X(0)}\right)}}{1 - r^{\left(\frac{X(0)+C(0)}{f \cdot X(0)}\right)}}$$

which we wish to minimize. Calculus quickly shows that since $G(f)$ has no critical values on the interval $0 \le f \le 1$, the minimum must occur at an endpoint, and this is easily shown to be minimized at $f = 0$.

This means, of course, that minimizing the chance of ruin means betting *nothing*! This is always a surefire way to avoid ruin, but if bets must be made, you will minimize your chance of going broke by betting the *minimum* bet size each time since you want f, the fraction of your money you wager each time, as close as possible to zero. The graph of $G(f)$, the probability of ruin given that fraction f, is given in Figure 10.13 for $r = 97/103$ (when $p = 0.515$ and $q = 0.485$), $X(0) = 1000$, and

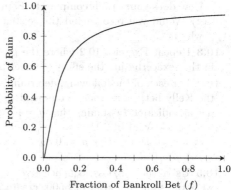

FIGURE 10.13: Probability of Ruin as a Function of f

$C(0) = 100000$. Note that $G(f)$ is indeed minimized when $f = 0$ (and maximized when $f = 1$, betting our entire bankroll at the start).

10.3 Exercises

10.1 Suppose that you are playing a game, starting with $500, using a base bet of $20. For the sequence of wins and losses

WWLLW LLWLW WLLLW WLLWW

create a table showing bet amounts and current money totals for a player using the indicated betting systems.

  (a) Martingale System   **(b)** Anti-Martingale System

10.2 Consider a modification of the Martingale system where instead of doubling your bet on a loss, you multiply your previous bet by $\sqrt{2}$ (this has the benefit of a slower exponential growth, but requires two wins in a row to recoup a loss). Assume a first wager of $100.

  (a) Create a table showing the wager amounts for the first five bets, assuming you lose each round. What is the total amount wagered (and lost) for these five rounds? How does this compare to the $3,100 lost with the standard Martingale system?

  (b) Modify Example 10.1 to account for this change in bet growth, noting that the net profit after a win is no longer a constant $100, and then graph your expected value function for $q = 0.45$ and values of n equal to 4, 6, 8, and 10. How does your graph compare to Figure 10.2 for $q = 0.45$?

  (c) Would you recommend this system over the Martingale system? Why or why not?

10.3 Repeat Exercise 10.2 where the previous bet is multiplied by 3 on a loss, further exacerbating the effect of losses but giving a higher reward on a win.

10.4 For each of the following, determine the percentage of your current bankroll the Kelly betting system recommends betting when the probability of winning p is as indicated (assuming that $q = 1 - p$ as usual). Do each for a payout of 1:1 and for a payout of 2:1.

  (a) $p = 0.60$ **(b)** $p = 0.80$ **(c)** $p = 0.52$ **(d)** $p = 0.50$
  (e) $p = 0.47$

10.5 Using the win/loss sequence given in Exercise 10.1, assuming that you start with the same bankroll of $500, create a table showing each bet amount and current money totals for a player using the Kelly System for each of the following values of the probability of winning p assuming a 1:1 payout.

  (a) $p = 0.60$ **(b)** $p = 0.80$ **(c)** $p = 0.52$

10.6 Algebraically derive the formula for k in Example 10.7 when $p \neq 0.50$.

11

Potpourri: Assorted Adventures in Probability

There's no way that a book could contain the entirety of probability knowledge developed since the days of the Chevalier de Méré (see Exercise 1.4). What we have aimed to do in this book is to give a friendly introduction to a lot of probability topics that come to mind when playing several games and, perhaps most importantly and somewhat hidden from plain sight, topics that can help you make better deci-

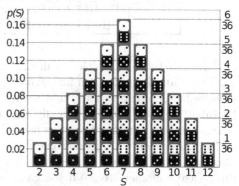

FIGURE 11.1: Density Function

sions in your life. Perhaps by author prerogative, this chapter contains some neat topics that are somewhat tangential to what we've seen so far in this book. In some cases, there's a direct connection between a section of this chapter and an earlier topic in this book; in other cases, topics appear only because I wanted to write something about that topic. Oftentimes our discussion will revolve around using functions, graphs, or images to convey complex ideas; an example of a simple, yet effective, image is shown in Figure 11.1, which shows the probabilities associated with Table 1.2.

11.1 True Randomness?

Virtually all of the games featured in this book involve elements of random chance; die are rolled, coins are flipped, and wheels are spun, giving the sense that the outcomes are truly random. What is meant, though, by the word random? The *Merriam-Webster* dictionary, [52], gives two definitions of this adjective. First, it can mean "lacking a definite plan, purpose, or pattern" and, perhaps more applicable to our use of the word, it also means "relating to, having, or being elements of events with definite probability of occurrence." It is this second version that we consider here.

TABLE 11.1: Fifty Coin Flip Outcomes

Round	Result
1	HHHTT TTTTT HHHHT HHHTT HHTHT
	TTHHH THHHH TTHTT THTTH THHHT
2	HTTHT HHHTT HTHHH HHTTH TTHTT
	TTTTT THHTH HTTHH THTTT HTHHT
3	HTHHT HTTTH THHTH HTTHT HHTTH
	HTHTH HTHTT HTHTH HTTHT HHTTH
4	HHTHH TTTHH THHTH THHTH THTTH
	THHTH TTHTH THHHT HTHTT HHTHT
5	TTHHT HHHHT HHHHH HHTHT THHHH
	THTTT TTHHH HHHTH TTTHT THTTH

Given in Table 11.1 is the result of five different rounds of fifty "coin flips" with a standard quarter. On first glance, it seems very plausible that the coin is indeed random, but how could we tell? In fact, of the five rounds shown in Table 11.1, *only one* is actually the result of flipping a coin fifty times! Can you tell which one of the five it is? Two students[1] of mine provided two of them, while one of my colleagues[2] provided another. The last sequence is from a random number generator on my computer that simulated fifty coin flips. Take a moment to see if you can pick out which sequences match these categories.

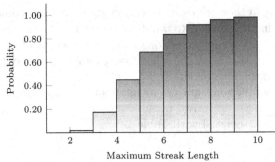

FIGURE 11.2: Maximum Streak Lengths

Each of these sequences could be considered random using the first definition, and as we discussed in this chapter, not only are each of these equally likely to occur, *any* of the 2^{50} possible sequences of heads and tails have the same chance to occur. However, two of the sequences, those created by my students, do not exhibit "typical" behavior of truly random coin flips; they do not contain larger streaks of either heads or tails and they switch from heads to tails or from tails to heads more than would be expected. If you identified sequences 3 and 4 as student-generated, congratulations! The other three sequences, however, are hard to connect with a particular way of creation; my colleague is aware of these "typical" properties and generated a sequence that looks just as good as one created either by a random number generator or by flipping a coin at my

---

[1]Thanks to Katia DeSimone and Andrew Feeney for these sequences.
[2]Thanks to Roland Minton for his sequence.

desk. In particular, to finally let the rest of the cats out of the bag, sequence 1 was the result of the random number generator, sequence 2 was created by my colleague, and sequence 5 is from flipping a coin.

These "typical" properties that we've alluded to are the number of runs and the lengths of runs (streaks). As we saw in Theorem 10.2, determining the probability that 50 coin flips will have a streak of at least k heads (or tails) in a row is fairly easy and can be computed recursively. Figure 11.2 shows the probability that a sequence of 50 coin flips results in runs of at most a certain length (the values for these probabilities are strictly increasing, since a sequence with a largest streak length of three, for example, is included in the value for a maximum of three, four, five, and so on).

From Table 11.1, the student sequences had streak lengths of at most 3; while each specific student sequence is equally likely among the sample space, the *type* of sequence, in terms of maximum streak length, is unlikely among the entire sample space (sequences with maximum

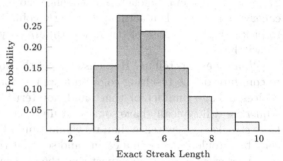

FIGURE 11.3: Exact Streak Lengths

streak length of three or less occur with probability 0.172586; see Figure 11.2). By comparison, the most likely streak length is four, with five just slightly behind. Figure 11.3 shows the probability of getting a maximum streak length of *exactly* the values indicated on the horizontal axis. Note that, as previously foreshadowed, the highest expected streak length is four, with a length of five not too far behind. By comparison, a maximum streak length of three occurs about half as often as a length of four! Using the idea of expected value, which was introduced in Chapter 2, the average highest streak length for all possible sequences of 50 coin flips turns out to be 5.006842.

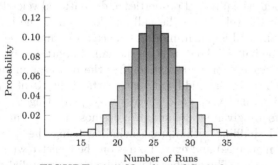

FIGURE 11.4: Number of Switches

The number of times that heads switches to tails or tails switches to heads in a sequence is another "test" to see how "realistic" the student sequences are compared to the others. While we saw that the student sequences were somewhat likely to occur based on the longest streak length, they are pretty far out of whack when it comes to the number of switches; again, without much justification, the probability for various numbers of switches, for 50 coin flips, is shown in Figure 11.4.

You'll notice that this distribution looks like a normal distribution (in other words, a bell-shaped curve); this is not a coincidence! The distribution for the number of switches can be thought of as a binomial distribution which you saw in Chapter 4, and binomial distributions can be approximated very nicely by normal distributions courtesy of Appendix C. Note that the student sequences, rounds 3 and 4 of Table 11.1, contain 34 and 36 switches, respectively; the probability that this number of switches occurs is very low (0.1199% and 0.0466%, respectively)! Indeed, the average number of switches for 50 coin flips is 24.5, and the other three sequences are very "close" to average with 21 from the randomly generated sequence, 26 for the sequence made by my colleague, and 22 when actually flipping a coin. The student sequences, hard to pick out just because of streak length, are easy to identify via the number of switches.

What does this tell us? Humans connect the notion of *random*, as it relates to coin flips or the resulting color on a spin at roulette, as switching between choices a large number of times and not terribly streaky; in fact, random sequences will typically have somewhat unexpected streaks and, as a result of large streaks, less switches than we would think. Don't be too surprised if you walk by a roulette table in a casino and see that the last ten numbers were all red! And, even more, don't expect that the next number can't be red; roulette wheels (along with coin flips and other "random" events) have no memory and still have the same probability for a red number being spun whether the last ten numbers were red or not!

Speaking of "random" as it applies in casinos, note that table games in modern casinos still use very physical methods of play; the behavior of the ball and wheel in roulette and the cards and shuffles of blackjack exhibit very random behavior. These methods, though, are not completely unpredictable. Given enough information about the starting state of a deck cards before it is shuffled, for example, and monitoring *exactly* how the shuffle is performed, gives us complete knowledge of which cards will come and when. With roulette, again, given information about the physical properties and rotational velocity at the start, combined with the velocity of the ball and location of the ball relative to the wheel when the ball is put in motion, one could perhaps know with high certainty where the ball will land; this is especially important given that players may still make wagers for a short time after the ball is spinning around the outside of the wheel! In the middle and late parts of the twentieth century, some mathematicians had very good success with predicting these processes and their accounts are given in [30] and [37]; casinos, however, have adapted by using advanced shuffling techniques (including using the more recent "continuous shuffle" machines) and extra bumps on the roulette wheel to add complexity to the processes. The number of variables needed to reliably predict the outcome quickly leaves the realm of possibility; while roulette and blackjack, for example, are *technically* predictable, they are still games whose outcomes are determined randomly as per our definition. It's worth noting here that some definitions of random include a statement about unpredictability and it's these two examples that demonstrate why I chose not to use definitions that include it.

On the flip side, modern slot machines and computer-based "table games" use random number generators. There are several different implementations of algorithms for using computers to generate sequences of numbers that appear random, but all of them share a common theme. To generate random numbers, a computer uses a specified iterative function combined with an initial value called the seed. For instance, as many versions use the current system time to start the iterative process, if the current time is t_0 and the function is $f(t)$, then the first random number would be $f(t_0)$, the second would be $f(f(t_0))$, the third would be $f(f(f(t_0)))$, and so on. One important observation here is that any of these *pseudo-random number generators* will give the *exact* same sequence of values if the seed, t_0, is the same.

Despite this limitation, modern computers have very good random number generators that pass the "diehard tests" developed by mathematician and computer scientist George Marsaglia in the mid-1990s. This collection of tests includes, among many others, the "runs test" that analyzes a long list of numbers between 0 and 1, counting the number of times the list moves from increasing to decreasing and vice versa; these counts should follow a known distribution. It also includes the "craps test" whereby hundreds of thousands of craps games are simulated with the number of wins and number of throws per game being counted and compared to known results for randomness. As another example, the "overlapping permutations" test looks at five consecutive random numbers, r_1, r_2, r_3, r_4, and r_5 and in a long sequence of randomly generated numbers, looks for other consecutive occurrences of $\{r_1, r_2, r_3, r_4, r_5\}$ (where the ordering may be different, such as in the order r_3, r_1, r_4, r_5, and r_2). Each of the $5! = 120$ possible orderings should occur equally often.

If by true randomness we mean something completely unpredictable, then it is extremely hard to find; flares from the sun, for instance, may be as random as we can observe but surely some chemistry and physics could be applied to predict them, albeit possibly outside the current realm of human knowledge. True randomness, then, is perhaps better thought of as in our second definition of the word random. Computers that use random number generators, roulette wheels that add so much complexity to the dynamical system involved, and shufflers that continuously move cards around are random "enough" for our purposes, and good enough for me to call them truly random. That is, as long as those random number generators pass randomness tests, of course!

11.2   Three Dice "Craps"

Game design for casinos can be very tough; in particular, the battle between needing a game to be enjoyable (so players want to play) and needing a that game to work mathematically (so the casino can make money) is the largest factor in making game design hard. For this section, we attempt to add one more die to the game of craps to see how the mathematics will play out; at the end, we also look at an actual game that uses three dice.

For our three dice "craps" we desire to stay as true as possible to the original game. Rather than focus on the copious amount of side bets available, we will examine just the pass line and don't pass line wagers and how they might be modified for a game using that one extra die. Since we will need the probability of obtaining each specific sum possible using three dice, we use Table 1.2 as inspiration to create Table 11.2; for each of the 36 possibilities given in Table 1.2, we could roll any of the numbers 1 through 6 on the third die. By considering all sums possible and grouping them together, we obtain the probabilities for the sums 3 through 18 given in our new table.

TABLE 11.2: Three Dice Sum Probabilities

Sum	Prob	Sum	Prob
3	1/216	11	27/216
4	3/216	12	25/216
5	6/216	13	21/216
6	10/216	14	15/216
7	15/216	15	10/216
8	21/216	16	6/216
9	25/216	17	3/216
10	27/216	18	1/216

How can we define a pass line win for three dice craps? Since an initial sum of 7 for regular craps results in an instant win on the come out roll, and 7 is the most popular sum for that version, it makes sense that our *new* special numbers for the pass line win should be 10 and 11 since, according to Table 11.2, these have the highest probability. Note that in the standard two dice craps, either a 7 or 11 on the come out roll is a win for the pass line, which happens with probability $8/36 = 48/216$ for that one roll. Since rolling a 10 or 11 with three dice happens with probability 54/216, our probability is already higher than we might perhaps like; for now, at least, let's consider a pass line win on the come out roll to happen only on a 10 or 11. Of course, we need to also decide what sums will determine an instant win for the don't pass line and thereby fixing which sums will establish a point. What goes into my thought process now is the desire to balance probability with symmetry and the hope that the "mathematics will work" down the line; we can always adjust the rules as we develop the game so that the casino still has the advantage.

For two dice craps, a 2, 3 , or 12 will be a win (or push) for the don't pass line on the come out roll, and the probability of obtaining these sums is $4/36 = 24/216$; using a sum of 3, 4, 5, 16, 17, or 18 for three dice craps gives a probability, using Table 11.2, of 20/216, which is close, and also keeps the possibilities for the point, the remaining sums of 6, 7, 8, 9, 12, 13, 14, and 15, symmetric. The worry here is that our pass line rules have a probability higher than the similar rules for regular craps and the don't pass line rules have a lower probability than normal. Is this a problem? We need to compute the probability of winning with each of those point totals. To be consistent with two dice craps, we say that, once a point is established, the pass line wins if the point is rolled before a 10 or 11 and the don't pass line wins otherwise.

This calculation proceeds much like we saw earlier; suppose that a point of 6 is established. At this point, there are 10 ways to roll a 6 and 54 ways to roll either a 10 or 11, so the probability, *once the point is established*, of winning on a pass line bet is 10/64 since there are 64 total outcomes that matter (all other sums are ignored) and 10 will have us win. Similarly, the probability of winning on the don't pass line with the point already established as 6 is then 54/64. The combined probability that we get a point total of 6 and then win with the pass line at least is

$$\underbrace{\left(\frac{10}{216}\right)}_{\substack{\text{establish 6} \\ \text{as the point}}} \cdot \underbrace{\left(\frac{10}{64}\right)}_{\substack{\text{roll 6 before} \\ \text{10 or 11}}} = \frac{25}{3456}$$

which isn't very high; maybe this will help balance the high initial probability of an initial pass line win.

You can verify that Table 11.3 contains the probabilities for establishing the indicated point on the come out roll and then winning with either a pass line bet or a don't pass line bet; note that the totals there do not add up to 100% since we (intentionally) left off the come out roll instant wins.

Combined with the $54/216 = 0.250000$ probability of winning with a pass line wager on the come out roll, this gives us a total win chance

TABLE 11.3: Point Win Probabilities

Point Total	Pass Line Win	Don't Pass Line Win
6	25/3456	5/128
7	25/1656	5/92
8	49/1800	7/100
9	625/17064	25/316
12	625/17064	25/316
13	49/1800	7/100
14	25/1656	5/92
15	25/3456	5/128
Total	0.172359	0.485049

for the pass line of 0.422359; too low to have any chance of a balanced game. If you're reading carefully, you'll remember that, since we have made no attempt to fix the game to have a house advantage yet, a pass line *loss* is a don't pass line win, which happens with probability $1 - 0.422359 = 0.577641$, too high to even modify the rules in a sane way to give the house the advantage it desires. We fix this now by increasing the chance of a pass line win on the come out roll; by trial and error, moving 5, 16, and 17 from a don't pass line initial win to a pass line initial win seems to do the job. The probability of getting a 5, 16, or 17 with three dice, when combined, is 15/216, and adding this to the 0.422359 probability gives 0.491803; this we can work with!

Thus far, we have a modified don't pass line win on a 3, 4, or 18 for the come out roll, and a probability of winning with the don't pass line of $1 - 0.491803 = 0.508197$; again, by trial and error, having the don't pass line *push* on a total of 3 or 4 seems to fix the problem, since the probability of winning is reduced by 4/216, yielding 0.489678. How do the expected values fare? We can compute the expected winnings for the pass line wager as

$$\text{Expected Value} = \underbrace{0.491803 \cdot \$1.00}_{\text{win}} + \underbrace{0.508197 \cdot (-\$1.00)}_{\text{lose}} = -\$0.016394$$

and that for the don't pass line wager of

$$\text{Expected Value} = \underbrace{0.489678 \cdot \$1.00}_{\text{win}} + \underbrace{0.018519 \cdot \$0.00}_{\text{push}} + \underbrace{0.491803 \cdot (-\$1.00)}_{\text{lose}}$$
$$= -\$0.002125$$

for house advantages of about 1.64% and 0.21% respectively. With two dice craps, the don't pass line was a slightly better wager than betting on the pass line; here, the don't pass line wager is even more attractive, but the game still works mathematically.

Of course, as we discussed earlier, the mathematical "soundness" of the game is only one component that goes into game design; is this game fun? A colleague of mine tried this game and decided "no" pretty quickly. Computing the total sum for three dice is not quite as natural as it is for two dice, so the game was considerably slower and awkward compared to the fast-pace that craps usually has; in addition, even though the game works to give the house a very small advantage for both the pass line and don't pass line wagers, the feeling that, once a point is established, don't pass was almost surely going to win wasn't fun either. Perhaps there's a good reason that you've never seen three dice "craps" anywhere in the world.

There *is* a three dice game, though, that can sometimes be found in casinos; chuck-a-luck is a game using three dice that uses a very different set of possible wagers than craps and, while modern casinos rarely carry it, the game is a favorite for carnivals and home game nights. The three dice are typically kept in a wire-frame cage rotated by a dealer, and some people use "birdcage" to refer to the game. Players have relatively few betting options, with a wager on one of the numbers 1 through 6 the most common. After the three dice are "rolled," for these single die bets, a payout of 1:1 is given if one of the dice matches the wagered number, with a 2:1 payout offered if two dice match, and 3:1 winnings are paid for three matching numbers. For example, if $1.00 were wagered on the number 4, then a result of ⚃ ⚄ ⚄ would result in winnings of $1.00 (while a wager on 6 for the same outcome would win $2.00). If none of the three dice matches the wagered die, the wager is lost.

How fair or unfair is this game? Of the 216 possible outcomes when rolling three dice, only 1 outcome will match the number wagered with all three dice, for a probability of 1/216. There are 15 possible outcomes which match the desired number on two dice, as there are five possibilities for the unmatched die and three possible dice on which that unmatched number could appear (recall the methods from Chapter 4 that help deal with counting scenarios such as this). Similarly, there are 75 outcomes where only one die will match the number we care about since there are 25 ways to number the two dice *not* matching our number and three ways to pick which die *does* match our number. The remaining 125 possible outcomes do not feature our number at all! The expected value of a $1.00 wager, then, is

$$\text{Expected Value} = \underbrace{\left(\frac{1}{216}\right) \cdot \$3.00}_{\text{match 3}} + \underbrace{\left(\frac{15}{216}\right) \cdot \$2.00}_{\text{match 2}} + \underbrace{\left(\frac{75}{216}\right) \cdot \$1.00}_{\text{match 1}}$$

$$+ \underbrace{\left(\frac{125}{216}\right) \cdot (-\$1.00)}_{\text{lose}} \approx -\$0.0787$$

giving a house advantage of almost 8%. Some carnivals increase the payout for a three dice match to 10:1; you can verify that this reduces the house advantage to about 4.63% and, if you *do* want this game to be fair, consider payouts of 1:1, 3:1, and 5:1 for a one die match, two dice match, and three dice match, respectively. This particular setup gives zero house (or player) advantage and, while fair, would never appear at a carnival, let alone a casino!

Chuck-a-luck, except for using dice, doesn't resemble the game of craps at all. Note that if we did want to continue development of craps with three dice, we could, of course, have developed the game in a different way than we did. You are certainly welcome to try, but I think, given the not-so-pretty version we have already seen combined with the not-so-fun game play, I'm not sure a "good" version of three dice craps is really out there. Game design can be tough, and for every new casino table game put into play by a real casino, you can be sure that there are several failed games that we have never seen and probably never well!

11.3   Counting "Fibonacci" Coins "Circularly"

The title for this section could be a little misleading. The focus here will be on counting the number of ways a particular type of event may happen. As you probably noticed, throughout this book the notion of counting is extremely important when it comes to calculating probabilities; these probabilities lead to better decision-making when playing games. The discussion that follows here is meant to be an exploration of "different" ways of counting that you would not normally see in a book such as this. While you may never need to use them again in your life, you never know when knowledge of "outside of the box" ways of counting will come in useful.

Question

What is the probability that flipping n coins results in no successive heads?

To answer this question, we'll need to count the number of sequences of length n involving heads and tails for which heads never appears twice in a row. If we let $F(n)$ represent the number of such sequences of length n, then we very quickly have $F(1) = 2$ since all sequences of length 1 do not have any heads in a row. Of the 4 sequences of length 2, only HH is disallowed, so we have $F(2) = 3$. We can continue and count by hand to see that $F(3) = 5$ and $F(4) = 8$ (you can look back at Table 1.1 for the $n = 4$ case and count them to verify that 8 is correct), but we would like a better way to count.

Note that, if flip n happens to be tails, then *any* sequence of $n - 1$ coin flips that has no consecutive heads will be fine; that is, any of the $F(n - 1)$ sequences with one less coin followed by a tails will still count for $F(n)$. On the other hand, if flip n is heads, then for our sequence to be permitted, flip $n - 1$ must have been tails and then *any* sequence of $n - 2$ flips before that with no consecutive heads is allowed. This means that any of the $F(n - 2)$ sequences can be made into a sequence of length n by appending a tails and then a heads, making $F(n)$ a recursive formula given by $F(n) = F(n-1) + F(n-2)$, and since there are 2^n possible sequences of length n overall, the probability becomes

$$\text{Probability No Successive Heads in } n \text{ Flips} = \frac{F(n)}{2^n}.$$

With the starting values of $F(n)$ given before as 2, 3, 5, and 8, this means we can easily compute $F(n)$ by adding the two previous terms to obtain more values, starting with 13 for $F(5)$ and continuing as 21, 34, 55, 89, and so on. This sequence of numbers may be familiar to you as they are quite famous in the world of mathematics. These are the Fibonacci numbers.[3]

Leonardo Fibonacci, in an early 1200s book of his, popularized this sequence of numbers; though the sequence was present in Indian mathematics, associated to a handful of mathematicians in that region during the period from 200 BC to 1150 AD, Fibonacci's presentation in terms of a "rabbit problem" caught on. Assume that a new pair of rabbits, one male and one female, are put into a field. A pair of rabbits is able to (and will always) mate at the age of one month and each month thereafter; a female rabbit always produces, after a one-month pregnancy, a single male and a single female as offspring. How many pairs of rabbits are present after a certain period of time, assuming that the rabbits never die?

---

[3]The indexing provided here is convenient for us, but the Fibonacci sequence traditionally starts with a pair of ones at the start, proceeding as 1, 1, 2, 3, 5, 8, and so on; the traditional labeling, f_n, relates to our labeling in that $f_n = F(n - 2)$.

At the start, there is only 1 pair of rabbits. After the first month, they mate, but there is still only 1 pair of rabbits. We reach 2 pairs of rabbits after the second month, since the first pair is still around, and the female gives birth to a new pair, and then mates again. With the passing of three months, there are 3 pairs of rabbits. The original pair is still around, and that female gives birth (and mates) again, and the newest pair will mate. This gives 5 pairs after month four since the original pair and pair born after month two give birth; the original pair, as well as the pair from month two and from month three, mate, which will give 8 pairs in the next month, and the process continues.

The Fibonacci numbers, along with the very related golden ratio (defined as the limiting value of $F(n)/F(n-1)$ as n goes to infinity) have a lot of neat mathematical connections but, more surprisingly, appear very often throughout many different biological settings (the spiral pattern on a seashell is but one example). Many books and articles have been written about the Fibonacci numbers (and other additive sequences like this); see [5] in the Bibliography for one such book that includes many other neat mathematical topics in addition to Fibonacci and his rabbits.

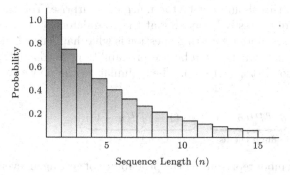

Sequence Length (n)

FIGURE 11.5: Non-Consecutive Heads Probability

The graph showing the probability of flipping a coin n times and getting no consecutive heads, for n between 1 and 15, is shown in Figure 11.5. As you can reason, as the number of flips involved increases, the probability of not getting two heads in a row decreases, and fairly quickly when small values of n are involved. For instance, when $n = 11$, this probability is

$$\text{No Consecutive Heads, 11 Flips} = \frac{F(11)}{2^{11}} = \frac{233}{2048} \approx 0.113770,$$

or a little bit more than a 10% chance. The value $n = 11$ is used only because it results in a numerator of 233, which happens to be the only three-digit prime Fibonacci number.

Question

What is the probability of an even number of heads in n flips of a coin?

On the surface, this sounds like a difficult question, but with a little bit of clever realization is how the probabilities for sequences that match this description will be very helpful. Define the probability of having an even number of heads on n as $P_e(n)$. If we again isolate flip n, the last flip, we will be able to proceed. One piece of the probability for getting an even number of heads in n flips is $1/2$ (the probability of getting tails on flip n) times the probability of getting an even number heads in $n-1$ flips, $P_e(n-1)$. Since the probability of getting an *odd* number of heads in $n-1$ flips is $1 - P_e(n-1)$, the other piece of $P_e(n)$ that we need is $1/2$ times $1 - P_e(n-1)$ which covers the case of getting heads on flip n and an odd number of heads beforehand. This gives us

$$P_e(n) = \underbrace{\frac{1}{2} \cdot P_e(n-1)}_{\text{flip } n \text{ is tails}} + \underbrace{\frac{1}{2} \cdot (1 - P_e(n-1))}_{\text{flip } n \text{ is heads}} = \frac{1}{2} \cdot P_e(n-1) + \frac{1}{2} - \frac{1}{2} \cdot P_e(n-1) = \frac{1}{2}$$

which, as you can see, is completely independent of the value of n. The probability of getting an even number of heads, no matter how many flips are involved, is always 50%.

After a bit of thought, this should not be too much of a surprise. You can verify this fact for $n = 4$ quite easily by looking at the 16 different outcomes for flips given in Table 1.1. A more interesting question is what happens when, perhaps, the coin is not fair; that is, let p be the probability of heads and q the probability of tails (so that $q = 1 - p$). Then, similar to the above, we have the relationship

$$^pP_e^q(n) = \underbrace{q \cdot {}^pP_e^q(n-1)}_{\text{flip } n \text{ is tails}} + \underbrace{p \cdot (1 - {}^pP_e^q(n-1))}_{\text{flip } n \text{ is heads}}$$

where $^pP_e^q(n)$ is a fancy symbol representing the probability of getting an even number of heads with n flips when the chance of heads and tails are p and q, respectively. Since this equality is true for any $n \geq 2$, this also means that

$$^pP_e^q(n-1) = \underbrace{q \cdot {}^pP_e^q(n-2)}_{\text{flip } n-1 \text{ is tails}} + \underbrace{p \cdot (1 - {}^pP_e^q(n-2))}_{\text{flip } n-1 \text{ is heads}}$$

which we can substitute into the first equation. This method for solving recursive relationships appeared in Chapter 10, and that technique produces a final value of

$$^pP_e^q(n) = \frac{1}{2} \left(1 + (q - p)^n\right)$$

which is no longer independent of the number of flips; however, as n increases, the $(q - p)^n$ term becomes closer and closer to zero, so the probability of an even number of heads gets closer to 50% as the number of flips increases *even* for an unfair coin.

Question

How many flips are expected in order to get n heads in a row?

By now, you might expect some sort of recursive way of thinking about this. Not intending to disappoint, let's start by setting $E(n)$ to be the expected number of flips needed to get n consecutive heads. Starting with $n = 1$, if we get a head on the first flip of the coin, we're all set, and this happens with probability $1/2$. Otherwise, with a tail coming up, we've used up a flip and we're essentially starting over, trying to get a head on the next flip. This give us the relationship

$$E(1) = \underbrace{1 \cdot \left(\frac{1}{2}\right)}_{\text{heads}} + \underbrace{\left(\ 1 + E(1)\ \right) \cdot \left(\frac{1}{2}\right)}_{\text{tails and try again}}$$

which can be algebraically solved to reach the value $E(1) = 2$. That is, the expected number of flips to get at 1 consecutive head is 2.

For $n = 2$, we need to think about flip $E(1) + 1$. Again, this has probability of $1/2$ for being heads, in which case the total number of expected flips in this case is $E(1) + 1$ itself (we expected to use $E(1)$ flips to get 1 consecutive head, and then we got the next head in a row on this flip). Otherwise, we effectively "start over," giving us the relationship

$$E(2) = \underbrace{\left(\frac{1}{2}\right) \cdot (E(1) + 1)}_{\text{heads this flip}} + \left(\frac{1}{2}\right) \cdot \left(\ \underbrace{E(1) + 1}_{\text{heads, then heads}} + \underbrace{E(2)}_{\text{try again}}\ \right)$$

which we can solve by substituting the known value $E(1) = 2$ and using algebra to get $E(2) = 6$ as the expected number of flips to get two heads in a row.

Since, for $n = 3$, we can write

$$E(3) = \underbrace{\left(\frac{1}{2}\right) \cdot (E(2) + 1)}_{\text{heads this flip}} + \left(\frac{1}{2}\right) \cdot \left(\ \underbrace{E(2) + 1}_{\text{2 heads, then heads}} + \underbrace{E(3)}_{\text{try again}}\ \right)$$

which looks virtually the same as above (using the same reasoning to get this form), we can abstract to n consecutive heads by writing

$$E(n) = \left(\frac{1}{2}\right) \cdot (E(n-1) + 1) + \left(\frac{1}{2}\right) \cdot (E(n-1) + 1 + E(n))$$

$$= E(n-1) + 1 + \frac{E(n)}{2}$$

which, using some nice algebra, gives us

$$E(n) = 2 \cdot (E(n-1) + 1)$$

as the recursively defined expected number of flips to get n consecutive heads in a row.

Table 11.4 contains the values of $E(n)$ for $1 \leq n \leq$ 10. Note how fast this value rises; the formula obtained for $E(n)$ effectively says that in order to get one additional head in a row, take the expected number of flips, add one, and double that value to get the new expected number of flips. You may notice that these numbers are very close to the powers of two; this is not a coincidence. A little bit of argument (using the principle of mathematical induction) can be used to show that this recursively defined expected value can be written as the closed-form

$$E(n) = 2^{n+1} - 2$$

which makes computing the values much easier than

TABLE 11.4: $E(n)$ Values

n	$E(n)$
1	2
2	6
3	14
4	30
5	62
6	126
7	254
8	510
9	1,022
10	2,046

having to compute $E(k)$ for all $1 \leq k \leq n-1$. Mathematicians like to take recursive formulas and find the equivalent closed-forms functions whenever possible.

Question

What is the probability that flipping n coins results in no streaks of m (or more) consecutive heads?

We end this section by reconsidering the question that started this discussion of counting that led to a study of recursion. Instead of sequences that have no consecutive heads when flipping n coins, we consider sequences that do not have m heads in a row; that earlier case where the Fibonacci numbers made an appearance was when $m = 2$, so let's discuss what happens with $m = 3$. Denote by $F(n, 3)$ the number of sequences of length n that do not have three heads in a row.

As before, if the last flip happens to be tails, *any* sequences of length $n-1$ that do not have 3 heads in a row can be combined with this last tails to give us a sequence of length n that we should count in our list. There are $F(n-1, 3)$ of these. Otherwise, if the last flip is heads, then we need to be careful. If the *prior* flip is tails (the sequence ends with TH), then we're in good shape. Any of the $F(n-2, 3)$ sequences of length $n-2$ that don't have three heads in a row can have TH appended to its end to get an acceptable sequence. If that prior flip is *heads* (the sequence ends with HH), then the flip immediately before must be tails for us to be happy. Taking any of the

$F(n - 3, 3)$ sequences of length $n - 3$ and attaching THH to the end is okay. This gives us the relationship

$$F(n, 3) = F(n - 1, 3) + F(n - 2, 3) + F(n - 3, 3)$$

where we can verify by hand the initial values of $F(1, 3) = 2$, $F(2, 3) = 4$, and $F(3, 3) = 7$. This means that we can compute $F(4, 3)$ as

$$F(4, 3) = F(3, 3) + F(2, 3) + F(1, 3) = 7 + 4 + 2 = 13$$

and you can verify by manual inspection of Table 1.1 that 13 is the correct number.

Abstracting to an arbitrary value of $m \geq 2$ then produces the expected

$$F(n, m) = F(n - 1, m) + F(n - 2, m) + \cdots + F(n - m, m)$$

so that we can give the probability of obtaining such a sequence as

$$\text{Probability No } m \text{ Consecutive Flips Heads} = \frac{F(n, m)}{2^n}.$$

As an interesting note, the sequences of values generated for $F(n, m)$ do have special names in addition to the Fibonacci sequence when $m = 2$. The sequence 2, 4, 7, 13, 24, 44, and so on, when $m = 3$, is called the Nebonacci sequence, and when $m = 4$ the resulting sequence (2, 4, 8, 15, 29, 56, and so forth) is the Pronebonacci sequence. This generalization works out nicely, and at this point you can compute some of the probabilities yourself.

11.4   Compositions and Probabilities

In Chapter 8, the numbers 1, 5, 10, 10, 5, and 1 came up as the number of different ways, in one, two, three, four, five, and six, respectively, additional cards, for the dealer to get a point total of 19 during a hand of Blackjack, assuming that the face up card is a ten-valued card. These particular values happen to be all of the binomial coefficients when $n = 5$ is involved, and are also the numbers that show up in the fifth row of Pascal's Triangle, pictured in Figure 11.6 (the numbers in Pascal's Triangle are themselves binomial coefficients). This is *not* a coincidence at all, and there are a lot of neat mathematical ideas used in constructing a general theorem for the number of ways the dealer can get a certain total. Many thanks to Roanoke College student Jonathan Marino who worked with me in developing the theory as his Honors in Mathematics project.

Determining the pattern involves looking at the possible ways the dealer can get from his or her current total, obtained from only the face up card present, to the target. In the example above, to get from a total of 10 to a total of 19, we need to consider all ways of adding cards to the dealer's hand that total 9 more points. There are, of course, some restrictions; getting a 7 followed by a 2 is not allowed, since the initial 10 plus another 7 will give the dealer 17 points, forcing the dealer to stand, due to the rules of Blackjack. How-

```
                1
             1     1
          1     2     1
       1     3     3     1
    1     4     6     4     1
 1     5    10    10     5     1
1    6    15    20    15     6    1
```

FIGURE 11.6: Pascal's Triangle

ever, the mathematical framework of compositions will be useful, and we introduce that here.

Definition 11.1 A *composition* of a positive integer n is a list of m positive integers $\lambda = (\lambda_1, \lambda_2, \ldots, \lambda_m)$ such that $\lambda_i > 0$ for each i and

$$\lambda_1 + \lambda_2 + \cdots + \lambda_m = n;$$

each of the λ_i terms is called a *part* of the composition. The length of λ is the number of parts, m, in the composition, and as a shorthand, we write $|\lambda| = n$ so that $|\lambda|$ always refers to the actual number being represented by the composition.

As an example, the possible compositions of $n = 5$, grouped into similar composition style, are given in Table 11.5. Note that there is 1 composition of length five, 4 compositions of length four, 6 compositions of length three, 4 compositions of length two, and 1 composition of length one. Again, numbers from

TABLE 11.5: All Compositions of $n = 5$

$1+1+1+1+1$		5
$2+1+1+1$	$3+1+1$	$2+2+1$
$1+2+1+1$	$1+3+1$	$2+1+2$
$1+1+2+1$	$1+1+3$	$1+2+2$
$1+1+1+2$		
$4+1$	$3+2$	
$1+4$	$2+3$	

Pascal's Triangle, Figure 11.6, appear; the numbers 1, 4, 6, 4, and 1 comprise the fourth row of the triangle. At this point, it should seem reasonable that our question about Blackjack and the notion of compositions *must* be related.

● ●|● ● ● ●|● ●|●

FIGURE 11.7: Composition

Why do these binomial coefficients show up when looking at compositions? Consider Figure 11.7 which shows the composition $\lambda = (2, 4, 2, 1)$ of $n = 9$. How could we picture a different composition of 9 that also has 4 parts? We could simply *move* the red bars to different positions. The only restriction is that the red bar cannot be placed to the left of the leftmost dot or to the right of the rightmost dot because we do not allow compositions to have parts equal to zero. For a composition of *nine* with *four* parts, there are *eight* choices of where to place *three* red bars. This reasoning gives us the following result.

Theorem 11.2 The number of compositions of n with m parts, denoted by $C(m, n)$, is given by the binomial coefficient

$$C(m, n) = \binom{n-1}{m-1}.$$

Returning to the Blackjack example where the dealer has a face up ten-valued card and we're counting the ways to get a final total of 19, it would seem that we need only count the compositions of 9, the difference between the current point total and desired final total. However, there are some restrictions we need to make due to the special rules involved in the game of Blackjack.

First, compositions that start with 1 are not allowed here; since the dealer currently has a 10, a "one" as the next card could only be obtained by getting an ace as the face down card, giving the dealer Blackjack (a total of 21, not 19). Note that the dealer would also have checked for an ace before any player would get to make a play choice. Therefore compositions of the form $(1, \lambda_2, \ldots, \lambda_m)$ need to be removed. In addition, our compositions also cannot *end* with a 1 or a 2, since ending with a 1 or a 2 means that the dealer's point total *before* drawing a 1 or a 2 would have been 18 or 17, causing the dealer to immediately stand and not have a chance of reaching 19. We remove compositions of the form $(\lambda_1, \cdots \lambda_{m-1}, 1)$ and $(\lambda_1, \cdots \lambda_{m-1}, 2)$. Our method of counting is now clear; we will count all compositions of $n = 9$, the difference between the desired total of 19 and the current total of 10, remove those that start with 1, remove those that *end* in a 1 or a 2, and then *add back* anything we accidentally removed twice. The adding back is required since compositions of the form $(1, \lambda_2, \ldots, \lambda_{m-1}, 1)$ and $(1, \lambda_2, \ldots, \lambda_{m-1}, 2)$ were removed courtesy of the two rounds of removals we did.

Assuming we want to reach 19 with m additional cards (or parts of a composition), we start with $C(m, 9)$ as our starting number of compositions. Each composition of nine with m parts that starts with a 1 corresponds identically with a composition of *eight* with $m - 1$ parts (removing the initial part equal to 1, decreasing the integer being composed and number of parts by 1), so the quantity that need removing are $C(m - 1, 8)$. We remove the compositions that *end* in a 1 exactly the same way, by removing $C(m - 1, 8)$ compositions. Those that *end* in a 2 correspond with compositions of *seven* with one less part, so we can remove those from our count by subtract $C(m - 1, 7)$ compositions as well. Note that $C(m, n)$ is defined to be zero if m happens to be negative.

Those that have been removed twice are easy to add back; compositions of nine with m parts of the form $(1, \lambda_2, \ldots, \lambda_{m-1}, 1)$ can be added back by counting compositions of *seven* with two fewer, $m - 2$, parts; similarly, those of the form $(1, \lambda_2, \ldots, \lambda_{m-1}, 2)$ can be found by looking at compositions of *six* with $m - 2$ parts. Therefore we add back $C(m - 2, 7)$ and $C(m - 2, 6)$ to reach our final total. The number of ways the dealer can reach 19, starting with 10, using m additional cards beyond that face up 10, is

$$\underbrace{C(m, 9)}_{\text{all}} - \underbrace{C(m - 1, 8)}_{\text{starts with 1}} \underbrace{-C(m - 1, 8) - C(m - 1, 7)}_{\text{ends with 1 or 2}}$$
$$+ \underbrace{C(m - 2, 7) + C(m - 2, 6)}_{\text{add back doubly removed}}$$

which using Theorem 11.2 to help with computation becomes

$$\binom{8}{m - 1} - \binom{7}{m - 2} - \binom{7}{m - 2} - \binom{6}{m - 2} + \binom{6}{m - 3} + \binom{5}{m - 3}.$$

For $m = 4$, this gives the number of ways the dealer can reach 19 with 4 additional cards as

$$\text{Ways to } 19 = 56 - 21 - 21 - 15 + 6 + 5 = 10,$$

the same as we found earlier; success!

This process can, of course, be generalized, but before we do, a small modification to a famous result of Pascal is needed. First, note from the triangle of Figure 11.6 that each number in the interior is found by adding the two numbers directly above (slightly to the left and to the right) of that number; this construction gives us the following relationship between binomial coefficients.

Theorem 11.3 (Pascal's Rule) For any positive integer n and $0 < k \leq n$, we have the relationship

$$\binom{n}{k} - \binom{n - 1}{k - 1} = \binom{n - 1}{k}.$$

You can note already that Theorem 11.3 would have been useful in simplifying the example with the dealer's face up card of ten!

The full generalization results in some rather messy formulas; to keep everything manageable while we explore how to take the case we've already seen and abstract to many arbitrary cases, we make one assumption on the values involved. Let s be the point total on which the dealer must stand (for normal Blackjack, $s = 17$), d be the point total of the dealer's face up card, and w the desired point total. The assumption we make is that $w - d \leq 11$ so that we need not worry about whether aces count as 1 or 11. We wish to count the number of ways to reach w with m cards.

Again, we start with the overall number of ways to get from d to w by using compositions; that is, we start with $C(m, w - d)$ compositions. As before, we need to eliminate compositions that start with 1, and there are $C(m - 1, w - d - 1)$ of these. The other compositions that must be removed (and then added back) depend on the values of s and w; for example, as before with $s = 17$ and $w = 19$, we disallowed compositions that ended with a 1 or a 2; if $w = 20$ instead, we would also need to remove those the end in 3. In addition, since we are generalizing our process, if we change the rules of Blackjack so that the dealer must stand on $s = 15$ and want $w = 20$, then compositions ending in 4 and 5 would need to be removed as well.

We remove compositions of the form $(\lambda_1, \ldots, \lambda_{m-1}, k)$ for all values of k from 1 to $w - s$ (removing *none* if $k \leq w - s$), and then add back those of the form $(1, \lambda_2, \ldots, \lambda_{m-1}, k)$ for the same values of k. These values are, respectively, $C(m - 1, w - d - k)$ and $C(m - 2, w - d - k - 1)$ where the $-k - 1$ in the second of these comes from removing the first part, equal to 1, and the last part, equal to k, from the integer we are composing. Overall, this gives us

$$\underbrace{C(m, w - d)}_{\text{all}} - \underbrace{C(m - 1, w - d - 1)}_{\text{starts with 1}}$$

$$- \sum_{k=1}^{w-s} \underbrace{(C(m - 1, w - d - k) - C(m - 2, w - d - k - 1))}_{\text{ends with } k, \text{ add back if starts with 1}}$$

using the summation notation you may have seen before (if not, it will be formally introduced in Definition 2.5).

Converting to binomial coefficients courtesy of Theorem 11.2, we can rewrite this as

$$\underbrace{\binom{w - d - 1}{m - 1} - \binom{w - d - 2}{m - 2}}_{\text{①}} - \sum_{k=1}^{w-s} \underbrace{\left(\binom{w - d - k - 1}{m - 2} - \binom{w - d - k - 2}{m - 3} \right)}_{\text{②}}$$

where we can use the nice result of Pascal's Rule in Theorem 11.3 to combine

(1) and (2) into single terms, obtaining

$$\text{Obtain } w \text{ in } m \text{ Cards} = \binom{w-d-2}{m-1} - \sum_{k=1}^{w-s} \binom{w-d-k-2}{m-2}.$$

Writing out the summation explicitly, we obtain the terms

$$\binom{w-d-2}{m-1} - \binom{w-d-3}{m-2} - \binom{w-d-4}{m-2} - \cdots$$
$$\cdots - \binom{w-d-(w-s-1)-2}{m-2} - \binom{w-d-(w-s)-2}{m-2}$$

which, using Pascal's Rule repeatedly, finally reduces to

$$\binom{w-d-(w-s)-2}{m-1} = \binom{s-d-2}{m-1},$$

giving us the following generalized result for Blackjack.

Theorem 11.4 If s is the point value upon which the dealer must stand, d is the value of the dealer's face up card, and w is the desired target point total, then as long as $w - d \leq 11$, the number of ways the dealer can reach w with m additional cards beyond the face up card, denoted as $N(m, s, d)$, is given by

$$N(m, s, d) = \binom{s-d-2}{m-1}.$$

Note that the result of Theorem 11.4 does *not* depend on the target point total w at all! The number of ways to reach a target only depends on the dealer's face up card along with the Blackjack rule that determines when the dealer must stand. Before we see how this can be used to help with computing probabilities, feel free to check out our work as [27] in the Bibliography for the most general results that relax the requirement that $w - d \leq 11$; this full generalization requires a few more intricacies that would needlessly make this section harder to follow.

How does Theorem 11.4 relate to the probability that the dealer obtains this total? Take a look back at Example 8.7 where the coefficients 1, 5, 10, 10, 5, and 1 appear in the calculation. We can combine our work here with that example, making use of the following well-known mathematical result.

Theorem 11.5 (Binomial Theorem) For positive integers n and any real numbers x and y, we have

$$(x+y)^n = \sum_{i=0}^{n} \binom{n}{i} x^i y^{n-i}$$

$$= \binom{n}{0} x^0 y^n + \binom{n}{1} x^1 y^{n-1} + \cdots + \binom{n}{n-1} x^{n-1} y^1 + \binom{n}{n} x^n y^0$$

(the name "binomial" in all of theorems we'll see comes from the fact that a binomial in mathematics is simply a polynomial with two terms).

Take a moment to expand the case when $n = 2$ by hand; you should see a familiar result, especially if you replace y by the number 1 and obtain the familiar $(x + 1)^2 = x^2 + 2x + 1$ from algebra class. Now, revisiting Example 8.7, we can use the Binomial Theorem and some algebra to inspire our last neat theorem.

Example 11.1 Replacing 4/52 by the equivalent fraction 1/13 in Example 8.7, we obtain that the probability that the dealer obtains a point total of 19 given that the face up card is of value 10 is

$$1 \cdot \left(\frac{1}{13}\right) + 5 \cdot \left(\frac{1}{13}\right)^2 + 10 \cdot \left(\frac{1}{13}\right)^3 + 10 \cdot \left(\frac{1}{13}\right)^4 + 5 \cdot \left(\frac{1}{13}\right)^5 + 1 \cdot \left(\frac{1}{13}\right)^6$$

$$= \left(\frac{1}{13}\right) \left(\binom{5}{0} \left(\frac{1}{13}\right)^0 + \binom{5}{1} \left(\frac{1}{13}\right)^1 + \binom{5}{2} \left(\frac{1}{13}\right)^2 + \binom{5}{3} \left(\frac{1}{13}\right)^3 \right.$$

$$\left. + \binom{5}{4} \left(\frac{1}{13}\right)^4 + \binom{5}{5} \left(\frac{1}{13}\right)^5 \right)$$

$$= \left(\frac{1}{13}\right) \cdot \sum_{i=0}^{5} \binom{5}{i} \cdot \left(\frac{1}{13}\right)^i \cdot (1)^{5-i} = \left(\frac{1}{13}\right) \cdot \left(\left(\frac{1}{13}\right) + 1 \right)^5 \approx 0.111424$$

where we have used Theorem 11.5, the Binomial Theorem, in the last equality by choosing a convenient value of $y = 1$.

All that remains is to combine the work of Example 11.1 with Theorems 11.4 and 11.5 into a concise nice result for probabilities. Again, the approximations provided are due to our use of the Infinite Deck Assumption of Theorem 8.1.

Theorem 11.6 If s is the point value upon which the dealer must stand,

d is the value of the dealer's face up card, and $w \neq 21$ is the desired target point total, then as long as $w - d \leq 9$, the probability of the dealer obtaining w is approximately

$$\text{Probability} = \frac{1}{13}\left(\frac{14}{13}\right)^{s-d-2}.$$

Note again that the probabilities given in Theorem 11.6 do not depend on w. The limitation that w not equal 21 and stricter control over $w - d$ are due to the fact that an ace as the first card is not allowed by our earlier theorem, slightly altering the probability, and that ten-valued cards themselves occur with a different probability. Nevertheless, we can still use Theorem 11.6 in interesting ways. Suppose a new version of Blackjack requires that the dealer stand with a total of *sixteen*. With a face up 10 and desiring to reach 19, that probability becomes approximately

$$\text{Probability of Dealer 19} \approx \frac{1}{13}\left(\frac{14}{13}\right)^{16-10-2} \approx 0.103465$$

using Theorem 11.6 with $s = 16$ and $d = 10$. This value is a bit less than we found in Examples 8.7 and 11.1 since, when standing on *seventeen*, the dealer has a chance to turn a 16 into a 19; this is no longer the case in this modified version!

11.5   Sicherman Dice

Rolling a pair of standard six-sided dice, as we've seen, gives the results that we saw back in Table 1.2. Six of the thirty-six outcomes give a sum of 7, while five of the outcomes give a sum of 6 and another five give a sum of 8. The other possible sums have probabilities that decrease in nature, and a full list of outcomes for the sums and the associated probabilities appears in Chapter 6; feel free to construct your own table for the sums based on the outcomes of Table 1.2 or refer to Table 6.4.

Is there another way of labeling two six-sided dice so that, when rolling them, the exact same probabilities for the sums as in Table 6.4 are obtained? This question was first answered by mathematician George Sicherman in the 1970s and you might be surprised to learn that the answer is yes. In fact, with reasonable assumptions on what numbers are allowed on the new dice, there is *only* one other way to make these six-sided dice!

We associate the standard six-sided die with faces $\{1, 2, 3, 4, 5, 6\}$ to the

polynomial
$$f(x) = x^1 + x^2 + x^3 + x^4 + x^5 + x^6$$

which will be called the *generating function* for the standard die. Verify that multiplying this function by itself yields

$$(f(x))^2 = (x^1 + x^2 + x^3 + x^4 + x^5 + x^6) \cdot (x^1 + x^2 + x^3 + x^4 + x^5 + x^6)$$
$$= 1x^2 + 2x^3 + 3x^4 + 4x^5 + 5x^6 + 6x^7 + 5x^8 + 4x^9 + 3x^{10} + 2x^{11} + 1x^{12}$$

as the result. This product is interesting because the exponents are the possible sums for rolling two dice *and* the coefficients are the number of outcomes (out of the thirty-six) that give that sum! Note that I've explicitly written coefficients and exponents of one in our function and product to illustrate this fact.

What we need is a pair of polynomials $g(x)$ and $h(x)$, representing the two new dice, that have the exact same product (there's no reason for us to assume at the start that the new dice will each have the same faces). That is, we want these new functions to have the property that

$$g(x) \cdot h(x) = (f(x))^2$$
$$= 1x^2 + 2x^3 + 3x^4 + 4x^5 + 5x^6 + 6x^7 + 5x^8 + 4x^9 + 3x^{10} + 2x^{11} + 1x^{12}.$$

It so happens that factoring the polynomial $f(x)$ gives a lot of useful information about how to proceed. Since

$$f(x) = x \cdot (x+1) \cdot (x^2 + x + 1) \cdot (x^2 - x + 1),$$

the two new dice given by generation functions $g(x)$ and $h(x)$ must satisfy

$$g(x) \cdot h(x) = x^2 \cdot (x+1)^2 \cdot (x^2 + x + 1)^2 \cdot (x^2 - x + 1)^2$$

so all that remains is for us to assign each of the terms on the right to the functions $g(x)$ and $h(x)$. We must, however, do so in such a way that

- the coefficients of $g(x)$ (and of $h(x)$) must be positive integers that add up to six since we want six faces on each of our new dice, and

- the exponents must be positive integers since we desire that each of the faces has at least one dot on them.

By trial and error, you can verify that the only assignment of terms to the new generating functions satisfying these rules is

$$g(x) = x^1 + 2x^2 + 2x^3 + x^4$$
$$h(x) = x^1 + x^3 + x^4 + x^5 + x^6 + x^8.$$

This means that if one die has faces $\{1, 2, 2, 3, 3, 4\}$ and another die has faces $\{1, 3, 4, 5, 6, 8\}$, rolling the two together gives the same probabilities for sums as rolling two standard six-sided dice. The outcomes are illustrated in Table 11.6 where you can verify by hand that six outcomes have a sum of 7, five outcomes have a sum of 6, and so on. Note that the dice with a face of 8 probably looks a good bit different than you're used to unless you've played a lot of dominoes.

At this point, you might wonder about possible other ways to label dice that have a different number of sides. For a standard four-sided die that typically has sides $\{1, 2, 3, 4\}$, we can repeat the same process. If we write

TABLE 11.6: Outcomes for Rolling Sicherman Dice

	⚀	⚂	⚃	⚄	⚅	8
⚀	⚀⚀	⚀⚂	⚀⚃	⚀⚄	⚀⚅	⚀8
⚁	⚁⚀	⚁⚂	⚁⚃	⚁⚄	⚁⚅	⚁8
⚁	⚁⚀	⚁⚂	⚁⚃	⚁⚄	⚁⚅	⚁8
⚂	⚂⚀	⚂⚂	⚂⚃	⚂⚄	⚂⚅	⚂8
⚂	⚂⚀	⚂⚂	⚂⚃	⚂⚄	⚂⚅	⚂8
⚃	⚃⚀	⚃⚂	⚃⚃	⚃⚄	⚃⚅	⚃8

$$f_4(x) = x^1 + x^2 + x^3 + x^4$$

as the generating function for the regular four-sided die, we have

$$(f_4(x))^2 = 1x^2 + 2x^3 + 3x^4 + 4x^5 + 3x^6 + 2x^7 + 1x^8$$

along with the knowledge that $f_4(x)$ factors into different pieces as

$$f_4(x) = x \cdot (x + 1) \cdot (x^2 + 1).$$

To find another pair of dice that have the same probabilities for the sum upon rolling two dice, we need two functions $g_4(x)$ and $h_4(x)$ that have positive integer exponents and positive integer coefficients that add to four so that

$$g_4(x) \cdot h_4(x) = 1x^2 + 2x^3 + 3x^4 + 4x^5 + 3x^6 + 2x^7 + 1x^8$$

where, again, trial and error produces one other possible pair of labels for four-sided dice that have the same probabilities for sums. One die would have faces $\{1, 2, 2, 3\}$ and the other would have faces $\{1, 3, 3, 5\}$.

The existence of a unique pair of dice that have the same sums as a pair of standard four or six-sided dice disappears for dice with more sides. You should take some time to explore the case where we start with a standard eight-sided die with sides $\{1, 2, 3, 4, 5, 6, 7, 8\}$ and try to find pairs for eight-sided dice with different sides that have the same probabilities for sums as rolling two standard dice with eight sides. For your convenience, the generating function for a single eight-sided die, $f_8(x)$, factors as

$$f_8(x) = x \cdot (x + 1) \cdot (x^2 + 1) \cdot (x^4 + 1)$$

and you should be able to find *three* pairs of dice with non-standard labels that give the usual probabilities for two eight-sided dice.

Continuing, there are *seven* pairs of non-standard labeled dice with twelve sides that give the same sum probabilities as two standard twelve-sided dice. There are also *seven* pairs for the solution when dice with twenty sides are involved; for more, see the work of Duane Broline found as [13] in the Bibliography. The mathematical theory involved lies in the areas of combinatorics and abstract algebra (the generating functions involved are *almost* what we call cyclotomic polynomials); I happen to find these dice rather fascinating, and, if you so desire, you can purchase Sicherman dice to use with your own board games that require two dice! Care must be taken, however, since the probability of rolling doubles is lower with Sicherman dice than standard dice. Don't use these for Monopoly unless, of course, everyone is aware of this change!

11.6 Traveling Salesmen

Imagine the following scenario. You're a salesman traveling around the country trying to sell the latest and greatest technology available. You are based in Harrisburg, the state capital of Pennsylvania, and your latest assignment is to travel to all of the other state capitals in the contiguous United States. Surely, to save money, gas, and time, finding the shortest distance you need to travel would be of interest. How can this be found?

The mathematical branch of graph theory provides the setup and model for a question such as this. Recall the probability trees and probability diagrams that appeared in this chapter and also in Chapter 8; here, the vertices for our *graph* will be the cities themselves and, since it is definitely possible to reach any city from any other city, there will be an edge between each possible pair of vertices (this type of graph is called a *complete* graph). We label each edge with a number indicating the distance between the cities represented by the two vertices for that edge (distance here will be "as the crow flies" or assuming that we have a private plane at our disposal that can fly from one city to another along a great circle of the earth). This type of graph to a mathematician is a weighted undirected complete graph (undirected because travel between cities can be in either direction).

TABLE 11.7: City Distances

	DE	MD	PA	VA	WV
DE	–	53.88	104.87	153.39	333.71
MD	53.88	–	91.79	111.84	280.18
PA	104.87	91.79	–	191.21	286.81
VA	153.39	111.84	191.21	–	234.15
WV	333.71	280.18	286.81	234.15	–

For our example, we will restrict ourselves to just five state capitals. Besides Harrisburg, PA, we'll use Dover, DE, Annapolis, MD, Richmond, VA, and Charleston, WV. The direct distances, in miles, between those cities is given in Table 11.7 and, to save space, only the state's initials are included. Note that the table is symmetric in that it can be flipped over the line created by the dashed entries since, for instance, the distance between Annapolis and Harrisburg is the same whether going from Maryland to Pennsylvania or from Pennsylvania to Maryland.

The graph version of this data appears in Figure 11.8. The vertices for the graph are the state capitals, again labeled by their respective states, while the distances between the cities are labeled along the edges. Finding the shortest distance, when visiting all five state state capitals, that starts and ends in Pennsylvania's state capital, is called finding a *Hamiltonian cycle* by graph theorists. For the graph of Figure 11.8, we can do this by brute-force; in other words, we can simply try *all* possible ways of visiting the five cities starting with PA. With the starting vertex fixed, there are only four choices here for which city to visit next, followed by only three choices. Using the permutations of Definition 4.4, we can quickly see that there are $4! = 24$ different cycles for our scenario, or $4!/2 = 12$ essentially different cycles; the cycle PA, WV, VA, MD, DE, PA is the same as the cycle PA, DE, MD, VA, WV, PA for purposes of computing distance since those distances between cities are the same in both directions. You should take a moment to write down all 24 (or 12) possible cycles along with their total distances to verify that the minimal distance of 791.55 miles is achieved by the example cycle mentioned.

For larger "tours" involving a lot more cities, the brute-force method *works* but is, however, *extremely* slow. For a graph with 11 vertices, there are 3,628,800 possible cycles to consider and that number grows to 39,916,800 cycles when just one more vertex is added (bringing the total to 12 vertices)![4] Finding a Hamiltonian cycle for a complete graph is

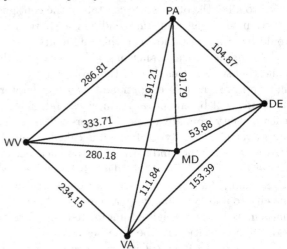

FIGURE 11.8: Graph of 5 State Capitals

among the class of problems in mathematics and computer science called "NP-complete" meaning that, while verifying a proposed solution's correctness is easy, *finding* such a solution in the first place takes a lot of time as the number of vertices grows larger and larger. In particular, the amount of time to solve even a medium-sized Hamiltonian cycle problem can take, using the best known algorithms today, millions or billions of years on a computer!

---

[4]The number of cycles given for 10 and 11 vertices are before dividing by two and therefore do not take into account the symmetry of tours; dividing by two, however, does not change how fast the number of cycles grows!

The fastest computer currently on the planet would take 20 years to perform a brute-force computation for a graph with 25 vertices and takes over 280,000,000 years when just 30 vertices are involved. Of course, there could be a much better algorithm for finding the exact solution to a traveling salesman problem, but such an algorithm's discovery has eluded mathematicians and computer scientists to date.

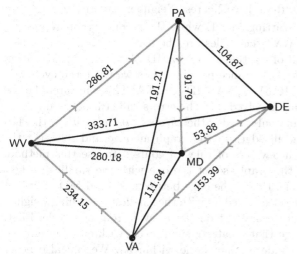

FIGURE 11.9: Nearest Neighbor Algorithm

A quicker method that solves such problems while introducing very little error is needed; we discuss two such methods here. First, the *nearest neighbor* algorithm focuses on choosing the next city based on its proximity to the current location. For our example, starting in PA, we look at all edges leaving the vertex labeled PA and choose the edge leading to MD because it is the location closest to PA; from MD, we consider all edges except the one leading back to PA and again take the shortest trip possible to our next city, ending up in DE. Verify that, by continuing, you obtain the cycle shown in Figure 11.9 that has length 820.02 miles. This very quick algorithm produced a cycle that is only about 3.6% above optimal; for a problem with many more cities, the simplicity of this method, combined with a low error, outweighs the time required to apply brute force!

Even better, we can use the *repetitive* nearest neighbor algorithm to possibly find improvements! This method considers each vertex as a possible starting point and applies the standard nearest neighbor algorithm; the resulting cycles can be adjusted afterwards to start and end in the desired vertex (such as PA).

The results, in terms of cycles and their distances, of applying the repetitive nearest neighbor algorithm to our example problem are shown in Table 11.8, while the resulting graphs are depicted in Figure 11.10; the first

TABLE 11.8: Repetitive Nearest Neighbor Results

Cycle	Distance
PA → MD → DE → VA → WV → PA	820.02
DE → MD → PA → VA → WV → DE	904.74
VA → MD → DE → PA → WV → VA	791.55
WV → VA → MD → DE → PA → WV	791.55
MD → DE → PA → VA → WV → MD	864.29

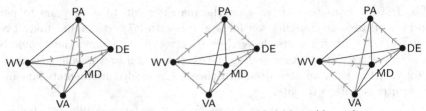

FIGURE 11.10: Repetitive Nearest Neighbor Algorithm

graph there comes from starting in DE while the second graph represents
starting in either VA or WV (the cycles are the same for these two cities).
The last graph is the result of starting at vertex MD. Note that this algorithm
successfully finds the shortest distance possible when we write the cycle VA,
MD, DE, PA, WV, VA as PA, WV, VA, MD, DE, PA! This will rarely be the
case when more vertices are involved, but the errors will still be small.

Another algorithm commonly used is the *cheapest link algorithm*; this has
the benefit of considering all edges in the graph at once and selecting the
smallest possible, within a few sets of rules of course. To use this method,
first, among all edges of the graph, select the one with the smallest weight
(distance). For our example, this is the edge between DE and MD of weight
53.88; we then continue by selecting edges with the smallest remaining weights,
subject to two rules, until we are just one edge away from our cycle. First,
we can never select an edge that creates a small cycle within our graph (no
Hamiltonian cycle will include a small cycle within it). We can also never
choose an edge that would be the third edge selected for any specific vertex
(once we arrive at and then leave a city, we will never return). Once all but
the last edge for our cycle has been selected, pick the edge that completes the
cycle.

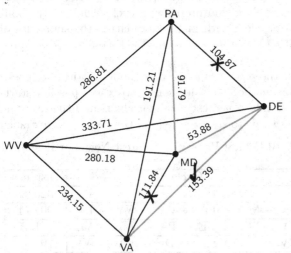

FIGURE 11.11: Cheapest Link Algorithm

Figure 11.11 shows
the result, part way into
the implementation of
the cheapest link al-
gorithm as it applies
to our example. After
selecting the DE-MD
edge of weight 53.88,
we choose the MD-PA
edge of weight 91.79. We
would *want* to choose
the edge between DE
and PA of weight 104.87
since it is the next
smallest, but this would
create a small cycle and
is therefore eliminated

(denoted by placing an X along that edge in our figure). Similarly, the edge

MD-VA of weight 111.84, while the next smallest weight, also receives an X but this time it's because of the three-edge rule. We are forced to select the DE-VA edge which is followed by choosing the VA-WV edge. Finally, since all but one edge remains for our tour, we must pick the PA-WV edge to finish the complete cycle. This algorithm gives us the cycle PA, MD, DE, VA, WV, PA which we saw earlier as the same given by the nearest neighbor algorithm. Note that both methods will not always give the same cycle, but will be similar in error compared to the optimal solution.

Very recently, evolutionary algorithms have been applied to several problems in mathematics with great success; in particular, for the traveling salesman problem explored in this section, one method has been very popular. Given a starting generation (set) of cycles (possibly randomly generated), "offspring" are created by randomly swapping pairs of cities appearing in the cycles of the first generation. Those "offspring" cycles that give a shorter distance than its parent survive while the offspring with larger distances die out (are removed). This process continues for many generations which yields cycles with distances within 1% or less of the optimal solution!

Research into ant colony behavior has also resulted in very good approximate solutions to traveling salesman problems. While the movement of a single ant away from its colony may seem random, observing many ants from the same colony foraging for food highlights how such an algorithm could be developed. After many seemingly random movements, the ants settle into one particular path to get from point A to point B; as these ants travel and explore, they lay down pheromone trails. Eventually, because ants prefer to do as little movement as possible, those trails that are frequently traveled find themselves rich in ant pheromones and will be preferred to those with less pheromones. We can model this using computers by programming artificial ants; each city becomes a colony of ants with ants of one colony wanting to visit another. Movement at first is mostly random with each ant having a preference for nearby cities. After several iterations of releasing the ants, the virtual pheromone trails give the approximate solution since those trails yielding shorter distances are traveled more, picking up more and more pheromones, and those trails with larger distances become less traveled, with those pheromone trails evaporating and becoming much less desired. These two "biological" solutions to the traveling salesman problem can give extremely good results!

A fuller version of the traveling salesman problem, which involved 49 cities from the United States (one from each state, not necessarily the capital, and Washington, DC), was first solved by mathematicians Dantzig, Fulkerson, and Johnson in 1954 (see [15] in the Bibliography); while their method does not use brute force, it does produce a candidate cycle for these 49 vertices and then proves using linear programming techniques that no cycle could have a shorter length. If you're curious, that length ended up at 12,345 miles for their selection of 49 cities. It is important to note, however, that even these techniques fail when the number of vertices grows bigger and bigger. Currently there is a large monetary prize for solving the "world traveling" problem which consists of over 1 million cities across the world; approximate solutions that are less than 1% off of the optimal solution are perfectly fine in practice for actually following the path, but a lot of time (and research) is needed to find those optimal solutions for large problems. Moreover, even as large problems are solved, larger problems will always remain. For more on graph theory, check out a very friendly introduction to the subject by my friend and colleague Karin Saoub in [44].

11.7   Random Walks and Generating Functions

The binomial distribution has appeared all throughout this book but was featured prominently in Chapter 10 without much explicit reference to it. When a particular situation allows for its use, we should celebrate; the power of this distribution lies in its simplicity, relative ease of calculation, and knowledge of how probabilities behave in the long run. Here we look at two more topics related to the binomial distribution and, as usual, we look at flipping coins as our primary motivation.

Random Walks

Imagine walking in a forest, following trails left by those who traveled there before you. Rather than taking the "road less traveled" as Robert Frost might, when we reach a fork in the road, we flip a coin to determine whether we should take the left or right branch. Figure 11.12 shows the result of a hypothetical "walk" through a "forest"

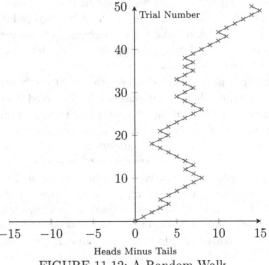

FIGURE 11.12: A Random Walk

using this method. Using a computer to simulate flipping coins, this graph shows 50 flips where each flip is spaced evenly along the y axis. The horizontal axis here keeps track of the difference between the number of heads and number of tails; in particular, each time we flip a coin, a head will move us one unit in the horizontal direction and one unit upwards, so at the start we move from the neutral position of $(0,0)$ to $(1,1)$ since a head was observed. Note that this particular sequence of random flips had four heads at the start, so it's not until we've reached the point $(4,4)$ that seeing a tail moves us to

$(3, 5)$ since a tail will move us one unit to the *left* along the horizontal axis while still moving one unit up.

This *random walk* graph *always* grows by one unit in the positive vertical direction since this axis keeps track of the number of flips that have been performed; the graph moves back and forth in the horizontal direction based on the result of the flip. Note that the particular example of Figure 11.12 appears somewhat unbalanced; after the 50 flips, we have already seen about 14 more heads than tails (with 32 heads and 18 tails to be exact). Is this a problem? Since we know that seeing this particular result, assuming a fair coin, happens with probability

$$\text{Probability 32 Heads, 50 Flips} = \binom{50}{32}\left(\frac{1}{2}\right)^{32}\left(\frac{1}{2}\right)^{18} \approx 0.0160$$

using Definition 4.1. Is this too low? If we revisit the idea of spread, first discussed in Chapter 4, we note that the expected number of heads is $n \cdot p$ where $n = 50$ and $p = 0.50$, for an expected number of 25; the standard deviation is $\sqrt{n \cdot p \cdot q}$ where q also equals 0.50, giving us 3.5355. About 95% of the time we would expect to see anywhere from 17.929 to 32.071 heads, so our result seems fine in that sense. More testing using techniques from statistics could help determine if this coin (or, in this case, random number generator) is actually fair, but that's for another course.

What behavior would we *expect* for this random walk graph? Assuming an even number, $2n$, of flips, we can see fairly easily that the expected number of heads is n and the expected number of tails is also n, for an expected *difference* of zero. For a fair coin, the random walk should mainly be centered around the vertical axis in the graph. While arbitrarily large deviations from this center axis are possible and even common, this particular walk, as should any with a fair coin, will "live" close to this axis while stepping away from it, on either side, quite often. For the curious, if you spent time reading the discussion of randomness that appears in Section 11.1, the streak lengths and switch numbers for this sequence do fit into our expectations!

These random walk graphs can provide a neat view of what happens for any binomial distribution situation; as a final note before moving onto the next topic, as you might guess, when other situations are considered that aren't balanced, such as rolling a die where $p = 1/6$ for any particular die value, staying centered around the center axis would *not* be the case. In fact, for this situation where, say, moving to the left happens when a 1 is rolled and, otherwise, we move to the right, we can determine the exact expected behavior. If $6n$ rolls are performed, we expect to see n rolls of a 1 and $5n$ rolls of something else; this difference would become $5n - n = 4n$ so, setting $y = n$, we should stay close to the line $x = 4y$, or $y = x/4$, much different than the even situation of coin flips!

Generating Functions

While we visited *generating functions* in Section 11.5, a binomial distribution situation provides another neat use of them to keep track of frequencies and variables. Again, consider the situation of flipping a coin where the probability of success, p, and probability of failure, q, are both the same. In this case, we use u and v to represent the *relative frequencies*, respectively, of success and failure. Here, since we expect the same number of successes as failures (since $p = q$), we set $u = 1$ and $v = 1$. Consider the function

$$f(x) = (ux^{-1} + vx)^n$$

which we call the generating function for n trials. What can this tell us? Let's consider

$$(f(x))^2 = (ux^{-1} + vx)^2 = (x^{-1} + x)^2 = 1 \cdot x^{-2} + 2 + 1 \cdot x^2$$

and discuss what each of these pieces means. First, since x^{-1} is keeping track of our successes, this squared generating function tells us there is one physical outcome that results in 2 successes; one outcome because the coefficient is one, and two successes because the power of x^{-1} is two (since $x^{-2} = (x^{-1})^2$). Similarly, with x keeping track of the failures, there is one physical outcome with 2 failures since the term $1 \cdot x^2$ appears in $(f(x))^2$. What does the middle term, 2, tell us? Since we could think of this term as $2 \cdot x^0$, the term indicates 2 physical outcomes where the number of successes and failures are the same.

This should be of no surprise; these coefficients directly correspond to the outcomes {HH, HT, TH, TT} when flipping a coin twice; two outcomes result in the same number of heads and tails, one outcome results in a pair of heads, and one outcome results in a pair of tails! Note that the cubed version of $f(x)$, given as

$$(f(x))^3 = (x^{-1} + x)^3 = 1 \cdot x^{-3} + 3 \cdot x^{-1} + 3x + x^3,$$

corresponds directly with the outcomes {HHH, HHT, HTH, THH, HTT, THT, TTH, TTT} for flipping a coin three times. There is one outcome with three successes, indicated by x^{-3}, three outcomes with one more success than failures because $3 \cdot x^{-1}$ is present, matching up with outcomes HHT, HTH, and THH, and the other terms have similar meanings.

Probabilities are also easy to determine using these generating functions. For the case of flipping three coins, the coefficients found in the expansion of $(f(x))^3$ add up to $1 + 3 + 3 + 1 = 8$, so to determine the probability of two heads (one more success than failures), we take the coefficient of x^{-1} and divide by 8, obtaining $3/8$ as that probability. This result should agree with the result you found in Exercise 1.1.

Of course, the strength of this method lies in the ability to work with any particular choices of p and q (as long as these two values add to one as we always require)! For instance, suppose $p = 1/3$ and $q = 2/3$; we must convert from p and q to u and v first, though, similar to how we converted odds with

Theorem 2.2. Since we expect to see, in this situation, twice as many failures as successes, the relative frequencies should be $u = 1$ and $v = 2$, giving us a new generating function

$$g(x) = (x^{-1} + 2x).$$

If, for example, this value of p modeled an unbalanced coin that landed on tails twice as often and we flipped this coin three times, then

$$(g(x))^3 = (x^{-1} + 2x)^2 = 1 \cdot x^{-3} + 6 \cdot x^{-1} + 12 \cdot x + 8 \cdot x^3$$

is the generating function for this new experiment. Very quickly we could determine the probability of getting one more success than failure as

$$\text{Probability 2 Successes, 1 Failure} = \frac{6}{1 + 6 + 12 + 8} = \frac{6}{27} \approx 0.2222,$$

much lower than that for a fair coin!

The reason that these generating functions work so well is hidden in the Binomial Theorem given earlier as Theorem 11.5. When expanding polynomials, the coefficients used happen to be binomial coefficients, the same ones that appear in the Binomial Distribution itself found in Definition 4.1; this neat connection allows us to abstract generating functions to be slightly more useful than just considering successes and failures!

Let us introduce a new parameter to a modified generating function $f(x)$ so that now

$$f(x) = (ux^{-1} + w + vx)^n$$

for n trials; this new variable w represents the relative frequency of a *tie*; that is, together, u, v, and w represent, respectively, the expected number of successes, failures, and ties in $u + v + w$ trials. For instance, suppose when rolling a five-sided die, a result of one or two is a success, a result of four or five is a failure, and a result of three results in a tie for that round. When rolling this die three times, we obtain the cubed generating function

$$(f(x))^3 = (2x^{-1} + 1 + 2x)^3 = 8 \cdot x^{-3} + 12 \cdot x^{-2} + 30 \cdot x^{-1} + 25 + 30 \cdot x + 12 \cdot x^2 + 8 \cdot x^3$$

which lets us now determine the probability again of having one more success than failures as

$$\text{Probability 2 Successes, 1 Failure} = \frac{30}{8 + 12 + 30 + 25 + 30 + 12 + 8} = \frac{30}{125}$$

which, at a value of 0.24, is much lower than the 3/8 we saw earlier in a fair situation where ties were not allowed (or didn't count toward the trial count). In this case, as well, we can also get two more successes than failures if the three trials consist of two success and one tie, a possibility that could not happen in prior circumstances.

Generating functions can be very helpful in many situations; their strength lies in doing *many* calculations at the same time by using variables to keep track of particular events. Note that never once did we substitute a value for x in any of these functions; the x is essentially a placeholder, and using the coefficients and powers of that placeholder becomes more important than the function evaluation you are probably more familiar with. As a final note, you might wonder why we used x^{-1} in our equation instead of a separate variable y (as we did back in the section 11.4). The simple answer here is that our exponents keep track of the *difference* between the number of successes and failures and while an $x \cdot y$ term *could* (and does) represent having one success and one failure, when talking about differences, the term $x \cdot x^{-1} = x^0$ is more useful since having a power of zero directly addresses differences; indeed, using x^{-1} allows simple function expansion to answer questions that we are already asking!

11.8 More Probability!

The main probability distributions covered in this book were the binomial and multinomial distributions covered in Chapters 4 and 9 along with our "counting" methods of Chapter 3. There are, as you might imagine, many different distributions we haven't had a chance to cover. This section discusses some of these. Note that we stick to *discrete* distributions here since we do not assume that you have a background in calculus. Continuous probability models take on values in a complete range of numbers, such as the interval $[0, 2]$ where *any* number in the range, including, for example, $\sqrt{2}$, could be an outcome to a process. Feel free to consult a standard probability textbook for continuous probability distribution models (and even more discrete distributions as well).

One distribution, related to our coverage of Chapter 3 and some of the lottery and bingo discussions in this chapter, is the hypergeometric distribution.

> **Definition 11.7** Suppose that we have a collection of $n = n_1 + n_2$ objects where n_1 of them belong to one type of object and n_2 to another; if a subset of r objects is selected at random from among the n total objects, the probability that exactly x of them belong to the first type of objects is
>
> $$\text{Probability } x \text{ Objects from Class 1} = \frac{\dbinom{n_1}{x} \cdot \dbinom{n_2}{r-x}}{\dbinom{n}{r}}$$
>
> and is called a *hypergeometric distribution*.

At this point, the formula should make sense; the total sample space here consists of all possible ways to choose r objects from the master collection of n items; to match our desire to have x items from the first class of objects, we need to choose x objects from that set of n_1 items and, therefore, the remaining $r - x$ objects must be chosen from the n_2 items that do *not* fit into that first class. Consider the following example that shows how the hypergeometric distribution can be used.

Example 11.2 Suppose that from a standard deck of playing cards, all of the spades are removed. This leaves, from among the remaining 39 cards, exactly 13 black cards and 26 red cards. What is the probability that, upon shuffling and choosing 9 random cards, we have 6 black cards? This question fits the setup of Definition 11.7 exactly! This hypergeometric distribution problem uses $n = 39$ where $n_1 = 13$ and $n_2 = 26$. With a random subset of size 9, we use $r = 9$ and want the probability of $x = 6$ black cards. Using that definition and formula, we obtain

$$\text{Probability 6 Black from 9 Cards} = \frac{\dbinom{13}{6} \cdot \dbinom{26}{3}}{\dbinom{39}{9}} = \frac{4,461,600}{211,915,132} \approx 0.0211$$

as that probability.

The next probability distribution to look at is called the *negative binomial distribution* which is closely related to the second part of the Streaks Theorem presented as Theorem 10.3. There we were looking for the probability, when presented with a binomial situation where the outcomes where either successes or failures, that k successes would be seen before l failures happened. The formula there was quite messy, and if we change the problem slightly, we can give a nice straightforward formula for this new distribution.

Definition 11.8 Consider an experiment for which the outcome is a success with probability p and a failure with probability $q = 1 - p$, repeated multiple times so that the probability of success and failure is the same for each repetition. The probability that success number r occurs on experiment x is given by

$$\text{Probability Success } r \text{ on Turn } x = \binom{x-1}{r-1} \cdot p^r q^{x-r}$$

and is called the *negative binomial distribution*.

Again, note that we can reason the formula from the scenario presented. For success number r to happen on experiment x, among the first $x - 1$ experiments, there must be exactly $r - 1$ failures, with probability p^{r-1}, so we choose $r - 1$ experiments from those $x - 1$ total on which the successes could occur; the rest of those $x - r$ experiments must be failures with probability q^{x-r}. We multiply by a final p to have success r happen on experiment x. Note here that the name stems from this distribution essentially being the "opposite" of the standard binomial distribution of Definition 4.1; there, we were concerned with looking at the number of successes in a *fixed* number of experiments, and here we are interested in the number of experiments needed to get a fixed number of *successes* instead! Let's apply this distribution to an example.

Example 11.3 Consider any of the standard 1:1 payout wagers at American roulette where $p = 18/38$ and $q = 20/38$ and let's determine the probability that we get our fifth win on the eighth wager. Using Definition 11.8 with $r = 5$ and $x = 8$, we obtain

$$\text{Probability Fifth Win on Wager 8} = \binom{7}{4} \cdot \left(\frac{18}{38}\right)^5 \left(\frac{20}{38}\right)^3 \approx 0.121689$$

as the desired probability.

For the curious, the probabilities associated to Example 11.3 for different numbers of wagers are given in Figure 11.13; note that this distribution is not quite symmetric (compared to the binomial distribution, which we know can be approximated by the bell-shaped curve of a normal distribution quite well); however, it does have a

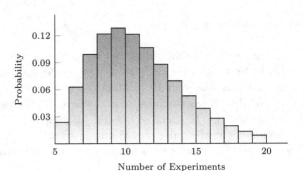

FIGURE 11.13: Experiments Needed for Win 5

nice shape with clearly defined peak. You might expect that the expected number of experiments needed to get that fifth win is somewhere between 8 and 12; indeed, it can be shown that the expected value for the negative binomial distribution of Definition 11.8 is r/p which, here, equals about 10.56 experiments.

As one final note, the special case of the negative binomial distribution when $r = 1$ is called the *geometric distribution*, named because its probability function is a geometric series that you might recall from advanced algebra in high school or a course in calculus. This distribution looks at the number of experiments needed to obtain the *first* success, with expected value $1/p$. Again, let's give this particular distribution a closer look via an example.

Example 11.4 For a single number bet at American roulette, the probability of success is $p = 1/38$ and the probability of failure is $q = 37/38$. Using our discussion above as a guide, the expected number of bets needed for us to see our first win on a chosen number is $1/p = 38$ which should come as no surprise. What is the probability that our first win actually happens on round 38? We use Definition 11.8 with $r = 1$ and $x = 38$ to determine that probability as

$$\text{Probability First Win on Round 38} = \binom{37}{0} \cdot \left(\frac{1}{38}\right)^1 \left(\frac{37}{38}\right)^{37} \approx 0.009810$$

or just slightly less than 1%, which seems very low for the average!

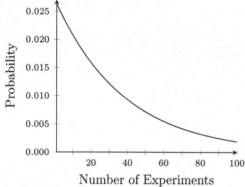

FIGURE 11.14: Experiments Needed for First Win

Since the binomial coefficient involved in a geometric distribution is always zero, as is the power on probability p, the formula shown in Example 11.4 can be written in a much nicer way, and in full generality, as the function

$$P(n) = \frac{37^{n-1}}{38^n}.$$

This function decreases *very* slowly as you can see in Figure 11.14. Since there is no guarantee that the first win will occur by the tenth, hundredth, or even millionth round, computing the expected value using Definition 2.4 here turns into an infinite sum; truncating this sum by using just the first 401 terms gets your expected value to within 0.01 of the real value of 38 though.

There are many other distributions out there for probability, and each generally has a specific type of question it was created to address; the examples above show that these two (or three, depending on how you count) have their own uses and benefits. Again, please consult a full probability textbook for more examples of probabilities distributions and their uses.

Appendices

A Probabilities with Infinity

The definition of probability given back in Definition 1.2 served us well throughout this book. In fact, the essence of that definition is very much correct, and the only case left out is when the sample space involved has an infinite number of possibilities. Thinking about infinity is never easy, and that notion indeed extends to probability.

If I asked you to select a random positive integer, which would you pick? Did you pick 10? How about 1,000,000? Or were you brazen enough to pick something off the wall such as $(5, 732, 652, 912)^{73}$? Whatever number you picked, the probability that, among all positive integers, your number would be randomly selected is zero. No, I'm not trying to fool you or be strange; the probability that any specific positive integer will be randomly selected given the sample space of *all* positive integers is exactly zero. It *has* to be zero. You might ask why, and so here is a discussion and explanation of how probabilities can be "strange" when infinity is involved.

A *uniform distribution* in probability gives each element of the sample space the same exact probability of being chosen. Rolling a (fair) six-sided results in a uniform distribution on the set $\{1, 2, 3, 4, 5, 6\}$ since each of the six numbers has a probability of being selected equal to $1/6$ when rolling the die. Switching to a twenty-sided die results in the same phenomenon where each of the numbers one through twenty has a $1/20$ chance of coming up when the die is rolled.

To formalize things a little more, let S be the set of elements (finite, for now) on which we are basing the uniform distribution, and denote by $|S|$ the number of elements of S. For example, with the standard six-sided die, we would have $S = \{1, 2, 3, 4, 5, 6\}$ and $|S| = 6$. As we just saw, if we have $|S|$ equally likely things, the probability of any specific one of them must be $1/|S|$. For the six-sided die, note that when we add up the probabilities for each possible outcome, we get

$$\underbrace{\frac{1}{6}}_{1} + \underbrace{\frac{1}{6}}_{2} + \underbrace{\frac{1}{6}}_{3} + \underbrace{\frac{1}{6}}_{4} + \underbrace{\frac{1}{6}}_{5} + \underbrace{\frac{1}{6}}_{6} = 6 \cdot \frac{1}{6} = 1$$

which is the same as 100%. We could also write this as

$$\left(\frac{1}{6}\right)\left(\underbrace{1}_{1}+\underbrace{1}_{2}+\underbrace{1}_{3}+\underbrace{1}_{4}+\underbrace{1}_{5}+\underbrace{1}_{6}\right)=\left(\frac{1}{6}\right)\cdot 6=1$$

by factoring out the 1/6 since each piece of the sum has the same probability.

This reasoning yields the following association between $|S|$ and p, since we can write

$$\underbrace{\left(\frac{1}{|S|}\right)}_{p}\cdot|S|=1.$$

Does this same reasoning apply when S is infinite? The short answer is *kind of*. While a good understanding of calculus is required to discuss infinity in any decent sense, for now it should be somewhat intuitive that *multiplying* infinity by any non-zero number still results in some form of infinity. If you think of "infinity" as a number that is growing and growing without bounds, then this should be clear. Taking any *fixed value*, say p, and multiplying by larger and larger numbers results in something that itself keeps getting larger and larger.

If S were an infinite set, say the set of all positive integers, and we continue working with a uniform distribution, we run into issues if $p \neq 0$. If instead the probability of any specific number were $0 < p < 1$, then the total probability when adding up all possibilities would be

$$p\cdot(1+1+\cdots 1\cdots)=p\cdot\infty=\infty>1$$

which cannot happen. Therefore, p must be identically zero. In fact, using $p = 1/|S|$ still works when $|S| = \infty$, as calculus says that $1/\infty = 0$ since dividing a constant non-zero number by larger and larger numbers produces values that get closer and closer (as close as you like) to zero.

This can be a bit upsetting! Indeed, if you look at the finite set of positive integers $S = \{1, 2, 3, \ldots, n\}$, the probability of selecting the number 1 at random is $p = 1/|S| = 1/n$ as you would expect. As the value of n increases, the probability of selecting the number *decreases* since the probability remains $1/n$. As n gets larger, $1/n$ gets as close to zero as desired, so in the limit, when we casually write $n = \infty$, the probability of selecting 1 at random becomes $1/\infty = 0$. Using notation borrowed from calculus, we would write

$$\lim_{n\to\infty}\frac{1}{n}=0$$

for which the left side is read as "the limit as n approaches infinity of $1/n$." Note that we should always be careful when quantities and ideas involving infinity are around, but here that *does* equal zero since it can be made arbitrarily as close to zero as desired.

Rather than determining the probability of the number 1 being selected, perhaps you want to know the probability of a number 1,000,000 or less being selected from among all positive integers? If $S = \{1, 2, 3, \ldots, n\}$, we can talk about this probability as long as $n \geq 1,000,000$ (otherwise the question doesn't really make too much sense, since we need $\{1, 2, \ldots, 1,000,000\}$ to be a subset of our sample space). For a fixed value of n, this probability is $1,000,000/n$ since there are 1,000,000 numbers that fit our event and a total of n possible things; we're still assuming a uniform distribution here. As n grows, however, we have a probability that again gets smaller and smaller, so in the limit we have

$$\lim_{n \to \infty} \frac{1,000,000}{n} = 0$$

since eventually, for large values of n, this fraction can still be made as small as desired. Yes, that means the probability of selecting any integer from one to a million from among *all* positive integers is zero.

Things change if we instead want the probability of selecting an *even* positive integer from among all of them. For any fixed value of n, we can consider our sample space to be $S = \{1, 2, 3, 4, \ldots, 2n - 1, 2n\}$ so that the set of *even* integers from S can be designated as $T = \{2, 4, \ldots, 2n\}$. Then from these finite sets, the total number of even integers is $|T| = n$ and the total size of the sample space is $|S| = 2n$, giving us a probability of choosing an even integer as $|T|/|S| = n/2n = 1/2$ as we might expect. However, letting n grow large in the limit now we have

$$\lim_{n \to \infty} \frac{n}{2n} = \lim_{n \to \infty} \frac{1}{2} = \frac{1}{2}$$

as we would like, since no matter how large n is, exactly half of the integers in the sample space are even. If we are truly using a uniform distribution, we would *want* this probability to be equal to one-half, and it is.

As we've seen, infinity can give interesting results as far as probabilities are concerned. Things can get even more interesting when continuous sets of numbers, such as all numbers between zero and two (including $1/7$ and $\sqrt{2}$), instead of discrete sets of numbers that we've used for S so far. That topic, though, I'll leave for a more traditional probability book that assumes prior study of calculus.

B   St. Petersburg Paradox

Suppose while walking through a casino, an attractive-looking dealer at a new table calls over you. "I've got an exciting new game using a simple fair coin for you to play and it can offer you limitless rewards for a small wager of just $10," he says. Being somewhat curious, you see no reason to not listen to him a little more.

The game being offered is quite simple. A coin is flipped, and should heads come up, you win \$1. Otherwise, the game continues and the coin is flipped again. Heads on this flip will net you \$2, and if tails is observed, the coin is flipped yet again. On this third flip, heads is worth \$4; as you can imagine, the game continues until heads is the outcome of the flip. On the fourth flip, it would net you \$8, on the fifth it would be \$16, and the possible reward doubles each time. If it takes until flip n to see the first heads once the game starts, your reward is 2^{n-1} dollars.

The question, as you might predict, concerns the expected value of this game. The probability of getting heads on the first flip is the usual $1/2$, and the probability of getting the first heads on the *second* flip is the same as getting a tails on the first flip and then the heads, for a probability of

$$\text{Tails, then Heads} = \underbrace{\left(\frac{1}{2}\right)}_{\text{tails}} \cdot \underbrace{\left(\frac{1}{2}\right)}_{\text{heads}} = \frac{1}{4} = \frac{1}{2^2}$$

as you should be able to compute quite quickly by now. Generalizing, the probability that you get the first heads on flip n is given (for $n \geq 1$) as

$$\text{First Heads on Flip } n = \underbrace{\left(\frac{1}{2}\right) \cdot \left(\frac{1}{2}\right) \cdots \left(\frac{1}{2}\right)}_{n-1 \text{ tails}} \cdot \underbrace{\left(\frac{1}{2}\right)}_{\text{heads}} = \frac{1}{2^n}$$

which will be quite handy for us to use in exploring expected value. Using Definition 2.4 as we have done many times, we get an expected value of

$$\text{Expected Value} = \underbrace{\left(\frac{1}{2}\right) \cdot \$1}_{\text{heads in 1}} + \underbrace{\left(\frac{1}{4}\right) \cdot \$2}_{\text{heads in 2}} + \underbrace{\left(\frac{1}{8}\right) \cdot \$4}_{\text{heads in 3}} + \cdots + \underbrace{\left(\frac{1}{2^n}\right) \cdot \$2^{n-1}}_{\text{heads in } n} + \cdots$$

where the extra dots at the end are included because, with this game, there is *no guarantee* that the game will stop after any fixed value k of flips! There are an *infinite* number of terms in this expected value calculation, and since each piece of the sum simplifies to $1/2$, we can get an expected value of (borrowing the summation notation of Definition 2.5)

$$\text{Expected Value} = \sum_{k=1}^{\infty} \left(\frac{1}{2^k}\right) \cdot \$2^{k-1} = \frac{1}{2} + \frac{1}{2} + \cdots + \frac{1}{2} + \cdots = \infty$$

where "equaling infinity" means that this quantity grows without bounds. How much should to be willing to pay to make this a fair game? Since the expected value grows without bounds, any finite "cost" to play this game (including the \$10 offered by our attractive dealer) makes this a *very* favorable game to the player, at least on the surface.

Would you pay \$10 to play this game? How about \$25? Since the expected value is infinite, simply relying on the idea that expected value should be our sole decision maker, would you be willing to pay \$1,000 to play? The answer to this last question for most people is no (and it's still no for most people even in the \$25 case); the idea of infinite expected value versus most people's unwillingness to play is why this scenario is a paradox (the specific name comes from the first exposition of the problem by Daniel Bernoulli, in 1738, in a St. Petersburg scientific journal; see [7] in the Bibliography for an English translation of Bernoulli's original paper).

There are a few ways to "deal" with the "paradox" involved in this situation that effectively make the naysayers (those unwilling to pay even \$25 to play) "correct" in the practical sense of the word.

The first involves using the idea of *utility*; in economics, utility theory is used to provide a more effective measure of how large sums of money can relate to smaller sums of money. While we will avoid introducing the full theory here, the intuitive reason that a utility function might be more useful than a raw amount of winnings is quickly demonstrated by noting that \$1 is effectively worth more to someone who currently has only \$100 to their name than \$1 is worth to someone who currently has \$100,000. For monetary amounts, a logarithmic function such as

$$U(w, T) = \ln\left(\frac{T + w}{T}\right) = \ln(T + w) - \ln(T)$$

is typically used, where T is how much money a person currently has and w is the amount of money to be won (or lost).

The $U(1, T)$ column of Table A.1 shows the utility value of \$1 for people who currently have particular amounts of money. Note that \$1 is definitely worth more to someone who only has \$100 than to the person with \$100,000 in the bank (by a factor of 1000)! To "solve" the St. Petersburg Paradox with utility theory, then, we replace the winnings for success on flip k (observing all tails

TABLE A.1: Values of U and EU

T	$U(1, T)$	$EU(T) = 0$
1	0.693147	\$1.67
2	0.405465	\$1.99
10	0.095310	\$2.88
100	0.009950	\$4.36
1000	0.000999	\$5.96
10000	0.000099	\$7.61
100000	0.000009	\$9.27
1000000	≈ 0	\$10.93

beforehand and seeing the first head on flip k); instead of the simple 2^{k-1} winnings from before, we will use $U(2^{k-1} - C, T)$ where C is the cost to play the game and T is the player's current net worth. This gives us an expected *utility* of

$$EU(T) = \sum_{k=1}^{\infty} \left(\frac{1}{2^k}\right) \cdot U(2^{k-1} - C, T)$$

which we consider a fair game when this expected utility is zero. The value of C (dependent on T) that yields an expected utility of zero is given in Table A.1

for various net worths. Note that not even the millionaire would (or should) be willing to pay \$25 for this game!

A second way to resolve the apparent paradox is to remember that the infinite expected value we saw earlier means that (paying \$0 to enter) the casino (or whoever the paying entity might be), on average, pays out an infinite amount of money to players of this game! Since this is very impossible and casinos (and other entities) have finite resources, at some point the ability of the casino to pay out the "required" winnings of 2^{k-1} is not possible. If the casino has T dollars available, then we modify the game slightly so that if the casino would be ordinarily required to pay out more than T, it instead pays out exactly T (and presumably goes out of business immediately). That is, when $2^{k-1} \geq T$, we are simply playing for T dollars. Algebra shows that this happens when $k \geq 1 + \log_2(T)$. The modified expected value computation becomes

$$\text{Modified Expected Value} = \sum_{k=1}^{L} \left(\frac{1}{2^k}\right) \cdot 2^{k-1} + \sum_{k=L+1}^{\infty} \left(\frac{1}{2^k}\right) \cdot T = \frac{L}{2} + \frac{T}{2^L}$$

where L is the cutoff point for which the winnings change from a power of two to the total opponent's net worth. Since $k \geq 1 + \log_2(T)$ means that the payoff will be T, values of k less than this still offer the exponential winnings; the largest possible value of L, then, is the integer part of $1 + \log_2(T)$. Infinite series results from calculus can be used to evaluate the right (infinite) part of the sum quickly.

TABLE A.2: Expected Value, Finite Resources

Opponent	T	EV
Friend	100	\$4.28
Millionaire	1000000	\$10.95
Billionaire	1000000000	\$15.93
Bill Gates (2013)	67000000000	\$18.98
World GDP (2012)	71830000000000	\$24.01

The expected values for different opponents is given in Table A.2; note that even if the opponent happens to have an amount of money equal to the entire world's output in 2012, the expected value is *still* less than \$25. Perhaps the naysayers from before really aren't naysayers at all! The probability of getting \$16 or less from this game is $31/32 \approx 0.96875$, so you have a small chance of reaching the part of the game that offers a payout of \$32 or more. Having a large streak of tails in a row definitely does happen, but keep in mind that the probability of getting a head on the first flip is *still* 1/2, so you have a 50% chance of walking away with \$1 right away. It would suck to pay \$25 to walk away with \$24 less, wouldn't it?

Other ways to eliminate the "paradox" exist, and every few years a new paper from a mathematician or economist surfaces offering another solution to this problem.

C Binomial Distribution versus Normal Distribution

As we have seen in Chapter 4 through Chapter 10, the binomial distribution is very important in the study of probability. Whenever we are concerned with the probability of results that can be divided into "success" or "failure" with more than one trial (round) involved, this distribution gives us that probability very quickly. Unfortunately, if the number of trials, n, is fairly large, the number of computations involved, using Definition 4.1, can also be very large. For example, for 100 trials, if we wanted to know the probability of 40 successes or fewer, we would have to compute the probability of zero successes, the probability of 1 success, the probability of 2 successes, continuing until we reach the probability of 40 successes, adding these probabilities together to obtain the final answer. Modern computers can usually make quick work of this but computations *can* get slow when even larger values of n are used.

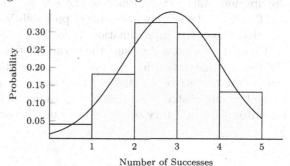

To solve this issue, many probability and statistics books will include tables of probabilities for common values of p, the probability of success, n, the number of trials, and k, the desired "maximum" number of successes (that is, the table values give the probability of zero

FIGURE A.1: Binomial versus Normal, $n = 5$

through k successes). This works well when the values of p are "nice" such as when $p = 0.25$, $p = 0.20$, or $p = 0.50$ (corresponding to randomly guessing on a multiple choice quiz with 4 answers per question, 5 answers per question, or 2 answers per question). Tables T.1 and T.2 that give probabilities for Liar's Dice, studied in Chapter 4, are examples of tables that these books include; note that Table T.2 indeed gives the probability of k through n successes, the complement of what books typically do have. However, if $p \approx 0.4737$ as it is for American roulette, tables are nowhere to be found!

Figure A.1 shows the binomial distribution for $n = 5$ and $p = 0.4737$ as a histogram with a normal distribution (bell curve) on top of it. Note that a normal distribution is given by the function

$$f(x) = \frac{1}{\sigma\sqrt{2\pi}}e^{-(1/2)\cdot((x-\mu)/\sigma)^2}$$

where μ is the mean and σ is the standard deviation; here, we use $\mu = n \cdot p$ and $\sigma = n \cdot p \cdot q$ where we last saw these equations for the mean and standard deviation of a binomial distribution in Theorems 4.2 and 4.3. While the normal distribution shown in Figure A.1 fits fairly well, it's not that good of an approximation.

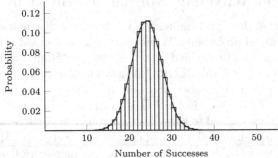

FIGURE A.2: Binomial versus Normal, $n = 50$

As n increases, as you can see in Figure A.2 when $n = 50$, the normal distribution approximates the binomial distribution quite well! You might wonder why this is important; after all, there is a unique normal distribution for each possible choice of μ and σ, giving us an infinite number of them. The key here is that *any* normal distribution can be transformed to the *standard* normal distribution (where $\mu = 0$ and $\sigma = 1$), so only *one* table of probabilities would be needed! It's this fact that makes the approximation useful.

Generally speaking, by using the normal distribution and its table of probabilities (or integration in calculus), as long as the interval $\mu \pm 3\sigma$ does *not* contain the values 0 or n, using $\mu = n \cdot p$ and $\sigma = n \cdot p \cdot q$ gives an approximation of the binomial distribution that is easy to use and very accurate.[1] For more, consult any introductory statistics book or a general probability book.

D   Matrix Multiplication Review

Matrix multiplication is not meant to be *hard* but it can be involved, especially when the size of the matrices involved gets somewhat large. First, suppose that A is an m-by-n matrix, meaning that it has m rows and n columns, and let B be an n-by-r matrix. We will use the symbol $a_{i,j}$ to represent the entry in row i and column j of matrix A, with the notation $b_{i,j}$ meaning the same, just for matrix B. Written out in the usual matrix form, we have

$$A = \begin{bmatrix} a_{1,1} & a_{1,2} & \cdots & a_{1,n} \\ a_{2,1} & a_{2,2} & \cdots & a_{2,n} \\ \vdots & \vdots & \ddots & \vdots \\ a_{m,1} & a_{m,2} & \cdots & a_{m,n} \end{bmatrix} \qquad B = \begin{bmatrix} b_{1,1} & b_{1,2} & \cdots & b_{1,r} \\ b_{2,1} & b_{2,2} & \cdots & b_{2,r} \\ \vdots & \vdots & \ddots & \vdots \\ b_{n,1} & b_{n,2} & \cdots & b_{n,r} \end{bmatrix}$$

---

[1]As a small technicality, there is a continuity correction that must be applied to the normal distribution to take care of the fact that we're using a continuous function to model a discrete distribution; this is done by changing x in the formula for the normal curve to $x - 0.5$ and is accounted for in Figures A.1 and A.2.

where the format looks the same; the only differences are in the variables used to represent the number of rows and columns. Note that in our example the number of columns for A is the same as the number of rows for B; because of the way we multiply matrices, if we wish to find $A \cdot B$, this relationship must always hold (we could not find the product $B \cdot A$ here because multiplying an n-by-r matrix by a m-by-n matrix does not exhibit this relationship; order matters)!

If we denote by $(AB)_{i,j}$ the entry in row i and column j for the product of A and B, then there's a nice formula for determining that value based on the values from A and B. While it may not seem pretty, the elements appearing in the formula are instructive; that formula is

$$(AB)_{i,j} = \sum_{k=1}^{n} a_{i,k} \cdot b_{k,j}$$

which indicates that, to get the (i, j) entry for the product, we take the values from row i of matrix A and multiply them, on a one-to-one matching basis, by the values from column j of matrix B. At this point, perhaps an example is in order. Suppose the matrices A and B are given as

$$A = \begin{bmatrix} 1 & 2 & 8 \\ 9 & 1 & 1 \\ 2 & 3 & 3 \\ 2 & 5 & 9 \end{bmatrix} \qquad B = \begin{bmatrix} -1 & 6 \\ 3 & 0 \\ 2 & -2 \end{bmatrix}$$

where we note that A is a 4-by-3 matrix and B is a 3-by-2 matrix so multiplication will make sense. To get $(AB)_{1,1}$, the first row first column entry of the product, we use the formula to get

$$(AB)_{1,1} = \sum_{k=1}^{3} a_{1,k} \cdot b_{k,1} = 1 \cdot (-1) + 2 \cdot 3 + 8 \cdot 2 = 21.$$

For those of you familiar with computing dot products, what we're doing to find $(AB)_{1,1}$ is taking the dot product of the first row of A and the first column of B as in

$$\langle 1, 2, 8 \rangle \cdot \langle -1, 3, 2 \rangle = 1 \cdot (-1) + 2 \cdot 3 + 8 \cdot 2 = 21.$$

How many rows and columns will be in $A \cdot B$? In general, multiplying an m-by-n matrix by an n-by-r matrix results in an m-by-r matrix; in our example, this means we will end up with a 4-by-2 matrix.

Thinking of matrix multiplication as a series of dot products makes the process very algorithmic. What I do is take the first row of matrix A and multiply[2] it by the first column of matrix B, resulting in the $(1, 1)$ entry of $A \cdot B$. I then continue using the first row of A by multiplying it by the second

---

[2]For the rest of this section of the Appendix, "multiply" is used for rows and columns meaning you should take the dot product of the row and column involved.

column of B to get $(AB)_{1,2}$, then by the third column of B to get $(AB)_{1,3}$, repeating this until the last column of B has been used. Then I take the second row of A and multiply by the first column of B to get $(AB)_{2,1}$, by the second column of B to get $(AB)_{2,2}$, and so on, using all columns of B until advancing to the next row of A. Once I've multiplied the last row of A by the last column of B, the process is over and we have $A \cdot B$.

For the example matrices A and B we saw earlier, we complete the dot products, getting

$$(AB)_{1,2} = \langle 1,2,8 \rangle \cdot \langle 6,0,-2 \rangle = 1 \cdot 6 + 2 \cdot 0 + 8 \cdot (-2) = -10$$
$$(AB)_{2,1} = \langle 9,1,1 \rangle \cdot \langle -1,3,2 \rangle = 9 \cdot (-1) + 1 \cdot 3 + 1 \cdot 2 = -4$$
$$(AB)_{2,2} = \langle 9,1,1 \rangle \cdot \langle 6,0,-2 \rangle = 9 \cdot 6 + 1 \cdot 0 + 1 \cdot (-2) = 52$$
$$(AB)_{3,1} = \langle 2,3,3 \rangle \cdot \langle -1,3,2 \rangle = 2 \cdot (-1) + 3 \cdot 3 + 3 \cdot 2 = 13$$
$$(AB)_{3,2} = \langle 2,3,3 \rangle \cdot \langle 6,0,-2 \rangle = 2 \cdot 6 + 3 \cdot 0 + 3 \cdot (-2) = 6$$
$$(AB)_{4,1} = \langle 2,5,9 \rangle \cdot \langle -1,3,2 \rangle = 2 \cdot (-1) + 5 \cdot 3 + 9 \cdot 2 = 31$$
$$(AB)_{4,2} = \langle 2,5,9 \rangle \cdot \langle 6,0,-2 \rangle = 2 \cdot 6 + 5 \cdot 0 + 9 \cdot (-2) = -6$$

which, in matrix form, results in

$$A \cdot B = \begin{bmatrix} 21 & -10 \\ -4 & 52 \\ 13 & 6 \\ 31 & -6 \end{bmatrix}.$$

One of the nice properties of square matrices that have the same number of rows as they do columns is that the matrices can be multiplied in *either* order; beware, however, since, generally, $A \cdot B \neq B \cdot A$ when multiplying matrices! The matrix multiplication in Chapter 6, though, can be fairly nice. Yes, repeated matrix multiplication is needed when taking powers (one multiplication when squaring, two when cubing, and so on), but since the same matrix is used, order doesn't matter. That *is* nice, isn't it?

Tables

This section of the book contains a few tables that are either too large or not necessary to include in their respective chapters. Several are referenced and used by the exercises appearing throughout this text.

You can find more details about these tables in the chapters that they support. The first two tables are useful probability tables for Liar's Dice; see the coverage of that game in Chapter 4 for more details. The Yahtzee Dice Roll table is explained in 4 as well, and its use is highlighted in Example 4.16 there. Following that appears the two Monopoly tables that are the correct steady-state vectors showing long-term behavior of where players' pieces end their turns (the probabilities here are slightly different than what appeared in 6 for the reasons explained there). The next four tables are used for Blackjack or Blackjack Switch and give the expected value of each possible player hand versus each possible dealer face up card; note that these expected values assume that you follow the basic strategy for each of those games, given in Chapter 8 as Tables 8.3, 8.4, and 8.5 for regular Blackjack and Tables 8.6, 8.7, and 8.8 for Blackjack Switch. Finally, the first table corresponding to Zombie Dice is Table T.10 which gives the appropriate dice configurations given a particular set of dice obtained from the cup (or leftover as footprints); see Example 9.33 for more details. While there, recall Example 9.35 which explains the use of the Decision Points table for Zombie Dice and how it can be used to play the game optimally. The list of tables appearing in this section is repeated here for convenience.

TABLE T.1: Liar's Dice Probability Table, Exact

n	μ	σ	1	2	3	4	5	6	7	8	9	10	11	12	13	14	15	16	17	18
1	0.33	0.47	0.33	–	–	–	–	–	–	–	–	–	–	–	–	–	–	–	–	–
2	0.66	0.66	0.44	0.11	–	–	–	–	–	–	–	–	–	–	–	–	–	–	–	–
3	1.00	0.81	0.44	0.22	0.03	–	–	–	–	–	–	–	–	–	–	–	–	–	–	–
4	1.33	0.94	0.39	0.29	0.09	0.01	–	–	–	–	–	–	–	–	–	–	–	–	–	–
5	1.66	1.05	0.32	0.32	0.16	0.04	0.00	–	–	–	–	–	–	–	–	–	–	–	–	–
6	2.00	1.15	0.26	0.32	0.21	0.08	0.01	0.00	–	–	–	–	–	–	–	–	–	–	–	–
7	2.33	1.24	0.20	0.30	0.25	0.12	0.03	0.00	0.00	–	–	–	–	–	–	–	–	–	–	–
8	2.66	1.33	0.15	0.27	0.27	0.17	0.06	0.01	0.00	0.00	–	–	–	–	–	–	–	–	–	–
9	3.00	1.41	0.11	0.23	0.27	0.20	0.10	0.03	0.00	0.00	0.00	–	–	–	–	–	–	–	–	–
10	3.33	1.49	0.08	0.19	0.26	0.22	0.13	0.05	0.01	0.00	0.00	0.00	–	–	–	–	–	–	–	–
11	3.66	1.56	0.06	0.15	0.23	0.23	0.16	0.08	0.02	0.00	0.00	0.00	0.00	–	–	–	–	–	–	–
12	4.00	1.63	0.04	0.12	0.21	0.23	0.19	0.11	0.04	0.01	0.00	0.00	0.00	0.00	–	–	–	–	–	–
13	4.33	1.69	0.03	0.10	0.18	0.22	0.2	0.13	0.06	0.02	0.00	0.00	0.00	0.00	0.00	–	–	–	–	–
14	4.66	1.76	0.02	0.07	0.15	0.21	0.21	0.16	0.09	0.04	0.01	0.00	0.00	0.00	0.00	0.00	–	–	–	–
15	5.00	1.82	0.01	0.05	0.12	0.19	0.21	0.17	0.11	0.05	0.02	0.01	0.00	0.00	0.00	0.00	0.00	–	–	–
16	5.33	1.88	0.01	0.04	0.10	0.17	0.20	0.19	0.13	0.07	0.03	0.01	0.00	0.00	0.00	0.00	0.00	0.00	–	–
17	5.66	1.94	0.00	0.03	0.08	0.15	0.19	0.19	0.15	0.09	0.04	0.01	0.00	0.00	0.00	0.00	0.00	0.00	0.00	–
18	6.00	2.00	0.00	0.02	0.06	0.12	0.18	0.19	0.16	0.11	0.06	0.02	0.01	0.00	0.00	0.00	0.00	0.00	0.00	0.00
19	6.33	2.05	0.00	0.01	0.05	0.10	0.16	0.19	0.17	0.13	0.08	0.04	0.01	0.00	0.00	0.00	0.00	0.00	0.00	0.00
20	6.66	2.10	0.00	0.01	0.04	0.09	0.14	0.18	0.18	0.14	0.09	0.05	0.02	0.00	0.00	0.00	0.00	0.00	0.00	0.00
21	7.00	2.16	0.00	0.01	0.03	0.07	0.12	0.16	0.18	0.15	0.11	0.06	0.03	0.01	0.00	0.00	0.00	0.00	0.00	0.00
22	7.33	2.21	0.00	0.01	0.02	0.06	0.10	0.16	0.17	0.16	0.12	0.08	0.04	0.02	0.00	0.00	0.00	0.00	0.00	0.00
23	7.66	2.26	0.00	0.00	0.01	0.04	0.09	0.14	0.17	0.17	0.14	0.09	0.05	0.02	0.01	0.00	0.00	0.00	0.00	0.00
24	8.00	2.30	0.00	0.00	0.01	0.03	0.07	0.12	0.16	0.17	0.15	0.11	0.07	0.03	0.01	0.00	0.00	0.00	0.00	0.00
25	8.33	2.35	0.00	0.00	0.01	0.03	0.06	0.10	0.14	0.16	0.15	0.12	0.08	0.05	0.02	0.01	0.00	0.00	0.00	0.00
26	8.66	2.40	0.00	0.00	0.00	0.02	0.05	0.09	0.13	0.16	0.16	0.13	0.09	0.06	0.03	0.01	0.00	0.00	0.00	0.00
27	9.00	2.44	0.00	0.00	0.00	0.01	0.04	0.08	0.12	0.15	0.16	0.14	0.11	0.07	0.04	0.02	0.00	0.00	0.00	0.00
28	9.33	2.49	0.00	0.00	0.01	0.01	0.03	0.06	0.10	0.14	0.15	0.15	0.12	0.08	0.05	0.02	0.01	0.00	0.00	0.00
29	9.66	2.53	0.00	0.00	0.00	0.01	0.02	0.05	0.09	0.13	0.15	0.15	0.13	0.09	0.06	0.03	0.01	0.01	0.00	0.00
30	10.00	2.58	0.00	0.00	0.00	0.00	0.02	0.04	0.08	0.11	0.14	0.15	0.13	0.11	0.07	0.04	0.02	0.01	0.00	0.00

Probability of EXACTLY This Number of 1s or Xs

TABLE T.2: Liar's Dice Probability Table, At Least

n	\multicolumn{18}{c}{Probability of AT LEAST This Number of 1s or Xs}																	
	1	2	3	4	5	6	7	8	9	10	11	12	13	14	15	16	17	18
1	0.33	-	-	-	-	-	-	-	-	-	-	-	-	-	-	-	-	-
2	0.55	0.11	-	-	-	-	-	-	-	-	-	-	-	-	-	-	-	-
3	0.70	0.25	0.03	-	-	-	-	-	-	-	-	-	-	-	-	-	-	-
4	0.80	0.40	0.11	0.01	-	-	-	-	-	-	-	-	-	-	-	-	-	-
5	0.86	0.53	0.20	0.04	0.00	-	-	-	-	-	-	-	-	-	-	-	-	-
6	0.91	0.64	0.31	0.10	0.01	0.00	-	-	-	-	-	-	-	-	-	-	-	-
7	0.94	0.73	0.42	0.17	0.04	0.00	0.00	-	-	-	-	-	-	-	-	-	-	-
8	0.96	0.80	0.53	0.25	0.08	0.01	0.00	0.00	-	-	-	-	-	-	-	-	-	-
9	0.97	0.85	0.62	0.34	0.14	0.04	0.00	0.00	0.00	-	-	-	-	-	-	-	-	-
10	0.98	0.89	0.70	0.44	0.21	0.07	0.01	0.00	0.00	0.00	-	-	-	-	-	-	-	-
11	0.98	0.92	0.76	0.52	0.28	0.12	0.03	0.00	0.00	0.00	0.00	-	-	-	-	-	-	-
12	0.99	0.94	0.81	0.60	0.36	0.17	0.06	0.01	0.00	0.00	0.00	0.00	-	-	-	-	-	-
13	0.99	0.96	0.86	0.67	0.44	0.24	0.10	0.03	0.00	0.00	0.00	0.00	0.00	-	-	-	-	-
14	0.99	0.97	0.89	0.73	0.52	0.31	0.14	0.05	0.01	0.00	0.00	0.00	0.00	0.00	-	-	-	-
15	0.99	0.98	0.92	0.79	0.59	0.38	0.20	0.08	0.03	0.01	0.00	0.00	0.00	0.00	0.00	-	-	-
16	0.99	0.98	0.94	0.83	0.66	0.45	0.26	0.12	0.04	0.01	0.00	0.00	0.00	0.00	0.00	0.00	-	-
17	0.99	0.99	0.95	0.86	0.71	0.52	0.32	0.17	0.07	0.02	0.00	0.00	0.00	0.00	0.00	0.00	0.00	-
18	0.99	0.99	0.96	0.89	0.76	0.58	0.39	0.22	0.10	0.04	0.01	0.00	0.00	0.00	0.00	0.00	0.00	0.00
19	0.99	0.99	0.97	0.92	0.81	0.64	0.45	0.27	0.14	0.06	0.02	0.00	0.00	0.00	0.00	0.00	0.00	0.00
20	0.99	0.99	0.98	0.93	0.84	0.70	0.52	0.33	0.19	0.09	0.03	0.01	0.00	0.00	0.00	0.00	0.00	0.00
21	0.99	0.99	0.98	0.95	0.87	0.75	0.58	0.39	0.23	0.12	0.05	0.02	0.00	0.00	0.00	0.00	0.00	0.00
22	0.99	0.99	0.99	0.96	0.90	0.79	0.63	0.45	0.29	0.16	0.07	0.03	0.01	0.00	0.00	0.00	0.00	0.00
23	0.99	0.99	0.99	0.97	0.92	0.83	0.68	0.51	0.34	0.20	0.10	0.04	0.01	0.00	0.00	0.00	0.00	0.00
24	0.99	0.99	0.99	0.98	0.94	0.86	0.73	0.57	0.40	0.25	0.14	0.06	0.02	0.01	0.00	0.00	0.00	0.00
25	0.99	0.99	0.99	0.98	0.95	0.88	0.77	0.62	0.46	0.30	0.17	0.09	0.04	0.01	0.00	0.00	0.00	0.00
26	0.99	0.99	0.99	0.98	0.96	0.90	0.81	0.67	0.51	0.35	0.22	0.12	0.05	0.02	0.00	0.00	0.00	0.00
27	0.99	0.99	0.99	0.99	0.97	0.92	0.84	0.72	0.57	0.41	0.26	0.15	0.07	0.03	0.01	0.00	0.00	0.00
28	0.99	0.99	0.99	0.99	0.97	0.94	0.87	0.76	0.62	0.46	0.31	0.19	0.10	0.05	0.02	0.00	0.00	0.00
29	0.99	0.99	0.99	0.99	0.98	0.95	0.89	0.80	0.67	0.51	0.36	0.23	0.13	0.06	0.03	0.01	0.00	0.00
30	0.99	0.99	0.99	0.99	0.98	0.96	0.91	0.83	0.71	0.56	0.41	0.27	0.16	0.08	0.04	0.01	0.00	0.00

TABLE T.3: Yahtzee Dice Roll Advice

Ending Dice	Categories	Ending Dice	Categories
⚀⚀⚀⚀⚀	Y6FTC12H3L45S	⚀⚀⚂⚃⚅	C1F632HY4TL5S
⚀⚀⚀⚀⚁	6FTC12HY35L4S	⚀⚀⚂⚄⚄	2C1F6HY4T3L5S
⚀⚀⚀⚀⚂	6FTC12HY43L5S	⚀⚀⚂⚄⚅	1C2F6HY4T3L5S
⚀⚀⚀⚀⚃	6FTC132HYL45S	⚀⚀⚂⚅⚅	2CF62HY4T3L5S
⚀⚀⚀⚀⚄	6FTC12HY3L45S	⚀⚀⚃⚃⚃	H3TC1F62YL45S
⚀⚀⚀⚀⚅	6FTC12HY3L45S	⚀⚀⚃⚃⚄	C312F6HYTL45S
⚀⚀⚀⚁⚁	TH6C1F52Y3L4S	⚀⚀⚃⚃⚅	31CF62HYTL45S
⚀⚀⚀⚁⚂	T6C1F2HY435LS	⚀⚀⚃⚄⚄	2C1F63HYTL45S
⚀⚀⚀⚁⚃	T6C1F32HY5L4S	⚀⚀⚃⚄⚅	1C2F63HYTL45S
⚀⚀⚀⚁⚄	T6C12FHY35L4S	⚀⚀⚃⚅⚅	1CF632HYTL45S
⚀⚀⚀⚁⚅	6TC1F2HY35L4S	⚀⚀⚄⚄⚄	H2TC1F6Y3L45S
⚀⚀⚀⚂⚂	HT6C41F2Y3L5S	⚀⚀⚄⚄⚅	21CF6HYT3L45S
⚀⚀⚀⚂⚃	T6C1F32HY4L5S	⚀⚀⚄⚅⚅	12CF6HYT3L45S
⚀⚀⚀⚂⚄	6TC12FHY43L5S	⚀⚀⚅⚅⚅	H1TFC62Y3L45S
⚀⚀⚀⚂⚅	6TC1F2HY43L5S	⚀⚁⚁⚁⚁	5FTC12HY36L4S
⚀⚀⚀⚃⚃	H6TC31F2YL45S	⚀⚁⚁⚁⚂	T5C1F2HY436L5
⚀⚀⚀⚃⚄	6TC12F3HYL45S	⚀⚁⚁⚁⚃	5TC1F32HY6L4S
⚀⚀⚀⚃⚅	6TC1F32HYL45S	⚀⚁⚁⚁⚄	5TC12FHY36L45
⚀⚀⚀⚄⚄	H6TC21FY3L45S	⚀⚁⚁⚁⚅	5TC1F2HY36L4S
⚀⚀⚀⚄⚅	6TC12FHY3L45S	⚀⚁⚁⚂⚂	C41F52HYT36LS
⚀⚀⚀⚅⚅	H6T1CF2Y3L45S	⚀⚁⚁⚂⚃	SC1F532HY4T6L
⚀⚀⚁⚁⚁	HT5C1F62Y3L4S	⚀⚁⚁⚂⚄	C12F5HY4T36LS
⚀⚀⚁⚁⚂	C1F562HY4T3LS	⚀⚁⚁⚂⚅	C1F52HY4T36LS
⚀⚀⚁⚁⚃	C1F5632HYTL4S	⚀⚁⚁⚃⚃	C31F52HYT6L4S
⚀⚀⚁⚁⚄	C12F56HYT3L4S	⚀⚁⚁⚃⚄	C12F53HYT6L4S
⚀⚀⚁⚁⚅	C1F562HYT3L4S	⚀⚁⚁⚃⚅	C1F532HYT6L4S
⚀⚀⚁⚂⚂	C41F62HYT35LS	⚀⚁⚁⚄⚄	2C1F5HYT36L4S
⚀⚀⚁⚂⚃	SC1F632HY4T5L	⚀⚁⚁⚄⚅	1C2F5HYT36L4S
⚀⚀⚁⚂⚄	C12F6HY4T35LS	⚀⚁⚁⚅⚅	1CF52HYT36L45
⚀⚀⚁⚂⚅	C1F62HY4T35LS	⚀⚁⚂⚂⚂	4TC1F2HY356LS
⚀⚀⚁⚃⚃	C31F62HYT5L4S	⚀⚁⚂⚂⚃	SC41F32HYT56L
⚀⚀⚁⚃⚄	C12F63HYT5L4S	⚀⚁⚂⚂⚄	C412FHYT356LS
⚀⚀⚁⚃⚅	C1F632HYT5L4S	⚀⚁⚂⚂⚅	C14F2HYT356LS
⚀⚀⚁⚄⚄	2C1F6HYT35L4S	⚀⚁⚂⚃⚃	SC31F2HY4T56L
⚀⚀⚁⚄⚅	C12F6HYT35L4S	⚀⚁⚂⚃⚄	LSC12F3HY4T56
⚀⚀⚁⚅⚅	2CF62HYT35L4S	⚀⚁⚂⚃⚅	S1CF32HY4T56L
⚀⚀⚂⚂⚂	HT4C1F62Y3L5S	⚀⚁⚂⚄⚄	2C1FHY4T356LS
⚀⚀⚂⚂⚃	C41F632HYTL5S	⚀⚁⚂⚄⚅	1C2FHY4T356LS
⚀⚀⚂⚂⚄	C412F6HYT3L5S	⚀⚁⚂⚅⚅	1CF2HY4T356LS
⚀⚀⚂⚂⚅	C14F62HYT3L5S	⚀⚁⚃⚃⚃	3TC1F2HY56L4S
⚀⚀⚂⚃⚃	C31F62HY4TL5S	⚀⚁⚃⚃⚄	3C12FHYT56L45
⚀⚀⚂⚃⚄	C12F63HY4TL5S	⚀⚁⚃⚃⚅	31CF2HYT56L4S

Ending Dice	Categories	Ending Dice	Categories
🎲🎲🎲🎲🎲	2C1F3HYT56L4S	🎲🎲🎲🎲🎲	5FTC132HYL4S6
🎲🎲🎲🎲🎲	1C2F3HYT56L4S	🎲🎲🎲🎲🎲	5FTC12HY3L4S6
🎲🎲🎲🎲🎲	1CF32HYT56L4S	🎲🎲🎲🎲🎲	5FTC12HY3L4S6
🎲🎲🎲🎲🎲	2TC1FHY356L4S	🎲🎲🎲🎲🎲	H5TC41F2Y3LS6
🎲🎲🎲🎲🎲	21CFHYT356L4S	🎲🎲🎲🎲🎲	5TC1F32HY4LS6
🎲🎲🎲🎲🎲	12FCHYT356L4S	🎲🎲🎲🎲🎲	5TC12FHY43LS6
🎲🎲🎲🎲🎲	1TFC2HY356L4S	🎲🎲🎲🎲🎲	5TC1F2HY43LS6
🎲🎲🎲🎲🎲	4FTC12HY36L5S	🎲🎲🎲🎲🎲	H5TC31F2YL4S6
🎲🎲🎲🎲🎲	4TC1F32HY6L5S	🎲🎲🎲🎲🎲	5TC12F3HYL4S6
🎲🎲🎲🎲🎲	4TC12FHY36L5S	🎲🎲🎲🎲🎲	5T1CF32HYL4S6
🎲🎲🎲🎲🎲	4T1CF2HY36L5S	🎲🎲🎲🎲🎲	H5T2C1FY3L4S6
🎲🎲🎲🎲🎲	C341F2HYT6L5S	🎲🎲🎲🎲🎲	5T1C2FHY3L4S6
🎲🎲🎲🎲🎲	C412F3HYT6L5S	🎲🎲🎲🎲🎲	H5T1CF2Y3L4S6
🎲🎲🎲🎲🎲	1C4F32HYT6L5S	🎲🎲🎲🎲🎲	H4TC1F52Y3LS6
🎲🎲🎲🎲🎲	2C41FHYT36L5S	🎲🎲🎲🎲🎲	C41F532HYTLS6
🎲🎲🎲🎲🎲	14C2FHYT36L5S	🎲🎲🎲🎲🎲	C412F5HYT3LS6
🎲🎲🎲🎲🎲	14CF2HYT36L5S	🎲🎲🎲🎲🎲	1C4F52HYT3LS6
🎲🎲🎲🎲🎲	3TC1F2HY46L5S	🎲🎲🎲🎲🎲	C31F52HY4TLS6
🎲🎲🎲🎲🎲	3C12FHY4T6L5S	🎲🎲🎲🎲🎲	SC12F53HY4TL6
🎲🎲🎲🎲🎲	31CF2HY4T6L5S	🎲🎲🎲🎲🎲	1CF532HY4TLS6
🎲🎲🎲🎲🎲	2C1F3HY4T6L5S	🎲🎲🎲🎲🎲	2C1F5HY4T3LS6
🎲🎲🎲🎲🎲	S12CF3HY4T6L5	🎲🎲🎲🎲🎲	1C2F5HY4T3LS6
🎲🎲🎲🎲🎲	1FC32HY4T6L5S	🎲🎲🎲🎲🎲	1CF52HY4T3LS6
🎲🎲🎲🎲🎲	2T1CFHY436L5S	🎲🎲🎲🎲🎲	H3TC1F52YL4S6
🎲🎲🎲🎲🎲	21FCHY4T36L5S	🎲🎲🎲🎲🎲	3C12F5HYTL4S6
🎲🎲🎲🎲🎲	12FCHY4T36L5S	🎲🎲🎲🎲🎲	31CF52HYTL4S6
🎲🎲🎲🎲🎲	1TFC2HY436L5S	🎲🎲🎲🎲🎲	2C1F53HYTL4S6
🎲🎲🎲🎲🎲	3FTC12HY6L45S	🎲🎲🎲🎲🎲	12CF53HYTL4S6
🎲🎲🎲🎲🎲	3TC12FHY6L45S	🎲🎲🎲🎲🎲	1FC532HYTL4S6
🎲🎲🎲🎲🎲	3T1CF2HY6L45S	🎲🎲🎲🎲🎲	H2T1CF5Y3L4S6
🎲🎲🎲🎲🎲	231CFHYT6L45S	🎲🎲🎲🎲🎲	21FC5HYT3L4S6
🎲🎲🎲🎲🎲	312FCHYT6L45S	🎲🎲🎲🎲🎲	12F5CHYT3L4S6
🎲🎲🎲🎲🎲	13FC2HYT6L45S	🎲🎲🎲🎲🎲	H1TF5C2Y3L4S6
🎲🎲🎲🎲🎲	2T1FC3HY6L45S	🎲🎲🎲🎲🎲	4FTC12HY35LS6
🎲🎲🎲🎲🎲	21FC3HYT6L45S	🎲🎲🎲🎲🎲	4TC1F32HY5LS6
🎲🎲🎲🎲🎲	12FC3HYT6L45S	🎲🎲🎲🎲🎲	4TC12FHY35LS6
🎲🎲🎲🎲🎲	1FTC32HY6L45S	🎲🎲🎲🎲🎲	4T1CF2HY35LS6
🎲🎲🎲🎲🎲	2FT1CHY36L45S	🎲🎲🎲🎲🎲	3C41F2HYT5LS6
🎲🎲🎲🎲🎲	21TFCHY36L45S	🎲🎲🎲🎲🎲	SC412F3HYT5L6
🎲🎲🎲🎲🎲	12FCHYT36L45S	🎲🎲🎲🎲🎲	14CF32HYT5LS6
🎲🎲🎲🎲🎲	1T2FCHY36L45S	🎲🎲🎲🎲🎲	24C1FHYT35LS6
🎲🎲🎲🎲🎲	1FT2CHY36L45S	🎲🎲🎲🎲🎲	142CFHYT35LS6
🎲🎲🎲🎲🎲	Y5FTC12H3L4S6	🎲🎲🎲🎲🎲	14FC2HYT35LS6
🎲🎲🎲🎲🎲	5FTC12HY43LS6	🎲🎲🎲🎲🎲	3TC1F2HY45LS6

Ending Dice	Categories	Ending Dice	Categories
⚅ ⚅ ⚃ ⚃ ⚀	S3C12FHY4T5L6	⚅ ⚅ ⚀ ⚃ ⚀	H2T41FCY3L5S6
⚅ ⚅ ⚃ ⚃ ⚀	31CF2HY4T5LS6	⚅ ⚅ ⚀ ⚃ ⚀	214FCHYT3L5S6
⚅ ⚅ ⚃ ⚀ ⚀	S21CF3HY4T5L6	⚅ ⚅ ⚀ ⚀ ⚀	142FCHYT3L5S6
⚅ ⚅ ⚃ ⚀ ⚀	LS12FC3HY4T56	⚅ ⚅ ⚀ ⚀ ⚀	H14TFC2Y3L5S6
⚅ ⚅ ⚃ ⚀ ⚀	1FC32HY4T5LS6	⚅ ⚃ ⚃ ⚃ ⚃	3FT1C2HY4L5S6
⚅ ⚅ ⚀ ⚀ ⚀	2T14CHY435LS6	⚅ ⚃ ⚃ ⚃ ⚀	3T12FCHY4L5S6
⚅ ⚅ ⚀ ⚀ ⚀	21FCHY4T35LS6	⚅ ⚃ ⚃ ⚃ ⚀	3T1FC2HY4L5S6
⚅ ⚅ ⚀ ⚀ ⚀	12FCHY4T35LS6	⚅ ⚃ ⚃ ⚀ ⚀	231FCHY4TL5S6
⚅ ⚅ ⚀ ⚀ ⚀	1TFC2HY435LS6	⚅ ⚃ ⚃ ⚀ ⚀	S312FCHY4TL56
⚅ ⚃ ⚃ ⚃ ⚃	3FTC12HY5L4S6	⚅ ⚃ ⚃ ⚀ ⚀	23FC2HY4TL5S6
⚅ ⚃ ⚃ ⚃ ⚀	3T12CFHY5L4S6	⚅ ⚃ ⚀ ⚀ ⚀	2T1FC3HY4L5S6
⚅ ⚃ ⚃ ⚃ ⚀	3T1FC2HY5L4S6	⚅ ⚃ ⚀ ⚀ ⚀	S21FC3HY4TL56
⚅ ⚃ ⚃ ⚀ ⚀	231FCHYT5L4S6	⚅ ⚃ ⚀ ⚀ ⚀	S12F3CHY4TL56
⚅ ⚃ ⚃ ⚀ ⚀	312FCHYT5L4S6	⚅ ⚃ ⚀ ⚀ ⚀	1TF32CHY4L5S6
⚅ ⚃ ⚃ ⚀ ⚀	13FC2HYT5L4S6	⚅ ⚀ ⚀ ⚀ ⚀	2FT1CHY43L5S6
⚅ ⚃ ⚀ ⚀ ⚀	2T1FC3HY5L4S6	⚅ ⚀ ⚀ ⚀ ⚀	21TFCHY43L5S6
⚅ ⚃ ⚀ ⚀ ⚀	21FC3HYT5L4S6	⚅ ⚀ ⚀ ⚀ ⚀	12FCHY4T3L5S6
⚅ ⚃ ⚀ ⚀ ⚀	12FC3HYT5L4S6	⚅ ⚀ ⚀ ⚀ ⚀	12FTCHY43L5S6
⚅ ⚃ ⚀ ⚀ ⚀	1TF3C2HY5L4S6	⚅ ⚀ ⚀ ⚀ ⚀	1FT2HYC43L5S6
⚅ ⚀ ⚀ ⚀ ⚀	2FT1CHY35L4S6	⚃ ⚃ ⚃ ⚃ ⚃	Y3FT1C2HL45S6
⚅ ⚀ ⚀ ⚀ ⚀	21TFCHY35L4S6	⚃ ⚃ ⚃ ⚃ ⚀	3FT12CHYL45S6
⚅ ⚀ ⚀ ⚀ ⚀	12FCHYT35L4S6	⚃ ⚃ ⚃ ⚃ ⚀	3F1TC2HYL45S6
⚅ ⚀ ⚀ ⚀ ⚀	12TFCHY35L4S6	⚃ ⚃ ⚃ ⚀ ⚀	H32T1FCYL45S6
⚅ ⚀ ⚀ ⚀ ⚀	1FT2CHY35L4S6	⚃ ⚃ ⚃ ⚀ ⚀	31T2FCHYL45S6
⚅ ⚅ ⚅ ⚅ ⚅	Y4FTC12H3L5S6	⚃ ⚃ ⚃ ⚀ ⚀	H31TFC2YL45S6
⚅ ⚅ ⚅ ⚅ ⚃	4FTC132HYL5S6	⚃ ⚃ ⚀ ⚀ ⚀	H23T1FCYL45S6
⚅ ⚅ ⚅ ⚅ ⚀	4FTC12HY3L5S6	⚃ ⚃ ⚀ ⚀ ⚀	231FCHYTL45S6
⚅ ⚅ ⚅ ⚅ ⚀	4FT1C2HY3L5S6	⚃ ⚃ ⚀ ⚀ ⚀	132FCHYTL45S6
⚅ ⚅ ⚅ ⚃ ⚃	H4T3C1F2YL5S6	⚃ ⚃ ⚀ ⚀ ⚀	H13FT2CYL45S6
⚅ ⚅ ⚅ ⚃ ⚀	4TC12F3HYL5S6	⚃ ⚀ ⚀ ⚀ ⚀	2FT13CHYL45S6
⚅ ⚅ ⚅ ⚃ ⚀	4T1CF32HYL5S6	⚃ ⚀ ⚀ ⚀ ⚀	21TF3CHYL45S6
⚅ ⚅ ⚅ ⚀ ⚀	H4T21CFY3L5S6	⚃ ⚀ ⚀ ⚀ ⚀	12F3CHYTL45S6
⚅ ⚅ ⚅ ⚀ ⚀	4T12FCHY3L5S6	⚃ ⚀ ⚀ ⚀ ⚀	12FT3HYCL45S6
⚅ ⚅ ⚅ ⚀ ⚀	H41TFC2Y3L5S6	⚃ ⚀ ⚀ ⚀ ⚀	1FT32HYCL45S6
⚅ ⚅ ⚃ ⚃ ⚃	H3T4C1F2YL5S6	⚀ ⚀ ⚀ ⚀ ⚀	Y2F1TCH3L45S6
⚅ ⚅ ⚃ ⚃ ⚀	3412CFHYTL5S6	⚀ ⚀ ⚀ ⚀ ⚀	2F1TCHY3L45S6
⚅ ⚅ ⚃ ⚀ ⚀	314FC2HYTL5S6	⚀ ⚀ ⚀ ⚀ ⚀	H21FTYC3L45S6
⚅ ⚅ ⚃ ⚀ ⚀	241FC3HYTL5S6	⚀ ⚀ ⚀ ⚀ ⚀	H12FTYC3L45S6
⚅ ⚅ ⚃ ⚀ ⚀	S142FC3HYTL56	⚀ ⚀ ⚀ ⚀ ⚀	1F2THYC3L45S6
⚅ ⚅ ⚃ ⚀ ⚀	14FC32HYTL5S6	⚀ ⚀ ⚀ ⚀ ⚀	Y1FT2H3CL45S6

TABLE T.4: Correct Steady-State Vector Values for Monopoly, Short Jail

Space	Prob		Space	Prob
Go	0.0309612		New York Ave	0.0308517
Mediterranean Ave	0.0213138		Free Parking	0.0288360
Comm Chest (South)	0.0188488		Kentucky Ave	0.0283584
Baltic Ave	0.0216240		Chance (North)	0.0104803
Income Tax	0.0232852		Indiana Ave	0.0273569
Reading RR	0.0296310		Illinois Ave	0.0318577
Oriental Ave	0.0226214		B & O RR	0.0306590
Chance (South)	0.0086505		Atlantic Ave	0.0270720
Vermont Ave	0.0232096		Ventnor Ave	0.0267886
Connecticut Ave	0.0230034		Water Works	0.0280742
Just Visiting	0.0226954		Marvin Gardens	0.0258605
In Jail	0.0394998		Pacific Ave	0.0267737
St. Charles Place	0.0270166		North Carolina Ave	0.0262517
Electric Company	0.0260404		Comm Chest (East)	0.0236605
States Ave	0.0237209		Pennsylvania Ave	0.0250063
Virginia Ave	0.0246489		Short Line	0.0243264
Pennsylvania RR	0.0291997		Chance (East)	0.0086687
St. James Place	0.0279242		Park Place	0.0218640
Comm Chest (West)	0.0259446		Luxury Tax	0.0217985
Tennessee Ave	0.0293559		Boardwalk	0.0262596

TABLE T.5: Correct Steady-State Vector Values for Monopoly, Long Jail

	Space	Prob		Space	Prob
	Go	0.0291826		New York Ave	0.0281156
	Mediterranean Ave	0.0201003		Free Parking	0.0282480
	Comm Chest (South)	0.0177751		Kentucky Ave	0.0261421
	Baltic Ave	0.0203976		Chance (North)	0.0104495
	Income Tax	0.0219659		Indiana Ave	0.0256728
	Reading RR	0.0280496		Illinois Ave	0.0299549
	Oriental Ave	0.0213457		B & O RR	0.0289285
	Chance (South)	0.0081636		Atlantic Ave	0.0254008
	Vermont Ave	0.0219034		Ventnor Ave	0.0251920
	Connecticut Ave	0.0217118		Water Works	0.0265475
	Just Visiting	0.0214223		Marvin Gardens	0.0243872
	In Jail	0.0938552		Pacific Ave	0.0252492
	St. Charles Place	0.0255955		North Carolina Ave	0.0247692
	Electric Company	0.0261548		Comm Chest (East)	0.0222919
	States Ave	0.0217600		Pennsylvania Ave	0.0235764
	Virginia Ave	0.0242527		Short Line	0.0229247
	Pennsylvania RR	0.0263657		Chance (East)	0.0081745
	St. James Place	0.0267892		Park Place	0.0206162
	Comm Chest (West)	0.0229513		Luxury Tax	0.0205584
	Tennessee Ave	0.0281968		Boardwalk	0.0248612

TABLE T.6: Expected Values for Blackjack Hard Hands

Hard Hand	Dealer's Up Card									
	2	3	4	5	6	7	8	9	10	A
10-9	+0.37797	+0.39479	+0.41389	+0.43718	+0.45118	+0.61528	+0.59178	+0.28483	+0.06785	+0.19096
10-8	+0.10984	+0.13789	+0.16484	+0.19539	+0.22106	+0.39820	+0.10453	-0.18468	-0.17549	-0.22397
10-7	-0.15700	-0.12041	-0.08149	-0.04607	-0.00890	-0.10835	-0.38367	-0.42242	-0.41877	-0.50000
9-8	-0.15445	-0.12037	-0.08385	-0.04618	-0.00901	-0.10901	-0.38610	-0.42175	-0.41621	-0.50000
10-6	-0.28707	-0.24644	-0.20377	-0.16167	-0.12366	-0.41017	-0.45447	-0.50000	-0.50000	-0.50000
9-7	-0.28458	-0.24625	-0.20376	-0.16414	-0.12386	-0.41003	-0.45484	-0.50000	-0.50000	-0.50000
10-5	-0.28669	-0.24607	-0.20341	-0.16115	-0.12073	-0.36914	-0.41688	-0.47203	-0.50000	-0.50000
9-6	-0.28420	-0.24588	-0.20324	-0.16119	-0.12327	-0.36900	-0.41725	-0.47250	-0.50000	-0.50000
8-7	-0.28407	-0.24331	-0.20315	-0.16359	-0.12338	-0.36436	-0.41241	-0.46818	-0.50000	-0.50000
10-4	-0.28855	-0.24637	-0.20367	-0.16141	-0.12081	-0.32374	-0.37019	-0.42891	-0.46391	-0.46888
9-5	-0.28381	-0.24551	-0.20288	-0.16067	-0.12032	-0.32437	-0.37164	-0.43036	-0.46473	-0.46870
8-6	-0.28367	-0.24294	-0.20263	-0.16065	-0.12276	-0.32443	-0.37159	-0.43159	-0.46615	-0.46988
10-3	-0.28864	-0.24826	-0.20400	-0.16171	-0.12108	-0.26922	-0.32422	-0.38383	-0.42151	-0.42729
9-4	-0.28568	-0.24579	-0.20314	-0.16093	-0.12039	-0.26960	-0.32540	-0.38454	-0.42215	-0.42595
8-5	-0.28330	-0.24256	-0.20227	-0.16013	-0.11982	-0.27585	-0.33100	-0.39026	-0.42779	-0.43184
7-6	-0.28331	-0.24242	-0.19969	-0.16003	-0.12222	-0.27606	-0.33188	-0.39085	-0.42824	-0.43281
10-2	-0.25261	-0.23226	-0.20589	-0.16202	-0.12137	-0.21269	-0.27189	-0.34053	-0.37702	-0.38355
9-3	-0.25530	-0.23647	-0.20346	-0.16123	-0.12067	-0.21688	-0.27687	-0.34615	-0.38203	-0.38658
8-4	-0.25448	-0.23350	-0.20252	-0.16039	-0.11989	-0.21665	-0.27717	-0.34655	-0.38167	-0.38650
7-5	-0.25365	-0.23265	-0.19932	-0.15952	-0.11928	-0.21787	-0.27727	-0.34581	-0.38150	-0.38667
9-2	+0.47630	+0.52008	+0.56969	+0.62198	+0.66777	+0.46213	+0.34791	+0.22604	+0.17538	+0.11252
8-3	+0.47749	+0.52605	+0.57284	+0.62452	+0.67027	+0.46412	+0.34823	+0.22626	+0.17762	+0.11490
7-4	+0.47882	+0.52851	+0.57958	+0.62731	+0.67305	+0.46648	+0.34950	+0.22734	+0.17871	+0.11752
6-5	+0.48167	+0.53060	+0.58164	+0.63434	+0.67605	+0.46805	+0.35277	+0.22948	+0.17877	+0.11881

| Hard Hand | Dealer's Up Card | | | | | | | | | |
	2	3	4	5	6	7	8	9	10	A
8-2	+0.36443	+0.41626	+0.46483	+0.52124	+0.56971	+0.39535	+0.28732	+0.14758	+0.02597	+0.03362
7-3	+0.36504	+0.41749	+0.47234	+0.52393	+0.57238	+0.39918	+0.29108	+0.14541	+0.02589	+0.03363
6-4	+0.36596	+0.41870	+0.47383	+0.53082	+0.57544	+0.40261	+0.29036	+0.14690	+0.02642	+0.03417
7-2	+0.07350	+0.12755	+0.19069	+0.25280	+0.30971	+0.17369	+0.09939	-0.05217	-0.15307	-0.12616
6-3	+0.07456	+0.12720	+0.19274	+0.25935	+0.31226	+0.17502	+0.10048	-0.05207	-0.15166	-0.12452
5-4	+0.07491	+0.12824	+0.19343	+0.26182	+0.31953	+0.17517	+0.09953	-0.05219	-0.15120	-0.12420
6-2	-0.02328	+0.00768	+0.04206	+0.07694	+0.10604	+0.08351	-0.05938	-0.20996	-0.24956	-0.26529
5-3	-0.02350	+0.00746	+0.04272	+0.07807	+0.10954	+0.08343	-0.05957	-0.21108	-0.24956	-0.26527
5-2	-0.10833	-0.07485	-0.03878	+0.00023	+0.03542	-0.06869	-0.21140	-0.28543	-0.31943	-0.35117
4-3	-0.11056	-0.07695	-0.03952	-0.00067	+0.03446	-0.06904	-0.21260	-0.28768	-0.32093	-0.35232
4-2	-0.13892	-0.10437	-0.06779	-0.02848	+0.00859	-0.15329	-0.21909	-0.29405	-0.33880	-0.34340
3-2	-0.12598	-0.09318	-0.05625	-0.01747	+0.01823	-0.11947	-0.18741	-0.26615	-0.31291	-0.31744

TABLE T.7: Expected Values for Blackjack Soft and Paired Hands

Other Hands	Dealer's Up Card									
	2	3	4	5	6	7	8	9	10	A
A-10	+1.50000	+1.50000	+1.50000	+1.50000	+1.50000	+1.50000	+1.50000	+1.50000	+1.50000	+1.50000
A-9	+0.63699	+0.64541	+0.65612	+0.66990	+0.67775	+0.77326	+0.79095	+0.75920	+0.55455	+0.60453
A-8	+0.38008	+0.39908	+0.41577	+0.43898	+0.46184	+0.61579	+0.59538	+0.28761	+0.06327	+0.18912
A-7	+0.11587	+0.17468	+0.24176	+0.29999	+0.35647	+0.40129	+0.10752	-0.09904	-0.14315	-0.15985
A-6	-0.00063	+0.05635	+0.12183	+0.19346	+0.25140	+0.05448	-0.07205	-0.14790	-0.19593	-0.22099
A-5	-0.02103	+0.00880	+0.06139	+0.12948	+0.20052	-0.00713	-0.06873	-0.15072	-0.20947	-0.20764
A-4	-0.00050	+0.02929	+0.06420	+0.13298	+0.19877	+0.03656	-0.02806	-0.11236	-0.17335	-0.17125
A-3	+0.02228	+0.05057	+0.08232	+0.13651	+0.20123	+0.07729	+0.01566	-0.07335	-0.13770	-0.13492
A-2	+0.04587	+0.07386	+0.10402	+0.13756	+0.20217	+0.12074	+0.05240	-0.03500	-0.10295	-0.1007
A-A	+0.60656	+0.65505	+0.70543	+0.75767	+0.80522	+0.62643	+0.49878	+0.36163	+0.29994	+0.22800
10-10	+0.63345	+0.64419	+0.65494	+0.66879	+0.67700	+0.77232	+0.79077	+0.75665	+0.55799	+0.60068
9-9	+0.19443	+0.25108	+0.31938	+0.39234	+0.45469	+0.39956	+0.23082	-0.08091	-0.17290	-0.22107
8-8	+0.07211	+0.14563	+0.21682	+0.29901	+0.37398	+0.31818	-0.02839	-0.38971	-0.47706	-0.50000
7-7	-0.12352	-0.03271	+0.06688	+0.15660	+0.24901	-0.04940	-0.37625	-0.43616	-0.47212	-0.47542
6-6	-0.19415	-0.09812	+0.00399	+0.10741	+0.19040	-0.21860	-0.27735	-0.34535	-0.38173	-0.38649
5-5	+0.36755	+0.41975	+0.47431	+0.53259	+0.58185	+0.40166	+0.29144	+0.14797	+0.02667	+0.03406
4-4	-0.02279	+0.00840	+0.04397	+0.10756	+0.20276	+0.08550	-0.05929	-0.20994	-0.24836	-0.26378
3-3	-0.13061	-0.04789	+0.04948	+0.14966	+0.24311	-0.05168	-0.21868	-0.29460	-0.33846	-0.34340
2-2	-0.08003	-0.00905	+0.07063	+0.17042	+0.26423	+0.00804	-0.15744	-0.23858	-0.28764	-0.29257

TABLE T.8: Expected Values for Blackjack Switch Hard Hands

Hard Hand	Dealer's Up Card									
	2	3	4	5	6	7	8	9	10	A
10-9	+0.22647	+0.29783	+0.31738	+0.34366	+0.36114	+0.55038	+0.53082	+0.22712	+0.01257	+0.13823
10-8	−0.04175	+0.03997	+0.07027	+0.10177	+0.13093	+0.33311	+0.04407	−0.24195	−0.23139	−0.27713
10-7	−0.30857	−0.21830	−0.17659	−0.13817	−0.09905	−0.17344	−0.44420	−0.47865	−0.47528	−0.50000
9-8	−0.30690	−0.21669	−0.17800	−0.13985	−0.09920	−0.17409	−0.44745	−0.47802	−0.47186	−0.50000
10-6	−0.43857	−0.34418	−0.29880	−0.25451	−0.21221	−0.43367	−0.47638	−0.50000	−0.50000	−0.50000
9-7	−0.43696	−0.34239	−0.29875	−0.25606	−0.21400	−0.43340	−0.47689	−0.50000	−0.50000	−0.50000
10-5	−0.43844	−0.34404	−0.29867	−0.25418	−0.21034	−0.39549	−0.44185	−0.49568	−0.50000	−0.50000
9-6	−0.43640	−0.34184	−0.29803	−0.25380	−0.21159	−0.39537	−0.44223	−0.49618	−0.50000	−0.50000
8-7	−0.43626	−0.34014	−0.29633	−0.25527	−0.21329	−0.38940	−0.43586	−0.49060	−0.50000	−0.50000
10-4	−0.44076	−0.34441	−0.29900	−0.25449	−0.21042	−0.35336	−0.39611	−0.45351	−0.48849	−0.49286
9-5	−0.43625	−0.34171	−0.29790	−0.25346	−0.20970	−0.35368	−0.39841	−0.45546	−0.48978	−0.49267
8-6	−0.43568	−0.33960	−0.29560	−0.25301	−0.21087	−0.35394	−0.39809	−0.45697	−0.49135	−0.49418
10-3	−0.40143	−0.34675	−0.29939	−0.25485	−0.21076	−0.30035	−0.35331	−0.40972	−0.44768	−0.45294
9-4	−0.40065	−0.34205	−0.29823	−0.25378	−0.20979	−0.30065	−0.35434	−0.41094	−0.44858	−0.45119
8-5	−0.40649	−0.33944	−0.29547	−0.25267	−0.20898	−0.30837	−0.36188	−0.41841	−0.45616	−0.45883
7-6	−0.40635	−0.33888	−0.29337	−0.25059	−0.21006	−0.30880	−0.36291	−0.41892	−0.45624	−0.46004
10-2	−0.35442	−0.29946	−0.27701	−0.25391	−0.21110	−0.24610	−0.30313	−0.37009	−0.40479	−0.41126
9-3	−0.35879	−0.30370	−0.28182	−0.25413	−0.21012	−0.25137	−0.30961	−0.37694	−0.41187	−0.41528
8-4	−0.35824	−0.30102	−0.27901	−0.25299	−0.20907	−0.25107	−0.30940	−0.37786	−0.41163	−0.41519
7-5	−0.35740	−0.29989	−0.27678	−0.25026	−0.20818	−0.25207	−0.30981	−0.37663	−0.41152	−0.41525
9-2	+0.17222	+0.32841	+0.37956	+0.43645	+0.48973	+0.33270	+0.22657	+0.11189	+0.07282	+0.06446
8-3	+0.17373	+0.33293	+0.38659	+0.43989	+0.49307	+0.33534	+0.22737	+0.11401	+0.07279	+0.06482
7-4	+0.17551	+0.33586	+0.39213	+0.44673	+0.49691	+0.33845	+0.23091	+0.11438	+0.07255	+0.06527
6-5	+0.17904	+0.33868	+0.39489	+0.45271	+0.50410	+0.34248	+0.23378	+0.11724	+0.07374	+0.06627

| Hard Hand | Dealer's Up Card | | | | | | | | | |
	2	3	4	5	6	7	8	9	10	A
8-2	+0.06081	+0.22326	+0.27841	+0.33652	+0.39231	+0.26722	+0.16770	+0.07218	-0.01790	-0.00862
7-3	+0.06163	+0.22484	+0.28470	+0.34323	+0.39597	+0.27240	+0.17262	+0.07293	-0.01870	-0.00866
6-4	+0.06284	+0.22614	+0.28662	+0.34900	+0.40320	+0.27696	+0.17354	+0.07228	-0.01784	-0.00789
7-2	-0.07231	-0.00853	+0.03934	+0.07166	+0.13252	+0.12443	+0.05353	-0.09577	-0.19640	-0.16847
6-3	-0.07101	-0.00983	+0.04181	+0.07697	+0.13905	+0.12624	+0.05496	-0.09492	-0.19482	-0.16616
5-4	-0.07090	-0.00986	+0.04195	+0.07949	+0.14460	+0.12672	+0.05413	-0.09594	-0.19526	-0.16593
6-2	-0.16877	-0.08682	-0.05063	-0.01475	+0.01896	+0.03522	-0.10451	-0.25249	-0.29161	-0.30649
5-3	-0.16952	-0.08751	-0.05041	-0.01365	+0.02155	+0.03543	-0.10503	-0.25353	-0.29303	-0.30665
5-2	-0.25381	-0.16959	-0.13203	-0.09169	-0.05275	-0.11590	-0.25608	-0.32641	-0.36028	-0.39073
4-3	-0.25626	-0.17234	-0.13331	-0.09290	-0.05408	-0.11700	-0.25710	-0.32870	-0.36276	-0.39236
4-2	-0.28352	-0.19915	-0.16138	-0.12115	-0.07998	-0.18994	-0.25283	-0.32611	-0.37052	-0.37436
3-2	-0.26962	-0.18770	-0.14958	-0.11003	-0.07057	-0.15703	-0.22209	-0.29826	-0.34512	-0.34896

TABLE T.9: Expected Values for Blackjack Switch Soft and Paired Hands

Other Hands	Dealer's Up Card									
	2	3	4	5	6	7	8	9	10	A
A-10[a]	+1.00000	+1.00000	+1.00000	+1.00000	+1.00000	+1.00000	+1.00000	+1.00000	+1.00000	+1.00000
A-10[b]	+0.72817	+0.78655	+0.79212	+0.79801	+0.80319	+0.86210	+0.87034	+0.88288	+0.90583	+0.85051
A-9	+0.48445	+0.54888	+0.56002	+0.57646	+0.58801	+0.70851	+0.73030	+0.70263	+0.49772	+0.55181
A-8	+0.22758	+0.30172	+0.32136	+0.34604	+0.36270	+0.55099	+0.53581	+0.23078	+0.00650	+0.13581
A-7	-0.03999	+0.04355	+0.07512	+0.11771	+0.17806	+0.33721	+0.04800	-0.14301	-0.18776	-0.20274
A-6	-0.14734	-0.06640	-0.03297	+0.00943	+0.07649	+0.00376	-0.11906	-0.19094	-0.23921	-0.26268
A-5	-0.16833	-0.08745	-0.05389	-0.01904	+0.02223	-0.04809	-0.10718	-0.18603	-0.24542	-0.24178
A-4	-0.14687	-0.06647	-0.03399	+0.00063	+0.03639	-0.00608	-0.06726	-0.14844	-0.20945	-0.20613
A-3	-0.12273	-0.04480	-0.01181	+0.02191	+0.05626	+0.03253	-0.02458	-0.10968	-0.17468	-0.17061
A-2	-0.08932	-0.02115	+0.01026	+0.04341	+0.07668	+0.07444	+0.00934	-0.07308	-0.14073	-0.13720
A-A	+0.17891	+0.33644	+0.39026	+0.44548	+0.49998	+0.34531	+0.23875	+0.12382	+0.06833	+0.01420
10-10	+0.48311	+0.54585	+0.55852	+0.57502	+0.58712	+0.70707	+0.73013	+0.69927	+0.50225	+0.54683
9-9	-0.04035	+0.04016	+0.09406	+0.16406	+0.22864	+0.33519	+0.10747	-0.19332	-0.22793	-0.27347
8-8	-0.31684	-0.12754	-0.05185	+0.02066	+0.09707	+0.09650	-0.19806	-0.50000	-0.50000	-0.50000
7-7	-0.43569	-0.29019	-0.20641	-0.12151	-0.03551	-0.20574	-0.40416	-0.46279	-0.49936	-0.50000
6-6	-0.35714	-0.29966	-0.25780	-0.16887	-0.07519	-0.25322	-0.30972	-0.37583	-0.41090	-0.41480
5-5	+0.06386	+0.22677	+0.28653	+0.35058	+0.40766	+0.27826	+0.17271	+0.07194	-0.01828	-0.00808
4-4	-0.16843	-0.08659	-0.04884	-0.01166	+0.02343	+0.03741	-0.10342	-0.25262	-0.29150	-0.30482
3-3	-0.28351	-0.20068	-0.16039	-0.10634	-0.01694	-0.17046	-0.25284	-0.32589	-0.37048	-0.37436
2-2	-0.25493	-0.17431	-0.13805	-0.08291	+0.00627	-0.11264	-0.19307	-0.27197	-0.32029	-0.32506

[a]This row refers to a non-switched Blackjack.
[b]This row refers to a Blackjack formed by switching.

TABLE T.10: Values of $T(\{g, y, r\}; x)$

x	Element of S	Element #	Green	Yellow	Red
3	$\{3,0,0\}$	1	(3,0,0)	(0,0,0)	(0,0,0)
3	$\{0,3,0\}$	1	(0,0,0)	(0,3,0)	(0,0,0)
3	$\{0,0,3\}$	1	(0,0,0)	(0,0,0)	(0,0,3)
3	$\{2,1,0\}$	1	(2,0,0)	(1,0,0)	(0,0,0)
3	$\{2,0,1\}$	1	(2,0,0)	(0,0,0)	(1,0,0)
3	$\{1,2,0\}$	1	(1,0,0)	(2,0,0)	(0,0,0)
3	$\{0,2,1\}$	1	(0,0,0)	(2,0,0)	(1,0,0)
3	$\{1,0,2\}$	1	(1,0,0)	(0,0,0)	(2,0,0)
3	$\{0,1,2\}$	1	(0,0,0)	(1,0,0)	(2,0,0)
3	$\{1,1,1\}$	1	(1,0,0)	(1,0,0)	(1,0,0)
2	$\{3,0,0\}$	1	(2,1,0)	(0,0,0)	(0,0,0)
		2	(2,0,1)	(0,0,0)	(0,0,0)
2	$\{0,3,0\}$	1	(0,0,0)	(2,1,0)	(0,0,0)
		2	(0,0,0)	(2,0,1)	(0,0,0)
2	$\{0,0,3\}$	1	(0,0,0)	(0,0,0)	(2,1,0)
		2	(0,0,0)	(0,0,0)	(2,0,1)
2	$\{2,1,0\}$	1	(2,0,0)	(0,1,0)	(0,0,0)
		2	(2,0,0)	(0,0,1)	(0,0,0)
		3	(1,1,0)	(1,0,0)	(0,0,0)
		4	(1,0,1)	(1,0,0)	(0,0,0)
2	$\{2,0,1\}$	1	(2,0,0)	(0,0,0)	(0,1,0)
		2	(2,0,0)	(0,0,0)	(0,0,1)
		3	(1,1,0)	(0,0,0)	(1,0,0)
		4	(1,0,1)	(0,0,0)	(1,0,0)
2	$\{1,2,0\}$	1	(0,1,0)	(2,0,0)	(0,0,0)
		2	(0,0,1)	(2,0,0)	(0,0,0)
		3	(1,0,0)	(1,1,0)	(0,0,0)
		4	(1,0,0)	(1,0,1)	(0,0,0)
2	$\{0,2,1\}$	1	(0,0,0)	(2,0,0)	(0,1,0)
		2	(0,0,0)	(2,0,0)	(0,0,1)
		3	(0,0,0)	(1,1,0)	(1,0,0)
		4	(0,0,0)	(1,0,1)	(1,0,0)
2	$\{1,0,2\}$	1	(0,1,0)	(0,0,0)	(2,0,0)
		2	(0,0,1)	(0,0,0)	(2,0,0)
		3	(1,0,0)	(0,0,0)	(1,1,0)
		4	(1,0,0)	(0,0,0)	(1,0,1)
2	$\{0,1,2\}$	1	(0,0,0)	(0,1,0)	(2,0,0)
		2	(0,0,0)	(0,0,1)	(2,0,0)
		3	(0,0,0)	(1,0,0)	(1,1,0)
		4	(0,0,0)	(1,0,0)	(1,0,1)
2	$\{1,1,1\}$	1	(1,0,0)	(1,0,0)	(0,1,0)

x	Element of S	Element #	Green	Yellow	Red
		2	(1,0,0)	(1,0,0)	(0,0,1)
		3	(1,0,0)	(0,1,0)	(1,0,0)
		4	(1,0,0)	(0,0,1)	(1,0,0)
		5	(0,1,0)	(1,0,0)	(1,0,0)
		6	(0,0,1)	(1,0,0)	(1,0,0)
1	$\{3,0,0\}$	1	(1,2,0)	(0,0,0)	(0,0,0)
		2	(1,0,2)	(0,0,0)	(0,0,0)
		3	(1,1,1)	(0,0,0)	(0,0,0)
1	$\{0,3,0\}$	1	(0,0,0)	(1,2,0)	(0,0,0)
		2	(0,0,0)	(1,0,2)	(0,0,0)
		3	(0,0,0)	(1,1,1)	(0,0,0)
1	$\{0,0,3\}$	1	(0,0,0)	(0,0,0)	(1,2,0)
		2	(0,0,0)	(0,0,0)	(1,0,2)
		3	(0,0,0)	(0,0,0)	(1,1,1)
1	$\{2,1,0\}$	1	(1,1,0)	(0,1,0)	(0,0,0)
		2	(1,0,1)	(0,1,0)	(0,0,0)
		3	(1,1,0)	(0,0,1)	(0,0,0)
		4	(1,0,1)	(0,0,1)	(0,0,0)
		5	(0,2,0)	(1,0,0)	(0,0,0)
		6	(0,0,2)	(1,0,0)	(0,0,0)
		7	(0,1,1)	(1,0,0)	(0,0,0)
1	$\{2,0,1\}$	1	(1,1,0)	(0,0,0)	(0,1,0)
		2	(1,0,1)	(0,0,0)	(0,1,0)
		3	(1,1,0)	(0,0,0)	(0,0,1)
		4	(1,0,1)	(0,0,0)	(0,0,1)
		5	(0,2,0)	(0,0,0)	(1,0,0)
		6	(0,0,2)	(0,0,0)	(1,0,0)
		7	(0,1,1)	(0,0,0)	(1,0,0)
1	$\{1,2,0\}$	1	(0,1,0)	(1,1,0)	(0,0,0)
		2	(0,1,0)	(1,0,1)	(0,0,0)
		3	(0,0,1)	(1,1,0)	(0,0,0)
		4	(0,0,1)	(1,0,1)	(0,0,0)
		5	(1,0,0)	(0,2,0)	(0,0,0)
		6	(1,0,0)	(0,0,2)	(0,0,0)
		7	(1,0,0)	(0,1,1)	(0,0,0)
1	$\{0,2,1\}$	1	(0,0,0)	(1,1,0)	(0,1,0)
		2	(0,0,0)	(1,0,1)	(0,1,0)
		3	(0,0,0)	(1,1,0)	(0,0,1)
		4	(0,0,0)	(1,0,1)	(0,0,1)
		5	(0,0,0)	(0,2,0)	(1,0,0)
		6	(0,0,0)	(0,0,2)	(1,0,0)
		7	(0,0,0)	(0,1,1)	(1,0,0)
1	$\{1,0,2\}$	1	(0,1,0)	(0,0,0)	(1,1,0)
		2	(0,1,0)	(0,0,0)	(1,0,1)

x	Element of S	Element #	Green	Yellow	Red
		3	(0,0,1)	(0,0,0)	(1,1,0)
		4	(0,0,1)	(0,0,0)	(1,0,1)
		5	(1,0,0)	(0,0,0)	(0,2,0)
		6	(1,0,0)	(0,0,0)	(0,0,2)
		7	(1,0,0)	(0,0,0)	(0,1,1)
1	$\{0,1,2\}$	1	(0,0,0)	(0,1,0)	(1,1,0)
		2	(0,0,0)	(0,1,0)	(1,0,1)
		3	(0,0,0)	(0,0,1)	(1,1,0)
		4	(0,0,0)	(0,0,1)	(1,0,1)
		5	(0,0,0)	(1,0,0)	(0,2,0)
		6	(0,0,0)	(1,0,0)	(0,0,2)
		7	(0,0,0)	(1,0,0)	(0,1,1)
1	$\{1,1,1\}$	1	(1,0,0)	(0,1,0)	(0,1,0)
		2	(1,0,0)	(0,0,1)	(0,1,0)
		3	(1,0,0)	(0,1,0)	(0,0,1)
		4	(1,0,0)	(0,0,1)	(0,0,1)
		5	(0,1,0)	(1,0,0)	(0,1,0)
		6	(0,0,1)	(1,0,0)	(0,1,0)
		7	(0,1,0)	(1,0,0)	(0,0,1)
		8	(0,0,1)	(1,0,0)	(0,0,1)
		9	(0,1,0)	(0,1,0)	(1,0,0)
		10	(0,0,1)	(0,1,0)	(1,0,0)
		11	(0,1,0)	(0,0,1)	(1,0,0)
		12	(0,0,1)	(0,0,1)	(1,0,0)
0	$\{3,0,0\}$	1	(0,2,1)	(0,0,0)	(0,0,0)
		2	(0,1,2)	(0,0,0)	(0,0,0)
		3	(0,3,0)	(0,0,0)	(0,0,0)
		4	(0,0,3)	(0,0,0)	(0,0,0)
0	$\{0,3,0\}$	1	(0,0,0)	(0,2,1)	(0,0,0)
		2	(0,0,0)	(0,1,2)	(0,0,0)
		3	(0,0,0)	(0,3,0)	(0,0,0)
		4	(0,0,0)	(0,0,3)	(0,0,0)
0	$\{0,0,3\}$	1	(0,0,0)	(0,0,0)	(0,2,1)
		2	(0,0,0)	(0,0,0)	(0,1,2)
		3	(0,0,0)	(0,0,0)	(0,3,0)
		4	(0,0,0)	(0,0,0)	(0,0,3)
0	$\{2,1,0\}$	1	(0,2,0)	(0,1,0)	(0,0,0)
		2	(0,2,0)	(0,0,1)	(0,0,0)
		3	(0,0,2)	(0,1,0)	(0,0,0)
		4	(0,0,2)	(0,0,1)	(0,0,0)
		5	(0,1,1)	(0,1,0)	(0,0,0)
		6	(0,1,1)	(0,0,1)	(0,0,0)
0	$\{2,0,1\}$	1	(0,2,0)	(0,0,0)	(0,1,0)
		2	(0,2,0)	(0,0,0)	(0,0,1)

x	Element of S	Element #	Green	Yellow	Red
		3	(0,0,2)	(0,0,0)	(0,1,0)
		4	(0,0,2)	(0,0,0)	(0,0,1)
		5	(0,1,1)	(0,0,0)	(0,1,0)
		6	(0,1,1)	(0,0,0)	(0,0,1)
0	$\{1,2,0\}$	1	(0,1,0)	(0,2,0)	(0,0,0)
		2	(0,0,1)	(0,2,0)	(0,0,0)
		3	(0,1,0)	(0,0,2)	(0,0,0)
		4	(0,0,1)	(0,0,2)	(0,0,0)
		5	(0,1,0)	(0,1,1)	(0,0,0)
		6	(0,0,1)	(0,1,1)	(0,0,0)
0	$\{0,2,1\}$	1	(0,0,0)	(0,2,0)	(0,1,0)
		2	(0,0,0)	(0,2,0)	(0,0,1)
		3	(0,0,0)	(0,0,2)	(0,1,0)
		4	(0,0,0)	(0,0,2)	(0,0,1)
		5	(0,0,0)	(0,1,1)	(0,1,0)
		6	(0,0,0)	(0,1,1)	(0,0,1)
0	$\{1,0,2\}$	1	(0,1,0)	(0,0,0)	(0,2,0)
		2	(0,0,1)	(0,0,0)	(0,2,0)
		3	(0,1,0)	(0,0,0)	(0,0,2)
		4	(0,0,1)	(0,0,0)	(0,0,2)
		5	(0,1,0)	(0,0,0)	(0,1,1)
		6	(0,0,1)	(0,0,0)	(0,1,1)
0	$\{0,1,2\}$	1	(0,0,0)	(0,1,0)	(0,2,0)
		2	(0,0,0)	(0,0,1)	(0,2,0)
		3	(0,0,0)	(0,1,0)	(0,0,2)
		4	(0,0,0)	(0,0,1)	(0,0,2)
		5	(0,0,0)	(0,1,0)	(0,1,1)
		6	(0,0,0)	(0,0,1)	(0,1,1)
0	$\{0,0,0\}$	1	(0,1,0)	(0,1,0)	(0,1,0)
		2	(0,1,0)	(0,1,0)	(0,0,1)
		3	(0,1,0)	(0,0,1)	(0,1,0)
		4	(0,1,0)	(0,0,1)	(0,0,1)
		5	(0,0,1)	(0,1,0)	(0,1,0)
		6	(0,0,1)	(0,1,0)	(0,0,1)
		7	(0,0,1)	(0,0,1)	(0,1,0)
		8	(0,0,1)	(0,0,1)	(0,0,1)

TABLE T.11: Decision Points for Zombie Dice

Dice in Cup			Current FPs			Decision	Decision	Decision
R	Y	G	R	Y	G	(SG 0)	(SG 1)	(SG 2)
3	4	6	0	0	0	132.869279	7.912293	0.434451
3	4	5	0	0	1	227.648386	11.892983	0.656155
3	4	5	0	0	0	111.462951	6.799302	0.371282
3	4	4	0	0	2	424.499389	19.869442	1.096177
3	4	4	0	0	1	196.926266	10.432803	0.573141
3	4	4	0	0	0	91.073496	5.704195	0.308864
3	4	3	0	0	3	874.975515	37.053477	2.164569
3	4	3	0	0	2	382.461145	18.088099	0.993253
3	4	3	0	0	1	166.642408	8.951067	0.488107
3	4	3	0	0	0	72.020787	4.642113	0.248171
3	4	2	0	0	3	831.236899	35.527396	2.102402
3	4	2	0	0	2	339.261889	16.212234	0.882011
3	4	2	0	0	1	137.242819	7.462848	0.401977
3	4	2	0	0	0	54.581356	3.630077	0.190447
3	4	1	0	0	3	784.584930	33.872163	2.034298
3	4	1	0	0	2	295.351036	14.246486	0.761831
3	4	1	0	0	1	109.102081	5.986013	0.316183
3	4	1	0	0	0	39.199877	2.693273	0.137318
3	4	0	0	0	3	735.557387	32.083285	1.959650
3	4	0	0	0	2	250.938240	12.194740	0.632242
3	4	0	0	0	1	83.037598	4.553092	0.233007
3	4	0	0	0	0	26.257331	1.860368	0.090756
3	3	6	0	1	0	113.804718	7.034921	0.388644
3	3	6	0	0	0	141.352547	8.257477	0.450169
3	3	5	0	1	1	203.650293	10.794322	0.597104
3	3	5	0	1	0	98.986100	6.256192	0.346129
3	3	5	0	0	1	235.047495	12.183941	0.669498
3	3	5	0	0	0	117.398321	7.023927	0.380281
3	3	4	0	1	2	402.232818	18.875666	1.025975
3	3	4	0	1	1	184.100678	9.879073	0.549625
3	3	4	0	1	0	84.257169	5.454928	0.301515
3	3	4	0	0	2	428.935831	20.005156	1.103410
3	3	4	0	0	1	201.219533	10.577507	0.578046
3	3	4	0	0	0	94.656818	5.811214	0.311230
3	3	3	0	1	2	382.665471	18.135648	0.999036
3	3	3	0	1	1	163.879742	8.905169	0.497375
3	3	3	0	1	0	69.827567	4.637279	0.255044
3	3	3	0	0	3	875.680013	36.990609	2.160524
3	3	3	0	0	2	383.153418	18.048915	0.989932
3	3	3	0	0	1	167.998829	8.946514	0.484150
3	3	3	0	0	0	73.447915	4.636873	0.244278

| Dice in Cup | | | Current FPs | | | Decision | Decision | Decision |
R	Y	G	R	Y	G	(SG 0)	(SG 1)	(SG 2)
3	3	2	0	1	2	361.681410	17.328765	0.969293
3	3	2	0	1	1	143.178272	7.872683	0.439708
3	3	2	0	1	0	55.882352	3.810553	0.207193
3	3	2	0	0	3	828.862814	35.317284	2.091528
3	3	2	0	0	2	336.303794	15.983964	0.866356
3	3	2	0	0	1	135.767038	7.307063	0.389037
3	3	2	0	0	0	54.320058	3.528783	0.181169
3	3	1	0	1	2	339.483108	16.451845	0.936381
3	3	1	0	1	1	122.089115	6.780698	0.375887
3	3	1	0	1	0	42.819560	2.990387	0.158901
3	3	1	0	0	3	779.658499	33.502648	2.015403
3	3	1	0	0	2	288.541988	13.810069	0.731805
3	3	1	0	0	1	105.440689	5.692407	0.294820
3	3	1	0	0	0	37.788143	2.519971	0.124145
3	3	0	0	1	2	315.960618	15.496730	0.899887
3	3	0	0	1	1	101.222929	5.640239	0.305246
3	3	0	0	1	0	31.057107	2.197458	0.111817
3	3	0	0	0	3	727.834434	31.529384	1.931296
3	3	0	0	0	2	241.318612	11.553586	0.585860
3	3	0	0	0	1	77.976092	4.145274	0.204871
3	3	0	0	0	0	23.537133	1.625230	0.075648
3	2	6	0	2	0	96.138053	6.158154	0.338135
3	2	6	0	1	0	119.522364	7.283383	0.400773
3	2	6	0	0	0	152.500560	8.690608	0.469010
3	2	5	0	2	1	181.823035	9.689944	0.527204
3	2	5	0	2	0	87.233123	5.690633	0.315121
3	2	5	0	1	1	207.981452	10.969123	0.605838
3	2	5	0	1	0	103.048293	6.423112	0.353825
3	2	5	0	0	1	244.700202	12.548300	0.685393
3	2	5	0	0	0	125.435330	7.310781	0.390879
3	2	4	0	2	1	173.268217	9.332162	0.514615
3	2	4	0	2	0	77.955568	5.185796	0.289455
3	2	4	0	1	2	404.308490	18.913920	1.026741
3	2	4	0	1	1	186.558912	9.961040	0.553516
3	2	4	0	1	0	86.699396	5.535885	0.304269
3	2	4	0	0	2	435.042774	20.178725	1.111954
3	2	4	0	0	1	207.235404	10.767826	0.583648
3	2	4	0	0	0	99.703109	5.952265	0.313650
3	2	3	0	2	1	164.039759	8.939572	0.500604
3	2	3	0	2	0	68.381398	4.641631	0.260678
3	2	3	0	1	2	383.284669	18.100717	0.996783
3	2	3	0	1	1	164.454276	8.883865	0.495354
3	2	3	0	1	0	70.660582	4.627588	0.252399

Dice in Cup			Current FPs			Decision	Decision	Decision
R	Y	G	R	Y	G	(SG 0)	(SG 1)	(SG 2)
3	2	3	0	0	3	878.171515	36.930674	2.155208
3	2	3	0	0	2	385.158725	18.017245	0.985597
3	2	3	0	0	1	170.362040	8.954134	0.478846
3	2	3	0	0	0	75.950699	4.644869	0.239108
3	2	2	0	2	1	154.208433	8.509927	0.484958
3	2	2	0	2	0	58.548106	4.054881	0.228212
3	2	2	0	1	2	361.016192	17.213154	0.963413
3	2	2	0	1	1	141.731535	7.734039	0.430424
3	2	2	0	1	0	55.368422	3.713454	0.198938
3	2	2	0	0	3	828.977594	35.096280	2.077842
3	2	2	0	0	2	333.863477	15.720654	0.846651
3	2	2	0	0	1	135.072098	7.139814	0.372811
3	2	2	0	0	0	54.845989	3.425920	0.169707
3	2	1	0	2	1	143.718361	8.038191	0.467402
3	2	1	0	2	0	48.718092	3.428090	0.191398
3	2	1	0	1	2	337.384749	16.241200	0.926094
3	2	1	0	1	1	119.020912	6.520838	0.357769
3	2	1	0	1	0	41.313168	2.814679	0.145305
3	2	1	0	0	3	777.129227	33.091837	1.991614
3	2	1	0	0	2	282.641571	13.312110	0.694078
3	2	1	0	0	1	102.477169	5.369143	0.268678
3	2	1	0	0	0	35.956796	2.307091	0.108258
3	2	0	0	2	1	133.194759	7.531904	0.447733
3	2	0	0	2	0	39.236490	2.766610	0.149502
3	2	0	0	1	2	314.011036	15.202864	0.884468
3	2	0	0	1	1	97.147844	5.260155	0.276468
3	2	0	0	1	0	28.044438	1.931484	0.093861
3	2	0	0	0	3	726.756657	30.959668	1.895864
3	2	0	0	0	2	233.417414	10.829316	0.527414
3	2	0	0	0	1	71.165328	3.641197	0.171127
3	2	0	0	0	0	22.530332	1.434012	0.059493
3	1	6	0	3	0	80.282335	5.269032	0.283297
3	1	6	0	2	0	99.383955	6.311605	0.345580
3	1	6	0	1	0	126.993170	7.594537	0.415502
3	1	6	0	0	0	167.756693	9.252114	0.491988
3	1	5	0	3	0	76.659433	5.087551	0.277182
3	1	5	0	2	1	183.657300	9.748891	0.528988
3	1	5	0	2	0	89.568339	5.796803	0.320305
3	1	5	0	1	1	213.612711	11.188052	0.616410
3	1	5	0	1	0	108.541417	6.636612	0.363124
3	1	5	0	0	1	257.813779	13.020956	0.704658
3	1	5	0	0	0	136.695222	7.688063	0.403486
3	1	4	0	3	0	72.725062	4.886993	0.270322

| Dice in Cup | | | Current FPs | | | Decision | Decision | Decision |
R	Y	G	R	Y	G	(SG 0)	(SG 1)	(SG 2)
3	1	4	0	2	1	174.434666	9.355163	0.514972
3	1	4	0	2	0	79.345429	5.237427	0.291777
3	1	4	0	1	2	407.422582	18.966426	1.027457
3	1	4	0	1	1	190.082734	10.070251	0.558188
3	1	4	0	1	0	90.151192	5.643074	0.307399
3	1	4	0	0	2	443.867768	20.413232	1.122243
3	1	4	0	0	1	215.761858	11.024718	0.590030
3	1	4	0	0	0	107.321271	6.151491	0.316002
3	1	3	0	3	0	68.500446	4.665775	0.262588
3	1	3	0	2	1	164.583748	8.922087	0.499210
3	1	3	0	2	0	68.737489	4.628601	0.259336
3	1	3	0	1	2	385.170529	18.071795	0.993761
3	1	3	0	1	1	165.655894	8.865135	0.492614
3	1	3	0	1	0	72.277297	4.627770	0.248759
3	1	3	0	0	3	884.177658	36.892780	2.148088
3	1	3	0	0	2	388.906322	17.997283	0.979735
3	1	3	0	0	1	174.757935	8.994502	0.471569
3	1	3	0	0	0	80.464035	4.683374	0.232099
3	1	2	0	3	0	63.957437	4.420638	0.253811
3	1	2	0	2	1	154.046150	8.443624	0.481372
3	1	2	0	2	0	58.004080	3.971258	0.222180
3	1	2	0	1	2	361.513329	17.086215	0.955743
3	1	2	0	1	1	140.947197	7.578898	0.418418
3	1	2	0	1	0	55.480835	3.611858	0.188291
3	1	2	0	0	3	832.477158	34.863142	2.060011
3	1	2	0	0	2	333.355976	15.433425	0.821251
3	1	2	0	0	1	136.061917	6.974715	0.351988
3	1	2	0	0	0	55.530482	3.289799	0.154825
3	1	1	0	3	0	59.333928	4.154282	0.243811
3	1	1	0	2	1	143.446897	7.926121	0.461130
3	1	1	0	2	0	47.518881	3.269297	0.179334
3	1	1	0	1	2	338.047481	16.024903	0.912776
3	1	1	0	1	1	116.846794	6.225099	0.334192
3	1	1	0	1	0	39.152138	2.588124	0.127915
3	1	1	0	0	3	782.062003	32.686357	1.960874
3	1	1	0	0	2	279.281618	12.755119	0.645315
3	1	1	0	0	1	97.846906	4.948883	0.235453
3	1	1	0	0	0	36.719544	2.142847	0.089842
3	1	0	0	3	0	55.140058	3.875175	0.232421
3	1	0	0	2	1	134.103286	7.387581	0.438199
3	1	0	0	2	0	36.404649	2.494916	0.129472
3	1	0	0	1	2	318.087572	14.927589	0.864366
3	1	0	0	1	1	90.914756	4.752783	0.238137

Dice in Cup			Current FPs			Decision	Decision	Decision
R	Y	G	R	Y	G	(SG 0)	(SG 1)	(SG 2)
3	1	0	0	1	0	27.225450	1.696999	0.072664
3	1	0	0	0	3	741.100852	30.448396	1.849808
3	1	0	0	0	2	220.118698	9.849348	0.450436
3	1	0	0	0	1	69.711302	3.189850	0.131779
3	1	0	0	0	0	19.982022	1.172725	0.041598
3	0	6	0	3	0	81.539805	5.324243	0.285016
3	0	6	0	2	0	103.564985	6.503897	0.354699
3	0	6	0	1	0	137.142905	7.996346	0.433750
3	0	6	0	0	0	189.509550	10.004539	0.520533
3	0	5	0	3	0	77.623545	5.124466	0.278209
3	0	5	0	2	1	186.092668	9.823195	0.531057
3	0	5	0	2	0	92.701337	5.933464	0.326697
3	0	5	0	1	1	221.207382	11.471840	0.629465
3	0	5	0	1	0	116.186399	6.916590	0.374539
3	0	5	0	0	1	276.089913	13.651267	0.728396
3	0	5	0	0	0	153.498428	8.210022	0.418651
3	0	4	0	3	0	73.400912	4.902780	0.270477
3	0	4	0	2	1	176.263112	9.388834	0.515259
3	0	4	0	2	0	81.301355	5.306328	0.294591
3	0	4	0	1	2	412.290983	19.044157	1.028004
3	0	4	0	1	1	195.050104	10.217070	0.563852
3	0	4	0	1	0	95.387830	5.794641	0.310917
3	0	4	0	0	2	456.303739	20.731151	1.134759
3	0	4	0	0	1	228.658538	11.393419	0.597227
3	0	4	0	0	0	119.589323	6.450806	0.317973
3	0	3	0	3	0	68.843528	4.655394	0.261620
3	0	3	0	2	1	165.715781	8.905547	0.497212
3	0	3	0	2	0	69.600018	4.620913	0.257430
3	0	3	0	1	2	388.683165	18.048377	0.989435
3	0	3	0	1	1	168.205220	8.859732	0.488733
3	0	3	0	1	0	75.364456	4.650669	0.243517
3	0	3	0	0	3	894.548894	36.876438	2.137907
3	0	3	0	0	2	396.127506	18.012160	0.971444
3	0	3	0	0	1	182.820874	9.100539	0.461131
3	0	3	0	0	0	87.022472	4.724978	0.221593
3	0	2	0	3	0	64.176565	4.383844	0.251387
3	0	2	0	2	1	155.032321	8.376989	0.476432
3	0	2	0	2	0	57.989980	3.879000	0.214002
3	0	2	0	1	2	365.070613	16.962805	0.945182
3	0	2	0	1	1	141.598962	7.409358	0.402189
3	0	2	0	1	0	55.343971	3.469406	0.173659
3	0	2	0	0	3	843.875250	34.647981	2.035468
3	0	2	0	0	2	336.556009	15.125474	0.787056

| Dice in Cup | | | Current FPs | | | Decision | Decision | Decision |
R	Y	G	R	Y	G	(SG 0)	(SG 1)	(SG 2)
3	0	2	0	0	1	136.371438	6.739134	0.323471
3	0	2	0	0	0	60.725255	3.211826	0.135416
3	0	1	0	3	0	59.923793	4.094891	0.239451
3	0	1	0	2	1	145.549562	7.817301	0.452308
3	0	1	0	2	0	45.468854	3.045497	0.162601
3	0	1	0	1	2	344.761451	15.817565	0.894054
3	0	1	0	1	1	112.527824	5.802033	0.301647
3	0	1	0	1	0	39.661062	2.386265	0.104933
3	0	1	0	0	3	801.910834	32.300952	1.917692
3	0	1	0	0	2	270.542202	11.934987	0.578494
3	0	1	0	0	1	99.361120	4.560385	0.191945
3	0	1	0	0	0	36.722285	1.879411	0.067232
3	0	0	0	3	0	54.921548	3.760358	0.225627
3	0	0	0	2	1	134.176368	7.166485	0.424510
3	0	0	0	2	0	37.214944	2.228861	0.100153
3	0	0	0	1	2	319.757593	14.475081	0.835439
3	0	0	0	1	1	93.242279	4.238269	0.182565
3	0	0	0	1	0	24.053724	1.356609	0.048445
3	0	0	0	0	3	748.493886	29.511502	1.783356
3	0	0	0	0	2	225.558661	8.798883	0.340534
3	0	0	0	0	1	61.956577	2.504455	0.086879
3	0	0	0	0	0	12.176259	0.698747	0.022125
2	4	6	1	0	0	75.224034	4.358624	0.188180
2	4	6	0	0	0	178.769353	10.142458	0.556630
2	4	5	1	0	1	129.473504	6.439814	0.268600
2	4	5	1	0	0	65.830827	3.908239	0.171001
2	4	5	0	0	1	282.450426	14.371293	0.792231
2	4	5	0	0	0	150.722869	8.758959	0.478828
2	4	4	1	0	2	244.273166	10.891791	0.415609
2	4	4	1	0	1	117.541669	5.920328	0.251114
2	4	4	1	0	0	56.404073	3.440623	0.152523
2	4	4	0	0	2	486.296431	22.374960	1.234854
2	4	4	0	0	1	244.624194	12.635585	0.694649
2	4	4	0	0	0	123.661052	7.379448	0.400614
2	4	3	1	0	2	232.841609	10.495781	0.406738
2	4	3	1	0	1	105.095554	5.362311	0.231497
2	4	3	1	0	0	47.068706	2.958651	0.132713
2	4	3	0	0	3	929.829454	38.849989	2.234049
2	4	3	0	0	2	437.696654	20.347443	1.119959
2	4	3	0	0	1	206.980527	10.852914	0.593019
2	4	3	0	0	0	97.978923	6.021275	0.323166
2	4	2	1	0	2	220.485551	10.061004	0.396836
2	4	2	1	0	1	92.230026	4.764464	0.209353

Dice in Cup			Current FPs			Decision	Decision	Decision
R	Y	G	R	Y	G	(SG 0)	(SG 1)	(SG 2)
2	4	2	1	0	0	37.949301	2.465940	0.111604
2	4	2	0	0	3	882.859712	37.183015	2.165118
2	4	2	0	0	2	387.481958	18.190751	0.993757
2	4	2	0	0	1	169.950190	9.036572	0.488248
2	4	2	0	0	0	74.283018	4.715007	0.248277
2	4	1	1	0	2	207.279646	9.584994	0.385745
2	4	1	1	0	1	79.004166	4.124808	0.184165
2	4	1	1	0	0	29.291412	1.970962	0.089402
2	4	1	0	0	3	833.229910	35.373549	2.088675
2	4	1	0	0	2	335.822535	15.900005	0.854887
2	4	1	0	0	1	134.447994	7.220673	0.382203
2	4	1	0	0	0	53.239404	3.498405	0.178379
2	4	0	1	0	2	193.161632	9.062771	0.373255
2	4	0	1	0	1	65.750314	3.447658	0.155347
2	4	0	1	0	0	21.379970	1.485797	0.066614
2	4	0	0	0	3	780.830032	33.405233	2.003670
2	4	0	0	0	2	284.006411	13.503327	0.702413
2	4	0	0	0	1	101.581247	5.450643	0.278049
2	4	0	0	0	0	34.649406	2.379859	0.115992
2	3	6	1	1	0	59.959047	3.625925	0.156466
2	3	6	1	0	0	80.530497	4.579415	0.196207
2	3	6	0	1	0	145.381174	8.589821	0.469805
2	3	6	0	0	0	196.932210	10.897721	0.592445
2	3	5	1	1	1	108.139681	5.385356	0.224987
2	3	5	1	1	0	54.668593	3.376951	0.147751
2	3	5	1	0	1	133.991769	6.623264	0.274667
2	3	5	1	0	0	69.977875	4.079170	0.177309
2	3	5	0	1	1	237.805166	12.326504	0.672798
2	3	5	0	1	0	126.955943	7.671750	0.421062
2	3	5	0	0	1	299.403247	15.066420	0.825783
2	3	5	0	0	0	164.812496	9.337398	0.504726
2	3	4	1	1	1	103.291732	5.204844	0.220663
2	3	4	1	1	0	49.104379	3.104708	0.137873
2	3	4	1	0	2	246.984499	10.969701	0.417225
2	3	4	1	0	1	120.825288	6.050016	0.255512
2	3	4	1	0	0	59.373318	3.557528	0.156790
2	3	4	0	1	2	430.925863	19.914266	1.062135
2	3	4	0	1	1	215.099245	11.292775	0.621014
2	3	4	0	1	0	108.395109	6.713785	0.368901
2	3	4	0	0	2	499.688195	22.874030	1.260641
2	3	4	0	0	1	257.022946	13.132575	0.716677
2	3	4	0	0	0	133.728499	7.774878	0.416165
2	3	3	1	1	1	98.016384	5.005150	0.215801

| Dice in Cup | | | Current FPs | | | Decision | Decision | Decision |
R	Y	G	R	Y	G	(SG 0)	(SG 1)	(SG 2)
2	3	3	1	1	0	43.299513	2.807052	0.126588
2	3	3	1	0	2	234.642080	10.534345	0.407391
2	3	3	1	0	1	107.086273	5.430405	0.233753
2	3	3	1	0	0	48.839768	3.017063	0.134577
2	3	3	0	1	2	409.886163	19.109509	1.032512
2	3	3	0	1	1	191.406021	10.178852	0.563009
2	3	3	0	1	0	89.904941	5.719813	0.313373
2	3	3	0	0	3	937.964153	39.040464	2.238368
2	3	3	0	0	2	446.285145	20.631803	1.132744
2	3	3	0	0	1	214.705788	11.135002	0.602286
2	3	3	0	0	0	104.344931	6.241734	0.328494
2	3	2	1	1	1	92.333277	4.784680	0.210310
2	3	2	1	1	0	37.275669	2.480891	0.113564
2	3	2	1	0	2	221.412122	10.055284	0.396310
2	3	2	1	0	1	92.815234	4.760674	0.208813
2	3	2	1	0	0	38.636913	2.465665	0.110736
2	3	2	0	1	2	387.441251	18.229719	0.999360
2	3	2	0	1	1	166.796508	8.978312	0.497644
2	3	2	0	1	0	71.900931	4.704790	0.254953
2	3	2	0	0	3	888.319959	37.227507	2.161330
2	3	2	0	0	2	390.829349	18.227296	0.990575
2	3	2	0	0	1	173.366413	9.106172	0.484069
2	3	2	0	0	0	77.483608	4.779592	0.244253
2	3	1	1	1	1	86.211347	4.540396	0.204061
2	3	1	1	1	0	31.166656	2.125835	0.098390
2	3	1	1	0	2	207.233739	9.526534	0.383737
2	3	1	1	0	1	78.338580	4.043669	0.180008
2	3	1	1	0	0	29.093019	1.915477	0.085587
2	3	1	0	1	2	363.539063	17.265504	0.962072
2	3	1	0	1	1	141.791783	7.700481	0.423743
2	3	1	0	1	0	54.937257	3.690820	0.194755
2	3	1	0	0	3	835.826887	35.248352	2.074959
2	3	1	0	0	2	334.509703	15.685694	0.832579
2	3	1	0	0	1	134.243122	7.092559	0.364787
2	3	1	0	0	0	52.942249	3.389278	0.166043
2	3	0	1	1	1	79.938315	4.275675	0.196963
2	3	0	1	1	0	25.175673	1.743005	0.080528
2	3	0	1	0	2	192.873677	8.957156	0.369515
2	3	0	1	0	1	64.155407	3.286546	0.146473
2	3	0	1	0	0	20.019561	1.364794	0.059695
2	3	0	0	1	2	339.502979	16.245422	0.920426
2	3	0	0	1	1	117.238852	6.363737	0.339989
2	3	0	0	1	0	38.649433	2.663658	0.134349

Dice in Cup			Current FPs			Decision	Decision	Decision
R	Y	G	R	Y	G	(SG 0)	(SG 1)	(SG 2)
2	3	0	0	0	3	783.787526	33.172880	1.978924
2	3	0	0	0	2	279.270737	13.052777	0.657459
2	3	0	0	0	1	96.176576	5.061866	0.247880
2	3	0	0	0	0	34.013566	2.284524	0.101175
2	2	6	1	2	0	46.370780	2.918755	0.124663
2	2	6	1	1	0	62.943327	3.758112	0.161097
2	2	6	1	0	0	87.519921	4.859872	0.206174
2	2	6	0	2	0	114.758520	7.088505	0.381937
2	2	6	0	1	0	157.540851	9.122736	0.495498
2	2	6	0	0	0	221.731061	11.896585	0.637974
2	2	5	1	2	0	44.403876	2.829304	0.122524
2	2	5	1	1	1	109.969821	5.448132	0.226434
2	2	5	1	1	0	57.075520	3.484796	0.151629
2	2	5	1	0	1	139.800390	6.855131	0.282180
2	2	5	1	0	0	75.550879	4.299324	0.185211
2	2	5	0	2	1	196.563879	10.290659	0.547564
2	2	5	0	2	0	104.388727	6.571226	0.357473
2	2	5	0	1	1	247.779106	12.754252	0.693001
2	2	5	0	1	0	136.677337	8.097426	0.441131
2	2	5	0	0	1	321.982115	15.978851	0.868206
2	2	5	0	0	0	184.385692	10.112890	0.537628
2	2	4	1	2	0	42.246673	2.729550	0.120100
2	2	4	1	1	1	104.722269	5.249880	0.221652
2	2	4	1	1	0	50.885813	3.183401	0.140747
2	2	4	1	0	2	250.544761	11.069265	0.419214
2	2	4	1	0	1	125.200889	6.218267	0.261046
2	2	4	1	0	0	63.451373	3.710204	0.162158
2	2	4	0	2	1	187.369486	9.904821	0.533885
2	2	4	0	2	0	93.446063	6.004342	0.329667
2	2	4	0	1	2	437.003413	20.113230	1.068028
2	2	4	0	1	1	222.627578	11.609787	0.635658
2	2	4	0	1	0	115.547982	7.019014	0.382298
2	2	4	0	0	2	517.314137	23.534606	1.293277
2	2	4	0	0	1	273.865290	13.799530	0.744588
2	2	4	0	0	0	148.178374	8.325614	0.435887
2	2	3	1	2	0	39.901020	2.618443	0.117340
2	2	3	1	1	1	99.043183	5.029744	0.216217
2	2	3	1	1	0	44.385955	2.849908	0.128098
2	2	3	1	0	2	237.351267	10.590174	0.408221
2	2	3	1	0	1	109.863716	5.522304	0.236633
2	2	3	1	0	0	51.475355	3.099113	0.136919
2	2	3	0	2	1	177.457478	9.479484	0.518434
2	2	3	0	2	0	81.955701	5.381448	0.297778

| Dice in Cup | | | Current FPs | | | Decision | Decision | Decision |
R	Y	G	R	Y	G	(SG 0)	(SG 1)	(SG 2)
2	2	3	0	1	2	414.566849	19.236000	1.034920
2	2	3	0	1	1	196.222859	10.363572	0.570453
2	2	3	0	1	0	94.536495	5.900427	0.319131
2	2	3	0	0	3	949.729148	39.332312	2.243892
2	2	3	0	0	2	457.977146	21.028909	1.149161
2	2	3	0	0	1	225.887513	11.548927	0.614110
2	2	3	0	0	0	114.060929	6.578758	0.335230
2	2	2	1	2	0	37.351474	2.494079	0.114166
2	2	2	1	1	1	92.902998	4.784249	0.209984
2	2	2	1	1	0	37.697976	2.482855	0.113232
2	2	2	1	0	2	223.171098	10.057966	0.395649
2	2	2	1	0	1	94.082608	4.767843	0.208168
2	2	2	1	0	0	39.985878	2.477809	0.109592
2	2	2	0	2	1	166.804606	9.009153	0.500858
2	2	2	0	2	0	70.107878	4.703234	0.260934
2	2	2	0	1	2	390.608332	18.270095	0.997369
2	2	2	0	1	1	168.997225	9.021940	0.495936
2	2	2	0	1	0	74.240807	4.763471	0.252323
2	2	2	0	0	3	897.219677	37.354269	2.156675
2	2	2	0	0	2	396.820258	18.349687	0.986976
2	2	2	0	0	1	179.362701	9.272958	0.478910
2	2	2	0	0	0	81.896068	4.861794	0.238065
2	2	1	1	2	0	34.697729	2.357532	0.110512
2	2	1	1	1	1	86.575856	4.516237	0.202835
2	2	1	1	1	0	31.034720	2.082459	0.095537
2	2	1	1	0	2	208.730591	9.480446	0.381282
2	2	1	1	0	1	78.377450	3.960454	0.174636
2	2	1	1	0	0	28.725840	1.839071	0.080312
2	2	1	0	2	1	155.876736	8.505588	0.480938
2	2	1	0	2	0	58.269127	3.974893	0.218022
2	2	1	0	1	2	366.296662	17.244384	0.954981
2	2	1	0	1	1	141.806403	7.602153	0.410379
2	2	1	0	1	0	54.253866	3.581876	0.182648
2	2	1	0	0	3	844.614887	35.273995	2.058629
2	2	1	0	0	2	335.780290	15.541714	0.804471
2	2	1	0	0	1	133.041222	6.912496	0.341190
2	2	1	0	0	0	55.384616	3.434466	0.153116
2	2	0	1	2	0	32.195718	2.213764	0.106328
2	2	0	1	1	1	80.746100	4.236849	0.194693
2	2	0	1	1	0	23.974744	1.627065	0.073839
2	2	0	1	0	2	195.798413	8.884817	0.365012
2	2	0	1	0	1	61.527357	3.062202	0.134023
2	2	0	1	0	0	20.004381	1.280366	0.051563

| Dice in Cup | | | Current FPs | | | Decision | Decision | Decision |
R	Y	G	R	Y	G	(SG 0)	(SG 1)	(SG 2)
2	2	0	0	2	1	145.893797	8.002584	0.458656
2	2	0	0	2	0	45.684628	3.149498	0.166770
2	2	0	0	1	2	344.641622	16.237468	0.907858
2	2	0	0	1	1	112.604317	6.013160	0.310022
2	2	0	0	1	0	38.120066	2.573621	0.116669
2	2	0	0	0	3	799.160431	33.275643	1.950325
2	2	0	0	0	2	269.496272	12.402178	0.595937
2	2	0	0	0	1	95.179447	4.906291	0.214704
2	2	0	0	0	0	31.915311	2.118415	0.082803
2	1	6	1	2	0	47.460695	2.966989	0.125795
2	1	6	1	1	0	66.809183	3.925829	0.166878
2	1	6	1	0	0	97.081821	5.227284	0.218867
2	1	6	0	3	0	87.665510	5.635625	0.295279
2	1	6	0	2	0	121.513236	7.411552	0.396743
2	1	6	0	1	0	174.004426	9.824999	0.528412
2	1	6	0	0	0	257.081111	13.268093	0.697530
2	1	5	1	2	0	45.329085	2.869046	0.123441
2	1	5	1	1	1	112.289080	5.527564	0.228236
2	1	5	1	1	0	60.260062	3.624421	0.156549
2	1	5	1	0	1	147.495818	7.157490	0.291726
2	1	5	1	0	0	83.325613	4.591303	0.195368
2	1	5	0	3	0	83.784045	5.442488	0.288748
2	1	5	0	2	1	200.542876	10.451295	0.552652
2	1	5	0	2	0	109.946517	6.841970	0.369923
2	1	5	0	1	1	260.789077	13.316159	0.718769
2	1	5	0	1	0	150.087331	8.667419	0.467053
2	1	5	0	0	1	353.055779	17.216181	0.923366
2	1	5	0	0	0	213.011264	11.207950	0.580819
2	1	4	1	2	0	42.996247	2.759217	0.120740
2	1	4	1	1	1	106.643667	5.309524	0.222905
2	1	4	1	1	0	53.292760	3.287422	0.144453
2	1	4	1	0	2	255.352433	11.201555	0.421733
2	1	4	1	0	1	131.122574	6.442125	0.268188
2	1	4	1	0	0	69.387154	3.919408	0.169125
2	1	4	0	3	0	79.549330	5.227557	0.281297
2	1	4	0	2	1	190.652003	10.030468	0.537392
2	1	4	0	2	0	97.651492	6.209696	0.338904
2	1	4	0	1	2	444.914940	20.384214	1.075496
2	1	4	0	1	1	232.619114	12.036762	0.654619
2	1	4	0	1	0	125.754276	7.446092	0.399797
2	1	4	0	0	2	540.641214	24.429749	1.335725
2	1	4	0	0	1	297.801279	14.746755	0.781241
2	1	4	0	0	0	170.153283	9.141756	0.461851

| Dice in Cup | | | Current FPs | | | Decision | Decision | Decision |
R	Y	G	R	Y	G	(SG 0)	(SG 1)	(SG 2)
2	1	3	1	2	0	40.448214	2.635409	0.117610
2	1	3	1	1	1	100.513110	5.064606	0.216741
2	1	3	1	1	0	46.000976	2.911893	0.130108
2	1	3	1	0	2	241.227509	10.669546	0.409265
2	1	3	1	0	1	113.989518	5.655464	0.240470
2	1	3	1	0	0	55.642041	3.222183	0.139979
2	1	3	0	3	0	74.951578	4.987535	0.272724
2	1	3	0	2	1	179.978031	9.562480	0.519879
2	1	3	0	2	0	84.753029	5.513183	0.302704
2	1	3	0	1	2	420.932704	19.423567	1.037972
2	1	3	0	1	1	203.067259	10.641485	0.580467
2	1	3	0	1	0	101.597886	6.181406	0.326805
2	1	3	0	0	3	965.584783	39.767357	2.250896
2	1	3	0	0	2	474.428377	21.630243	1.171323
2	1	3	0	0	1	242.755216	12.199647	0.629961
2	1	3	0	0	0	128.652432	7.044701	0.342455
2	1	2	1	2	0	37.771254	2.498191	0.113964
2	1	2	1	1	1	94.133309	4.794517	0.209586
2	1	2	1	1	0	38.595979	2.496919	0.112848
2	1	2	1	0	2	226.691609	10.085931	0.394843
2	1	2	1	0	1	96.607409	4.800643	0.207433
2	1	2	1	0	0	41.778832	2.485984	0.107547
2	1	2	0	3	0	70.133886	4.727022	0.262854
2	1	2	0	2	1	168.885716	9.058148	0.499787
2	1	2	0	2	0	71.599033	4.756536	0.260050
2	1	2	0	1	2	396.244536	18.396703	0.995077
2	1	2	0	1	1	172.928427	9.145498	0.494195
2	1	2	0	1	0	77.265009	4.834254	0.247842
2	1	2	0	0	3	912.123763	37.688979	2.151346
2	1	2	0	0	2	406.889945	18.656146	0.983388
2	1	2	0	0	1	186.855462	9.486774	0.470005
2	1	2	0	0	0	91.860614	5.208164	0.232017
2	1	1	1	2	0	35.220317	2.352152	0.109723
2	1	1	1	1	1	88.181286	4.509568	0.201303
2	1	1	1	1	0	30.614960	2.015470	0.091302
2	1	1	1	0	2	213.477134	9.475837	0.378231
2	1	1	1	0	1	77.668834	3.828497	0.166646
2	1	1	1	0	0	30.260693	1.812557	0.073673
2	1	1	0	3	0	65.556554	4.462093	0.251561
2	1	1	0	2	1	158.528703	8.552783	0.476926
2	1	1	0	2	0	57.456514	3.882173	0.208219
2	1	1	0	1	2	373.646283	17.385601	0.946541
2	1	1	0	1	1	140.277736	7.436229	0.391021

| Dice in Cup | | | Current FPs | | | Decision | Decision | Decision |
R	Y	G	R	Y	G	(SG 0)	(SG 1)	(SG 2)
2	1	1	0	1	0	56.074572	3.645350	0.168329
2	1	1	0	0	3	864.290970	35.687114	2.039350
2	1	1	0	0	2	333.133404	15.261499	0.763773
2	1	1	0	0	1	137.400778	7.090538	0.313786
2	1	1	0	0	0	58.311758	3.437054	0.134701
2	1	0	1	2	0	32.305277	2.179268	0.104513
2	1	0	1	1	1	81.312995	4.171950	0.191162
2	1	0	1	1	0	24.741190	1.549621	0.064667
2	1	0	1	0	2	197.980220	8.751159	0.357955
2	1	0	1	0	1	63.841597	2.909935	0.117161
2	1	0	1	0	0	18.824373	1.116446	0.039449
2	1	0	0	3	0	60.387716	4.151540	0.238124
2	1	0	0	2	1	146.755601	7.958370	0.449826
2	1	0	0	2	0	46.311798	3.132139	0.146499
2	1	0	0	1	2	347.714037	16.188183	0.889229
2	1	0	0	1	1	114.248292	6.011436	0.271301
2	1	0	0	1	0	35.465285	2.365778	0.093249
2	1	0	0	0	3	808.730718	33.286240	1.907602
2	1	0	0	0	2	273.213861	12.525908	0.517816
2	1	0	0	0	1	88.331265	4.472522	0.170053
2	1	0	0	0	0	23.950185	1.533982	0.054175
2	0	6	1	2	0	48.822212	3.028017	0.127220
2	0	6	1	1	0	71.968965	4.144746	0.174282
2	0	6	1	0	0	110.769281	5.724848	0.235507
2	0	6	0	3	0	90.053613	5.758359	0.299183
2	0	6	0	2	0	130.397129	7.835741	0.415776
2	0	6	0	1	0	197.279608	10.782259	0.571865
2	0	6	0	0	0	310.347495	15.254792	0.778360
2	0	5	1	2	0	46.516676	2.920581	0.124609
2	0	5	1	1	1	115.266268	5.630586	0.230529
2	0	5	1	1	0	64.583311	3.809861	0.162960
2	0	5	1	0	1	157.931883	7.562650	0.304190
2	0	5	1	0	0	94.817856	4.995401	0.208859
2	0	5	0	3	0	85.842840	5.548033	0.291928
2	0	5	0	2	1	205.475965	10.662159	0.559128
2	0	5	0	2	0	117.370298	7.204636	0.386189
2	0	5	0	1	1	278.096699	14.074448	0.752557
2	0	5	0	1	0	169.538085	9.468133	0.501732
2	0	5	0	0	1	397.911724	18.981054	0.997790
2	0	5	0	0	0	257.674333	12.854166	0.639844
2	0	4	1	2	0	43.983589	2.798457	0.121554
2	0	4	1	1	1	109.172737	5.388347	0.224496
2	0	4	1	1	0	56.696379	3.431218	0.149396

| Dice in Cup | | | Current FPs | | | Decision | Decision | Decision |
R	Y	G	R	Y	G	(SG 0)	(SG 1)	(SG 2)
2	0	4	1	0	2	261.676861	11.376204	0.424928
2	0	4	1	0	1	139.477857	6.753409	0.277724
2	0	4	1	0	0	78.612178	4.220726	0.178505
2	0	4	0	3	0	81.245374	5.311546	0.283494
2	0	4	0	2	1	194.796525	10.200629	0.541848
2	0	4	0	2	0	103.424424	6.499980	0.351268
2	0	4	0	1	2	454.830991	20.751952	1.084974
2	0	4	0	1	1	246.236515	12.643742	0.680076
2	0	4	0	1	0	141.160754	8.078926	0.423592
2	0	4	0	0	2	572.192223	25.707126	1.393044
2	0	4	0	0	1	333.641280	16.174999	0.831434
2	0	4	0	0	0	205.786216	10.345459	0.495274
2	0	3	1	2	0	41.283607	2.661240	0.117943
2	0	3	1	1	1	102.731348	5.117184	0.217385
2	0	3	1	1	0	48.496168	3.006390	0.132890
2	0	3	1	0	2	247.001121	10.787820	0.410542
2	0	3	1	0	1	120.300200	5.857781	0.245785
2	0	3	1	0	0	61.867502	3.381011	0.143481
2	0	3	0	3	0	76.345053	5.052323	0.273633
2	0	3	0	2	1	183.491781	9.697938	0.521703
2	0	3	0	2	0	88.841671	5.724304	0.309607
2	0	3	0	1	2	429.627545	19.727367	1.041806
2	0	3	0	1	1	212.895600	11.085830	0.594536
2	0	3	0	1	0	112.091769	6.565748	0.336062
2	0	3	0	0	3	986.821757	40.464650	2.259657
2	0	3	0	0	2	497.613334	22.589117	1.202604
2	0	3	0	0	1	267.491832	13.117118	0.649014
2	0	3	0	0	0	156.100250	8.002941	0.352737
2	0	2	1	2	0	38.654802	2.512617	0.113654
2	0	2	1	1	1	96.566747	4.825284	0.208973
2	0	2	1	1	0	39.539067	2.504469	0.111940
2	0	2	1	0	2	233.242292	10.158246	0.393599
2	0	2	1	0	1	99.179769	4.817725	0.205631
2	0	2	1	0	0	46.736050	2.561485	0.104430
2	0	2	0	3	0	71.549584	4.786427	0.262105
2	0	2	0	2	1	172.564271	9.189073	0.498259
2	0	2	0	2	0	73.059748	4.813012	0.258215
2	0	2	0	1	2	405.596103	18.706107	0.991810
2	0	2	0	1	1	176.657836	9.274034	0.490430
2	0	2	0	1	0	84.538313	5.160140	0.241723
2	0	2	0	0	3	935.512367	38.437096	2.143753
2	0	2	0	0	2	416.172914	18.969955	0.975277
2	0	2	0	0	1	203.828966	10.257663	0.458049

Dice in Cup			Current FPs			Decision	Decision	Decision
R	Y	G	R	Y	G	(SG 0)	(SG 1)	(SG 2)
2	0	2	0	0	0	110.014635	5.684696	0.219869
2	0	1	1	2	0	35.682105	2.333507	0.108242
2	0	1	1	1	1	89.567729	4.473216	0.198381
2	0	1	1	1	0	32.069662	1.981405	0.084793
2	0	1	1	0	2	217.491437	9.397259	0.372303
2	0	1	1	0	1	81.397280	3.744639	0.154413
2	0	1	1	0	0	31.708175	1.706759	0.061343
2	0	1	0	3	0	66.255408	4.471799	0.248106
2	0	1	0	2	1	160.511794	8.586368	0.469871
2	0	1	0	2	0	58.976134	3.989075	0.193630
2	0	1	0	1	2	379.082164	17.492884	0.931429
2	0	1	0	1	1	143.804226	7.651829	0.362393
2	0	1	0	1	0	57.531408	3.670846	0.147247
2	0	1	0	0	3	878.794121	36.012831	2.004140
2	0	1	0	0	2	341.242640	15.756497	0.704021
2	0	1	0	0	1	139.723805	7.184207	0.271960
2	0	1	0	0	0	61.128401	2.846840	0.097440
2	0	0	1	2	0	37.213124	2.228752	0.100148
2	0	0	1	1	1	93.237717	4.238062	0.182556
2	0	0	1	1	0	24.053724	1.356609	0.048445
2	0	0	1	0	2	225.547625	8.798453	0.340517
2	0	0	1	0	1	61.956577	2.504455	0.086879
2	0	0	1	0	0	12.176259	0.698747	0.022125
2	0	0	0	3	0	67.529274	4.605739	0.227277
2	0	0	0	2	1	163.111839	8.860542	0.427658
2	0	0	0	2	0	44.378347	2.938875	0.116424
2	0	0	0	1	2	384.372458	18.057849	0.841677
2	0	0	0	1	1	108.981726	5.635394	0.213327
2	0	0	0	1	0	23.950185	1.533982	0.054175
2	0	0	0	0	3	891.999859	37.086498	1.796636
2	0	0	0	0	2	258.895915	11.840472	0.400016
2	0	0	0	0	1	61.128401	2.846840	0.097440
1	4	6	2	0	0	36.495937	2.006041	0.073846
1	4	6	1	0	0	99.330279	5.454976	0.227699
1	4	6	0	0	0	254.905934	13.737302	0.752707
1	4	5	2	0	1	62.600833	2.761407	0.099488
1	4	5	2	0	0	33.468211	1.888967	0.070506
1	4	5	1	0	1	154.869202	7.509470	0.302578
1	4	5	1	0	0	87.536491	4.924985	0.208802
1	4	5	0	0	1	367.870744	18.188431	0.999802
1	4	5	0	0	0	216.827108	11.982499	0.655714
1	4	4	2	0	1	59.970297	2.681092	0.097977
1	4	4	2	0	0	30.248353	1.758900	0.066669

Dice in Cup			Current FPs			Decision	Decision	Decision
R	Y	G	R	Y	G	(SG 0)	(SG 1)	(SG 2)
1	4	4	1	0	2	264.642429	11.587235	0.430765
1	4	4	1	0	1	140.967595	6.928063	0.284226
1	4	4	1	0	0	75.480244	4.364551	0.188019
1	4	4	0	0	2	576.446125	26.084254	1.435155
1	4	4	0	0	1	319.938082	16.098411	0.884570
1	4	4	0	0	0	179.324530	10.193572	0.555658
1	4	3	2	0	1	57.078491	2.591319	0.096260
1	4	3	2	0	0	26.845744	1.614054	0.062217
1	4	3	1	0	2	252.359952	11.160161	0.421257
1	4	3	1	0	1	126.267588	6.292775	0.263235
1	4	3	1	0	0	63.296737	3.774925	0.165158
1	4	3	0	0	3	1003.357582	41.450776	2.336347
1	4	3	0	0	2	519.225513	23.767854	1.307584
1	4	3	0	0	1	271.372834	13.906431	0.761554
1	4	3	0	0	0	143.008968	8.397856	0.454058
1	4	2	2	0	1	53.922116	2.491046	0.094302
1	4	2	2	0	0	23.271231	1.451982	0.056984
1	4	2	1	0	2	239.072456	10.687425	0.410498
1	4	2	1	0	1	110.814165	5.598042	0.238979
1	4	2	1	0	0	51.214865	3.162396	0.140144
1	4	2	0	0	3	952.828296	39.651483	2.260913
1	4	2	0	0	2	459.196326	21.255572	1.164171
1	4	2	0	0	1	222.898510	11.635863	0.631693
1	4	2	0	0	0	108.830244	6.638355	0.353349
1	4	1	2	0	1	50.484345	2.378574	0.092048
1	4	1	2	0	0	19.584172	1.271043	0.050751
1	4	1	1	0	2	224.769014	10.163233	0.398237
1	4	1	1	0	1	94.843247	4.844306	0.210720
1	4	1	1	0	0	39.589644	2.538554	0.113088
1	4	1	0	0	3	899.183804	37.677227	2.175915
1	4	1	0	0	2	397.091365	18.558635	1.002867
1	4	1	0	0	1	175.804038	9.330065	0.497177
1	4	1	0	0	0	77.144941	4.922076	0.255500
1	4	0	2	0	1	46.880072	2.254845	0.089461
1	4	0	2	0	0	15.894266	1.070092	0.043212
1	4	0	1	0	2	210.026302	9.597136	0.384392
1	4	0	1	0	1	78.839207	4.037521	0.177484
1	4	0	1	0	0	28.334033	1.898347	0.084040
1	4	0	0	0	3	844.692639	35.582192	2.081595
1	4	0	0	0	2	334.707277	15.716352	0.821648
1	4	0	0	0	1	129.643807	6.953637	0.359474
1	4	0	0	0	0	50.691656	3.461757	0.171078
1	3	6	2	1	0	25.763503	1.534180	0.055951

| Dice in Cup | | | Current FPs | | | Decision | Decision | Decision |
R	Y	G	R	Y	G	(SG 0)	(SG 1)	(SG 2)
1	3	6	2	0	0	39.028261	2.100776	0.076567
1	3	6	1	1	0	73.608192	4.239740	0.176990
1	3	6	1	0	0	110.278027	5.907606	0.243200
1	3	6	0	1	0	196.321294	11.029345	0.593560
1	3	6	0	0	0	293.191737	15.371179	0.833660
1	3	5	2	1	0	24.760756	1.494111	0.055205
1	3	5	2	0	1	64.183203	2.810058	0.100403
1	3	5	2	0	0	35.655352	1.973681	0.073008
1	3	5	1	1	1	118.406559	5.768594	0.233991
1	3	5	1	1	0	67.453212	3.971043	0.168160
1	3	5	1	0	1	164.216230	7.896281	0.314527
1	3	5	1	0	0	96.767986	5.312837	0.222434
1	3	5	0	1	1	288.862101	14.628626	0.782472
1	3	5	0	1	0	172.792075	9.941115	0.538436
1	3	5	0	0	1	403.364299	19.711163	1.076685
1	3	5	0	0	0	248.401947	13.350659	0.722234
1	3	4	2	1	0	23.647804	1.448931	0.054352
1	3	4	2	0	1	61.329488	2.722488	0.098755
1	3	4	2	0	0	32.042244	1.830881	0.068856
1	3	4	1	1	1	113.218595	5.577018	0.229456
1	3	4	1	1	0	60.857373	3.671490	0.157946
1	3	4	1	0	2	270.380641	11.785834	0.435056
1	3	4	1	0	1	148.683197	7.252352	0.294536
1	3	4	1	0	0	82.862532	4.677376	0.199234
1	3	4	0	1	2	469.580742	21.433040	1.116260
1	3	4	0	1	1	262.010973	13.465985	0.727197
1	3	4	0	1	0	148.572673	8.778231	0.477751
1	3	4	0	0	2	605.160521	27.294615	1.498647
1	3	4	0	0	1	348.388606	17.344489	0.946135
1	3	4	0	0	0	203.873542	11.275678	0.606283
1	3	3	2	1	0	22.419014	1.397982	0.053371
1	3	3	2	0	1	58.192031	2.624013	0.096862
1	3	3	2	0	0	28.193225	1.669551	0.063942
1	3	3	1	1	1	107.536263	5.362749	0.224281
1	3	3	1	1	0	53.834015	3.336357	0.145983
1	3	3	1	0	2	257.138797	11.318633	0.424565
1	3	3	1	0	1	132.151632	6.539827	0.271253
1	3	3	1	0	0	68.742113	4.005092	0.173355
1	3	3	0	1	2	446.764421	20.569385	1.084464
1	3	3	0	1	1	233.490801	12.184195	0.663720
1	3	3	0	1	0	123.902140	7.546304	0.411281
1	3	3	0	0	3	1020.218528	42.094028	2.359570
1	3	3	0	0	2	541.081874	24.702374	1.355842

| Dice in Cup | | | Current FPs | | | Decision | Decision | Decision |
R	Y	G	R	Y	G	(SG 0)	(SG 1)	(SG 2)
1	3	3	0	0	1	292.294448	14.846586	0.805346
1	3	3	0	0	0	160.605859	9.187021	0.487930
1	3	2	2	1	0	21.066401	1.340252	0.052232
1	3	2	2	0	1	54.756726	2.512833	0.094669
1	3	2	2	0	0	24.152278	1.487436	0.058045
1	3	2	1	1	1	101.355746	5.122618	0.218325
1	3	2	1	1	0	46.448639	2.962694	0.131822
1	3	2	1	0	2	242.829050	10.797213	0.412524
1	3	2	1	0	1	114.770634	5.756018	0.243886
1	3	2	1	0	0	54.764500	3.306185	0.144672
1	3	2	0	1	2	422.274751	19.612019	1.048249
1	3	2	0	1	1	203.461666	10.779645	0.590475
1	3	2	0	1	0	99.352055	6.262979	0.339256
1	3	2	0	0	3	966.214705	40.131125	2.275241
1	3	2	0	0	2	473.661947	21.879589	1.193208
1	3	2	0	0	1	236.312240	12.254811	0.655657
1	3	2	0	0	0	119.443517	7.089816	0.368400
1	3	1	2	1	0	19.622344	1.275909	0.050911
1	3	1	2	0	1	51.120765	2.389610	0.092136
1	3	1	2	0	0	20.026448	1.282861	0.050847
1	3	1	1	1	1	94.849208	4.859452	0.211526
1	3	1	1	1	0	38.890325	2.549823	0.114838
1	3	1	1	0	2	227.910111	10.229601	0.398828
1	3	1	1	0	1	96.994075	4.904644	0.211378
1	3	1	1	0	0	40.876337	2.570941	0.112686
1	3	1	0	1	2	396.763538	18.578761	1.007577
1	3	1	0	1	1	172.636245	9.263031	0.505572
1	3	1	0	1	0	75.009142	4.905873	0.260954
1	3	1	0	0	3	910.523213	38.030219	2.180926
1	3	1	0	0	2	404.481499	18.857691	1.007940
1	3	1	0	0	1	180.438204	9.519597	0.496468
1	3	1	0	0	0	82.551855	5.201810	0.260204
1	3	0	2	1	0	18.201010	1.207253	0.049404
1	3	0	2	0	1	47.605880	2.259494	0.089262
1	3	0	2	0	0	15.651366	1.041812	0.041641
1	3	0	1	1	1	88.616598	4.587202	0.203958
1	3	0	1	1	0	30.857523	2.072558	0.093520
1	3	0	1	0	2	213.914142	9.650026	0.383682
1	3	0	1	0	1	77.947474	3.940068	0.171002
1	3	0	1	0	0	28.964551	1.907006	0.080391
1	3	0	0	1	2	372.725868	17.541384	0.963563
1	3	0	0	1	1	139.820542	7.546798	0.403109
1	3	0	0	1	0	53.376718	3.660564	0.185252

Dice in Cup			Current FPs			Decision	Decision	Decision
R	Y	G	R	Y	G	(SG 0)	(SG 1)	(SG 2)
1	3	0	0	0	3	859.078910	35.957568	2.079613
1	3	0	0	0	2	330.390519	15.450046	0.789356
1	3	0	0	0	1	130.202639	7.008153	0.347293
1	3	0	0	0	0	49.836786	3.436737	0.161457
1	2	6	2	1	0	26.614243	1.569508	0.056618
1	2	6	2	0	0	42.363332	2.222452	0.080013
1	2	6	1	2	0	51.365838	3.150508	0.130149
1	2	6	1	1	0	79.563358	4.497656	0.185504
1	2	6	1	0	0	125.331949	6.514054	0.263469
1	2	6	0	2	0	143.414453	8.495363	0.445875
1	2	6	0	1	0	221.527079	12.150970	0.647632
1	2	6	0	0	0	347.839255	17.649754	0.944380
1	2	5	2	1	0	25.528411	1.526188	0.055815
1	2	5	2	0	1	66.162654	2.872309	0.101580
1	2	5	2	0	0	38.576294	2.084455	0.076239
1	2	5	1	2	0	49.279419	3.057759	0.127975
1	2	5	1	1	1	121.853890	5.903991	0.237212
1	2	5	1	1	0	72.655523	4.204406	0.176083
1	2	5	1	0	1	176.489497	8.407594	0.330056
1	2	5	1	0	0	109.703758	5.840216	0.240551
1	2	5	0	2	1	216.578279	11.179028	0.578586
1	2	5	0	2	0	131.161166	7.934081	0.420992
1	2	5	0	1	1	308.877978	15.549755	0.825509
1	2	5	0	1	0	194.371173	10.922308	0.586146
1	2	5	0	0	1	452.107527	21.791147	1.180748
1	2	5	0	0	0	294.129976	15.292307	0.814584
1	2	4	2	1	0	24.317809	1.476927	0.054885
1	2	4	2	0	1	63.066228	2.776848	0.099781
1	2	4	2	0	0	34.479236	1.926944	0.071743
1	2	4	1	2	0	46.960445	2.952779	0.125471
1	2	4	1	1	1	116.213664	5.695065	0.232258
1	2	4	1	1	0	65.194247	3.872674	0.164966
1	2	4	1	0	2	277.478312	12.045361	0.440730
1	2	4	1	0	1	158.983906	7.689440	0.308229
1	2	4	1	0	0	93.420122	5.112528	0.214484
1	2	4	0	2	1	206.542035	10.769730	0.564310
1	2	4	0	2	0	117.959916	7.303132	0.391955
1	2	4	0	1	2	480.286079	21.912703	1.133427
1	2	4	0	1	1	278.632608	14.254706	0.764711
1	2	4	0	1	0	166.021003	9.594279	0.517386
1	2	4	0	0	2	642.435044	28.917396	1.584025
1	2	4	0	0	1	387.975540	19.086005	1.031339
1	2	4	0	0	0	240.164005	12.848513	0.677967

| Dice in Cup | | | Current FPs | | | Decision | Decision | Decision |
R	Y	G	R	Y	G	(SG 0)	(SG 1)	(SG 2)
1	2	3	2	1	0	22.976455	1.420703	0.053797
1	2	3	2	0	1	59.654408	2.668266	0.097681
1	2	3	2	0	0	30.087769	1.746914	0.066306
1	2	3	1	2	0	44.408381	2.833794	0.122560
1	2	3	1	1	1	110.044260	5.459162	0.226512
1	2	3	1	1	0	57.186138	3.497714	0.151694
1	2	3	1	0	2	263.192608	11.531898	0.429079
1	2	3	1	0	1	140.204919	6.886786	0.282335
1	2	3	1	0	0	76.788717	4.338292	0.184893
1	2	3	0	2	1	195.640570	10.310700	0.547868
1	2	3	0	2	0	103.774789	6.594478	0.357811
1	2	3	0	1	2	455.612189	20.967183	1.098106
1	2	3	0	1	1	246.156091	12.812421	0.693781
1	2	3	0	1	0	136.938153	8.178722	0.440998
1	2	3	0	0	3	1040.139695	42.941966	2.391360
1	2	3	0	0	2	569.687535	26.000231	1.423456
1	2	3	0	0	1	322.057700	16.211167	0.868176
1	2	3	0	0	0	187.812414	10.331383	0.534513
1	2	2	2	1	0	21.525067	1.357430	0.052526
1	2	2	2	0	1	55.992083	2.546696	0.095236
1	2	2	2	0	0	25.493872	1.542133	0.059615
1	2	2	1	2	0	41.657602	2.701217	0.119198
1	2	2	1	1	1	103.444474	5.197783	0.219897
1	2	2	1	1	0	48.783989	3.077514	0.135645
1	2	2	1	0	2	248.045317	10.966489	0.415711
1	2	2	1	0	1	120.511577	5.999977	0.251286
1	2	2	1	0	0	59.898798	3.504447	0.150682
1	2	2	0	2	1	183.982059	9.805460	0.529166
1	2	2	0	2	0	88.837955	5.806992	0.317356
1	2	2	0	1	2	429.428161	19.933162	1.058072
1	2	2	0	1	1	211.962269	11.226617	0.610649
1	2	2	0	1	0	107.637143	6.653080	0.354939
1	2	2	0	0	3	982.861681	40.837414	2.298200
1	2	2	0	0	2	493.128883	22.816551	1.237950
1	2	2	0	0	1	255.386840	13.117302	0.688051
1	2	2	0	0	0	137.106536	7.907914	0.397716
1	2	1	2	1	0	20.069304	1.289338	0.051068
1	2	1	2	0	1	52.376613	2.417131	0.092446
1	2	1	2	0	0	20.544476	1.296544	0.050900
1	2	1	1	2	0	38.909791	2.561024	0.115412
1	2	1	1	1	1	96.944586	4.924354	0.212493
1	2	1	1	1	0	39.761244	2.584350	0.115142
1	2	1	1	0	2	233.375832	10.382087	0.400837

Dice in Cup			Current FPs			Decision	Decision	Decision
R	Y	G	R	Y	G	(SG 0)	(SG 1)	(SG 2)
1	2	1	1	0	1	99.249333	4.976793	0.211976
1	2	1	1	0	0	44.183450	2.715082	0.114421
1	2	1	0	2	1	172.425763	9.282878	0.508667
1	2	1	0	2	0	72.952697	4.892927	0.266915
1	2	1	0	1	2	403.798359	18.876828	1.014484
1	2	1	0	1	1	175.505898	9.407103	0.508312
1	2	1	0	1	0	79.388416	5.175955	0.267629
1	2	1	0	0	3	927.643637	38.719306	2.197464
1	2	1	0	0	2	411.233215	19.180670	1.013834
1	2	1	0	0	1	190.395904	10.107129	0.510643
1	2	1	0	0	0	90.761209	5.558196	0.264929
1	2	0	2	1	0	18.456181	1.207848	0.049197
1	2	0	2	0	1	48.368623	2.262080	0.088871
1	2	0	2	0	0	16.420494	1.053715	0.040145
1	2	0	1	2	0	35.913847	2.398081	0.110681
1	2	0	1	1	1	89.861198	4.607142	0.203257
1	2	0	1	1	0	31.901935	2.111151	0.090935
1	2	0	1	0	2	217.360495	9.704886	0.382320
1	2	0	1	0	1	80.673467	4.018402	0.166560
1	2	0	1	0	0	28.818615	1.893445	0.073081
1	2	0	0	2	1	160.175660	8.683242	0.483902
1	2	0	0	2	0	58.368393	4.018596	0.210469
1	2	0	0	1	2	376.732214	17.665171	0.961952
1	2	0	0	1	1	141.678750	7.693781	0.396774
1	2	0	0	1	0	52.163799	3.617357	0.173873
1	2	0	0	0	3	869.463832	36.286704	2.076331
1	2	0	0	0	2	334.421964	15.806673	0.777074
1	2	0	0	0	1	126.405479	6.876649	0.324057
1	2	0	0	0	0	43.945983	2.984539	0.129952
1	1	6	2	1	0	27.675562	1.614872	0.057483
1	1	6	2	0	0	46.897292	2.383097	0.084506
1	1	6	1	2	0	53.292457	3.246312	0.132474
1	1	6	1	1	0	87.539269	4.838952	0.196654
1	1	6	1	0	0	147.081038	7.359006	0.290910
1	1	6	0	3	0	97.786135	6.184387	0.313889
1	1	6	0	2	0	156.727286	9.143542	0.475218
1	1	6	0	1	0	257.383154	13.704121	0.721651
1	1	6	0	0	0	429.801576	20.981901	1.103271
1	1	5	2	1	0	26.493831	1.568086	0.056622
1	1	5	2	0	1	68.642333	2.953577	0.103137
1	1	5	2	0	0	42.622915	2.233666	0.080548
1	1	5	1	2	0	51.006858	3.145798	0.130137
1	1	5	1	1	1	126.033836	6.079383	0.241481

Dice in Cup R	Y	G	Current FPs R	Y	G	Decision (SG 0)	Decision (SG 1)	Decision (SG 2)
1	1	5	1	1	0	79.754679	4.519343	0.186688
1	1	5	1	0	1	193.140474	9.105618	0.350908
1	1	5	1	0	0	128.781592	6.589676	0.265662
1	1	5	0	3	0	93.531189	5.983340	0.307347
1	1	5	0	2	1	223.035992	11.514444	0.590829
1	1	5	0	2	0	142.917083	8.529509	0.448643
1	1	5	0	1	1	335.474475	16.795568	0.883736
1	1	5	0	1	0	225.565942	12.305281	0.653011
1	1	5	0	0	1	522.150045	24.748960	1.328499
1	1	5	0	0	0	364.348838	18.196802	0.949925
1	1	4	2	1	0	25.173891	1.514213	0.055606
1	1	4	2	0	1	65.276098	2.849182	0.101170
1	1	4	2	0	0	37.926939	2.060329	0.075720
1	1	4	1	2	0	48.472539	3.030840	0.127396
1	1	4	1	1	1	119.891719	5.850706	0.236053
1	1	4	1	1	0	71.195685	4.152167	0.174678
1	1	4	1	0	2	286.105214	12.386725	0.448413
1	1	4	1	0	1	173.134533	8.303177	0.327229
1	1	4	1	0	0	109.436610	5.748453	0.236300
1	1	4	0	3	0	88.847893	5.754860	0.299721
1	1	4	0	2	1	212.057069	11.066103	0.575158
1	1	4	0	2	0	127.731121	7.827623	0.417053
1	1	4	0	1	2	492.530467	22.528706	1.156702
1	1	4	0	1	1	300.783800	15.345792	0.817208
1	1	4	0	1	0	191.829306	10.772893	0.574762
1	1	4	0	0	2	691.753231	31.159157	1.704509
1	1	4	0	0	1	446.194384	21.629315	1.156155
1	1	4	0	0	0	300.384520	15.277443	0.782454
1	1	3	2	1	0	23.717809	1.452713	0.054406
1	1	3	2	0	1	61.588438	2.730517	0.098853
1	1	3	2	0	0	32.864626	1.860227	0.069734
1	1	3	1	2	0	45.682863	2.900752	0.124193
1	1	3	1	1	1	113.174906	5.593166	0.229728
1	1	3	1	1	0	61.935542	3.733610	0.160036
1	1	3	1	0	2	270.645481	11.827269	0.435582
1	1	3	1	0	1	151.496333	7.398134	0.298571
1	1	3	1	0	0	89.550420	4.823770	0.201167
1	1	3	0	3	0	83.695005	5.497451	0.290911
1	1	3	0	2	1	200.050112	10.563361	0.557108
1	1	3	0	2	0	111.245015	7.032033	0.379279
1	1	3	0	1	2	465.474891	21.495796	1.117908
1	1	3	0	1	1	263.139060	13.715138	0.738344
1	1	3	0	1	0	157.641608	9.099844	0.483605

Dice in Cup			Current FPs			Decision	Decision	Decision
R	Y	G	R	Y	G	(SG 0)	(SG 1)	(SG 2)
1	1	3	0	0	3	1061.991865	44.052494	2.437501
1	1	3	0	0	2	607.742584	27.854153	1.524436
1	1	3	0	0	1	369.124272	18.234621	0.958839
1	1	3	0	0	0	232.864390	12.223450	0.612717
1	1	2	2	1	0	22.211534	1.385513	0.053018
1	1	2	2	0	1	57.820769	2.601918	0.096186
1	1	2	2	0	0	27.351888	1.616901	0.061798
1	1	2	1	2	0	42.791116	2.760654	0.120557
1	1	2	1	1	1	106.281561	5.318270	0.222583
1	1	2	1	1	0	51.956013	3.237927	0.141033
1	1	2	1	0	2	254.964012	11.235964	0.421165
1	1	2	1	0	1	128.139912	6.340087	0.261638
1	1	2	1	0	0	69.191439	3.889127	0.161986
1	1	2	0	3	0	78.314735	5.221767	0.281108
1	1	2	0	2	1	187.606384	10.029198	0.537136
1	1	2	0	2	0	93.764544	6.104256	0.331518
1	1	2	0	1	2	437.691014	20.407912	1.075236
1	1	2	0	1	1	223.254323	11.832899	0.639597
1	1	2	0	1	0	121.502155	7.355608	0.386047
1	1	2	0	0	3	1001.838892	41.852328	2.338390
1	1	2	0	0	2	518.645901	24.065138	1.302123
1	1	2	0	0	1	287.054487	14.665641	0.753090
1	1	2	0	0	0	170.186251	9.191560	0.438574
1	1	1	2	1	0	20.577777	1.304387	0.051186
1	1	1	2	0	1	53.748275	2.446471	0.092665
1	1	1	2	0	0	22.176980	1.364888	0.051869
1	1	1	1	2	0	39.734863	2.598432	0.115917
1	1	1	1	1	1	99.045771	5.000779	0.213471
1	1	1	1	1	0	42.096423	2.732936	0.118293
1	1	1	1	0	2	238.617553	10.554688	0.402781
1	1	1	1	0	1	104.808327	5.277810	0.218299
1	1	1	1	0	0	49.153159	2.913817	0.113338
1	1	1	0	3	0	72.811154	4.908483	0.269094
1	1	1	0	2	1	175.016724	9.425025	0.512687
1	1	1	0	2	0	75.649725	5.142326	0.277272
1	1	1	0	1	2	409.886998	19.184256	1.023044
1	1	1	0	1	1	181.470493	9.888814	0.529878
1	1	1	0	1	0	86.326654	5.534329	0.275320
1	1	1	0	0	3	942.162105	39.395164	2.217253
1	1	1	0	0	2	424.766623	20.148991	1.060927
1	1	1	0	0	1	205.001154	10.870958	0.523632
1	1	1	0	0	0	107.109567	5.671926	0.238807
1	1	0	2	1	0	20.617233	1.290693	0.050305

| Dice in Cup | | | Current FPs | | | Decision | Decision | Decision |
R	Y	G	R	Y	G	(SG 0)	(SG 1)	(SG 2)
1	1	0	2	0	1	54.073494	2.428239	0.091082
1	1	0	2	0	0	16.748579	1.010729	0.035347
1	1	0	1	2	0	39.268217	2.582615	0.114374
1	1	0	1	1	1	97.983884	4.986202	0.210777
1	1	0	1	1	0	31.915311	2.118415	0.082803
1	1	0	1	0	2	236.172717	10.560664	0.398038
1	1	0	1	0	1	80.086610	3.985454	0.150712
1	1	0	1	0	0	23.950185	1.533982	0.054175
1	1	0	0	3	0	70.866556	4.851658	0.267244
1	1	0	0	2	1	170.382281	9.324419	0.509980
1	1	0	0	2	0	57.290355	4.005658	0.200382
1	1	0	0	1	2	399.590228	18.984732	1.019579
1	1	0	0	1	1	138.047847	7.604019	0.375516
1	1	0	0	1	0	43.945983	2.984539	0.129952
1	1	0	0	0	3	922.561430	38.953206	2.214718
1	1	0	0	0	2	323.545805	15.563918	0.727856
1	1	0	0	0	1	107.109567	5.671926	0.238807
1	0	6	2	1	0	28.993399	1.674009	0.058632
1	0	6	2	0	0	53.350108	2.601915	0.090558
1	0	6	1	2	0	55.608444	3.370082	0.135568
1	0	6	1	1	0	98.703361	5.304993	0.211749
1	0	6	1	0	0	180.448764	8.594429	0.329714
1	0	6	0	3	0	101.491618	6.421861	0.322481
1	0	6	0	2	0	175.009355	10.023356	0.515209
1	0	6	0	1	0	311.372744	15.947550	0.827817
1	0	6	0	0	0	561.070265	26.155421	1.345662
1	0	5	2	1	0	27.705629	1.623513	0.057714
1	0	5	2	0	1	71.733448	3.060876	0.105245
1	0	5	2	0	0	48.485795	2.441718	0.086524
1	0	5	1	2	0	53.100694	3.261043	0.133066
1	0	5	1	1	1	131.034566	6.308164	0.247262
1	0	5	1	1	0	89.812708	4.958980	0.201468
1	0	5	1	0	1	216.521241	10.095910	0.380073
1	0	5	1	0	0	159.019048	7.712903	0.302314
1	0	5	0	3	0	96.786184	6.202354	0.315446
1	0	5	0	2	1	230.326246	11.941248	0.607418
1	0	5	0	2	0	159.156765	9.351719	0.487446
1	0	5	0	1	1	371.893072	18.531386	0.965984
1	0	5	0	1	0	273.977748	14.344202	0.752018
1	0	5	0	0	1	630.487972	29.155525	1.551305
1	0	5	0	0	0	488.608265	23.003454	1.159533
1	0	4	2	1	0	26.245863	1.564543	0.056609
1	0	4	2	0	1	68.016160	2.946450	0.103102

Dice in Cup			Current FPs			Decision	Decision	Decision
R	Y	G	R	Y	G	(SG 0)	(SG 1)	(SG 2)
1	0	4	2	0	0	43.039011	2.253162	0.081479
1	0	4	1	2	0	50.262419	3.134684	0.130086
1	0	4	1	1	1	124.163608	6.056556	0.241348
1	0	4	1	1	0	79.818036	4.555604	0.188817
1	0	4	1	0	2	295.913465	12.835152	0.459122
1	0	4	1	0	1	193.227110	9.203123	0.354982
1	0	4	1	0	0	137.297617	6.731595	0.268478
1	0	4	0	3	0	91.468562	5.948566	0.307158
1	0	4	0	2	1	217.872085	11.442515	0.590352
1	0	4	0	2	0	141.302492	8.570031	0.453886
1	0	4	0	1	2	505.054053	23.298186	1.189259
1	0	4	0	1	1	331.249849	16.900550	0.894756
1	0	4	0	1	0	236.849017	12.594185	0.661121
1	0	4	0	0	2	759.423417	34.335215	1.884617
1	0	4	0	0	1	547.683379	25.639203	1.346887
1	0	4	0	0	0	403.449260	19.334573	0.955290
1	0	3	2	1	0	24.664064	1.498280	0.055310
1	0	3	2	0	1	64.020959	2.818631	0.100591
1	0	3	2	0	0	37.112504	2.022749	0.074682
1	0	3	1	2	0	47.166425	2.994173	0.126643
1	0	3	1	1	1	116.712120	5.778488	0.234544
1	0	3	1	1	0	69.156511	4.080112	0.172209
1	0	3	1	0	2	278.797101	12.231512	0.445307
1	0	3	1	0	1	168.472757	8.158392	0.322202
1	0	3	1	0	0	112.420744	5.641857	0.228250
1	0	3	0	3	0	85.622246	5.665999	0.297765
1	0	3	0	2	1	204.249219	10.889567	0.571097
1	0	3	0	2	0	122.695202	7.669107	0.411207
1	0	3	0	1	2	474.462939	22.159413	1.147854
1	0	3	0	1	1	289.076375	15.037378	0.804733
1	0	3	0	1	0	193.220721	10.573245	0.556677
1	0	3	0	0	3	1081.760697	45.393185	2.507193
1	0	3	0	0	2	665.851582	30.554542	1.676134
1	0	3	0	0	1	450.386723	21.538541	1.116712
1	0	3	0	0	0	324.840791	15.699034	0.736521
1	0	2	2	1	0	23.015843	1.418584	0.053551
1	0	2	2	0	1	59.889204	2.664416	0.097184
1	0	2	2	0	0	30.677308	1.761947	0.065881
1	0	2	1	2	0	44.050081	2.834992	0.122194
1	0	2	1	1	1	109.311417	5.464501	0.225739
1	0	2	1	1	0	56.989949	3.544210	0.151523
1	0	2	1	0	2	262.070790	11.552483	0.427403
1	0	2	1	0	1	139.812831	6.990359	0.281990

| Dice in Cup | | | Current FPs | | | Decision | Decision | Decision |
R	Y	G	R	Y	G	(SG 0)	(SG 1)	(SG 2)
1	0	2	1	0	0	88.841718	4.516633	0.174748
1	0	2	0	3	0	79.961729	5.354276	0.286298
1	0	2	0	2	1	191.276651	10.284715	0.547570
1	0	2	0	2	0	100.613515	6.622165	0.360320
1	0	2	0	1	2	445.812783	20.928043	1.097212
1	0	2	0	1	1	238.562076	12.865281	0.699085
1	0	2	0	1	0	152.253868	8.496675	0.436622
1	0	2	0	0	3	1020.378167	42.912114	2.388657
1	0	2	0	0	2	553.679932	26.102696	1.435299
1	0	2	0	0	1	355.919027	17.275076	0.854061
1	0	2	0	0	0	252.715066	11.812252	0.449392
1	0	1	2	1	0	22.020287	1.362846	0.051970
1	0	1	2	0	1	57.517788	2.560835	0.094183
1	0	1	2	0	0	24.227945	1.404291	0.050507
1	0	1	1	2	0	41.754908	2.731632	0.118679
1	0	1	1	1	1	103.840291	5.269706	0.218963
1	0	1	1	1	0	44.749072	2.967051	0.119791
1	0	1	1	0	2	249.571126	11.151949	0.413998
1	0	1	1	0	1	110.014635	5.684696	0.219869
1	0	1	1	0	0	61.128401	2.846840	0.097440
1	0	1	0	3	0	74.993082	5.129113	0.278749
1	0	1	0	2	1	179.696000	9.850187	0.532659
1	0	1	0	2	0	78.567685	5.487187	0.296787
1	0	1	0	1	2	420.094029	20.038493	1.066370
1	0	1	0	1	1	186.646111	10.450435	0.566275
1	0	1	0	1	0	107.109567	5.671926	0.238807
1	0	1	0	0	3	966.997899	41.075360	2.319429
1	0	1	0	0	2	433.656132	21.065638	1.131019
1	0	1	0	0	1	252.715066	11.812252	0.449392
1	0	0	2	1	0	24.052547	1.356543	0.048443
1	0	0	2	0	1	61.953546	2.504333	0.086875
1	0	0	2	0	0	12.176259	0.698747	0.022125
1	0	0	1	2	0	44.376175	2.938731	0.116418
1	0	0	1	1	1	108.976394	5.635118	0.213317
1	0	0	1	1	0	23.950185	1.533982	0.054175
1	0	0	1	0	2	258.883249	11.839892	0.399997
1	0	0	1	0	1	61.128401	2.846840	0.097440
1	0	0	0	3	0	77.864046	5.433151	0.294361
1	0	0	0	2	1	184.679420	10.350248	0.561860
1	0	0	0	2	0	43.945983	2.984539	0.129952
1	0	0	0	1	2	428.009084	20.863432	1.122692
1	0	0	0	1	1	107.109567	5.671926	0.238807
1	0	0	0	0	3	978.432280	42.331689	2.434881

Dice in Cup			Current FPs			Decision	Decision	Decision
R	Y	G	R	Y	G	(SG 0)	(SG 1)	(SG 2)
1	0	0	0	0	2	252.715066	11.812252	0.449392
0	4	6	3	0	0	13.509839	0.786900	0.025239
0	4	6	2	0	0	46.361011	2.370412	0.083776
0	4	6	1	0	0	139.572511	7.244731	0.288004
0	4	6	0	0	0	387.208860	19.950380	1.099840
0	4	5	3	0	0	13.046340	0.770039	0.024985
0	4	5	2	0	1	69.617663	2.982327	0.103619
0	4	5	2	0	0	42.872539	2.250344	0.080610
0	4	5	1	0	1	194.020108	9.164027	0.352034
0	4	5	1	0	0	124.404337	6.625330	0.268169
0	4	5	0	0	1	506.134792	24.451704	1.345834
0	4	5	0	0	0	333.666960	17.685479	0.979979
0	4	4	3	0	0	12.524179	0.750764	0.024691
0	4	4	2	0	1	66.837496	2.899990	0.102110
0	4	4	2	0	0	39.061422	2.113585	0.076886
0	4	4	1	0	2	292.431392	12.630105	0.453935
0	4	4	1	0	1	177.575150	8.525173	0.333697
0	4	4	1	0	0	108.499730	5.951004	0.245645
0	4	4	0	0	2	709.945730	31.866372	1.748052
0	4	4	0	0	1	443.492945	21.909080	1.212055
0	4	4	0	0	0	279.102131	15.295107	0.851417
0	4	3	3	0	0	11.936324	0.728651	0.024348
0	4	3	2	0	1	63.730808	2.806162	0.100360
0	4	3	2	0	0	34.932696	1.957232	0.072445
0	4	3	1	0	2	279.028304	12.178591	0.444138
0	4	3	1	0	1	159.831896	7.807795	0.312132
0	4	3	1	0	0	91.903166	5.219072	0.219978
0	4	3	0	0	3	1094.310488	45.268226	2.499486
0	4	3	0	0	2	640.873971	29.214371	1.610211
0	4	3	0	0	1	378.082568	19.157548	1.063925
0	4	3	0	0	0	224.439889	12.807168	0.715221
0	4	2	3	0	0	11.279407	0.703229	0.023944
0	4	2	2	0	1	60.301846	2.699334	0.098317
0	4	2	2	0	0	30.483617	1.777651	0.067062
0	4	2	1	0	2	264.450749	11.669997	0.432794
0	4	2	1	0	1	140.726027	7.001379	0.286456
0	4	2	1	0	0	74.906474	4.434175	0.190750
0	4	2	0	0	3	1037.933225	43.318236	2.419064
0	4	2	0	0	2	566.693107	26.262570	1.450280
0	4	2	0	0	1	310.864454	16.213910	0.901225
0	4	2	0	0	0	172.247368	10.276068	0.572704
0	4	1	3	0	0	10.556639	0.674183	0.023468
0	4	1	2	0	1	56.580469	2.578883	0.095932

| Dice in Cup | | | Current FPs | | | Decision | Decision | Decision |
R	Y	G	R	Y	G	(SG 0)	(SG 1)	(SG 2)
0	4	1	2	0	0	25.787887	1.571743	0.060414
0	4	1	1	0	2	248.828825	11.104342	0.419710
0	4	1	1	0	1	120.556822	6.102252	0.255496
0	4	1	1	0	0	57.901002	3.595560	0.157300
0	4	1	0	0	3	978.175763	41.178870	2.327479
0	4	1	0	0	2	488.388465	23.014289	1.264056
0	4	1	0	0	1	243.937697	13.078995	0.722557
0	4	1	0	0	0	121.620142	7.784467	0.432513
0	4	0	3	0	0	9.807705	0.642033	0.022912
0	4	0	2	0	1	52.814849	2.448637	0.093196
0	4	0	2	0	0	20.796873	1.328543	0.051859
0	4	0	1	0	2	233.316928	10.507469	0.405006
0	4	0	1	0	1	99.142057	5.075220	0.216678
0	4	0	1	0	0	41.759404	2.768833	0.121099
0	4	0	0	0	3	919.421654	38.973666	2.226848
0	4	0	0	0	2	406.102977	19.357035	1.040110
0	4	0	0	0	1	178.196865	9.930455	0.541307
0	4	0	0	0	0	76.483214	5.428062	0.297674
0	3	6	3	0	0	14.109858	0.810631	0.025609
0	3	6	2	1	0	29.030858	1.671526	0.058523
0	3	6	2	0	0	51.397179	2.547587	0.088561
0	3	6	1	1	0	95.416938	5.189253	0.207170
0	3	6	1	0	0	162.258686	8.200424	0.318753
0	3	6	0	1	0	282.492793	15.140776	0.799557
0	3	6	0	0	0	468.724290	23.533620	1.294273
0	3	5	3	0	0	13.612112	0.792822	0.025345
0	3	5	2	1	0	27.991395	1.632059	0.057818
0	3	5	2	0	1	72.425323	3.077999	0.105465
0	3	5	2	0	0	47.481340	2.419898	0.085333
0	3	5	1	1	1	132.589431	6.350577	0.248010
0	3	5	1	1	0	88.220913	4.912479	0.198975
0	3	5	1	0	1	212.342664	9.958631	0.375372
0	3	5	1	0	0	144.655049	7.505550	0.297592
0	3	5	0	1	1	367.230611	18.285135	0.954514
0	3	5	0	1	0	251.867357	13.863984	0.740990
0	3	5	0	0	1	578.861857	27.706135	1.524752
0	3	5	0	0	0	403.668000	20.867705	1.156139
0	3	4	3	0	0	13.043304	0.772118	0.025033
0	3	4	2	1	0	26.807331	1.586336	0.056989
0	3	4	2	0	1	69.373741	2.989004	0.103857
0	3	4	2	0	0	43.165698	2.272217	0.081454
0	3	4	1	1	1	126.911524	6.152986	0.243544
0	3	4	1	1	0	80.322640	4.594843	0.189210

Dice in Cup			Current FPs			Decision	Decision	Decision
R	Y	G	R	Y	G	(SG 0)	(SG 1)	(SG 2)
0	3	4	1	0	2	301.693969	13.043849	0.463709
0	3	4	1	0	1	193.790723	9.255925	0.356013
0	3	4	1	0	0	125.758172	6.731883	0.272956
0	3	4	0	1	2	517.773953	23.689583	1.203994
0	3	4	0	1	1	334.979791	17.001476	0.899232
0	3	4	0	1	0	219.039560	12.443096	0.673681
0	3	4	0	0	2	765.307122	34.443782	1.893224
0	3	4	0	0	1	504.888606	24.778978	1.373819
0	3	4	0	0	0	336.158604	17.995663	1.004436
0	3	3	3	0	0	12.402159	0.748102	0.024663
0	3	3	2	1	0	25.482324	1.533609	0.056010
0	3	3	2	0	1	65.978313	2.886694	0.101962
0	3	3	2	0	0	38.395965	2.099794	0.076708
0	3	3	1	1	1	120.648550	5.927297	0.238308
0	3	3	1	1	0	71.602514	4.227123	0.177384
0	3	3	1	0	2	287.043298	12.549643	0.453055
0	3	3	1	0	1	173.348878	8.449620	0.332680
0	3	3	1	0	0	105.710175	5.876696	0.244150
0	3	3	0	1	2	491.982845	22.760637	1.171347
0	3	3	0	1	1	299.530633	15.534656	0.833440
0	3	3	0	1	0	184.198893	10.871795	0.596306
0	3	3	0	0	3	1116.253054	46.556855	2.562872
0	3	3	0	0	2	685.958467	31.441706	1.740985
0	3	3	0	0	1	426.437669	21.553235	1.202686
0	3	3	0	0	0	270.924862	15.024905	0.840270
0	3	2	3	0	0	11.682987	0.720245	0.024222
0	3	2	2	1	0	24.003849	1.472863	0.054849
0	3	2	2	0	1	62.214848	2.769302	0.099720
0	3	2	2	0	0	33.199507	1.898372	0.070785
0	3	2	1	1	1	113.734235	5.670029	0.232158
0	3	2	1	1	0	62.096913	3.802340	0.162841
0	3	2	1	0	2	271.003221	11.989006	0.440572
0	3	2	1	0	1	151.108216	7.528498	0.304176
0	3	2	1	0	0	85.379801	4.948037	0.210309
0	3	2	0	1	2	463.823596	21.712682	1.133383
0	3	2	0	1	1	260.962056	13.866881	0.754479
0	3	2	0	1	0	149.322038	9.176981	0.507214
0	3	2	0	0	3	1054.092749	44.399585	2.474095
0	3	2	0	0	2	599.880455	28.051718	1.560192
0	3	2	0	0	1	347.846135	18.083832	1.008877
0	3	2	0	0	0	202.157402	11.848379	0.666704
0	3	1	3	0	0	10.914526	0.688808	0.023699
0	3	1	2	1	0	22.433784	1.405099	0.053492

| Dice in Cup | | | Current FPs | | | Decision | Decision | Decision |
R	Y	G	R	Y	G	(SG 0)	(SG 1)	(SG 2)
0	3	1	2	0	1	58.263768	2.639296	0.097109
0	3	1	2	0	0	27.635176	1.659678	0.063117
0	3	1	1	1	1	106.479561	5.388066	0.225085
0	3	1	1	1	0	52.024376	3.307168	0.144318
0	3	1	1	0	2	254.386947	11.379797	0.426279
0	3	1	1	0	1	127.533621	6.468006	0.268154
0	3	1	1	0	0	64.230152	3.962956	0.171365
0	3	1	0	1	2	434.503995	20.580898	1.090510
0	3	1	0	1	1	220.784848	11.973310	0.656712
0	3	1	0	1	0	112.201118	7.342687	0.407984
0	3	1	0	0	3	990.286228	42.094147	2.374306
0	3	1	0	0	2	510.218853	24.232314	1.339395
0	3	1	0	0	1	263.799641	14.321002	0.798042
0	3	1	0	0	0	139.056606	8.748515	0.490717
0	3	0	3	0	0	10.098245	0.652788	0.023063
0	3	0	2	1	0	20.795855	1.328478	0.051856
0	3	0	2	0	1	54.161736	2.492773	0.093969
0	3	0	2	0	0	22.143652	1.395851	0.053153
0	3	0	1	1	1	99.137206	5.074972	0.216667
0	3	0	1	1	0	41.759404	2.768833	0.121099
0	3	0	1	0	2	237.701875	10.705752	0.409299
0	3	0	1	0	1	103.453836	5.336208	0.223730
0	3	0	1	0	0	43.643299	2.942979	0.127400
0	3	0	0	1	2	406.083109	19.356088	1.040059
0	3	0	0	1	1	178.196865	9.930455	0.541307
0	3	0	0	1	0	76.483214	5.428062	0.297674
0	3	0	0	0	3	929.049466	39.618204	2.257063
0	3	0	0	0	2	414.993345	20.146844	1.084591
0	3	0	0	0	1	181.159451	10.318799	0.567811
0	3	0	0	0	0	76.856179	5.473756	0.302118
0	2	6	3	0	0	14.863567	0.841458	0.026097
0	2	6	2	1	0	30.489718	1.738255	0.059820
0	2	6	2	0	0	58.342826	2.784038	0.094895
0	2	6	1	2	0	58.353676	3.507177	0.138857
0	2	6	1	1	0	107.249177	5.692308	0.223188
0	2	6	1	0	0	195.266880	9.568012	0.361531
0	2	6	0	2	0	189.355767	10.741039	0.547113
0	2	6	0	1	0	335.581979	17.547125	0.918669
0	2	6	0	0	0	591.110555	28.790466	1.586732
0	2	5	3	0	0	14.321350	0.822655	0.025826
0	2	5	2	1	0	29.342400	1.696182	0.059092
0	2	5	2	0	1	75.822632	3.202099	0.107931
0	2	5	2	0	0	53.989630	2.651103	0.091737

Dice in Cup			Current FPs			Decision	Decision	Decision
R	Y	G	R	Y	G	(SG 0)	(SG 1)	(SG 2)
0	2	5	1	2	0	56.073205	3.414977	0.136859
0	2	5	1	1	1	137.960288	6.613786	0.254828
0	2	5	1	1	0	99.227883	5.400221	0.215051
0	2	5	1	0	1	237.437458	11.057568	0.407300
0	2	5	1	0	0	174.529164	8.784882	0.339501
0	2	5	0	2	1	241.831186	12.515368	0.629721
0	2	5	0	2	0	174.973262	10.165049	0.524850
0	2	5	0	1	1	406.282985	20.176818	1.045926
0	2	5	0	1	0	299.508615	16.100991	0.856232
0	2	5	0	0	1	683.857352	32.334253	1.787823
0	2	5	0	0	0	510.669450	25.593156	1.426794
0	2	4	3	0	0	13.703344	0.800581	0.025500
0	2	4	2	1	0	28.044934	1.647046	0.058221
0	2	4	2	0	1	72.478404	3.106263	0.106239
0	2	4	2	0	0	49.029757	2.492520	0.087821
0	2	4	1	2	0	53.517581	3.307911	0.134480
0	2	4	1	1	1	131.686649	6.399110	0.250097
0	2	4	1	1	0	90.120666	5.054103	0.205045
0	2	4	1	0	2	312.196647	13.576613	0.477017
0	2	4	1	0	1	216.112336	10.282947	0.387322
0	2	4	1	0	0	151.637520	7.884970	0.312935
0	2	4	0	2	1	230.355670	12.082375	0.615873
0	2	4	0	2	0	158.721376	9.486235	0.497709
0	2	4	0	1	2	531.345345	24.581085	1.244576
0	2	4	0	1	1	369.131902	18.746782	0.987846
0	2	4	0	1	0	259.800028	14.441489	0.781900
0	2	4	0	0	2	839.628839	37.956887	2.100863
0	2	4	0	0	1	595.144558	28.910399	1.617672
0	2	4	0	0	0	433.262616	22.277003	1.248788
0	2	3	3	0	0	12.990824	0.774362	0.025104
0	2	3	2	1	0	26.556399	1.588970	0.057166
0	2	3	2	0	1	68.665301	2.993343	0.104191
0	2	3	2	0	0	43.406629	2.302396	0.082867
0	2	3	1	2	0	50.599674	3.181960	0.131612
0	2	3	1	1	1	124.573412	6.147450	0.244402
0	2	3	1	1	0	79.812339	4.642602	0.192536
0	2	3	1	0	2	295.601191	13.024418	0.465397
0	2	3	1	0	1	192.059748	9.369159	0.362467
0	2	3	1	0	0	128.473152	6.901712	0.281241
0	2	3	0	2	1	217.420303	11.577191	0.599291
0	2	3	0	2	0	140.374523	8.683759	0.464205
0	2	3	0	1	2	501.894026	23.529026	1.208719
0	2	3	0	1	1	327.365512	17.067900	0.916536

| Dice in Cup | | | Current FPs | | | Decision | Decision | Decision |
R	Y	G	R	Y	G	(SG 0)	(SG 1)	(SG 2)
0	2	3	0	1	0	220.340547	12.656007	0.694797
0	2	3	0	0	3	1136.131225	48.055979	2.650049
0	2	3	0	0	2	746.917530	34.502970	1.933751
0	2	3	0	0	1	507.077847	25.246667	1.420568
0	2	3	0	0	0	343.256767	18.336617	1.041267
0	2	2	3	0	0	12.193294	0.743755	0.024624
0	2	2	2	1	0	24.896565	1.521688	0.055899
0	2	2	2	0	1	64.451683	2.863210	0.101737
0	2	2	2	0	0	37.414382	2.080163	0.076498
0	2	2	1	2	0	47.350042	3.036953	0.128197
0	2	2	1	1	1	116.721717	5.859324	0.237638
0	2	2	1	1	0	69.016032	4.169524	0.176721
0	2	2	1	0	2	277.489562	12.396171	0.451635
0	2	2	1	0	1	166.960700	8.324709	0.331229
0	2	2	1	0	0	101.551382	5.744186	0.241750
0	2	2	0	2	1	203.143299	11.000764	0.579778
0	2	2	0	2	0	121.558173	7.779706	0.422576
0	2	2	0	1	2	469.749227	22.336624	1.166647
0	2	2	0	1	1	284.727395	15.190079	0.828486
0	2	2	0	1	0	173.726692	10.511961	0.588756
0	2	2	0	0	3	1066.147342	45.600043	2.551416
0	2	2	0	0	2	652.419762	30.675673	1.728988
0	2	2	0	0	1	403.120825	20.844686	1.184887
0	2	2	0	0	0	259.713072	14.419665	0.820688
0	2	1	3	0	0	11.381688	0.709780	0.024054
0	2	1	2	1	0	23.242741	1.448021	0.054406
0	2	1	2	0	1	60.284200	2.721139	0.098851
0	2	1	2	0	0	30.606578	1.809797	0.067855
0	2	1	1	2	0	44.198341	2.880720	0.124211
0	2	1	1	1	1	109.193478	5.550156	0.229751
0	2	1	1	1	0	56.407176	3.597338	0.155861
0	2	1	1	0	2	260.356655	11.725075	0.435600
0	2	1	1	0	1	137.571085	7.077858	0.290496
0	2	1	1	0	0	75.180807	4.543592	0.195648
0	2	1	0	2	1	189.915685	10.397981	0.557210
0	2	1	0	2	0	99.022830	6.672854	0.369303
0	2	1	0	1	2	440.410276	21.101904	1.118003
0	2	1	0	1	1	233.582627	12.907383	0.717213
0	2	1	0	1	0	128.541138	8.290870	0.466760
0	2	1	0	0	3	1003.075287	43.091610	2.437395
0	2	1	0	0	2	539.504246	26.037135	1.473975
0	2	1	0	0	1	299.622497	16.197292	0.916367
0	2	1	0	0	0	181.168978	10.342497	0.573561

Dice in Cup			Current FPs			Decision	Decision	Decision
R	Y	G	R	Y	G	(SG 0)	(SG 1)	(SG 2)
0	2	0	3	0	0	10.893648	0.684945	0.023541
0	2	0	2	1	0	22.142568	1.395782	0.053150
0	2	0	2	0	1	57.648829	2.624513	0.096472
0	2	0	2	0	0	23.682871	1.501725	0.055719
0	2	0	1	2	0	41.757361	2.768697	0.121093
0	2	0	1	1	1	103.448775	5.335947	0.223719
0	2	0	1	1	0	43.643299	2.942979	0.127400
0	2	0	1	0	2	247.547858	11.277657	0.423620
0	2	0	1	0	1	107.025246	5.631030	0.235088
0	2	0	1	0	0	43.945983	2.984539	0.129952
0	2	0	0	2	1	178.188147	9.929969	0.541280
0	2	0	0	2	0	76.483214	5.428062	0.297674
0	2	0	0	1	2	414.973042	20.145858	1.084537
0	2	0	0	1	1	181.159451	10.318799	0.567811
0	2	0	0	1	0	76.856179	5.473756	0.302118
0	2	0	0	0	3	951.909316	41.138215	2.360983
0	2	0	0	0	2	419.438122	20.752732	1.133775
0	2	0	0	0	1	181.168978	10.342497	0.573561
0	1	6	3	0	0	15.801231	0.881819	0.026752
0	1	6	2	1	0	32.252698	1.824866	0.061561
0	1	6	2	0	0	68.498526	3.110295	0.103578
0	1	6	1	2	0	61.295434	3.685887	0.143580
0	1	6	1	1	0	124.184558	6.386705	0.245275
0	1	6	1	0	0	245.982062	11.627882	0.423697
0	1	6	0	3	0	110.733432	7.020250	0.345089
0	1	6	0	2	0	216.345314	12.021618	0.606934
0	1	6	0	1	0	415.964661	21.078355	1.094763
0	1	6	0	0	0	787.427933	36.878387	2.056884
0	1	5	3	0	0	15.214477	0.862334	0.026484
0	1	5	2	1	0	30.991928	1.780579	0.060833
0	1	5	2	0	1	79.916839	3.364854	0.111301
0	1	5	2	0	0	63.667382	2.974881	0.100736
0	1	5	1	2	0	58.752939	3.587163	0.141558
0	1	5	1	1	1	144.062712	6.952440	0.264129
0	1	5	1	1	0	115.191825	6.082009	0.237721
0	1	5	1	0	1	273.382655	12.628560	0.452522
0	1	5	1	0	0	221.529890	10.748088	0.402077
0	1	5	0	3	0	105.824618	6.813589	0.339294
0	1	5	0	2	1	249.999197	13.113159	0.656578
0	1	5	0	2	0	200.106591	11.405970	0.585622
0	1	5	0	1	1	461.380720	22.808015	1.176633
0	1	5	0	1	0	373.434040	19.430791	1.030936
0	1	5	0	0	1	847.372028	39.196852	2.197619

| Dice in Cup | | | Current FPs | | | Decision | Decision | Decision |
R	Y	G	R	Y	G	(SG 0)	(SG 1)	(SG 2)
0	1	5	0	0	0	704.643960	33.541325	1.888457
0	1	4	3	0	0	14.511092	0.838258	0.026146
0	1	4	2	1	0	29.490902	1.726055	0.059916
0	1	4	2	0	1	76.038807	3.257943	0.109509
0	1	4	2	0	0	57.901165	2.805556	0.097042
0	1	4	1	2	0	55.752333	3.466144	0.139016
0	1	4	1	1	1	136.688341	6.708420	0.259046
0	1	4	1	1	0	104.534894	5.703406	0.227972
0	1	4	1	0	2	322.700083	14.238265	0.495151
0	1	4	1	0	1	248.564513	11.773545	0.432813
0	1	4	1	0	0	197.815838	9.808583	0.376666
0	1	4	0	3	0	100.091121	6.561754	0.332021
0	1	4	0	2	1	236.410334	12.610305	0.641480
0	1	4	0	2	0	181.025061	10.646280	0.558296
0	1	4	0	1	2	543.113797	25.610252	1.299543
0	1	4	0	1	1	418.124614	21.195864	1.117434
0	1	4	0	1	0	333.111623	17.718105	0.957004
0	1	4	0	0	2	947.284358	42.860352	2.409751
0	1	4	0	0	1	758.540587	35.683774	2.023060
0	1	4	0	0	0	594.318999	28.930800	1.658303
0	1	3	3	0	0	13.669355	0.808429	0.025716
0	1	3	2	1	0	27.704226	1.658782	0.058753
0	1	3	2	0	1	71.458491	3.126439	0.107236
0	1	3	2	0	0	51.990673	2.618174	0.092499
0	1	3	1	2	0	52.202719	3.317463	0.135803
0	1	3	1	1	1	128.045898	6.409760	0.252624
0	1	3	1	1	0	93.876480	5.291012	0.216191
0	1	3	1	0	2	302.635528	13.579600	0.481965
0	1	3	1	0	1	223.959278	10.839864	0.409077
0	1	3	1	0	0	165.836919	8.505949	0.340507
0	1	3	0	3	0	93.358575	6.253749	0.322862
0	1	3	0	2	1	220.617971	11.998475	0.622477
0	1	3	0	2	0	162.431063	9.838530	0.525795
0	1	3	0	1	2	507.495984	24.329809	1.258183
0	1	3	0	1	1	376.307555	19.487802	1.047145
0	1	3	0	1	0	278.431213	15.316555	0.852597
0	1	3	0	0	3	1147.067354	49.511506	2.764738
0	1	3	0	0	2	855.560804	39.337618	2.241738
0	1	3	0	0	1	639.113448	30.785697	1.779352
0	1	3	0	0	0	493.765525	24.401681	1.404817
0	1	2	3	0	0	12.869885	0.777444	0.025235
0	1	2	2	1	0	26.040168	1.589705	0.057463
0	1	2	2	0	1	67.228093	2.991637	0.104714

Dice in Cup			Current FPs			Decision	Decision	Decision
R	Y	G	R	Y	G	(SG 0)	(SG 1)	(SG 2)
0	1	2	2	0	0	43.891328	2.348882	0.085695
0	1	2	1	2	0	48.981875	3.167012	0.132263
0	1	2	1	1	1	120.307877	6.108589	0.245541
0	1	2	1	1	0	79.080025	4.698751	0.198754
0	1	2	1	0	2	284.969483	12.918248	0.467401
0	1	2	1	0	1	189.894352	9.514559	0.374103
0	1	2	1	0	0	135.923981	7.176171	0.298645
0	1	2	0	3	0	87.425693	5.949004	0.312827
0	1	2	0	2	1	206.918028	11.398165	0.601626
0	1	2	0	2	0	136.355943	8.669373	0.478215
0	1	2	0	1	2	477.095779	23.088040	1.212722
0	1	2	0	1	1	317.922657	17.037248	0.944794
0	1	2	0	1	0	227.803002	12.887366	0.732797
0	1	2	0	0	3	1081.909338	46.965656	2.656944
0	1	2	0	0	2	728.713716	34.311553	1.999005
0	1	2	0	0	1	524.727631	25.769455	1.497825
0	1	2	0	0	0	417.484188	20.681047	1.139477
0	1	1	3	0	0	11.865740	0.737036	0.024576
0	1	1	2	1	0	23.899771	1.499849	0.055723
0	1	1	2	0	1	61.867462	2.817701	0.101324
0	1	1	2	0	0	35.990221	2.076884	0.077071
0	1	1	1	2	0	44.697727	2.969604	0.127563
0	1	1	1	1	1	110.125172	5.716360	0.236160
0	1	1	1	1	0	64.593711	4.086796	0.177626
0	1	1	1	0	2	262.108294	12.064601	0.448160
0	1	1	1	0	1	155.567025	8.086290	0.331450
0	1	1	1	0	0	107.109567	5.671926	0.238807
0	1	1	0	3	0	79.277596	5.539442	0.299655
0	1	1	0	2	1	188.275054	10.594747	0.574320
0	1	1	0	2	0	111.029017	7.457810	0.421314
0	1	1	0	1	2	436.428105	21.434863	1.153320
0	1	1	0	1	1	259.713072	14.419665	0.820688
0	1	1	0	1	0	181.168978	10.342497	0.573561
0	1	1	0	0	3	997.522563	43.600802	2.516390
0	1	1	0	0	2	596.988284	28.840743	1.699605
0	1	1	0	0	1	417.484188	20.681047	1.139477
0	1	0	3	0	0	11.957629	0.741422	0.024547
0	1	0	2	1	0	23.681712	1.501652	0.055716
0	1	0	2	0	1	60.897870	2.807210	0.101066
0	1	0	2	0	0	23.950185	1.533982	0.054175
0	1	0	1	2	0	43.641163	2.942835	0.127394
0	1	0	1	1	1	107.020010	5.630755	0.235076
0	1	0	1	1	0	43.945983	2.984539	0.129952

| Dice in Cup | | | Current FPs | | | Decision | Decision | Decision |
R	Y	G	R	Y	G	(SG 0)	(SG 1)	(SG 2)
0	1	0	1	0	2	253.947796	11.803823	0.444356
0	1	0	1	0	1	107.109567	5.671926	0.238807
0	1	0	0	3	0	76.479472	5.427796	0.297660
0	1	0	0	2	1	181.150588	10.318294	0.567783
0	1	0	0	2	0	76.856179	5.473756	0.302118
0	1	0	0	1	2	419.417602	20.751717	1.133719
0	1	0	0	1	1	181.168978	10.342497	0.573561
0	1	0	0	0	3	958.433527	41.997513	2.456915
0	1	0	0	0	2	417.484188	20.681047	1.139477
0	0	6	3	0	0	16.972299	0.935433	0.027651
0	0	6	2	1	0	34.379839	1.938567	0.063948
0	0	6	2	0	0	84.452380	3.571766	0.115888
0	0	6	1	2	0	64.659750	3.916192	0.150041
0	0	6	1	1	0	150.155915	7.369267	0.276707
0	0	6	1	0	0	329.776185	14.957185	0.518417
0	0	6	0	3	0	115.459703	7.433207	0.363132
0	0	6	0	2	0	257.075594	13.803858	0.692635
0	0	6	0	1	0	547.692858	26.581883	1.368700
0	0	6	0	0	0	1165.744406	51.244065	2.924494
0	0	5	3	0	0	16.279723	0.914078	0.027386
0	0	5	2	1	0	32.860710	1.888572	0.063207
0	0	5	2	0	1	84.448248	3.571591	0.115883
0	0	5	2	0	0	79.369503	3.436431	0.113804
0	0	5	1	2	0	61.549486	3.801431	0.147922
0	0	5	1	1	1	150.148569	7.368906	0.276694
0	0	5	1	1	0	140.492316	7.052978	0.270606
0	0	5	1	0	1	329.776185	14.957185	0.518417
0	0	5	1	0	0	311.454819	14.312726	0.505898
0	0	5	0	3	0	109.418046	7.185785	0.356882
0	0	5	0	2	1	257.063017	13.803183	0.692602
0	0	5	0	2	0	239.422526	13.140558	0.673946
0	0	5	0	1	1	547.692858	26.581883	1.368700
0	0	5	0	1	0	516.037206	25.334665	1.327940
0	0	5	0	0	1	1165.744406	51.244065	2.924494
0	0	5	0	0	0	1047.708435	46.429681	2.720086
0	0	4	3	0	0	15.371545	0.884729	0.027010
0	0	4	2	1	0	30.897235	1.820242	0.062153
0	0	4	2	0	1	79.365620	3.436263	0.113799
0	0	4	2	0	0	75.075486	3.311903	0.111765
0	0	4	1	2	0	57.605500	3.645951	0.144908
0	0	4	1	1	1	140.485442	7.052633	0.270593
0	0	4	1	1	0	132.508516	6.764101	0.264622
0	0	4	1	0	2	329.760051	14.956453	0.518392

Dice in Cup			Current FPs			Decision	Decision	Decision
R	Y	G	R	Y	G	(SG 0)	(SG 1)	(SG 2)
0	0	4	1	0	1	311.454819	14.312726	0.505898
0	0	4	1	0	0	278.359669	13.042845	0.479864
0	0	4	0	3	0	101.909868	6.855700	0.347985
0	0	4	0	2	1	239.410812	13.139915	0.673913
0	0	4	0	2	0	225.169174	12.546038	0.655557
0	0	4	0	1	2	547.666063	26.580583	1.368633
0	0	4	0	1	1	516.037206	25.334665	1.327940
0	0	4	0	1	0	459.969760	22.964647	1.243092
0	0	4	0	0	2	1165.744406	51.244065	2.924494
0	0	4	0	0	1	1047.708435	46.429681	2.720086
0	0	4	0	0	0	951.598076	42.060973	2.499797
0	0	3	3	0	0	14.591838	0.857736	0.026646
0	0	3	2	1	0	29.227890	1.757496	0.061129
0	0	3	2	0	1	75.071813	3.311741	0.111759
0	0	3	2	0	0	66.902037	3.061478	0.107512
0	0	3	1	2	0	54.308924	3.503917	0.141969
0	0	3	1	1	1	132.502033	6.763770	0.264609
0	0	3	1	1	0	117.776520	6.191331	0.252173
0	0	3	1	0	2	311.439582	14.312026	0.505873
0	0	3	1	0	1	278.359669	13.042845	0.479864
0	0	3	1	0	0	252.380266	11.893159	0.452811
0	0	3	0	3	0	95.763586	6.558184	0.339284
0	0	3	0	2	1	225.158158	12.545424	0.655525
0	0	3	0	2	0	199.500790	11.398940	0.617267
0	0	3	0	1	2	516.011960	25.333426	1.327875
0	0	3	0	1	1	459.969760	22.964647	1.243092
0	0	3	0	1	0	416.383869	20.840712	1.153923
0	0	3	0	0	3	1165.687374	51.241558	2.924350
0	0	3	0	0	2	1047.708435	46.429681	2.720086
0	0	3	0	0	1	951.598076	42.060973	2.499797
0	0	3	0	0	0	949.010700	41.631582	2.456211
0	0	2	3	0	0	13.063199	0.802717	0.025880
0	0	2	2	1	0	26.019073	1.630988	0.058992
0	0	2	2	0	1	66.898764	3.061328	0.107506
0	0	2	2	0	0	60.722973	2.843406	0.102971
0	0	2	1	2	0	48.132182	3.220928	0.135855
0	0	2	1	1	1	117.770758	6.191028	0.252161
0	0	2	1	1	0	106.503911	5.685186	0.239595
0	0	2	1	0	2	278.346050	13.042207	0.479841
0	0	2	1	0	1	252.380266	11.893159	0.452811
0	0	2	1	0	0	252.715066	11.812252	0.449392
0	0	2	0	3	0	84.523969	5.977927	0.321160
0	0	2	0	2	1	199.491030	11.398382	0.617237

| Dice in Cup | | | Current FPs | | | Decision | Decision | Decision |
R	Y	G	R	Y	G	(SG 0)	(SG 1)	(SG 2)
0	0	2	0	2	0	179.981573	10.387024	0.578087
0	0	2	0	1	2	459.947256	22.963523	1.243031
0	0	2	0	1	1	416.383869	20.840712	1.153923
0	0	2	0	1	0	417.484188	20.681047	1.139477
0	0	2	0	0	3	1047.657177	46.427410	2.719953
0	0	2	0	0	2	951.598076	42.060973	2.499797
0	0	2.	0	0	1	949.010700	41.631582	2.456211
0	0	1	3	0	0	11.978630	0.753866	0.024989
0	0	1	2	1	0	23.658007	1.523947	0.056774
0	0	1	2	0	1	60.720002	2.843267	0.102966
0	0	1	2	0	0	61.128401	2.846840	0.097440
0	0	1	1	2	0	43.500536	2.977028	0.129859
0	0	1	1	1	1	106.498700	5.684908	0.239584
0	0	1	1	1	0	107.109567	5.671926	0.238807
0	0	1	1	0	2	252.367918	11.892577	0.452789
0	0	1	1	0	1	252.715066	11.812252	0.449392
0	0	1	0	3	0	76.083000	5.475230	0.303133
0	0	1	0	2	1	179.972768	10.386516	0.578058
0	0	1	0	2	0	181.168978	10.342497	0.573561
0	0	1	0	1	2	416.363498	20.839693	1.153867
0	0	1	0	1	1	417.484188	20.681047	1.139477
0	0	1	0	0	3	951.551520	42.058915	2.499675
0	0	1	0	0	2	949.010700	41.631582	2.456211
0	0	0	3	0	0	12.175663	0.698713	0.022124
0	0	0	2	1	0	23.949013	1.533907	0.054172
0	0	0	2	0	1	61.125410	2.846701	0.097435
0	0	0	1	2	0	43.943833	2.984393	0.129946
0	0	0	1	1	1	107.104327	5.671649	0.238795
0	0	0	1	0	2	252.702702	11.811674	0.449370
0	0	0	0	3	0	76.852419	5.473488	0.302103
0	0	0	0	2	1	181.160115	10.341991	0.573533
0	0	0	0	1	2	417.463763	20.680035	1.139421
0	0	0	0	0	3	948.964271	41.629545	2.456091

Answers and Selected Solutions

Chapter 1: Games, Gambling, and Probability

1.1 The outcomes are {HHH, HHT, HTH, THH, HTT, THT, TTH, TTT}.

  (a) With only one out of the eight outcomes having zero heads, the probability is 1/8.
  (b) Similar to the first part, we get 3/8.
  (c) 3/8
  (d) 1/8

1.2 **(a)** The outcomes are {1,2,3,4,5,6,7,8}.
  (b) 1/8
  (c) The outcomes that match this event are just 3 or 6, so the probability is 2/8.
  (d) The outcomes that match this event are 2, 4, 6, 7, and 8, so the probability is 5/8.
  (e) Since the two rolls are completely independent, the probability is 1/8.

1.3 For the first part, with 10 outcomes per die, there are $10 \cdot 10 = 100$ possible outcomes. When adding four coins, we have $10 \cdot 10 \cdot 2 \cdot 2 \cdot 2 \cdot 2 = 1,600$ possible outcomes.

1.4 Using Theorem 1.5, we look at the probability of getting no sixes on the four dice; since the dice are independent, this happens with probability $(\frac{5}{6})^4 = 625/1296$ so the probability of getting at least 1 six is $1 - 625/1296 = 671/1296 \approx 0.5177$.

1.5 Using Theorem 1.5, we look at the probability of getting no sixes on the three dice; since the dice are independent, this happens with probability $(\frac{5}{6})^3 = 125/216$ so the probability of getting at least 1 six is $1 - 125/216 = 91/216 \approx 0.4213$. For five dice, the probability is $4651/7776 \approx 0.5981$; the probabilities change a lot, so four dice makes "sense" as an almost fair game.

1.6 From Table 1.2, there is 1 way to roll a total of two, 2 ways to roll a three, and 1 way to roll a twelve, so the probability of a player "crapping out" is $4/36 \approx 0.1111$.

1.7   (a)

	⚀	⚁	⚂	⚃	⚄	⚅
⚀	⚀⚀	⚀⚁	⚀⚂	⚀⚃	⚀⚄	⚀⚅
⚁	⚁⚀	⚁⚁	⚁⚂	⚁⚃	⚁⚄	⚁⚅
⚂	⚂⚀	⚂⚁	⚂⚂	⚂⚃	⚂⚄	⚂⚅
⚃	⚃⚀	⚃⚁	⚃⚂	⚃⚃	⚃⚄	⚃⚅
⚄	⚄⚀	⚄⚁	⚄⚂	⚄⚃	⚄⚄	⚄⚅
⚅	⚅⚀	⚅⚁	⚅⚂	⚅⚃	⚅⚄	⚅⚅

(b) 3/36

(c) 6/36   (d) 20/36   (e) 3/36

1.8 (a) The table is given as Table 11.6 on page 360.
(b) 3/36
(c) 6/36
(d) 25/36
(e) 3/36

   The results are the same except for part (d); these dice are called the Sicherman Dice, which have the property that the probability of getting any particular sum is the same as for a pair of standard six-sided dice, but probabilities involving dice *faces* may change (see Section 11.5 for more).

1.9 (a) There are $6^3 = 216$ elements of the sample space.
(b) The only way to roll a total of three is by rolling (1,1,1) on the dice. For four, we can roll a (1,1,2), a (1,2,1), or a (2,1,1), for a probability of a total sum of 3 or 4 of $4/216 \approx 0.0185$.
(c) To roll a sum of five, we can have the outcome of (1,1,3), (1,3,1), (3,1,1), (2,2,1), (2,1,2), or (1,2,2) for a probability of $6/216 \approx 0.0278$.
(d) Using Theorem 1.5, the probability of getting a sum of 6 or more is $1 - \frac{4}{216} - \frac{6}{216} \approx 0.9537$.

1.10 (a) $1/38 \approx 0.0263$
(b) $18/38 \approx 0.4737$
(c) $12/38 \approx 0.3158$
(d) Using Theorem 1.3, the probability of odd or the first dozen is the probability of odd plus the probability of the first dozen minus the probability of the overlap; since the overlap consists of the numbers 1, 3, 5, 7, 9, and 11, we get $18/38 + 12/38 - 6/38 = 24/38 \approx 0.6316$.

1.11 (a) $5/8 \approx 0.6250$
(b) Since the marble is replaced, we can treat the draws as independent, getting $(\frac{3}{8})^2 = 9/64 \approx 0.1406$.
(c) We can get a red marble then a blue marble, or a blue marble followed by a red marble. This happens with probability $(\frac{3}{8})(\frac{5}{8}) + (\frac{5}{8})(\frac{3}{8}) = 30/64 \approx 0.4688$.
(d) Since there is one less marble for the second draw (and one less blue marble for the blue quantity), we get $(\frac{5}{8})(\frac{4}{7}) = 20/56 \approx 0.3571$.

1.12 (a) There is a total of $m + n$ marbles, so the probability for getting a blue marble is $\frac{n}{m+n}$ and the probability of a red marble is $\frac{m}{m+n}$.

(b) Both probabilities get closer and closer to zero since the quantity of the opposite color marble is getting larger.

(c) Without replacement, there is one fewer marble (and one fewer blue marble) for the second draw, resulting in a probability of $(\frac{n}{m+n})(\frac{n-1}{m+n-1}) = \frac{n(n-1)}{(m+n)(m+n-1)}$.

(d) With replacement, the two draws are independent, so the probability is $(\frac{n}{m+n})^2$.

1.13 **(a)** Since the flips are independent, we get $(0.7)(0.3) = 0.21$ as the probability.

(b) Since the flips are independent, we get $(0.7)(0.3) = 0.21$ as the probability.

(c) Yes, since the probability of HT is the same as the probability of TH.

(d) Yes, assuming that the coin is not two-headed or two-tailed (or so unbalanced that the probability of getting a head (or a tail) is 0%).

Chapter 2: Games, Gambling, and Probability

2.1 For a single number bet, the probability is $1/38$. For the adjacent number bet, it's $2/38$, with the probability being $3/38$ for the three number bet, $4/38$ for the four number bet, $5/38$ for the five number bet, and $6/38$ for the six number bet. The dozen bet and the column bet each have a probability of $12/38$, and any of the even-money wagers (first 18, last 18, red, black, even, and odd) have probabilities of $18/38$.

2.2 For a single number bet, where 1 number wins and 37 lose, the true mathematical odds are $37 : 1$. For the adjacent number bet, where 2 numbers wins and 36 numbers lose, those odds are $36 : 2$ or $18 : 1$. Similarly, the odds are $35 : 3$ for the three number bet, $17 : 2$ for the four number bet, $33 : 5$ for the five number bet, and $16 : 3$ for the six number bet. For the dozen bet and column bet, the odds are $13 : 6$ and any of the even money wagers (first 18, last 18, red, black, even, and odd) have true mathematical odds of $10 : 9$.

2.3 Each of these expected winnings calculations assumes a wager of $1.00.

All even money wagers were covered by Example 2.5.

$$\text{Adjacent} \quad = \frac{2}{38}(\$17) + \frac{36}{38}(-\$1) = -\frac{2}{38} \approx -0.0526$$

$$\text{Three Numbers} \quad = \frac{3}{38}(\$11) + \frac{35}{38}(-\$1) = -\frac{2}{38} \approx -0.0526$$

$$\text{Four Numbers} \quad = \frac{4}{38}(\$8) + \frac{34}{38}(-\$1) = -\frac{2}{38} \approx -0.0526$$

$$\text{Six Numbers} \quad = \frac{6}{38}(\$5) + \frac{32}{38}(-\$1) = -\frac{2}{38} \approx -0.0526$$

$$\text{Dozen/Column} \quad = \frac{12}{38}(\$2) + \frac{26}{38}(-\$1) = -\frac{2}{38} \approx -0.0526$$

2.4 **(a)** With 8 ways to win and 30 ways to lose, the true mathematical odds are 30 : 8 which is equivalent to 15 : 4 or 3.75 : 1.
(b) Offering odds of X : 1 where $X < 3.75$ will work; for example, odds of 3.5 : 1 (equivalent to 7 : 2) would work.
(c) Using a payout of 3.5 : 1 gives expected winnings of

$$\frac{8}{38}(\$3.50) + \frac{30}{38}(-\$1.00) = -\frac{2}{38} \approx -0.0526.$$

2.5 **(a)** Since the house advantage comes from the presence of 0 and 00, the expected winnings should be \$0.00.
(b) $\frac{1}{36}(\$35) + \frac{35}{36}(-\$1) = 0$
(c) $\frac{18}{36}(\$1) + \frac{18}{36}(-\$1) = 0$

2.6 **(a)** The expected winnings of $\frac{10W-28}{38}$ given in Example 2.6, when set equal to -0.0526, gives $W = 2.60012$ when algebraically solved; odds of 2.60012 : 1 as a payout makes the prime number bet have the same house advantage as most of the other wagers.
(b) Solving $\frac{5}{38}W + \frac{33}{38}(-\$1) = \frac{5W-33}{38} = -0.0526$ gives $W = 6.2002$, or 6.2002 : 1 as the appropriate payout to have the same house advantage; using 6 : 1 is easier to work with and properly pay out winning players.

2.7 **(a)** $\frac{1}{39}(\$35) + \frac{38}{39}(-\$1) = -\$0.076923$
(b) $\frac{18}{39}(\$1) + \frac{21}{39}(-\$1) = -\$0.076923$
(c) The house advantage has increased by about 50%.

2.8 For each of the following, the true mathematical odds are given in parentheses after the bet name. The dollar signs are left out of the calculations.

$$\text{Single Number (36:1)} = \frac{1}{37}(35) + \frac{36}{37}(-1) = -\frac{1}{37} \approx -0.0270$$

$$\text{Adjacent Numbers (35:2)} = \frac{2}{37}(17) + \frac{35}{37}(-1) = -\frac{1}{37} \approx -0.0270$$

$$\text{Three Numbers (34:3)} = \frac{3}{37}(11) + \frac{34}{37}(-1) = -\frac{1}{37} \approx -0.0270$$

$$\text{Four Numbers (33:4)} = \frac{1}{37}(8) + \frac{33}{37}(-1) = -\frac{1}{37} \approx -0.0270$$

$$\text{Five Numbers (32:5)} = \frac{5}{37}(6) + \frac{32}{37}(-1) = -\frac{2}{37} \approx -0.0540$$

$$\text{Six Numbers (31:6)} = \frac{6}{37}(5) + \frac{31}{37}(-1) = -\frac{1}{37} \approx -0.0270$$

$$\text{Dozen/Column (25:12)} = \frac{12}{37}(2) + \frac{25}{37}(-1) = -\frac{1}{37} \approx -0.0270$$

$$\text{Even Money (37:18)} = \frac{18}{37}(1) + \frac{19}{37}(-1) = -\frac{1}{37} \approx -0.0270$$

2.9 Solving $\frac{10}{37}W + \frac{27}{37}(-1) = \frac{10W-27}{37} = -0.0270$ for W gives $W = 2.6001$ so a payout of $2.6001 : 1$ would be appropriate.

2.10 Since with probability $1/37$ we result in a spin of 0, giving us half of our \$1.00 assumed wager back, we get

$$\frac{18}{37}(1) + \frac{1}{37}(-0.50) + \frac{18}{37}(-1) = -0.0135.$$

2.11 Here assuming a \$1 wager, with probability $1/37$ a result of 0 happens, at which point the wager is left on the board, with a payout of \$0 for a win and the loss of \$1 on a loss. This gives

$$\frac{18}{37}(1) + \frac{1}{37}\left(\frac{18}{37}(0) + \frac{19}{37}(-1)\right) + \frac{18}{37}(-1) \approx -0.0139$$

which is a good bit better than the usual 2.7% house advantage for European roulette.

2.12 Establishing a point total of 5 on the come out roll happens with probability $4/36$, after which only rolling a 5 (with 4 possibilities) or 7 (with 6 possibilities) matters, so the probability of the pass line winning after a point total of 5 is made (rolling a 5 before a 7) is $(\frac{4}{36})(\frac{4}{10}) = 2/45$. For a point total of 6, we have $(\frac{5}{36})(\frac{5}{11}) = 25/396$ and for a point total of 10 we get $(\frac{3}{36})(\frac{3}{9}) = 1/36$ (since totals of 5 and 9 have equal probabilities with two dice, the calculation for a total of 9 is the same as it is for 5).

2.13 Establishing a point total of 4 on the come out roll happens with probability $3/36$, after which only rolling a 4 (with 3 possibilities) or 7 (with 6 possibilities) matters, so the probability of the don't pass line winning after a point total of 4 is made (rolling a 7 before a 4) is $(\frac{3}{36})(\frac{6}{9}) = 1/18$. For a total of 5, we get $(\frac{4}{36})(\frac{6}{10}) = 1/15$ and for a total of 6 we have $(\frac{5}{36})(\frac{6}{11}) = 5/66$. The calculations for totals of 8, 9, and 10 are the same, respectively, as those for 6, 5, and 4.

2.14 Modifying Example 2.11, we subtract 2/36 from the raw don't pass line win probability of 251/495 to get 149/330 as the probability of the don't pass line winning. This results in the (lousy) expected winnings of

$$\frac{149}{330}(1) + \frac{2}{36}(0) + \frac{244}{495}(-1) = -\frac{41}{990} \approx -0.0414.$$

2.15 Remembering that 12 for an initial roll is not a win or loss for the don't pass line player, we have

$$\text{Expected Value} = \underbrace{\left(\frac{3}{36}\right) \cdot \$6}_{\substack{\text{win come} \\ \text{out roll}}} + \underbrace{\left(\frac{8}{36}\right) \cdot -\$6}_{\substack{\text{lose come} \\ \text{out roll}}} + \underbrace{\left(\frac{1}{18}\right) \cdot \$9}_{\substack{\text{win point} \\ \text{of four}}} + \underbrace{\left(\frac{1}{36}\right) \cdot -\$12}_{\substack{\text{lose point} \\ \text{of four}}}$$

$$+ \underbrace{\left(\frac{1}{15}\right) \cdot \$10}_{\substack{\text{win point} \\ \text{of five}}} + \underbrace{\left(\frac{2}{45}\right) \cdot -\$12}_{\substack{\text{lose point} \\ \text{of five}}} + \underbrace{\left(\frac{5}{66}\right) \cdot \$11}_{\substack{\text{win point} \\ \text{of six}}} + \underbrace{\left(\frac{25}{396}\right) \cdot -\$12}_{\substack{\text{lose point} \\ \text{of six}}}$$

$$+ \underbrace{\left(\frac{5}{66}\right) \cdot \$11}_{\substack{\text{win point} \\ \text{of eight}}} + \underbrace{\left(\frac{25}{396}\right) \cdot -\$12}_{\substack{\text{lose point} \\ \text{of eight}}} + \underbrace{\left(\frac{1}{15}\right) \cdot \$10}_{\substack{\text{win point} \\ \text{of nine}}} + \underbrace{\left(\frac{2}{45}\right) \cdot -\$12}_{\substack{\text{lose point} \\ \text{of nine}}}$$

$$+ \underbrace{\left(\frac{1}{18}\right) \cdot \$9}_{\substack{\text{win point} \\ \text{of ten}}} + \underbrace{\left(\frac{1}{36}\right) \cdot -\$12}_{\substack{\text{lose point} \\ \text{of ten}}} = -\$0.0818$$

for the "raw" expected value of a \$6.00 wager, but since the expected wager (see Example 2.14) is \$10, the expected winnings are −\$0.00818.

2.16 For 5x odds, an initial \$10 wager turns into an additional \$50 wager for free odds; using the same calculation as Example 2.14 with modified payouts and losses for the odds, we have:

$$\text{Expected Value} = \underbrace{\left(\frac{8}{36}\right) \cdot \$10}_{\substack{\text{win come} \\ \text{out roll}}} + \underbrace{\left(\frac{4}{36}\right) \cdot -\$10}_{\substack{\text{lose come} \\ \text{out roll}}} + \underbrace{\left(\frac{1}{36}\right) \cdot \$110}_{\substack{\text{win point} \\ \text{of four}}} + \underbrace{\left(\frac{1}{18}\right) \cdot -\$60}_{\substack{\text{lose point} \\ \text{of four}}}$$

$$+ \underbrace{\left(\frac{2}{45}\right) \cdot \$85}_{\substack{\text{win point} \\ \text{of five}}} + \underbrace{\left(\frac{1}{15}\right) \cdot -\$60}_{\substack{\text{lose point} \\ \text{of five}}} + \underbrace{\left(\frac{25}{396}\right) \cdot \$70}_{\substack{\text{win point} \\ \text{of six}}} + \underbrace{\left(\frac{5}{66}\right) \cdot -\$60}_{\substack{\text{lose point} \\ \text{of six}}}$$

$$+ \underbrace{\left(\frac{25}{396}\right) \cdot \$70}_{\substack{\text{win point} \\ \text{of eight}}} + \underbrace{\left(\frac{5}{66}\right) \cdot -\$60}_{\substack{\text{lose point} \\ \text{of eight}}} + \underbrace{\left(\frac{2}{45}\right) \cdot \$85}_{\substack{\text{win point} \\ \text{of nine}}} + \underbrace{\left(\frac{1}{15}\right) \cdot -\$60}_{\substack{\text{lose point} \\ \text{of nine}}}$$

$$+ \underbrace{\left(\frac{1}{36}\right) \cdot \$110}_{\substack{\text{win point} \\ \text{of ten}}} + \underbrace{\left(\frac{1}{18}\right) \cdot -\$60}_{\substack{\text{lose point} \\ \text{of ten}}} = -\$0.1414$$

and with an expected wager of \$43.33, the house advantage is 0.00326. For 100x odds, the \$10 wager turns into \$1000 on free odds, giving an expected value of:

$$\underbrace{\left(\frac{8}{36}\right) \cdot \$10}_{\substack{\text{win come} \\ \text{out roll}}} + \underbrace{\left(\frac{4}{36}\right) \cdot -\$10}_{\substack{\text{lose come} \\ \text{out roll}}} + \underbrace{\left(\frac{1}{36}\right) \cdot \$2010}_{\substack{\text{win point} \\ \text{of four}}} + \underbrace{\left(\frac{1}{18}\right) \cdot -\$1010}_{\substack{\text{lose point} \\ \text{of four}}}$$

$$+ \underbrace{\left(\frac{2}{45}\right) \cdot \$1510}_{\substack{\text{win point} \\ \text{of five}}} + \underbrace{\left(\frac{1}{15}\right) \cdot -\$1010}_{\substack{\text{lose point} \\ \text{of five}}} + \underbrace{\left(\frac{25}{396}\right) \cdot \$1210}_{\substack{\text{win point} \\ \text{of six}}} + \underbrace{\left(\frac{5}{66}\right) \cdot -\$1010}_{\substack{\text{lose point} \\ \text{of six}}}$$

$$+ \underbrace{\left(\frac{25}{396}\right) \cdot \$1210}_{\substack{\text{win point} \\ \text{of eight}}} + \underbrace{\left(\frac{5}{66}\right) \cdot -\$1010}_{\substack{\text{lose point} \\ \text{of eight}}} + \underbrace{\left(\frac{2}{45}\right) \cdot \$1510}_{\substack{\text{win point} \\ \text{of nine}}} + \underbrace{\left(\frac{1}{15}\right) \cdot -\$1010}_{\substack{\text{lose point} \\ \text{of nine}}}$$

$$+ \underbrace{\left(\frac{1}{36}\right) \cdot \$2010}_{\substack{\text{win point} \\ \text{of ten}}} + \underbrace{\left(\frac{1}{18}\right) \cdot -\$1010}_{\substack{\text{lose point} \\ \text{of ten}}} = -\$0.1414$$

but an expected wager of \$676.67 gives us a house advantage of 0.000209. Note that the *unweighted* expected *value* computations are the same since the house's advantage comes *entirely* from the initial wager only.

2.17 With this structure for odds and the payouts given in Table 2.3, an initial \$10 wager turns into an additional wager, for a point total of 4 or 10, of

$30; a win gives us $70 (with 1:1 payout on the initial wager and 2:1 payout on that additional $30 wager). For a point total of 5 or 9, the odds wager is an additional $40, but is paid at 3:2, for a net total on a win of $70, and the same $70 appears for the total win on a point total of 6 or 8 as well. The ease of payouts is why casinos would use them! To do the expected value, we have

$$\text{Expected Value} = \underbrace{\left(\frac{8}{36}\right) \cdot \$10}_{\substack{\text{win come} \\ \text{out roll}}} + \underbrace{\left(\frac{4}{36}\right) \cdot -\$10}_{\substack{\text{lose come} \\ \text{out roll}}} + \underbrace{\left(\frac{1}{36}\right) \cdot \$70}_{\substack{\text{win point} \\ \text{of four}}} + \underbrace{\left(\frac{1}{18}\right) \cdot -\$40}_{\substack{\text{lose point} \\ \text{of four}}}$$

$$+ \underbrace{\left(\frac{2}{45}\right) \cdot \$70}_{\substack{\text{win point} \\ \text{of five}}} + \underbrace{\left(\frac{1}{15}\right) \cdot -\$50}_{\substack{\text{lose point} \\ \text{of five}}} + \underbrace{\left(\frac{25}{396}\right) \cdot \$70}_{\substack{\text{win point} \\ \text{of six}}} + \underbrace{\left(\frac{5}{66}\right) \cdot -\$60}_{\substack{\text{lose point} \\ \text{of six}}}$$

$$+ \underbrace{\left(\frac{25}{396}\right) \cdot \$70}_{\substack{\text{win point} \\ \text{of eight}}} + \underbrace{\left(\frac{5}{66}\right) \cdot -\$60}_{\substack{\text{lose point} \\ \text{of eight}}} + \underbrace{\left(\frac{2}{45}\right) \cdot \$70}_{\substack{\text{win point} \\ \text{of nine}}} + \underbrace{\left(\frac{1}{15}\right) \cdot -\$50}_{\substack{\text{lose point} \\ \text{of nine}}}$$

$$+ \underbrace{\left(\frac{1}{36}\right) \cdot \$70}_{\substack{\text{win point} \\ \text{of ten}}} + \underbrace{\left(\frac{1}{18}\right) \cdot -\$40}_{\substack{\text{lose point} \\ \text{of ten}}} = -\$0.1414$$

but the expected wager is now

$$\left(\frac{12}{36}\right) \cdot \$10 + \left(\frac{10}{36}\right) \cdot \$60 + \left(\frac{8}{36}\right) \cdot \$50 + \left(\frac{6}{36}\right) \cdot \$40$$

which is $37.78, for a net house advantage of 0.003743.

2.18 (a) Two out of the 36 outcomes result in a total die roll of 2 or 12, at a payout of 2:1, and 14 other outcomes are paid out at 1:1, so the expected value is

$$\frac{2}{36}(2) + \frac{14}{36}(1) + \frac{20}{36}(-1) \approx -\$0.0556.$$

(b) $\frac{4}{36}(7) + \frac{32}{36}(-1) \approx -\$0.1111$

(c) Here, only an 8 or 7 will resolve the wager, there are 11 total ways to roll a 7 or 8, and only one way (⚃⚃) results in a win. This gives us

$$\frac{1}{11}(9) + \frac{10}{11}(-1) \approx -\$0.0909.$$

(d) $\frac{1}{36}(30) + \frac{35}{36}(-1) \approx -\$0.1389$

2.19 (a) For the $1 space, we get true mathematical odds of 30:24 (since 24 spaces are wins and 30 spaces are losses), equivalent to 5:4. For $2, we get 39:15 (or 13:5), for $5 it's 47:7, for $10 it's 50:4 (or 25:2), for $20 it's 52:2 (or 26:1), and for either of the special spaces, it's 53:1.

(b) Using the following expected values, the \$1 wager is the best, if you want to wager at all.

- \$1: $\frac{24}{54}(1) + \frac{30}{54}(-1) \approx -\$0.1111$
- \$2: $\frac{15}{54}(2) + \frac{39}{54}(-1) \approx -\$0.1667$
- \$5: $\frac{7}{54}(5) + \frac{47}{54}(-1) \approx -\$0.2222$
- \$10: $\frac{4}{54}(10) + \frac{50}{54}(-1) \approx -\$0.1852$
- \$20: $\frac{2}{54}(20) + \frac{52}{54}(-1) \approx -\$0.2222$
- Special: $\frac{1}{54}(40) + \frac{53}{54}(-1) \approx -\$0.2407$

(c) It's now a good bit better, as we get $\frac{1}{54}(45) + \frac{53}{54}(-1) \approx -\$0.1401$

(d) With the following, the \$5 wager becomes the best!

- \$1: $\frac{23}{54}(1) + \frac{31}{54}(-1) \approx -\$0.1401$
- \$5: $\frac{8}{54}(5) + \frac{46}{54}(-1) \approx -\$0.1111$

Chapter 3: Games, Gambling, and Probability

3.1 Any 2 or any 7, drawn from the remaining 47 cards, will complete your straight, for a probability of $8/47 \approx 0.1702$.

3.2 This is different because now only a 9 will complete the straight, for a probability of $4/47 \approx 0.0851$.

3.3 (a) There are now 5 more cards we know are not in the deck, letting us draw from the remaining 42 (instead of 47 as before)! Again, any 2 or any 7 will complete the straight for a probability of $8/42 \approx 0.1905$.
(b) Since you know your opponent has one of the four 2s, there are now only 7 cards that complete our straight left in the deck for a probability of $7/42 \approx 0.1667$.
(c) With two 2s and one 7 used, only 5 cards are left that help, so the probability is $5/42 \approx 0.1190$.

3.4 (a) Choose a rank, pick all four cards, then choose a rank for the unmatched card, and select one out of the four of those to get $\binom{13}{1}\binom{4}{4}\binom{12}{1}\binom{4}{1} = 624$.
(b) Choose a rank for the triple, then three of those four cards, then choose a rank for the pair, and two of those four cards, obtaining $\binom{13}{1}\binom{4}{3}\binom{12}{1}\binom{4}{2} = 3,744$.
(c) Choose a rank for the high card of the straight (in one of 10 ways), then for each of the five cards (now fixed in rank), select one of the four possible suits to get $\binom{10}{1}\binom{4}{1}\binom{4}{1}\binom{4}{1}\binom{4}{1}\binom{4}{1} = 10,240$. Subtract the 40 straight flushes to get 10,200 straights.

(d) Choose a rank for the pair and pick two of those four cards, then select three of the remaining ranks for the singletons, and one of of the four possible cards for each to get $\binom{13}{1}\binom{4}{2}\binom{12}{3}\binom{4}{1}\binom{4}{1}\binom{4}{1} = 1,098,240$.

3.5 (a) We select the color (one choice among either red or black) followed by choosing five cards from the twenty-six cards of the chosen color. Afterwards, we subtract the number of raw flushes, 5,148, to get the final count, so $\binom{2}{1}\binom{26}{5} - 5,148 = 126,412$ is our answer.
(b) It's pretty common and ranks near two pair; adding this to the game could add a neat twist for players to think about and wouldn't be present *too* often such as with a single pair.

3.6 From the 52 cards, we choose 4, to get $\binom{52}{4} = 270,725$ possible four card hands.

3.7 Follow the explanations of the standard five card poker; the answers are given here (with computations). Note that the probabilities for each just take the number of hands for each type and divide by 270,725, the total number of four card poker hands.
 (a) $\binom{13}{1}\binom{4}{4} = 13$; probability ≈ 0.000048
 (b) $\binom{4}{1}\binom{11}{1} = 44$; probability ≈ 0.000163
 (c) $\binom{13}{1}\binom{4}{3}\binom{12}{1}\binom{4}{1} = 2,496$; probability ≈ 0.009220
 (d) $\binom{4}{1}\binom{13}{4} - 44 = 2,816$; probability ≈ 0.010402
 (e) $\binom{11}{1}\binom{4}{1}\binom{4}{1}\binom{4}{1}\binom{4}{1} - 44 = 2,772$; probability ≈ 0.010239
 (f) $\binom{13}{2}\binom{4}{2}\binom{4}{2} = 2,808$; probability ≈ 0.010372
 (g) $\binom{13}{1}\binom{4}{2}\binom{12}{2}\binom{4}{1}\binom{4}{1} = 82,368$; probability ≈ 0.304250
 (h) Subtract all others from 270,725 to get 177,408; probability ≈ 0.655307

3.8 Again, consult the five card hands for explanations on how to count; calculations and answers only are given here; probabilities are found by taking the count and dividing by 22,100.
 (a) $\binom{4}{1}\binom{12}{1} = 48$; probability ≈ 0.0022
 (b) $\binom{13}{1}\binom{4}{3} = 52$; probability ≈ 0.0024
 (c) $\binom{12}{1}\binom{4}{1}\binom{4}{1}\binom{4}{1} - 48 = 720$; probability ≈ 0.0326
 (d) $\binom{4}{1}\binom{13}{3} - 48 = 1,096$; probability ≈ 0.0496
 (e) $\binom{13}{1}\binom{4}{2}\binom{12}{1}\binom{4}{1} = 3,744$; probability ≈ 0.1694
 (f) Subtract all others from 22,100 to get 16,440; probability ≈ 0.7439

3.9 A few of the full calculations appear here, but all answers are present. To get the probability, divide the number of hands by $\binom{52}{6} = 20,358,520$.
 (a) 36; probability ≈ 0.000002
 (b) $\binom{13}{1}\binom{4}{4}\binom{12}{1}\binom{4}{2} = 936$; probability ≈ 0.000046
 (c) 1248; probability ≈ 0.000061
 (d) 6864; probability ≈ 0.000337
 (e) 13,728; probability ≈ 0.000674

(f) $\binom{9}{1}\binom{4}{1}^6 = 36,864$; probability ≈ 0.001811

(g) $\binom{13}{3}\binom{4}{2}^3 = 61,775$; probability ≈ 0.003034

(h) 164,736; probability ≈ 0.008092

(i) $\binom{13}{1}\binom{4}{3}\binom{12}{3}\binom{4}{1}^3 = 732,160$; probability $\approx= 0.035963$

(j) $\binom{13}{2}\binom{4}{2}^2\binom{11}{2}\binom{4}{1}^2 = 2,471,040$; probability ≈ 0.121376

(k) 9,884,160; probability ≈ 0.485505

(l) 6,984,972; probability ≈ 0.343098

3.10 (a) Here, we select one rank for the four of a kind, pick all four of those cards, and then of the remaining 48 cards, pick any 3, to get $\binom{13}{1}\binom{4}{4}\binom{48}{3} = 224,848$; divide by 133,784,560, the total number of seven-card poker hands, to get a probability of about 0.0017.

(b) For a flush, we need either 5, 6, or 7 cards of a single suit; for 7, we pick one of the four suits, then any 7 of those 13 cards to get $\binom{4}{1}\binom{13}{7}$. For 6, we pick a suit for the 6 cards, then pick 6 of those cards, then finish by picking a suit for the remaining card and one card from that suit for $\binom{4}{1}\binom{13}{6}\binom{3}{1}\binom{13}{1}$. For 5, we pick a suit, 5 cards from that suit, and then the other two cards are selected from any of the remaining 39 (cards from the other suits) to get $\binom{4}{1}\binom{13}{5}\binom{39}{2}$. After subtracting the straight flushes (41,584 of them), we get 4,047,644 and a probability of about 0.0303.

(c) We must have 5 distinct ranks (otherwise, the hand is a "better" hand, for example, a triple and a pair, would be a full house). There are $\binom{13}{5}$ to select the ranks, and after subtracting the 10 possible straights, we have 1,277 ways to choose five distinct ranks for the hand; of these 5 ranks, choose 1 for the three of a kind, and then pick 3 of those 4 cards, for $\binom{5}{1}\binom{4}{3} = 20$ ways to form the triple given the set of ranks chosen. Finally, for the other four ranks, there are $\binom{4}{1}^4$ ways to select the suits for those cards, but we must subtract $\binom{3}{1}$ in case *all* four of these cards match in suit *and* match one of the tree suits represented by the triple. This gives us 253 possibilities for the suits of those four cards. Thus we have $1,277 \cdot 20 \cdot 253 = 6,461,620$ possible three of a kind hands with probability approximately 0.0483.

(d) A pair must consist of 6 distinct ranks, chosen in $\binom{13}{6}$ ways, but straights must be removed; there are 9 straights consisting of all six cards in order (6 through Ace as the high card), 14 straights with five cards with the 5 or Ace as the high card (the other card, not part of the straight, cannot be a 6 or 9, respectively, otherwise we have six cards in order, so for each of these that sixth card has only seven choices, for $2 \cdot 7$ possibilities), and 48 straights with five cards with 6 through king as the high card (the other card, not part of the straight, cannot be a card immediately above or below the straight we've selected, so from these 8 possible straights, there are only 6 possibilities for that last card, for $8 \cdot 6$ total). This gives us 1,645 choices for the ranks. Once ranks are selected, there are $\binom{6}{1}\binom{4}{2} = 36$ ways to select which rank is for the pair and then pick those two cards. Finally, there are $\binom{4}{1}^5 - 34 = 990$ ways to choose the suits for the unmatched five cards (there are 34 ways to get a flush out of the suit selection, so these must be removed). This gives us $1,645 \cdot 36 \cdot 990 = 58,627,800$ hands with one pair, with a probability of about 0.4382.

3.11 (a) There are $\binom{4}{1}^7 = 16,384$ ways to select the suits for the seven cards of a distinct rank; of these, we have a flush if all seven are from one suit, happening by choosing that suit in $\binom{4}{1}$ ways, or if six of the seven share a suit, happening by choosing that suit, which six cards of the seven match the suit, and picking a different suit for the last card, in $\binom{4}{1}\binom{7}{6}\binom{3}{1}$ ways, or finally if five of the seven share a suit, where we pick the suit, which five cards match, then a different suit for the next card and for the card thereafter, in $\binom{4}{1}\binom{7}{5}\binom{3}{1}\binom{3}{1}$, for a total of 844 flushes, so there are only 15,540 distinct suit combinations for seven distinct ranks that do not give a flush.
(b) For an ace-high straight, the other two cards can be of any rank, for a total of $\binom{8}{2}$ possibilities; for the other 9 possible high cards, we must avoid the rank immediately above the high card, so we first pick the high card in 1 of 9 ways, and then pick the remaining two ranks from the seven allowed to get $\binom{9}{1}\binom{7}{2}$; combining these together, there are 217 straights allowed, so we multiply by our answer from part (a), getting $217 \cdot 15,540 = 3,372,180$ possible non-flush straights with seven distinct ranks.
(c) For possible sets of ranks for the straight, if we have an ace-high straight, we can choose the card not part of the straight in $\binom{8}{1}$ ways; otherwise, for the other 9 possible high cards, there are only 7 choices for the extra card, for $\binom{9}{1}\binom{7}{1}$ possibilities, for a total of 71. Now, for forming the pair, we pick one of the six ranks to be the rank for the pair, and then choose 2 cards of that rank, totaling $\binom{6}{1}\binom{4}{2} = 36$ ways. For suits, there are initially $\binom{4}{1}^5$ possibilities (the pair takes up two of the seven cards already), but some must be removed. If all five are in the same suit, we must exclude them, which happens in $\binom{4}{1}$ ways; otherwise, if *four* of these match *either* suit for the pair, we must also exclude them; this happens by selecting 4 of the 5 cards to share 1 of the 2 suits for the pair, and choosing one of the other 3 suits for the other card of the pair, for $\binom{5}{4}\binom{2}{1}\binom{3}{1}$ possibilities; combined we remove 34 options, leaving 990. Finally, taking the sets of ranks, pair choices, and suit sets together gives us $71 \cdot 36 \cdot 990 = 2,530,440$ straights with six distinct cards.
(d) For three of a kind possibilities that are actually straights, there are 10 choices for the high card; of the five distinct ranks, choose 1 for the triple and pick 3 of those 4 cards for $\binom{5}{1}\binom{4}{3}$ possible triples. There are $\binom{4}{1}^4$ possible suit sets for the non-triple cards, and we only need to remove the cases when all four cards match any of the $\binom{3}{1}$ suits of that triple, for a total of 253 suit sets. Combined, this gives us $10 \cdot 20 \cdot 253 = 50,600$ straights with a triple. For two pair, again we have 10 choices for the high card, and $\binom{5}{2} = 10$ ways to choose the ranks for the pairs. For suits, consider that we have a hand of the form AABBCDE where each distinct letter matches one of the ranks. If AABB has two suits, chosen from the four possible, we get a flush if the $\binom{4}{1}^3$

other cards are all in either of the suits for AA, so there are $\binom{4}{2}\left(\binom{4}{1}^{3} - \binom{2}{1}\right)$ possibilities here. If AABB has three suits, happening by selecting one suit for the overlap, another suit for the other A, and another different suit for the other B, we get a flush only if the extra cards share suit with the overlapped suit, for $\binom{4}{1}\binom{3}{1}\binom{2}{1}\left(\binom{4}{1}^{3} - 1\right)$ situations. Finally, if AABB has all four suits, no flush is possible; choose two suits for AA, which fixes the suits for BB, and the other cards can be anything, giving us $\binom{4}{2}\binom{4}{1}^{3}$ situations. This gives 2,268 non-flush suit sets, giving us $10 \cdot 10 \cdot 2,268 = 226,800$ straights that contain two pairs.

(e) Finally, adding it all together, we have 6,180,020 straights, with a probability of about 0.0462.

Chapter 4: Games, Gambling, and Probability

4.1 Using the binomial distribution, the probability of flipping 6 coins and getting x heads is

$$\binom{6}{x}\left(\frac{1}{2}\right)^{x}\left(\frac{1}{2}\right)^{6-x} = \binom{6}{x}\left(\frac{1}{2}\right)^{6}$$

so for 0, 1, 2, 3, 4, 5, and 6 heads, we get, respectively, probabilities of $1/64$, $6/64$, $15/64$, $20/64$, $15/64$, $6/64$, and $1/64$.

4.2 The probability of zero heads is $\binom{20}{0}\left(\frac{1}{2}\right)^{0}\left(\frac{1}{2}\right)^{20} = \left(\frac{1}{2}\right)^{20}$ which is what is discussed and exhibited in Example 1.11.

4.3 **(a)** Using the binomial distribution, we get $\binom{100}{15}\left(\frac{1}{6}\right)^{15}\left(\frac{5}{6}\right)^{85} \approx 0.1002$.
(b) The formula $n \cdot p$ for expected value gives $100 \cdot \frac{1}{6} \approx 16.67$.
(c) The formula $\sqrt{n \cdot p \cdot q}$ for standard deviation gives $\sqrt{100 \cdot \frac{1}{6} \cdot \frac{5}{6}} \approx 3.726$.
(d) We get 16.67 ± 3.726 or $(12.944, 20.396)$.

4.4 DO THIS

4.5 **(a)** Answer A
(b) For this, we use $1/2$ in place of both $2/3$ and $1/3$ in the general skill test probability calculation, to get

Successes Needed	Dice Rolled			
	1	2	3	4
1	0.500000	0.750000	0.875000	0.937500
2		0.250000	0.500000	0.687500
3			0.125000	0.312500
4				0.062500

Successes Needed	Dice Rolled			
	5	6	7	8
1	0.968750	0.984375	0.992188	0.996094
2	0.812500	0.890625	0.937500	0.964844
3	0.500000	0.656250	0.773438	0.855469
4	0.187500	0.343750	0.500000	0.636719

4.6 (a) Answer A

(b) For this, we use 5/6 in place of 2/3 and 1/6 in place of 1/3 in the general skill test probability calculation, to get

Successes Needed	Dice Rolled			
	1	2	3	4
1	0.166667	0.305556	0.421296	0.517747
2		0.027778	0.074074	0.131944
3			0.004630	0.016204
4				0.000772

Successes Needed	Dice Rolled			
	5	6	7	8
1	0.598122	0.665102	0.720918	0.767432
2	0.196245	0.263224	0.330204	0.395323
3	0.035494	0.062286	0.095776	0.134847
4	0.003344	0.008702	0.017633	0.030656

4.7 (a) DO THIS

(b) DO THIS

(c) DO THIS

4.8 DO THIS

4.9 DO THIS

4.10 For two pairs, we select three of the six die values to represent the three different face types we'll have, and then choose two of those three to select which represent the two pairs, to get $\binom{6}{3}\binom{3}{2} = 60$ rolls of this type. For one pair and three unmatched dice, select four of the six values for those we'll need, and select one of those four to be the pair, getting $\binom{6}{4}\binom{4}{1} = 60$ rolls of this type, for a combined 120 "pairs" at Yahtzee.

4.11 For a full house, select one of the thirty possible and select which three of the five dice get the triple; the probability for this specific arrangement is $\left(\frac{1}{6}\right)^5$ and the probability for matching on the reroll is $\left(\frac{1}{6}\right)^2$, for $\binom{30}{1}\binom{5}{3}\left(\frac{1}{6}\right)^5\left(\frac{1}{6}\right)^2 = \frac{25}{23,328}$. For a non-full house, we select one of the sixty possible, followed by which three of the five dice are for the triple, then use 2! to order the unmatched dice, again multiplying by the same probabilities as before to get $\binom{60}{1}\binom{5}{3}2!\left(\frac{1}{6}\right)^5\left(\frac{1}{6}\right)^2 = \frac{25}{5,832}$ which we combine with the other to get about 0.005358.

4.12 We can start with a full house, reroll the pair, obtain another pair, which we reroll to complete our Yahtzee, getting

$$\underbrace{\binom{30}{1}\binom{5}{3,2}\left(\frac{1}{6}\right)^5}\underbrace{\binom{5}{1}\binom{2}{2}\left(\frac{1}{6}\right)^2}\underbrace{\left(\frac{1}{6}\right)^2} = \frac{125}{839,808}$$

where the separated terms represent the first roll, first reroll, and second reroll, respectively. The following three calculations are for full house, reroll two non-matching dice, and then finish the Yahtzee, followed by two that represent both reroll situations starting with a non-full house.

$$\underbrace{\binom{30}{1}\binom{5}{3,2}\left(\frac{1}{6}\right)^5}\underbrace{\binom{5}{2}\binom{2}{1,1}\left(\frac{1}{6}\right)^2}\underbrace{\left(\frac{1}{6}\right)^2} = \frac{125}{209,952}$$

$$\underbrace{\binom{60}{1}\binom{5}{3,1,1}\left(\frac{1}{6}\right)^5}\underbrace{\binom{5}{1}\binom{2}{2}\left(\frac{1}{6}\right)^2}\underbrace{\left(\frac{1}{6}\right)^2} = \frac{125}{209,952}$$

$$\underbrace{\binom{60}{1}\binom{5}{3,1,1}\left(\frac{1}{6}\right)^5}\underbrace{\binom{5}{2}\binom{2}{1,1}\left(\frac{1}{6}\right)^2}\underbrace{\left(\frac{1}{6}\right)^2} = \frac{125}{52,488}$$

Combining these together, the probability is about 0.003721.

4.13 We can start with two pair, keep one of the pairs and reroll the remaining three dice, obtain another pair along with a matching die of interest, then reroll that pair to obtain a Yahtzee on roll three. We get

$$\underbrace{\binom{60}{1}\binom{5}{2,2,1}\left(\frac{1}{6}\right)^5}\underbrace{\binom{5}{1}\binom{3}{2,1}\left(\frac{1}{6}\right)^3}\underbrace{\left(\frac{1}{6}\right)^2} = \frac{125}{279,936}$$

where the separated terms refer to the first roll, first reroll, and second reroll. The next calculation is for *not* getting a pair on that first reroll, and the two that follow are the same as the first two for the case of only having one pair at the start.

$$\underbrace{\binom{60}{1}\binom{5}{2,2,1}\left(\frac{1}{6}\right)^{5}\binom{5}{2}\binom{3}{1,1,1}\left(\frac{1}{6}\right)^{3}\left(\frac{1}{6}\right)^{2}} = \frac{125}{69,984}$$

$$\underbrace{\binom{60}{1}\binom{5}{2,1,1,1}\left(\frac{1}{6}\right)^{5}\binom{5}{1}\binom{3}{2,1}\left(\frac{1}{6}\right)^{3}\left(\frac{1}{6}\right)^{2}} = \frac{125}{139,968}$$

$$\underbrace{\binom{60}{1}\binom{5}{2,1,1,1}\left(\frac{1}{6}\right)^{5}\binom{5}{2}\binom{3}{1,1,1}\left(\frac{1}{6}\right)^{3}\left(\frac{1}{6}\right)^{2}} = \frac{125}{34,992}$$

Combining yields a probability of about 0.006698.

4.14 $\binom{6}{1} \cdot 5! \cdot \left(\frac{1}{6}\right)^{5} \cdot \left(\frac{1}{6}\right)^{4} \approx 0.000071$, the same

4.15 The end values for modifying Figure 4.4 are {1,8,27,64,125,216} when cubing, with average value 73.5 and they are {3,6,9,12,15,18} for tripling, with average value 10.5. For an initial roll of 1, when tripling, that score of 3 gets added to the expected 73.5 (for cubing on the second roll) for an expected value of 76.5 and when cubing, that score of 1 gets added to the expected triple value 10.5 for an expected value of 11.5; we should definitely triple an initial 1. The full table is

Initial Roll	1	2	3	4	5	6
Triple	76.5	79.5	82.5	85.5	88.5	91.5
Cube	11.5	18.5	37.5	74.5	135.5	226.5

which means we should cube only fives and sixes on the first roll.

Chapter 5: Games, Gambling, and Probability

5.1 (a) Row B, Column C is a saddle point since 0 is the smallest number in the row and largest in the column, so the row player should choose B and the column player should choose C.

(b) There's no saddle point, so use the probabilities given in Theorem 5.4 to find that the row player should play $\frac{1}{2}$A and $\frac{1}{2}$B, and the column player should play $\frac{1}{6}$C and $\frac{5}{6}$D.

(c) Again, there's no saddle point, so Theorem 5.4 gives $\frac{1}{2}$A and $\frac{1}{2}$B for the row player and $\frac{2}{5}C$ and $\frac{3}{5}D$.

5.2 Using Theorem 5.4 on the numbers of Table 5.1 for the column player gives that the probability that Bob calls should be

$$p = \frac{|50.00 - 4.50|}{|50.00 - 4.50| + |-36.50 - 9.00|} = 0.50$$

as needed.

5.3 Essentially, no. If the row (or column) player had a pure strategy, he or she will always play it, so the column (or row) player, knowing this, should always pick the best for him or her (sometimes both entries for the pure strategy might be the same, so the opposing player would be ambivalent to their choice, but this isn't a mixed strategy).

5.4 (a) No dominance or saddle points exist in this game, so we'll need to use oddments or Theorem 5.4 to find the solution. The result is $(\frac{5}{7}A, \frac{2}{7}B)$ for the row player and $(\frac{5}{7}C, \frac{2}{7}D)$ for the column player, with a value of $-\frac{3}{7}$.
(b) Using dominance, we can remove row B (since it is dominated by row C) and column F (since it is dominated by column D). Solving the resulting two-by-two game using oddments or Theorem 5.4 gives $(\frac{3}{11}A, \frac{8}{11}C)$ for the row player and $(\frac{5}{11}D, \frac{6}{11}E)$ for the column player. The value is $\frac{4}{11}$.
(c) In this game, entry AE is a saddle point, so the row player should play A, the column player should play E, and the value is 2.
(d) The game itself has no obvious dominance or saddle points; being a 2×4 game, we plot the column player's choices on an axis and observe that the highest point of the lower envelope occurs when columns E and F meet, so we ignore columns C and D, proceeding to solve the resulting two-by-two game. We get $(\frac{8}{13}A, \frac{5}{13}B)$ for the row player and $(\frac{5}{13}E, \frac{8}{13}F)$ for the column player, with a value of $\frac{12}{13}$.

5.5 (a) For row A to dominate row B, we need to make sure that the row player prefers the value 0 in column C at least as much as the value of t; this will happen when $t \leq 0$.
(b) For t to be a saddle point, it must be the smallest entry in its row (so $t \leq 3$) and largest in its column (meaning $t \geq 0$). Combining these, we get $0 \leq t \leq 3$ for the initial range. Note that when $t = 0$, the BC entry is no longer a saddle point (since AC is the saddle point), so the final range is $0 < t \leq 3$.
(c) From the work on the two other parts, the range would be anything not included by (a) or (b), so $t > 3$ is the answer.

5.6 We use Rose for the row player and Colin for the column player. The resulting matrix is given by the following; you may have a slightly different look to your matrix, as you could swap columns, swap rows, or swap both rows and columns.

	Red 3	Black 8
Black 2	-3	2
Red 9	9	-8

Solving the game, we get $(\frac{17}{22}B2, \frac{5}{22}R9)$ for Rose and $(\frac{10}{22}R3, \frac{12}{22}B8)$ for Colin, and the value to Rose is $-\frac{6}{22}$.

5.7 Dominance is not affected, since if $a \geq b$ for any two matrix entries a and b, then $a + d \geq b + d$. For the same reason, if saddle point status does not change because the inequalities used for maximin and minimax are preserved. For a mixed strategy in the two-by-two case, note that in the term $\frac{|z-w|}{|x-y|+|z-w|}$ that appears in Theorem 5.4, if we add d to all entries, we get $\frac{|(z+d)-(w+d)|}{|(x+d)-(y+d)|+|(z+d)-(w+d)|} = \frac{|z-w|}{|x-y|+|z-w|}$ so that the probability of playing row A stays the same (and similarly for any other row or column or size since whatever formula gives the probability, the d terms will subtract out). Recall that if the column player uses probabilities $p_1, p_2, \ldots, p_n$ for their n columns, and row A had matrix entries $a_1, a_2, \ldots, a_n$, the normal value is $\sum_{i=1}^{n} a_i \cdot p_i$; adding d to each entry results in a value of $\sum_{i=1}^{n}(a_i + d) \cdot p_i = \sum_{i=1}^{n} a_i \cdot p_i + \sum_{i=1}^{n} d \cdot p_i = \sum_{i=1}^{n} a_i \cdot p_i + d \cdot \sum_{i=1}^{n} p_i = \sum_{i=1}^{n} a_i \cdot p_i + d$ since the sum of all of the p_i terms sum to 1, so the new value is the original with d added.

5.8 Dominance is not affected, since if $a \geq b$ for any two matrix entries a and b, then $ad \geq bd$ for a positive value of d. For the same reason, if saddle point status does not change because the inequalities used for maximin and minimax are preserved. For a mixed strategy in the two-by-two case, note that in the term $\frac{|z-w|}{|x-y|+|z-w|}$ that appears in Theorem 5.4, if we multiply all entries by a positive value d, we get $\frac{|(zd)-(wd)|}{|(xd)-(yd)|+|(zd)-(wd)|} = \frac{|z-w|}{|x-y|+|z-w|}$ by factoring and canceling so that the probability of playing row A stays the same (and similarly for any other row or column or size since whatever formula gives the probability, the d terms will subtract out). Recall that if the column player uses probabilities $p_1, p_2, \ldots, p_n$ for their n columns, and row A had matrix entries $a_1, a_2, \ldots, a_n$, the normal value is $\sum_{i=1}^{n} a_i \cdot p_i$; multiplying each entry by d results in a value of $\sum_{i=1}^{n}(a_i \cdot d) \cdot p_i = d \cdot \sum_{i=1}^{n} a_i \cdot p_i$ by factoring, so the new value is the original multiplied by d.

5.9 Multiplying by a negative number changes the direction of the inequality; a row or column A that dominates row or column B will now be dominated by B instead. Saddle points can move around (instead of the saddle point being at the smallest in the row, largest in the column, the location in the original matrix would be given by the largest in the row and smallest in the column). Because mixed strategy recommendations in Theorem 5.4 come from relative differences, the probabilities for rows and columns will stay the same, but the value will be multiplied by $-d$. In short, nothing changes for mixed strategies, while pure strategies may change a great deal.

5.10 Suppose that the probability for Row A is p and that for Row B is $1-p$. Then the expected value for column C is $px + (1-p)z$ and that for column D is $py + (1-p)w$. Equate these, move all terms with p to one side and factor to obtain $p(x - z + w - y) = w - z$. By grouping terms and adding absolute value bars (to assume that p is a positive quantity as required for a probability), we get

$$p = \frac{|z - w|}{|x - y| + |z - w|} \qquad 1 - p = \frac{|x - y|}{|x - y| + |z - w|}$$

for the probability the row player should play rows A and B.

Very similarly, set p to be the probability column C should be played and $1 - p$ that for column D. The expected value for row A is then $px + (1 - p)y$ and for row B it's $pz + (1 - p)w$; set equal, move all terms with a p to one side, and factor to get $p(x - y - z + w) = w - y$ which, as above, yields

$$p = \frac{|y - w|}{|x - z| + |y - w|} \qquad 1 - p = \frac{|x - z|}{|x - z| + |y - w|}$$

for the probabilities used for columns C and D.

5.11 Subtract 10 from each payoff to the row player to get the game

	C	D
A	$(3, -3)$	$(-6, 6)$
B	$(-2, 2)$	$(1, -1)$

which is a zero sum game (the equivalence comes from the fact that all ordered pairs line on a straight line of negative slope). The solution is $(\frac{1}{4}A, \frac{3}{4}B)$ for the row player and $(\frac{7}{12}C, \frac{5}{12}D)$ for the column player. The value of the new game is $-\frac{3}{4}$, which means, by adding 10 back for the row player, the value is $\frac{37}{4}$ for the row player and $\frac{3}{4}$ to the column player.

5.12 Rewriting the relationship as $y = -x + d$ means that the outcomes plotted would be of the form $(x, -x + d)$ which clearly lie on a straight line of negative slope, so the game must be equivalent to a zero sum game.

5.13 (a) No, since the four outcomes, when graphed, do not like on a straight line.
(b) Yes, since the four outcomes line on a straight line.

5.14 (a) $(0, 0)$ and $(-1, -1)$ are not Pareto optimal because, for example, $(2, 1)$ is better for both players. Both outcomes $(2, 1)$ and $(1, 2)$ are Pareto optimal because there is are no other outcomes that are better for one player and at least as good for the other player. Both $(2, 1)$ and $(1, 2)$ are also pure strategy equilibrium points. The movement diagram is given by

	C	D
A	$(0, 0) \quad \rightarrow$	$(1, 2)$
	$\downarrow$	$\uparrow$
B	$(2, 1) \quad \leftarrow$	$(-1, -1)$

(b) The outcome $(1, 1)$ is not Pareto optimal; all other outcomes are Pareto optimal. Outcome $(4, 4)$ is the only pure strategy equilibrium point. The movement diagram is given by

	C	D
A	$(1, 1) \quad \rightarrow$	$(0, 5)$
	$\downarrow$	$\downarrow$
B	$(5, 0) \quad \rightarrow$	$(4, 4)$

5.15 $(\frac{2}{5}A, \frac{3}{5}B)$, $(\frac{1}{2}C, \frac{1}{2}D)$, value $(\frac{3}{2}, \frac{9}{5})$

5.16 For the row player, the prudential strategy is $(\frac{6}{7}A, \frac{1}{7}B)$ with security level $\frac{11}{7}$. For the column player, the prudential strategy is $(\frac{4}{5}C, \frac{1}{5}D)$ and security level $\frac{8}{5}$. The row player's counter-prudential strategy is B, and the column player's counter-prudential strategy is D.

5.17 (a) BCF $= (4, 3, 3)$
   (b) The result is BDF $= (1, 2, 7)$
   (c) The result is BCF $= (4, 3, 3)$
   (d) The result is ACF $= (3, 5, 2)$
   (e) Larry prefers Colin, Colin prefers Rose, and Rose prefers Larry.

Chapter 6: Games, Gambling, and Probability

6.1 (a) For this, we need two nodes and four edges as shown in the following diagram.

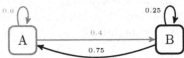

   (b) The matrix is $P = \begin{bmatrix} 0.6 & 0.4 \\ 0.75 & 0.25 \end{bmatrix}$.

   (c) The matrix P^2 is $\begin{bmatrix} 0.66 & 0.34 \\ 0.6375 & 0.3625 \end{bmatrix}$ so by looking at the upper-right entry, the probability of starting at A and ending at B with two moves is 0.34.

   (d) We compute P^3 to answer this, getting $\begin{bmatrix} 0.651 & 0.349 \\ 0.654375 & 0.345625 \end{bmatrix}$ so the requested probability is 0.349.

   (e) $[0.652174, 0.347826]$

6.2 (a) For this, we again need two nodes and four edges as shown in the following diagram.

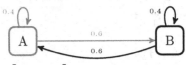

   (b) The matrix is $P = \begin{bmatrix} 0.4 & 0.6 \\ 0.6 & 0.4 \end{bmatrix}$.

   (c) The matrix P^2 is $\begin{bmatrix} 0.52 & 0.48 \\ 0.48 & 0.52 \end{bmatrix}$ so by looking at the upper-right entry, the probability of starting at A and ending at B with two moves is 0.48.

   (d) We compute P^3 to answer this, getting $\begin{bmatrix} 0.496 & 0.504 \\ 0.504 & 0.496 \end{bmatrix}$ so the requested probability is 0.504.

(e) $[0.5, 0.5]$

These results should not be surprising since this is an "equal" game.

6.3 (a) This matrix is $\begin{bmatrix} 0 & 3/4 & 1/4 \\ 1/4 & 1/4 & 1/2 \\ 1/2 & 1/2 & 0 \end{bmatrix}$.

(b) We get $P^2 = \begin{bmatrix} 5/16 & 5/16 & 5/8 \\ 5/16 & 1/2 & 3/16 \\ 1/8 & 1/2 & 3/8 \end{bmatrix}$ and by looking at row 2, column 3, we get $3/16$ as the probability requested.

(c) With $P^3 = \begin{bmatrix} 17/64 & 1/2 & 15/64 \\ 7/32 & 29/64 & 21/64 \\ 5/16 & 13/32 & 9/32 \end{bmatrix}$ the desired probability is $5/16$.

(d) $[0.258065, 0.451613, 0.290323]$ so I would want space B!

6.4 We would need to change the probability on the edge going from D to D to be $3/6$ and the probability on the edge going from D to "End" would increase by $1/6$ to be $2/6$. The matrix would be:

$$\begin{bmatrix} 0 & 1/6 & 1/6 & 1/6 & 1/6 & 1/6 & 1/6 \\ 0 & 1/6 & 1/6 & 1/6 & 1/6 & 1/6 & 1/6 \\ 0 & 0 & 2/6 & 1/6 & 1/6 & 1/6 & 1/6 \\ 0 & 0 & 0 & 3/6 & 1/6 & 1/6 & 1/6 \\ 0 & 0 & 0 & 0 & 3/6 & 1/6 & 2/6 \\ 0 & 0 & 0 & 0 & 0 & 5/6 & 1/6 \\ 0 & 0 & 0 & 0 & 0 & 0 & 1 \end{bmatrix}$$

6.5 (a) Each space would have six edges coming out of it to distinct spaces, each labeled with a probability of $1/6$. For example, from space C, edges would connect to D, E, "End," "Start," A, and B.

(b) The new matrix looks like

$$\begin{bmatrix} 0 & 1/6 & 1/6 & 1/6 & 1/6 & 1/6 & 1/6 \\ 1/6 & 0 & 1/6 & 1/6 & 1/6 & 1/6 & 1/6 \\ 1/6 & 1/6 & 0 & 1/6 & 1/6 & 1/6 & 1/6 \\ 1/6 & 1/6 & 1/6 & 0 & 1/6 & 1/6 & 1/6 \\ 1/6 & 1/6 & 1/6 & 1/6 & 0 & 1/6 & 1/6 \\ 1/6 & 1/6 & 1/6 & 1/6 & 1/6 & 0 & 1/6 \\ 0 & 0 & 0 & 0 & 0 & 0 & 1 \end{bmatrix}$$

(c) We get the same steady-state vector of $[0, 0, 0, 0, 0, 0, 1]$ since, long-term, we always get stuck at "End" once we arrive there.

6.6 (a) For spaces 9, 17, 20, and 30, we add up the probabilities to get $0.029750 + 0.030422 + 0.030308 + 0.031691 = 0.122171$. Similar to Example 6.4, "Buyer" in 8 has a probability of $1 - (1 - 0.122171)^8 \approx 0.647402$.

(b) We gain $300 profit if we are able to sell the ring, and lose $300 if we can't, so the expected value is $300p + (-300)(1 - p) = 600p - 300$ which is greater than 0 when $p > 0.5$.

(c) We want $1 - (1 - 0.122171)^x \geq 0.5$; solving for x yields $x \geq 5.31$, so six rolls, on average, should be sufficient.

6.7 (a) Using Table 11.6, there are only four outcomes where doubles are rolled, ⊡ ▪, ⊡ ▪, ⊡ ▪, and ⊡ ▪, so the probability is $4/36 = 1/9$.

(b) $\left(\frac{1}{9}\right)^3 = \frac{1}{729} \approx 0.001372$ which is much smaller than the $\frac{1}{216} \approx 0.004630$ for regular dice; interestingly, though the Sicherman Dice will provide the same general flow around the board for pretty much anything else since the dice sums have the same probability distribution as standard dice, players will end up in jail a little less.

6.8 (a) We did the dark blue, yellow, and red groups in the example. The others are here.

Brown	$(0.0212824)(250) + (0.0215936)(450) = \$15.04$
Light Blue	$(0.0225527)(550) + (0.0231383)(550) + (0.0229617)(600) = \$38.91$
Light Purple	$(0.0269635)(750) + (0.0237101)(750) + (0.0246382)(900) = \$60.18$
Orange	$(0.0279591)(950) + (0.0294002)(950) + (0.0308864)(1000) = \$85.37$
Green	$(0.0267389)(1275) + (0.0262337)(1275) + (0.0249801)(1400) = \$102.51$

(b) Using similar calculations, we get the table below. Note that "base rent" does not include the fact that owning all three properties entitles the owner to double the rent (for comparisons, a factor of two for each does not matter). Dollar signs are omitted.

Group	Base	1	2	3	4
Brown	0.13	0.64	1.93	5.80	10.32
Light Blue	0.46	2.29	6.41	19.23	28.61
Light Purple	0.80	4.01	12.04	35.12	48.92
Orange	1.30	6.49	18.27	50.08	67.73
Red	1.64	8.20	23.48	62.89	78.21
Yellow	1.80	9.02	27.07	65.03	78.98
Green	2.08	10.63	31.90	72.66	88.25
Dark Blue	2.08	9.07	26.65	60.73	72.96

(c) The yellow group remains better than the red group until hotels are added!

6.9 The best contains Reading Railroad and the B & O Railroad, with combined probability of 0.0602389 and expected value of $3.01; the worse contains Pennsylvania Railroad and the Short Line, with probability 0.0535154 and value $2.68; note that you can find these subsets easily by choosing the two railroads with highest probability for the best and the two with the lowest for the worst.

6.10 We list the results in order of color around the board.

Group	EV
Brown	$14.16
Light Blue	$36.66
Light Purple	$57.22
Orange	$80.37
Red	$87.22
Yellow	$87.28
Green	$96.52
Dark Blue	$80.40

6.11 To reach 14 on 4 dice, we can consider two pairs of dice; there are 6 ways to roll a 7, so there are 6 · 6 ways to *combine* a seven on the first pair with a seven on the second, giving 36 here. We can also reach 14 by 8 + 6 where we get 8 on the first pair and 6 on the second pair, or 6 + 8 where we get 6 on the first pair and 8 on the second, giving 2 · 5 · 5 = 50 possibilities. For 9 + 5 and 5 + 9, there are 32, for 10 + 4 and 4 + 10 there are 18, for 11 + 3 and 3 + 11 there are 8, and finally there are 2 possibilities for 12 + 2 and 2 + 12, giving 146 total.

6.12 (a) These values are included in the table presented for part (b).
(b) Dollar signs are left off, and "base rent" does not include doubling the value for owning all three properties.

Group	Base	1	2	3	4	H
Brown	0.13	0.65	1.94	5.81	10.33	15.06
Light Blue	0.46	2.30	6.43	19.28	28.68	39.01
Light Purple	0.80	4.02	12.05	35.16	48.97	60.24
Orange	1.30	6.48	18.24	50.02	67.64	85.27
Red	1.64	8.20	23.49	62.89	78.22	93.54
Yellow	1.81	9.03	27.08	65.07	79.02	92.97
Green	2.08	10.64	31.93	72.73	88.34	102.62
Dark Blue	2.08	9.08	26.69	60.81	73.06	85.32

(c) The best three-railroad subset contains the Reading Railroad, the Pennsylvania Railroad, and the B & O Railroad, with expected value $8.95; the worst three contains Reading Railroad, the Pennsylvania Railroad, and the Short Line, with value $8.32. The best two-railroad subset has the Reading Railroad and the B & O Railroad with value $3.01, and the worst has the Pennsylvania Railroad and the Short Line, with value $2.68.
(d) The results are almost the same, and the relative ranking of each color group remains the same.

6.13 (a) Again, these values are included in the table presented for part (b).
(b) Again, dollar signs have been left off, and "base rent" does not include doubling the value for owning all three properties.

Group	Base	1	2	3	4	H
Brown	0.12	0.61	1.83	5.48	9.74	14.20
Light Blue	0.43	2.17	6.06	18.19	27.07	36.81
Light Purple	0.76	3.82	11.47	33.44	46.57	57.34
Orange	1.22	6.10	17.18	47.11	63.73	80.35
Red	1.53	7.66	21.94	58.74	73.05	87.36
Yellow	1.70	8.49	25.48	61.20	74.32	87.45
Green	1.96	10.04	30.12	68.59	83.31	96.78
Dark Blue	1.96	8.58	25.22	57.48	69.07	80.65

(c) The best three railroad subset contains the Reading Railroad, the Pennsylvania Railroad, and the B & O Railroad, with expected value $8.33; the worst three contains Reading Railroad, the Pennsylvania Railroad, and the Short Line, with value $7.73. The best two railroad subset has the Reading Railroad and the B & O Railroad with value $2.85, and the worse has the Pennsylvania Railroad and the Short Line, with value $2.46.

(d) Again, the results are very close, and the relative ranking is unchanged!

Chapter 7: Games, Gambling, and Probability

7.1 For each of these, use Theorem 7.1 to get the results.

(a) Blue Jays 100 : 210, Padres 210 : 100, service charge $20, and expected winnings −$6.45

(b) Rays 100 : 155, Astros 155 : 100, service charge $10, and expected winnings −$3.92

(c) Phillies 100 : 135, Giants 135 : 100, service charge $20, and expected winnings −$8.51

7.2 Using Theorem 7.1, the service charge for this wager would be $S = 0$ and the casino or book maker would not make any profit off of the moneyline payouts.

7.3 For Flip Ahead, we need Double Six to lose (with probability $\frac{35}{36}$) and Bi-Nomial to lose (with probability $\frac{3}{4}$ since the probability of getting 3 heads is $\frac{1}{4}$), and Flip Ahead wins with probability $\frac{1}{2}$, so the probability of Flip Ahead winning is $\left(\frac{35}{36}\right)\left(\frac{3}{4}\right)\left(\frac{1}{2}\right) = \frac{35}{96}$ which translates to odds of 61 : 35. For Roll Four, we need the first three horses to lose, and roll a four with probability $\frac{1}{6}$ so we get $\left(\frac{35}{36}\right)\left(\frac{3}{4}\right)\left(\frac{1}{2}\right)\left(\frac{1}{6}\right) = \frac{35}{576}$ or odds of 541 : 35. Finally, for None Above, we need the first four horses to lose, giving us $\left(\frac{35}{36}\right)\left(\frac{3}{4}\right)\left(\frac{1}{2}\right)\left(\frac{5}{6}\right) = \frac{175}{576}$ which means the odds are 401 : 175.

7.4 Double six has a payout of $20.50 and breakage of $6.25. For Binomial, it's $5.00 and no breakage. For Flip Ahead, we get $4.60 and $110 breakage, and that for None Above is $9.20 and $87.

	Horse	Win	Place	Show
7.5	Double Six	$20.50	$5.00	$2.88
	Flip Ahead	—	$2.75	$2.11
	Roll Four	—	—	$7.33

7.6 Using Theorem 5.4, we get $(\frac{4}{11}KL, \frac{7}{11}KR)$ for the kicker and $(\frac{5}{11}GL, \frac{6}{11}GR)$ for the goalie, for a value of 79.1; the values of KL and GL match the laboratory results exactly.

7.7 (a) $P =$
$$\begin{bmatrix}
0.1 & 0.2 & 0.7 & 0.0 & 0.0 & 0.0 & 0.0 \\
0.1 & 0.2 & 0.7 & 0.0 & 0.0 & 0.0 & 0.0 \\
0.0 & 0.0 & 0.1 & 0.2 & 0.7 & 0.0 & 0.0 \\
0.0 & 0.0 & 0.1 & 0.2 & 0.7 & 0.0 & 0.0 \\
0.0 & 0.0 & 0.0 & 0.0 & 0.1 & 0.2 & 0.7 \\
0.0 & 0.0 & 0.0 & 0.0 & 0.1 & 0.2 & 0.7 \\
0.0 & 0.0 & 0.0 & 0.0 & 0.0 & 0.0 & 1.0
\end{bmatrix}$$

(b) $P =$
$$\begin{bmatrix}
0.03 & 0.06 & 0.28 & 0.14 & 0.49 & 0.00 & 0.00 \\
0.03 & 0.06 & 0.28 & 0.14 & 0.49 & 0.00 & 0.00 \\
0.00 & 0.00 & 0.03 & 0.06 & 0.28 & 0.14 & 0.49 \\
0.00 & 0.00 & 0.03 & 0.06 & 0.28 & 0.14 & 0.49 \\
0.00 & 0.00 & 0.00 & 0.00 & 0.03 & 0.06 & 0.91 \\
0.00 & 0.00 & 0.00 & 0.00 & 0.03 & 0.06 & 0.91 \\
0.00 & 0.00 & 0.00 & 0.00 & 0.00 & 0.00 & 1.00
\end{bmatrix}$$

(c) For the first six states, in usual order, the expected number of runs scored from those states are 0.6857, 0.9857, 0.4571, 0.7571, 0.2286, and 0.5286
(d) 4.2857

7.8 The new formula is $g_1 = p_1^4 + 4(1 - p_1)p_1^4 + 10(1 - p_1)^2 p_1^4 + 20(1 - p_1)^3 p_1^4$ which results in a probability of 0.4943 of the first player winning if $p_1 = 0.56$. This is much lower than in Example 7.6 and explains why the "win by two" is used in tennis games. The modified graph (with the original graph of Figure 7.2 in light gray) is below.

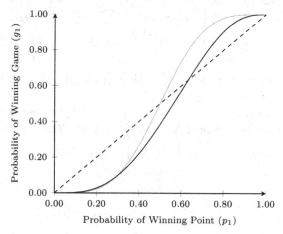

7.9 The new formula is $g_1 = p_1^5 + 5(1-p_1)p_1^5 + 15(1-p_1)^2 p_1^5 + 35(1-p_1)^3 p_1^5 + \frac{70p_1^4(1-p_1)^4 p_1^2}{2p_1^2 - 2p_1 + 1}$ which results in a probability of the first player winning of 0.6599 when $p_1 = 0.56$ which is higher than in the example problem, which makes sense since more points played should favor the player with a higher chance of winning a point. The modified figure, with the original graph from the example in light gray, is below.

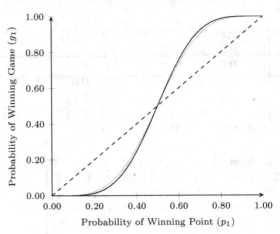

7.10 **(a)** For this, we modify the t_1 calculation on page 209 by removing the "win 7-6" option, and *replacing* the "win 7-5" with $\sum_{b=0}^{5} \binom{5}{b} g_1^{5-b}(1 - g_1)^b \binom{5}{5-b} g_2^{5-b}(1-g_2)^b \left(\sum_{j=0}^{\infty} (g_1 g_2 + (1-g_1)(1-g_2))^j \right) g_1(1-g_2)$ to complete the formula.

(b) The probability that the first player wins the seven-point tie-break game, denoted b_1, is given by the messy formula $b_1 = p_1^3(1-p_2)^4 + \sum_{b=0}^{1} \binom{4}{b}(1-p_2)^{4-b} p_2^b \binom{3}{1-b}(1-p_1)^b p_1^{3-b} p_1 + \sum_{b=0}^{2} \binom{4}{b}(1-p_2)^{4-b} p_2^b \binom{4}{2-b}(1-p_1)^b p_1^{4-b} p_1 + \sum_{b=0}^{3} \binom{4}{b}(1-p_2)^{4-b} p_2^b \binom{5}{2-b}(1-p_1)^b p_1^{5-b}(1-p_2) + \sum_{b=0}^{4} \binom{5}{b}(1-p_2)^{5-b} p_2^b \binom{5}{4-b}(1-p_1)^b p_1^{5-b}(1-p_2) + \sum_{b=0}^{5} \binom{6}{b}(1-p_2)^{6-b} p_2^b \binom{5}{5-b}(1-p_1)^b p_1^{5-b} p_1 + \sum_{b=0}^{6} \binom{6}{b}(1-p_2)^{6-b} p_2^b \binom{6}{6-b}(1-p_1)^b p_1^{6-b} \left(\sum_{j=0}^{\infty} (p_1 p_2 + (1-p_1)(1-p_2))^j \right) p_1(1-p_2)$. With this, on page 209, remove the "win 7-6" option and replace "win 7-5" with $\sum_{b=0}^{5} \binom{5}{b} g_1^{5-b}(1-g_1)^b \binom{5}{5-b} g_2^{5-b}(1-g_2)^b (g_1 g_2 + (1-g_1)(1-g_2)) \cdot b_1$.

(c) With b_1 as in (b), edit t_1 on page 209 by removing the "win 7-6" and replacing "win 7-5" by $\sum_{b=0}^{5} \binom{5}{b} g_1^{5-b}(1 - g_1)^b \binom{5}{5-b} g_2^{5-b}(1 - g_2)^b \left[\left(\sum_{j=0}^{6} (g_1 g_2 + (1-g_1)(1-g_2))^j \right) + (g_1 g_2 + (1-g_1)(1-g_2))^7 b_1 \right]$.

Chapter 8: Games, Gambling, and Probability

8.1 The probability of getting two 10-valued cards is $\frac{16}{52} \cdot \frac{15}{51}$, the probability of getting an ace followed by a 9 is $\frac{4}{52} \cdot \frac{4}{51}$, and the probability of getting a 9 followed by an ace is also $\frac{4}{52} \cdot \frac{4}{51}$ for a total probability of about 0.1026. Alternatively, we could choose 2 of the 16 ten-valued cards, or one of the aces and one of the nines, and divide by the total number of two-card hands and get

$$\frac{\binom{16}{2} + \binom{4}{1}\binom{4}{1}}{\binom{52}{2}} \approx 0.1026.$$

8.2 For the first four parts, we use the combinations method from Exercise 8.1.

(a) $\frac{\binom{32}{2}+\binom{8}{1}\binom{8}{1}}{\binom{104}{2}} \approx 0.1046$ (b) $\frac{\binom{64}{2}+\binom{16}{1}\binom{16}{1}}{\binom{208}{2}} \approx 0.1055$

(c) $\frac{\binom{96}{2}+\binom{24}{1}\binom{24}{1}}{\binom{312}{2}} \approx 0.1059$ (d) $\frac{\binom{128}{2}+\binom{32}{1}\binom{32}{1}}{\binom{416}{2}} \approx 0.1060$

(e) Using the non-combinations method of Exercise 8.1, we get $\left(\frac{16}{52}\right)^2 + \frac{4}{52} \cdot \frac{4}{52} + \frac{4}{52} \cdot \frac{4}{52} \approx 0.1065$

The probability of getting a hand valued at 20 points increases as the number of decks increases.

8.3 Since there are 49 unknown cards, we just need to count the number of ten-valued cards available.

(a) $\frac{16}{49}(2) + \frac{33}{49}(-1) \approx -\$0.0204$ (b) $\frac{15}{49}(2) + \frac{34}{49}(-1) \approx -\$0.0816$

(c) $\frac{14}{49}(2) + \frac{35}{49}(-1) \approx -\$0.1429$

8.4 Now, there are only 45 unknown cards, changing our denominators and the numerators for the losing part of the expected value computations.

(a) $\frac{16}{45}(2) + \frac{29}{45}(-1) \approx +\$0.0612$ (b) $\frac{15}{45}(2) + \frac{30}{45}(-1) \approx \$0.00$

(c) $\frac{14}{45}(2) + \frac{31}{45}(-1) \approx -\$0.0612$

Knowing that, among fewer unknown cards, the same amount of ten-valued cards remaining gives us higher expected values which, for some situations, is player-favorable!

8.5 The tree is similar to Figure 8.3 with different labels and numbers; see the tree labeled in this section of the answers.

(a) 3/7 (b) 4/7 (c) 5/7 (d) 15/56 (e) 20/56

8.6 Here each node has three edges coming off of it, but the result is very similar to before; see the tree labeled in this section of the answers.

8.7 There is only one type of change that needs to be made to Figure 8.4. Each of the right-most numbers in a path need to be increased by 1 (including the 9 at the top, which becomes a 10). Because of that 10, the probability of

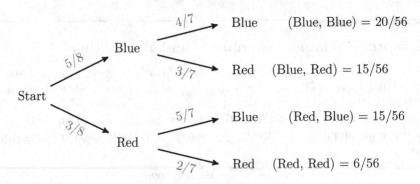

Probability Tree for Exercise 8.5

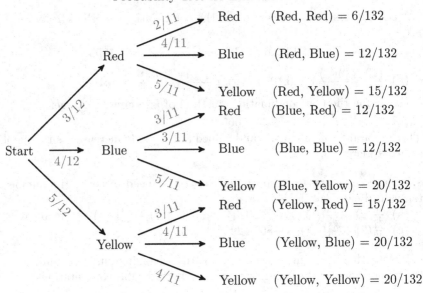

Probability Tree for Exercise 8.6

reaching 20 with one card (the face down card) is now 16/52 rather than 4/52, so the approximate probability is now

$$\left(\frac{16}{52}\right) + 5\left(\frac{4}{52}\right)^2 + 10\left(\frac{4}{52}\right)^3 + 10\left(\frac{4}{52}\right)^4 + 5\left(\frac{4}{52}\right)^5 + \left(\frac{4}{52}\right)^6 \approx 0.342194.$$

For reaching a point total of 21, take Figure 8.4, *delete* the top branch (the 9 would normally turn into an ace, but because a round immediately ends when the dealer has blackjack, we exclude this case) and then add 2 to the right-most numbers in each path. The possibility of reaching 21 with 1 card (the face down card) is now gone, giving an approximate probability of

$$5\left(\frac{4}{52}\right)^2 + 10\left(\frac{4}{52}\right)^3 + 10\left(\frac{4}{52}\right)^4 + 5\left(\frac{4}{52}\right)^5 + \left(\frac{4}{52}\right)^6 \approx 0.034501.$$

8.8 The expected value calculation for not surrendering is now $p \cdot \$1.00 + r \cdot$ $\$0.00 + (1 - p - r)(-\$1.00) = 2p + r - 1$ and that for surrendering is still $-\$0.50$. Equating these, obtaining $2p + r - 1 = -0.5$ gives $p = 0.25 - r/2$ so we should surrender if the probability of winning is less than a quarter minus half of the chance of a tie.

8.9 (a) The values in Table T.6 for a dealer 6 are all larger than those for a dealer 7 as long as our hand total is 17 or below, so we want to see a dealer 6. This makes sense, since with a dealer 6 we can reasonable expect the face down card to have value ten, making the dealer need to draw and possibly bust (versus a probable hand of 17 for the dealer on which the dealer would stand).

(b) Since for a point total of 18, the entry for a dealer 7 is higher than that for a dealer 6, we prefer to see the dealer having a 7 face up.

(c) No hands are 100% in the player's favor!

8.10 We would most prefer to see the dealer have a face up 4, 5, or 6 since all of the expected value entries are positive; between 3 and 4, we prefer 4 because the entries are the highest among those two.

8.11 (a) The 54 hands are $\{$(A,A), (A,9), (A,8), (A,7), (A,6), (A,5), (A,4), (A,3), (A,2), (10,10), (10,9), (10,8), (10,7), (10,6), (10,5), (10,4), (10,3), (10,2), (9,9), (9,8), (9,7), (9,6), (9,5), (9,4), (9,3), (9,2), (8,8), (8,7), (8,6), (8,5), (8,4), (8,3), (8,2), (7,7), (7,6), (7,5), (7,4), (7,3), (7,2), (6,6), (6,5), (6,4), (6,3), (6,2), (5,5), (5,4), (5,3), (5,2), (4,4), (4,3), (4,2), (3,3), (3,2), (2,2)$\}$.

(b) For any of the hands without a ten-valued card that's also not a pair, the probability is $2\left(\frac{4}{52}\right)\left(\frac{4}{52}\right) = 2/169 \approx 0.0118$. For hands without a ten-valued card that are also pairs, it's $\left(\frac{4}{52}\right)\left(\frac{4}{52}\right) = 1/169 \approx 0.0059$. On the other hand, if the hand is not paired but contains a ten-valued card, we get $2\left(\frac{16}{52}\right)\left(\frac{4}{52}\right) = 8/169 \approx 0.0473$ and for (10,10) we get $\left(\frac{16}{52}\right)\left(\frac{16}{52}\right) = 16/169 \approx$ 0.0948. Note that the factor of 2 for non-paired hands is due to the two possible orders.

(c) Counting the types of hands from part (a) that fit our situation and matching these with the probability types of part (b), we get

$$11\left(\frac{2}{169}\right) + 5\left(\frac{8}{169}\right) + 3\left(\frac{1}{169}\right) \approx 0.3846$$

where note that the hands in part (a) we care about are $\{$(10,6), (10,5), (10,4), (10, 3), (10,2), (9,7), (9,6), (9,5), (9,4), (9,3), (8,8), (8,7), (8,6), (8,5), (8,4), (7,7), (7,6), (7,5), (6,6)$\}$.

(d) Good! Though the probability is just slightly more than a third, it's important to note that for many other hands, the dealer may end up busting anyways if we can know exactly where a ten-valued card lies; the probability that we found is the chance that, say, if we were the only player at the table, if a ten-valued card is fifth down (first to be used as a hit), if we know nothing about the top four cards and simply stand our hand, we'll win with probability 38.46%. Of course, if we had knowledge of one card, we could have knowledge of others, which is exactly how the MIT team in the 1990s worked the casinos!

Chapter 9: Games, Gambling, and Probability

9.1 For the Virginia Lottery, "exact" pays $5,000 on a $1 wager; the expected winnings are

$$\frac{1}{10000}(\$4999) + \frac{9999}{10000}(-\$1) = -\$0.50.$$

9.2 For the Virginia Lottery, a "four way" bet (on a number such as 1112, coming up in any of the four ways) pays $1,200 on a $1 wager. The "24 way" bet (winning when any permutation of a number like 1234 comes up) pays $200 on a $1 wager. These expected values, respectively, are

$$\frac{4}{10000}(\$1199) + \frac{9996}{10000}(-\$1) = -\$0.52$$

$$\frac{24}{10000}(\$199) + \frac{9976}{10000}(-\$1) = -\$0.52.$$

9.3 As of July 2020, the Powerball has players select five numbers from 1 to 69 and one special number from 1 to 26. The jackpot is won when 5+1 wins, otherwise, payouts are $1,000,000 for 5+0, $50,000 for 4+1, $100 for 4+0 and 3+1, $7 for 3+0 and 2+1, and $4 for 1+1 and 0+1. The raw probabilities are

$$5 + 1 \quad \left(\binom{5}{5} / \binom{69}{5} \right) \cdot \frac{1}{26} = 1/292,201,338$$

$$5 + 0 \quad \left(\binom{5}{5} / \binom{69}{5} \right) \cdot \frac{25}{26} = 25/292,201,338$$

$$4 + 1 \quad \left(\binom{5}{4}\binom{64}{1} / \binom{69}{5} \right) \cdot \frac{1}{26} = 160/146,100,669$$

$$4 + 0 \quad \left(\binom{5}{4}\binom{64}{1} / \binom{69}{5} \right) \cdot \frac{25}{26} = 4,000/146,100,669$$

$$3 + 1 \quad \left(\binom{5}{3}\binom{64}{2} / \binom{69}{5} \right) \cdot \frac{1}{26} = 3,360/48,700,223$$

$$3 + 0 \quad \left(\binom{5}{3}\binom{64}{2} / \binom{69}{5} \right) \cdot \frac{25}{26} = 84,000/48,700,223$$

$$2+1 \quad \left(\binom{5}{2}\binom{64}{3} \Big/ \binom{69}{5} \right) \cdot \frac{1}{26} = 69,440/48,700,223$$

$$1+1 \quad \left(\binom{5}{1}\binom{64}{4} \Big/ \binom{69}{5} \right) \cdot \frac{1}{26} = 529,480/48,700,223$$

$$0+1 \quad \left(\binom{5}{0}\binom{64}{5} \Big/ \binom{69}{5} \right) \cdot \frac{1}{26} = 1,270,752/48,700,223$$

so we can do the expected value computation as we did in the book for Mega Millions, getting

$$EV = \underbrace{\left(\frac{1}{292,201,338} \right) \cdot (J-2)}_{5+1} + \underbrace{\left(\frac{25}{292,201,338} \right) \cdot 999,998}_{5+0}$$

$$+ \underbrace{\left(\frac{160}{146,100,669} \right) \cdot 49,998}_{4+1} + \underbrace{\left(\frac{4,000}{146,100,669} \right) \cdot 98}_{4+0}$$

$$+ \underbrace{\left(\frac{3,360}{48,700,223} \right) \cdot 98}_{3+1} + \underbrace{\left(\frac{84,000}{48,700,223} \right) \cdot 5}_{3+0} + \underbrace{\left(\frac{69,440}{48,700,223} \right) \cdot 5}_{2+1}$$

$$+ \underbrace{\left(\frac{529,480}{48,700,223} \right) \cdot 2}_{1+1} + \underbrace{\left(\frac{1,270,752}{48,700,223} \right) \cdot 2}_{0+1} + \underbrace{\left(\frac{46,741,800}{48,700,223} \right) \cdot (-2)}_{\text{completely lose}}$$

$$= \frac{J-2}{292,201,338} - \frac{81,822,771}{48,700,223}.$$

Setting this greater than zero gives $J > \$490,936,628$ as the point when playing Powerball gives a positive expected value (assuming this is the lump-sum payment and the prize is not split).

9.4 $\binom{15}{5}\binom{15}{1}\binom{15}{1}\binom{15}{5} = 2,029,052,025$

9.5 $\binom{15}{2}^4\binom{15}{4} = 165,916,603,125$

9.6 $\binom{15}{5}\binom{15}{5} = 9,019,009$

9.7 $PBB(50) = \frac{\binom{24}{1}\binom{51}{50-24}}{\binom{75}{50-1}}\left(\frac{1}{76-50} \right) \approx 0.00000226$

9.8 (a) $\binom{10}{2}^4 = 4,100,625$

(b) $\binom{10}{5}^4\binom{10}{4} = 846,879,183,360$

(c) $\binom{10}{5}\binom{10}{5}\binom{10}{1}\binom{10}{1} = 6,350,400$

(d) $\binom{10}{2}^4\binom{10}{4} = 861,131,250$

(e) $\binom{10}{5}\binom{10}{5} = 63,504$

(f) $(_{10}P_5)^4 \cdot {}_{10}P_4/2 = 2,107,306,409,538,355,200,000$

9.9 (a) $1/\binom{50}{16} = \frac{1}{62,852,101,650} \approx 0.000000000016$

(b) $PBB(n) = \frac{\binom{16}{1}\binom{24}{n-16}}{\binom{40}{n-1}}\left(\frac{1}{41-n}\right)$

(c) $656/17 \approx 38.5882$

9.10 (a) $1/\binom{24}{8} = \frac{1}{735,471} \approx 0.00000136$

(b) $PBB(n) = \frac{\binom{8}{1}\binom{16}{n-8}}{\binom{24}{n-1}}\left(\frac{1}{25-n}\right)$

(c) $200/9 \approx 22.22$

9.11 (a) $3,365,856/27,883,218,168 \approx 0.00012071$

(b) $480,700/27,883,218,168 \approx 0.00001724$

(c) $4,151,918,628/27,883,218,168 \approx 0.148904$

(d) $2,186,189,400/27,883,218,168 \approx 0.07840520$

(e) $8/27,883,218,168 \approx 0.0000000000287$

(f) Using inclusion-exclusion and combining like terms, we get

$$\binom{108}{7} - 4 \cdot \binom{83}{7} + 6 \cdot \binom{58}{7} - 4 \cdot \binom{33}{7} + \binom{8}{7}$$

which gives $13,062,500,000/27,883,218,168 \approx 0.468472$

9.12 The seven cards you drew, plus the "draw four plus wild" card your opponent played are known to you, so there are a total of 100 unknown cards to you, and 6 unknown cards in your opponent's hand. There are 22 outstanding red cards (since you drew three), so this becomes $\binom{78}{6}/\binom{100}{6} \approx 0.21547$ for about a 22% chance of an illegal play.

9.13 $(0.095155968)(8) + (0.904844032)(-1) = -\$0.1436$

9.14 $(0.458597423)(1) + (0.446246609)(-1) = +\$0.0124$, and a positive expected value is bad for the casino!

9.15 (a) $(0.436063423)(1) + (0.446246609)(-1) = -\$0.0102$

(b) $(0.022534)(40) + (0.977466)(-1) = -\$0.0761$

(c) $(0.034543)(25) + (0.965457)(-1) = -\$0.1019$

9.16 $(0.053863716)(0.50)+(0.404733707)(1)+(0.446246609)(-1) = -\$0.0146$

9.17 For two pair with five dice, we get

$$\frac{\binom{4}{2,1}\binom{5}{2,2,1}}{6^5} = \frac{5}{108} \approx 0.046296$$

and for one pair with five dice, we get

$$\frac{\binom{4}{1,3}\binom{5}{2,1,1,1}}{6^5} = \frac{5}{162} \approx 0.030864$$

and after combining we get the expected 0.077160.

9.18 The 15 pairs of kinds of triples are $\{(6,5), (6,4), (6,3), (6,2), (6,1), (5,4),$ $(5,3), (5,2), (5,1), (4,3), (4,2), (4,1), (3,2), (3,1), (2,1)\}$ with values, respectively, $\{1100, 1000, 900, 800, 1600, 900, 800, 700, 1500, 700, 600, 1400, 500,$ $1300, 1200\}$. Each of the 15 possibilities has $300/15 = 20$ ways out of the 46,656 possible six-dice rolls to be represented, so we multiply each of these 15 values by $20/46,656$ and sum up to obtain a contribution of about 6.4300.

9.19 Of the possible three of a kinds discussed in Example 9.15, 7,200 of them match the pattern AAABBC and 7,200 match the pattern AAABCD. For AAABBC and a triple of twos, for example, B and C can be chosen in one of $5 \cdot 4 = 20$ possible ways. Choosing (1,5) for (B,C) gives a score of 450, (5,1) gives a score of 400, and (1,X), where X is either a 3, 4, or 6, gives 400 points. Continuing, we have 300 points for (5,X), 300 points for (X,1), 250 points for (X,5), and 200 points for (X,Y) where the pair (X,Y) can be chosen in 6 possible ways. The weighted average score for a triple of 2s, counting possible 1s and 5s, is therefore 290. For 3s, 4s, and 6s, the average score can be similarly computed 390, 490, and 690. For a triple of 1s, the pair (5,X), where X can be a 2, 3, 4, or 6, gives a score of 1,100, the pair (X,5) gives a score of 1,050, and the pair (X,Y), with 12 possibilities, gives 1,000 for an average score of 1,030; similar analysis gives a average score for a triple of 5s as 560. Since each of 1 through 6 is an equal possibility for the value of A, each of the these values for A gives 1,200 possibilities for AAABBC out of all rolls possible. The contribution to your first roll in Farkle for this pattern is then $\left(\frac{1,200}{46,656}\right) \cdot (1030 + 290 + 390 + 490 + 560 + 690) \approx 88.7346$. Considering all possibilities for the pattern AAABCD in a similar way gives the same expectec contribution, so for all three of a kinds with only one three of a kind, the expected contribution is about 177.469.

9.20 The scoring system we'll use is additive; a four of a kind of 4s, for instance, is worth 800 points since the additional rolled 4 counts as an "additional three of a kind" here.

(a) For pattern AAAABB, we count $\binom{6}{1,1,4}\binom{6}{4,2} = 450$ possibilities and for AAAABC we find $\binom{6}{1,2,3}\binom{6}{4,1,1} = 1,800$ possibilities, for a combined probability of $\frac{2,250}{46,656} \approx 0.04822$.

(b) $\left(\frac{375}{46,656}\right) \cdot (2000 + 400 + 600 + 800 + 1000 + 1200) \approx 48.225$

(c) Proceeding like we did for Exercise 9.19, we get 50.3955 as the expected contribution to the first roll for all four of a kinds.

9.21 The scoring system we use gives, for example, 600 points for a five of a kind of 2s (200 for each additional die beyond the three required for a three of a kind), 800 points for a six of a kind of 2s, 750 points for a straight, and 750 points for three pairs.

(a) Probability is about 0.00386, contribution for just the five of a kind is about 5.787, and considering the sole pattern AAAAAB yields an expected overall contribution of about 5.8835.

(b) Probability is about 0.000129 and the expected contribution is about 0.257202.

(c) Probability is about 0.0154 and the expected contribution is about 11.57407.

(d) Probability is about 0.03858 and the expected contribution is about 28.93519.

(e) The probability can be obtained by subtracting all "named" rolls from 1, as well as subtracting the probability of having a Farkle with six dice. For the two patterns AABBCD and AABCDE, the expected contribution is about 86.8055.

9.22 Adding together the expected (overall) contributions from all of the Farkle calculations, for our scoring system, we get an expected value of that first roll, not including possible rerolls, of 367.75.

9.23   (a) 1,837,734,087; 0.114804
(b) 278,040,882; 0.017369
(c) 347,737,768; 0.021723
(d) 493,897,888; 0.030854
(e) 38,248,840; 0.002389
(f) 88,929,720; 0.005555

9.24   (a) 189,106,190; 0.012703
(b) 246,645,784; 0.016568
(c) 82,173,456; 0.005940

9.25 The probability of getting a hand to make DOG *decreases* a little when a D is missing, *increases* a little when an E is missing, and *decreases* substantially, reducing the probability by about a third, when two of the three Gs are missing.

9.26   (a) 28,165,960,690; 0.012232; the probability is slightly lower in Super Scrabble.
(b) 61,398,980,188; 0.026883; the probability is a good bit higher in Super Scrabble.

9.27

2	4	5	8	9	10	11
12/36	15/36	15/36	6/36	5/36	3/36	2/36
12	15	16	18	20	24	
3/36	1/36	1/36	1/36	1/36	1/36	

Other values are not possible, since, above 12, doubles need to be rolled in order to hit a piece that far away; only two, three, or four times the values 4, 5, or 6 will be possible this high.

9.28 Repeating Example 9.21, we get

$$\text{Expected Value, Concede} = -s$$
$$\text{Expected Value, Accept} = p \cdot 3s + (1 - p)(-3s) = s(6p - 3)$$

and equating these gives $p > 1/3$ as when you should accept. For Example 9.22, we have

$$\text{EV, Opponent Concedes} = s$$
$$\text{EV, Opponent Accepts} = p \cdot 3s + (1 - p)(-3s) = s(6p - 3)$$
$$\text{EV, No Double} = p \cdot s + (1 - p)(-s) = s(2p - 1)$$

where equating the first two yields $p > 2/3$ and equating the last two we get $p > 1/2$, so we want the opponent to accept if $p > 1/2$ and definitely hope they accept when $p > 2/3$.

9.29 (a) This is included in the table for part (b).
  (b)

Layout	Length	Layout	Length	Layout	Length
11223344	4	11232344	5	11322344	5
11223434	5	11232434	6	11322434	6
11224334	5	11234234	6	11324234	6
11242334	5	11243234	6	11342234	5
11422334	5	11423234	6	11432234	5
14122334	5	14123234	6	14132234	6
41122334	5	41123234	6	41132234	6
13122344	5	31122344	5	12123344	5
13122434	6	31122434	6	12123434	6
13124234	6	31124234	6	12124334	6
13142234	6	31142234	6	12142334	6
13412234	6	31412234	6	12412334	6
14312234	6	34112234	5	14212334	6
41312234	6	43112234	5	41212334	6
12132344	6	12312344	6	13212344	6
12132434	7	12312434	7	13212434	7
12134234	7	12314234	7	13214234	7
12143234	7	12341234	6	13241234	6
12413234	7	12431234	6	13421234	6
14213234	7	14231234	6	14321234	6
41213234	7	41231234	6	41321234	6
31212344	6	12213344	5	12231344	6
31212434	7	12213434	6	12231434	7
31214234	7	12214334	6	12234134	7
31241234	6	12241334	6	12243134	7
31421234	6	12421334	6	12423134	7
34121234	6	14221334	5	14223134	6
43121234	6	41221334	5	41223134	6
12321344	6	13221344	5	31221344	5
12321434	7	13221434	6	31221434	6

12324134	7	13224134	6	31224134	6
12342134	6	13242134	6	31242134	6
12432134	6	13222134	6	31422134	6
14232134	6	14322134	6	34122134	6
41232134	6	41322134	6	43122134	6

(c) $\frac{1}{105}(4) + \frac{20}{105}(5) + \frac{66}{105}(6) + \frac{18}{105}(7) \approx 5.962$

(d) From the formula of Theorem 9.3, we get about 5.944 which is very close! Given the amount of work to get the exact value, the formula is very useful.

9.30 (a) For $n = 2$ we have the number of essentially different layouts as $\binom{6}{3,3}/2! = 20/2 = 10$ and for $n = 3$ we get $\binom{9}{3,3,3}/3! = 1,680$.

(b) $\frac{\binom{3n}{3,3,\ldots,3}}{n!} = \frac{(3n)!}{6^n n!}$

(c) If the locations for three matching cards are known, flip them over. Otherwise, flip over an unknown card; if the other two positions for that card are known, complete the triple, and, otherwise, flip over two unknown cards.

(d) The 10 possible layouts are {111222, 112122, 121122, 211122, 112212, 121212, 211212, 122112, 212112, 221112} that have game lengths {2, 3, 3, 3, 4, 4, 4, 4, 4, 4} respectively. The expected game length ends up being 3.5.

9.31 Rather than use the $P_{\text{green}}^{(0,0,3)}$ and such notation, we use the actual probabilities in the answers below.

$$GGG = \frac{\binom{6}{3}\binom{4}{0}\binom{3}{0}}{\binom{13}{3}}\left(\frac{1}{6}\right)^3\left(\frac{1}{3}\right)^0\left(\frac{1}{2}\right)^0 \approx 0.000324$$

$$YYY = \frac{\binom{6}{0}\binom{4}{3}\binom{3}{0}}{\binom{13}{3}}\left(\frac{1}{6}\right)^0\left(\frac{1}{3}\right)^3\left(\frac{1}{2}\right)^0 \approx 0.000518$$

$$GGY = \frac{\binom{6}{2}\binom{4}{1}\binom{3}{0}}{\binom{13}{3}}\left(\frac{1}{6}\right)^2\left(\frac{1}{3}\right)^1\left(\frac{1}{2}\right)^0 \approx 0.001943$$

$$GGR = \frac{\binom{6}{2}\binom{4}{0}\binom{3}{1}}{\binom{13}{3}}\left(\frac{1}{6}\right)^2\left(\frac{1}{3}\right)^0\left(\frac{1}{2}\right)^1 \approx 0.002185$$

$$YYG = \frac{\binom{6}{1}\binom{4}{2}\binom{3}{0}}{\binom{13}{3}}\left(\frac{1}{6}\right)^1\left(\frac{1}{3}\right)^2\left(\frac{1}{2}\right)^0 \approx 0.002331$$

$$RRG = \frac{\binom{6}{1}\binom{4}{0}\binom{3}{2}}{\binom{13}{3}}\left(\frac{1}{6}\right)^1\left(\frac{1}{3}\right)^0\left(\frac{1}{2}\right)^2 \approx 0.002622$$

$$RRY = \frac{\binom{6}{0}\binom{4}{1}\binom{3}{2}}{\binom{13}{3}}\left(\frac{1}{6}\right)^3\left(\frac{1}{3}\right)^1\left(\frac{1}{2}\right)^2 \approx 0.003497$$

9.32 (a) 0.022181, 0.027062

(b) 0.025234, 0.023317
(c) 0.028287, 0.020286
(d) Since the regular probability is 0.02435, and 0.023317 is lower (for removing a yellow), that seems better.

9.33 **(a)** With a table value of 2.715082, stop if you have 3 brains.
(b) A table value of 2.076331 means to reroll if you have two or less brains.
(c) With a table value of 0.777074, reroll only if you have no brains.
(d) With a table value of 19.334573, roll again!
(e) With the notation XYZ/PQR representing the entry in the table for dice quantities $(r_c, y_c, g_c, r_f, y_f, g_f)$, the only rows where no shotguns has an entry less than 13 are {300/000, 200/100, 100/200, 044/300, 043/300, 042/300, 041/300, 040/300, 033/300, 032/300, 031/300, 030/300, 023/300, 022/300, 021/300, 020/300, 012/300, 011/300, 010/300, 000/300}. A fairly safe rule would be to simply reroll no matter what, given this small number of entries.

9.34 **(a)** Stop at 2 ($y_c > g_c$).
(b) Reroll 0 or 1 only ($g_f = 3$).
(c) Reroll 0 only ($g_f \neq 3$).
(d) If you have 3, stop ($g_c > y_c$).
Some are different, especially for (d), since the simple table is more cautious, generally, than the full table.

Chapter 10: Games, Gambling, and Probability

Round	Result	(a) Bet	(a) Total	(b) Bet	(b) Total
1	W	$20	$520	$20	$520
2	W	$20	$540	$40	$560
3	L	$20	$520	$80	$480
4	L	$40	$480	$20	$460
5	W	$80	$560	$20	$480
6	L	$20	$540	$40	$440
7	L	$40	$500	$20	$420
8	W	$80	$580	$20	$440
9	L	$20	$560	$40	$400
10	W	$40	$600	$20	$420
11	W	$20	$620	$40	$460
12	L	$20	$600	$80	$380
13	L	$40	$560	$20	$360
14	L	$80	$480	$20	$340
15	W	$160	$620	$20	$360

16	W	$20	$640	$40	$400
17	L	$20	$620	$80	$320
18	L	$40	$580	$20	$300
19	W	$80	$660	$20	$320
20	W	$20	$680	$40	$360

10.1 Starting with $500 and using $20 "base" bets, (a) and (b) are presented in the previous table.

10.2 (a) This is a good bit better than standard Martingale since we get

Bet	1	2	3	4	5	Total
Amount	$100.00	$141.42	$200.00	$282.84	$400.00	$1,124.26

(b) The expected amount won, weighted for that winning occurring on turn 1, 2, 3, and so on, is $(1 - q) \sum_{i=0}^{n-1} q^i \left((\sqrt{2})^i B - \sum_{j=0}^{i-1} (\sqrt{2})^j B \right)$ which after repeated application of Theorem 10.1 and some algebra, becomes $\frac{B(1-q)}{\sqrt{2}-1} \left(\frac{(\sqrt{2}q)^n - 1}{\sqrt{2}q - 1} (\sqrt{2} - 2) + \frac{q^n - 1}{q - 1} \right)$. The amount lost is $\sum_{i=1}^{n} B \cdot (\sqrt{2})^{i-1} = B \left(\frac{(\sqrt{2})^n - 1}{\sqrt{2}-1} \right) = \frac{1}{\sqrt{2}-1} B((\sqrt{2})^n - 1)$ so the expected value becomes

$$\frac{B(1-q)}{\sqrt{2}-1} \left(\frac{(\sqrt{2}q)^n - 1}{\sqrt{2}q - 1} (\sqrt{2} - 2) + \frac{q^n - 1}{q - 1} \right) - \frac{1}{\sqrt{2}-1} Bq^n((\sqrt{2})^n - 1).$$

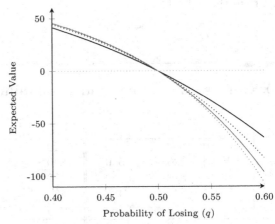

(c) This is better perhaps for losses since they do not grow as fast, but neither do the winnings! It's harder to get back afloat once losses start.

10.3 (a) Much worse!

Bet	1	2	3	4	5	Total
Amount	$100	$300	$900	$2,700	$8,100	$12,100

(b) The expected amount won, weighted for that winning occurring on turn 1, 2, 3, and so on, is $(1 - q) \sum_{i=0}^{n-1} q^i \left(3^i B - \sum_{j=0}^{i-1} 3^j B \right)$ which after repeated application of Theorem 10.1 and some algebra, becomes $\frac{1}{2} B(1 -$

$q\left(\frac{(3q)^n-1}{3q-1} + \frac{q^n-1}{q-1}\right)$. The amount lost is $\sum_{i=1}^{n} B \cdot 3^{i-1} = B\left(\frac{3^n-1}{3-1}\right) = \frac{1}{2}B(3^n - 1)$ so the expected value becomes

$$\frac{1}{2}B(1-q)\left(\frac{(3q)^n - 1}{3q - 1} + \frac{q^n - 1}{q - 1}\right) - \frac{1}{2}Bq^n(3^n - 1).$$

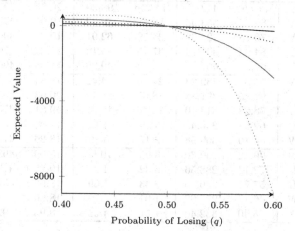

(c) As n increases, the expected losses become huge!

10.4 We use a fraction of $p - q$ for 1:1 wagers and $\frac{bp-q}{b} = \frac{2p-q}{2}$ for 2:1 wagers.

  (a) 0.2, 0.4   (b) 0.6, 0.7   (c) 0.04, 0.28   (d) 0, 0.25
  (e) −0.06, 0.205

10.5 Here we put all three in one table; dollar signs are left off to save space. Note that for (a) we use a 20% bankroll bet, for (b) we use a 60% bankroll bet, and for (c) we use a 4% bankroll bet.

Res	(a) Bet	(a) Tot	(b) Bet	(b) Tot	(c) Bet	(c) Tot
W	100.00	600.00	300.00	800.00	20.00	520.00
W	120.00	720.00	160.00	960.00	20.80	540.80
L	144.00	576.00	576.00	384.00	21.63	519.17
L	115.20	460.80	230.40	153.60	20.77	498.40
W	92.16	552.96	92.16	245.76	19.94	518.34
L	110.59	442.37	147.46	98.30	20.73	497.61
L	88.47	353.90	58.98	39.32	19.90	477.71
W	70.78	424.68	23.59	62.91	19.11	496.82
L	84.94	339.74	37.75	25.16	19.87	476.95
W	67.95	407.69	15.10	40.26	19.08	496.03
W	81.54	489.23	24.16	64.42	19.84	515.87
L	97.85	391.38	38.65	25.77	20.63	495.24
L	78.28	313.10	15.46	10.31	19.81	475.13
L	62.62	250.48	6.19	4.12	19.02	456.41
W	50.10	300.58	2.47	6.59	18.26	474.67
W	60.12	360.70	3.95	10.54	18.99	493.66
L	72.14	288.56	6.32	4.22	19.75	473.91
L	57.71	230.85	2.53	1.69	18.96	454.95
W	46.17	277.02	1.01	2.70	18.20	473.15
W	55.40	332.40	1.62	4.32	18.93	492.08

10.6 The equation $\frac{1-\left(\frac{q}{p}\right)^k}{1-\left(\frac{q}{p}\right)^T} = \frac{1}{2}$ yields $\frac{1}{2}\left(1-\left(\frac{q}{p}\right)^T\right) = 1-\left(\frac{q}{p}\right)^k$ and by moving terms around, we get $\left(\frac{q}{p}\right)^k = 1-\frac{1}{2}\left(1-\left(\frac{q}{p}\right)^T\right)$. Take the natural logarithm of both sides and divide by $\ln(q/p)$ to get the answer in Example 10.7.

Bibliography

[1]   Morton Abramson. "Restricted Combinations and Compositions." In: *Fibonacci Quart.* 14.5 (1976), pp. 439–452. ISSN: 0015-0517.

[2]   George E. Andrews. *The Theory of Partitions.* Cambridge Mathematical Library. Reprint of the 1976 original. Cambridge: Cambridge University Press, 1998. ISBN: 978-0-5216-3766-4.

[3]   Arrow International, Inc. *Game Patterns/Pattern Probabilities.* Dec. 2013. URL: http : / / www . arrowinternational . com / images / docs / bingogamepatterns.pdf.

[4]   Robert Axelrod. *The Evolution of Cooperation: Revised Edition.* New York, NY: Basic Books, 1984. ISBN: 978-0-465-00564-2.

[5]   Keith Ball. *Strange Curves, Counting Rabbits, & Other Mathematical Explorations.* Princeton, NJ: Princeton University Press, 2003. ISBN: 978-0-6911-1321-0.

[6]   Michael Barnsley. *Fractals Everywhere.* Burlington, MA: Morgan Kaufmann Publishers, 1993. ISBN: 978-0120790616.

[7]   Daniel Bernoulli. "Exposition of a New Theory on the Measurement of Risk." English. In: *Econometrica* 22.1 (1954), pp. 23–36. ISSN: 00129682.

[8]   Jörg Bewersdorff. *Luck, Logic, & White Lies: The Mathematics of Games.* Wellesley, MA: A K Peters, 2005. ISBN: 978-1-5688-1210-6.

[9]   Gunnar Blom, Lars Holst, and Dennis Sandell. *Problems and Snapshots from the World of Probability.* New York, NY: Springer-Verlag, 1994. ISBN: 978-0-3879-4161-5.

[10] Mark Bollman. *Mathematics of Keno and Lotteries.* Boca Raton, FL: CRC Press/Taylor & Francis Group, 2018. ISBN: 978-1-138-72372-6.

[11] Eric Bonabeau and Théraulaz. "Swarm Smarts." In: *Scientific American* (Mar. 2000).

[12] Maxine Brady. *The Monopoly Book: Strategy and Tactics of the World's Most Popular Game.* New York, NY: David McKay Company, Inc., 1974. ISBN: 978-0-6792-0292-9.

[13] Duane M. Broline. "Renumbering of the Faces of Dice." English. In: *Mathematics Magazine* 52.5 (1979), pp. 312–315. ISSN: 0025570X.

[14] Heather L. Cook and David G. Taylor. *Zombie Dice: An Optimal Play Strategy.* 2014. eprint: arXiv:1406.0351.

[15] G. Dantzig, R. Fulkerson, and S. Johnson. "Solution of a large-scale traveling-salesman problem." In: *J. Operations Res. Soc. Amer.* 2 (1954), pp. 393–410. ISSN: 0160-5682.

[16] Richard A. Epstein. *The Theory of Gambling and Statistical Logic.* 2nd ed. Burlington, MA: Elsevier/Academic Press, 2009. ISBN: 978-0-1237-4940-6.

[17] Joseph A. Gallian. *Contemporary Abstract Algebra.* 7th ed. Stamford, CT: Cengage Learning, 2009. ISBN: 978-0-5471-6509-7.

[18] James Glenn. *An Optimal Strategy for Yahtzee.* Tech. rep. CS-TR-0002. Loyola College in Maryland, 2006.

[19] Steven Levitt, Austan Goolsbee, and Chad Syverson. *Microeconomics.* 3rd ed. New York, NY: Worth Publishers, 2019. ISBN: 978-1-319-10556-3.

[20] Ronald J. Gould. *Mathematics in Games, Sports, and Gambling: The Games People Play.* 2nd ed. Boca Raton, FL: CRC Press/Taylor & Francis Group, 2016. ISBN: 978-1-4987-1952-0.

[21] Anders Hald. *A History of Probability and Statistics and their Applications Before 1750.* New York, NY: John Wiley & Sons, 1990. ISBN: 978-0-4715-0230-2.

[22] Robert V. Hogg and Elliot A. Tanis. *Probability and Statistical Inference.* 6th ed. Upper Saddle River, NJ: Prentice Hall, 2001. ISBN: 978-0-1302-7294-2.

[23] Tom Jolly. *Using Game Theory to Design Games.* Sept. 2014. URL: http://www.leagueofgamemakers.com/using-game-theory-to-design-games/.

[24] J. L. Kelly Jr. "A new interpretation of information rate." In: *Bell. System Tech. J.* 35 (1956), pp. 917–926. ISSN: 0005-8580.

[25] Jim Kiley, Jim Fox, and Anthony F. Lucas. *Casino Operations Management.* 2nd ed. Hoboken, NJ: John Wiley & Sons, Inc., 2004. ISBN: 978-0-4712-6632-7.

[26] Raph Koster. *A Theory of Fun for Game Design.* Scottsdale, AZ: Paraglyph Press, 2005. ISBN: 978-1-9321-1197-2.

[27] Jonathan Marino and David G. Taylor. "Integer Compositions Applied to the Probability Analysis of Blackjack and the Infinite Deck Assumption." English. In: *Topics in Recreational Mathematics* 2017.1 (2017), pp. 74–91.

[28] George Marsaglia. *The Marsaglia Random Number CDROM including tje Diehard Battery of Tests of Randomness.* [CD-ROM]. 1995.

[29] James T. McClave and Terry Sincich. *A First Course in Statistics.* 10th ed. Upper Saddle River, NJ: Pearson, 2009. ISBN: 978-0-1360-0047-1.

[30] Ben Mezrich. *Busting Vega$: The MIT Whiz Kid Who Brought the Casinos to Their Knees*. New York, NY: HarperCollins Publishers, 2005. ISBN: 978-0-0605-7511-3.

[31] Roland Minton. *Sports Math: An Introductory Course in the Mathematics of Sports Science and Sports Analytics*. Boca Raton, FL: CRC Press/Taylor & Francis Group, 2017. ISBN: 978-1-4987-0626-1.

[32] Paul K. Newton and Joseph B. Keller. "Probability of Winning at Tennis I. Theory and Data." In: *Stud. in Appl. Math.* 114 (2005), pp. 241–269.

[33] Edward Packel. *The Mathematics of Games and Gambling*. 2nd ed. Washington, DC: Mathematical Association of America, 2006. ISBN: 978-0-8835-5646-8.

[34] Ignacio Palacios-Huerta and Oscar Volji. "Experientia Docet: Professionals Play Minimax in Laboratory Experiments." In: *Econometrica* 76.1 (2008), pp. 71–115.

[35] Mark Pankin. *Retrosheet*. Sept. 2020. URL: https://www.retrosheet.org/.

[36] Alfred Posamentier. *The Fabulous Fibonacci Numbers*. Buffalo, NY: Prometheus Publishing, 2007. ISBN: 978-1591024750.

[37] William Poundstone. *Fortune's Formula: The Untold Story of the Scientific Betting System That Beat the Casinos and Wall Street*. New York, NY: Hill and Wang, 2006. ISBN: 978-0-8090-4599-0.

[38] Larry Rabinowitz. *Elementary Probability with Applications*. 2nd ed. Boca Raton, FL: CRC Press/Taylor & Francis Group, 2017. ISBN: 978-1-4987-7132-0.

[39] Hannah Robbins. *Functional Linear Algebra*. Boca Raton, FL: CRC Press/Taylor & Francis Group, 2021. ISBN: 978-0-367-48687-7.

[40] Bill Robertie. *Backgammon for Winners*. Third. Las Vegas, NV: Cardoza Books, 2002. ISBN: 978-1-5804-2043-3.

[41] Jason Rosenhouse. *The Monty Hall Problem: The Remarkable Story of Math's Most Contentious Brain Teaser*. New York, NY: Oxford University Press, 2009. ISBN: 978-0-19-536789-8.

[42] Ken Ross. *A Mathematician at the Ballpark: Odds and Probabilities for Baseball Fans*. New York, NY: Plume, 2007. ISBN: 978-0-452-28782-2.

[43] Louis M. Rotando and Edward O. Thorp. "The Kelly Criterion and the Stock Market." In: *Amer. Math. Monthly* 99.10 (1992), pp. 922–931. ISSN: 0002-9890. DOI: 10.2307/2324484.

[44] Karin Saoub. *A Tour Through Graph Theory*. Boca Raton, FL: CRC Press/Taylor & Francis Group, 2018. ISBN: 978-1-138-19780-0.

[45] Marilyn vos Savant. "Ask Marilyn." In: *Parade Magazine* 16 (1990).

[46] Frederic Paik Schoenberg. *Introduction to Probability with Texas Hold'Em Examples*. 2nd ed. Boca Raton, FL: CRC Press/Taylor & Francis Group, 2017. ISBN: 978-1-4987-7618-9.

[47] Thomas Severini. *Analytic Methods in Sports: Using Mathematics and Statistics to Understand Data from Baseball, Football, Basketball, and Other Sports*. Boca Raton, FL: CRC Press/Taylor & Francis Group, 2014. ISBN: 978-1-4822-3701-6.

[48] Arnold Snyder. *The Big Book of Blackjack*. New York, NY: Cardoza Publishing, 2006. ISBN: 978-1-5804-2155-3.

[49] Philip D. Straffin. *Game Theory and Strategy*. Washington, DC: Mathematical Association of America, 1993. ISBN: 978-0-8838-5637-6.

[50] Peter Tannenbaum. *Excursions in Modern Mathematics*. 8th ed. Boston, MA: Pearson, 2014. ISBN: 978-0-3218-2573-5.

[51] Courtney Taylor. *What is the Probability of Rolling a Yahtzee?* Oct. 2013. URL: http://statistics.about.com/od/ProbHelpandTutorials/a/What-Is-The-Probability-Of-Rolling-A-Yahtzee.htm.

[52] *The Merriam Webster Dictionary*. Springfield, MA: Merriam Webster, 2005. ISBN: 978-0-8777-9636-7.

[53] Edward O. Thorp. *Beat the Dealer: A Winning Strategy for The Game of Twenty-One*. New York, NY: Blaisdell Publishing Company, 1962. ISBN: 978-0-3947-0310-7.

[54] Edward O. Thorp. *The Mathematics of Gambling*. Fort Lee, NJ: Lyle Stuart, 1985. ISBN: 978-0-8974-6019-4.

[55] J. V. Uspensky. *Introduction to Mathematical Probability*. New York, NY: McGraw-Hill Book Company, 1937. ISBN: 978-0-0706-6735-8.

[56] Daniel J. Velleman and Gregory S. Warrington. "What to Expect in a Game of Memory." In: *Amer. Math. Monthly* 120.9 (2013), pp. 787–805. ISSN: 0002-9890. DOI: 10.4169/amer.math.monthly.120.09.787.

[57] Mark Walker and John Wooders. "Minimax Play at Winbledon." In: *Amer. Econ. Review* 91.5 (2001), pp. 1521–1538.

[58] WikiBooks. *Poker/Bluffing*. Oct. 2013. URL: http://en.wikibooks.org/wiki/Poker/Bluffing.

[59] J.D. Williams. *The Compleat Strategyst*. Revised. Santa Monica, CA: Rand Corporation, 2007. ISBN: 978-0-8330-4222-4.

[60] Phil Woodward. "Yahtzee®: The Solution." In: *Chance* 16.1 (2003), pp. 18–22. ISSN: 0933-2480.

Image Credits

Most of the figures that appear in this book were created electronically by the author or his close friends and colleagues; in particular, special thanks are due to Robert Allen and Roland Minton for their assistance in creating the wonderful figures and taking some of the pictures that you find throughout this text.

For the other images that were used, either with explicit permission or via public domain use, credit is given here, organized by order of appearance.

- Figure 1.1 is a public domain image, courtesy of author Ant'onio Miguel de Campos of Wikimedia Commons, file *Fractal_fern_explained.png*.

- Figure 1.2 is a public domain image, courtesy of author Metoc of Wikimedia Commons, file *10-sided dice 250.png*, released under the Creative Commons Attribution-ShareAlike 3.0 Unported license.

- Figures 1.4 and 1.5 are public domain images, courtesy of author Cepheus of Wikimedia Commons, files *Monty closed doors.svg* and *Monty open door chances.svg*.

- Figure 1.6 is a public domain image, courtesy of author Evan-Amos of Wikimedia Commons, file *Plain-M&Ms-Pile.jpg*.

- Figure 2.1 is a public domain image, courtesy of author Antoine Taveneaux of Wikimedia Commons, file *Sahara Hotel and Casino 2.jpg*, released under the Creative Commons Attribution-ShareAlike 3.0 Unported license.

- Figure 2.2 is courtesy of Jackpot Productions and Bernie Stevens & Associates, used with permission.

- Figure 2.4 is edited from a public domain image, courtesy of Betzaar.com on Wikimedia Commons, file *Craps table layout.svg*, released under the Creative Commons Attribution-ShareAlike 3.0 Unported license.

- Figure 3.1 is a public domain image, courtesy of author Moltovivace of Wikimedia Commons, file *Cards royalflushhearts.jpg*.

- Figure 4.1 is a public domain image, courtesy of author PierreSelim of Wikimedia Commons, file *Dice - 1-2-4-5-6.jpg*, released under the Creative Commons Attribution-ShareAlike 3.0 Unported license.

- Figure 5.1 is a public domain image, courtesy of author Todd Klassy on Wikipedia, file *Holdem.jpg*, used under the Creative Commons Attribution 2.5 Generic license.

- Figure 6.1 is a public domain image, courtesy of author Peter Griffin of Public Domain Pictures, file *board-games.jpg*.

- Figure 7.1 is a public domain image, courtesy of author Razzle-Dazzle of English Wikipedia, file *Centre Court.jpg*.

- Figure 8.1 is a public domain image, courtesy of author Rickyar of Wikipedia, file *BlackJackGame.jpg*.

- Figure 9.1 is a public domain image, courtesy of author Chris 73 of Wikimedia Commons, file *Old Backgammon Vasa.jpg*, released under the Creative Commons Attribution-ShareAlike 3.0 Unported license.

- Figures 9.2, 9.3, and 9.4 are courtesy of the Virginia Lottery, used with permission.

- Figure 9.8 is a public domain image, courtesy of author Kokoo of Wikimedia Commons, file *Baraja_de_UNO.JPG*.

- Figure 9.9 is a public domain image, courtesy of author Magnus Manske of Wikimedia Commons, file *ScrabbleDuplicate.jpg*, released under the Creative Commons Attribution-General 2.5 license.

- Figure 9.10 is a public domain image, courtesy of author Ptkfgs of Wikimedia Commons, file *Backgammon_lg.png*.

- Figure 9.11 is a public domain image, courtesy of author KF of Wikimedia Commons, file *Memory (Game).jpg*.

- Figure 9.12 is a public domain image, courtesy of author Casey Fiesler of Wikimedia Commons, file *Zombie Dice.jpg*, released under the Creative Commons Attribution-ShareAlike 3.0 Unported license.

- Figure 10.1 is a public domain image, courtesy of author Jamie Adams of Wikimedia Commons, file *Gambling chips.jpg*, released under the Creative Commons Attribution-ShareAlike 2.0 license.

- Figure 11.1 is a public domain image, courtesy of author Tim Stellmach of Wikimedia Commons, file *Dice Distribution (bar).svg*, released into the public domain.

Index

Printed in the United States
by Baker & Taylor Publisher Services

Printed in the United States
by Baker & Taylor Publisher Services